NUMERICAL METHODS THAT WORK

NUMERICAL METHODS THAT WORK

Forman S. Acton

PROFESSOR EMERITUS OF COMPUTER SCIENCE
PRINCETON UNIVERSITY

THE MATHEMATICAL ASSOCIATION
OF AMERICA

WASHINGTON D.C.

Published in the United States of America by
The Mathematical Association of America.
Copyright © 1990 by Forman S. Acton
All rights reserved under International and Pan-American Copyright
Conventions.

This book was updated and revised from the 1970 edition
published by Harper & Row, Publishers.

Library of Congress Catalog Card Number 90-62538
ISBN 0-88385-450-3

Manufactured in the United States of America

Cover by Forman S. Acton

To My Parents

CONTENTS

solved. We then solve a slightly nonlinear differential equation—and the student finally learns how nasty the world really is. We offer him a Hobson's choice between linearized algebraic iteration and a retreat to initial value integration techniques (iterated).

A discussion of some of the better methods, emphasizing both the need for an overall plan and the futility of trying to solve everybody's polynomial with a single system. We prefer Newton's method for refining isolated real roots, isolated complex roots, and isolated quadratic factors. Stepping searches, division of one polynomial by another, the Forward-Backward division algorithm, Lin's method for quadratic factors of difficult quartics, and Laguerre's method receive various amounts of attention. Root-squaring is not recommended. After discussing the *desiderata* for a public root-finding computer package, we finally raise the spectre of Wilkinson's pathologic polynomial.

An introductory discussion of the power method for finding the extreme eigenvalues of symmetric matrices. Three vibrating masses briefly introduce the subject. Shifting an eigenvalue into prominence and the art of running orthogonal to an extreme eigenvector are treated. A final section points to the unsymmetric eigenproblem but does not discuss it in realistic detail. In this chapter a brief exposure to the ideas and typography of vectors and matrices is helpful, although a lack of it should not stymie the student.

We stress the practical evaluation of Fourier series by recurrence relations, especially the finite Fourier series with its summation orthogonalities. Prior removal of singularities, suppression of the Gibb's phenomenon, and the role of various kinds of symmetry appear. We conclude with an extension of the recurrence scheme to polynomials orthogonal over a finite irregular set of points and a mere mention of the exponential formalism of the finite Fourier transform.

PREFACE-90

As I think about the reprinting of this numerical methods book, now twenty years old, I consider what Time and the PC have done to its message. In 1970 computers were 'large'; scientists and engineers used them in laboratories; numerical software for solving various types of equations could be borrowed from a friend but was not very reliable and was poorly documented. In 1989 computers have moved into the office and into the spare bedroom of the engineer's home. Numerical software is widely available but is not very reliable and is poorly documented. Of course Newton's Method is 20 years older but it seems to be bearing up well. Indeed, with the advent of programs that do *analytic* differentiation, his method has become even more attractive.

The big change has been to place the capacities of the large 1970 computer in the hands of many who previously would have had to seek far and perhaps borrow time on someone else's machine. But access to bigger memories and faster cycle times have not changed the useful algorithms; they have merely increased the sizes of the problems that now seem feasible—almost always *linear* problems. And linear algebraic algorithms and software were, and still are, the one reasonably reliable group of 'black-boxes'. They were good enough in 1970 so that most computer users were not going to write their own; one simply had to know why one *needed* to solve a system of 100 linear equations—a question all too often ignored then, and now! But that important question lies outside the scope of this book (and the competence of its author). He has here merely raised the minatory finger—and having writ, moves on.

New tools have appeared: symbolic mathematical packages like Maple and Mathematica* that can reduce the drudgery (and increase the accuracy) of expanding complicated expressions into Taylor series. But they do not address questions about whether expansion will be a useful tactic—and if you have never before used one of these systems you will spend much more time learning how than you would expanding the immediate problem by hand. (Don't buy a chainsaw if you only have one sapling to cut! I assume that anyone reading this book is not about to tackle a forest.)

Since 1970 the Bulirsch-Stoer algorithm, mentioned briefly herein, has achieved a small but firm place among the ordinary differential equation integrators but its enthusiastic promoters have yet to produce an exposition that could fairly be included in a book at this level. Indeed, even to more sophisticated audiences they recommend it as a black-box, saying, in effect, "Try it—you'll like it!" I have been burnt by too many black boxes to take that position. Of course I regularly use other peoples' software, but if I can't fix a numerical algorithm, I won't use it. And since I haven't used B-S, I'm in no position to urge it on the public, however felicitous it may be.

Partial differential equations at a realistic algorithmic level require a book of their own. I only attempt to show the 'flavor' of the classical, still useful Finite Difference technique and point out some of the issues that must be addressed. For some partial differential equations Finite Elements, not even mentioned before, have certainly established themselves as a serious alternative to Finite Differences (tho, in my opinion, will never eliminate them) but to add Finite Elements now would have considerably lengthened the book and raised the expository level without enabling the reader to use them even on simple problems.

My doleful conclusion is that REAL algorithmic progress during the last twenty years has been principally in specialized areas that the student will wisely avoid in a first course. But from the student's point of view that means he really doesn't have to master a lot of new esoteric material. It's the best of worlds; it's the worst of worlds.

---

*Trademark

TO THE INSTRUCTOR

This book discusses efficient numerical methods for the solution of equations—algebraic, transcendental, and differential—assuming an electronic computer is available to perform the bulk of the arithmetic. I wrote it for upper-class students in engineering and the physical sciences—men who have had calculus and a first exposure to differential equations. More importantly, I wrote it for students whose motivations lie in the physical world, who would get answers to "real" problems. This intended audience shapes both the content of the book and its expository method.

My principal concern has been the proper matching of the tool to the job. Conversely, it has not been the indoctrination of the student into the beauties of analysis—numerical or otherwise. Students with motivations from the physical world are best led from the specific example to the general method. They rapidly become impatient with the development of a logical superstructure for which they have seen no practical use. An average junior in engineering at Princeton will follow an unmotivated mathematical derivation or proof for about 20 minutes before writing it off as "some more mathematical Mickey Mouse."

When a student's principal thrust is to solve problems, the author and instructor should talk principally about how problems are solved. Methods can be introduced with geometric, sometimes heuristic, arguments. Initially at least, their justification is that they work. The student uses these methods, achieving a sense of power in being able to solve problems that his mathematics courses seem to have ignored. Then he tries a problem on which the method falters: a singularity appears, or too many significant figures disappear, or a debilitating instability arises. Now, *and not before*, is the student willing to delve into the deeper structure of the numerical method—convinced that its established utility can be retrieved and broadened only by such an effort. At this point his calculus training can be called into necessary play; at this point he can be shown the proof, especially a constructive proof. But at no point can he be expected to enthuse over, or even to tolerate, the systematic derivation of 17 quadrature formulas via some finite difference operator calculus. He doesn't need the 17 quadrature formulas, and he doesn't need to know how to derive them—elegantly or at all.

The book is divided into two parts, either of which could form the basis for a one-semester course in numerical methods. Part I discusses most of the standard techniques: roots of transcendental equations, roots of polynomials, eigenvalues of symmetric matrices and so on. This material can profitably be learned at the sophomore or junior level and, indeed, with the increasing availability of automatic computing equipment on American campuses, much of it is already diffusing into the standard freshman and sophomore courses in mathematics, physics, and engineering.

Part II cuts across the basic tools, stressing such common problems as instabilities in extrapolation, removal of singularities, loss of significant figures, and so on. It also introduces some of the methods that are useful with the larger problems associated with partial differential equations. At Princeton many students take the material of Part II as their first formal course in numerical methods—having absorbed the earlier material largely by osmosis. Some remedial reference to Part I is occasionally needed, but surprisingly little has been necessary in the last two years. The material of Part II is normally taken at Princeton by juniors and seniors.

I have tried to write a readable book. Clarity in presenting major points often requires the suppression of minor ones, at least temporarily. Thus the trained mathematician will encounter statements that are "incorrect" in the sense that all the qualifications and exceptions necessary to make them precise are not set forth nearby. The instructor may wish to warn the students about these inaccuracies and even to supply the lacunae, but here I would recommend caution. An excess of fleshy detail before the student has a firm skeleton on which to hang it is a burden that frequently will bring down the entire structure, leaving a heap of rubble that has neither beauty nor utility. Exceptional information ought to be supplied the *second* time through the subject, or when the student asks for it. The curious student who wants more complete discussions may be referred to the book by Ralston with a warning that it is written for the first-year graduate student.

One topic is largely missing: formal error analysis with its emphasis on inequalities leading to bounds on approximations. I firmly believe such an analysis should be delayed until at least the third time through a numerical method. The student oriented toward results quickly discovers that the expedient way to test his methods is to run them again at half the interval and with slightly perturbed input data. Not that this expediency will catch all inadequacies in his finite difference approximations, but it will deal with a far greater percentage of them than will any formal error analysis.

I have not usually tried to prove convergence of any iterative process. It is a commonplace that numerical processes that are efficient usually cannot be proven to converge, while those amenable to proof are inefficient. Again my plea is insufficient pertinence to the student's purpose. The best demonstration of convergence is convergence itself. Judicious introduction of these two topics at the few places where they produce insight may be desirable, but they are so formidable in typographic appearance that all but the most docile students usually skip them. I prefer to leave their formal presentation to the instructor's discretion.

Finally, pedagogical expedience strongly suggests that the teaching of programming of digital computers be clearly separated from the teaching of numerical methods for solving problems, at least at the elementary levels. Any person who has mixed the two realizes how hopelessly the two sets of difficulties become intertwined both in the minds and in the practice of the students. At the advanced level, to be sure, the interaction between programming techniques and numerical algorithms is a fruitful study, but the sophomore should be exposed to fruitful studies in homeopathic doses! We recommend that the ability to write simple programs in FORTRAN, PL/1 or ALGOL be a prerequisite to this course, while access to a reasonable computer is almost a necessity.

FORMAN S. ACTON

TO THE STUDENT

I hope you like to formulate physical phenomena into equations and then solve those equations to see how well your formulations actually predict experimental results. It is for such people that I wrote this book. But if you have tried this fascinating game, you have realized that solving the equation is frequently harder than getting it in the first place.

The complexities of theoretical formulations soon frustrate our ability to solve their equations analytically, and numerical solution methods—when performed by hand—often require days of laborious arithmetic. Thus, in the past, engineers, always under pressure to produce working devices quickly, have tended to avoid theoretical formulations. To solve anything quickly before the advent of the electronic computer meant to solve it analytically, and this usually required analytic simplification to the point where the answers had only a remote connection with the original problem. Experiment with an actual device was much quicker.

In the 1940s a big change occurred in some sectors of Engineering. Atomic explosions and jet aircraft required experiments that were no longer simple, cheap, or quick. Five years and millions of dollars were involved before a new idea for a jet engine could be pronounced successful, and the costs of an atomic test were not measured merely in money. Theoretical investigation suddenly had become a necessity for the engineer instead of a pleasant pastime. In 1950, Von Neumann's computer opportunely appeared, solidly establishing the theoretical investigation as a practical developmental tool, sometimes *the* practical tool, and nothing has been the same since.

The incorporation of numerical methods into the engineering curriculum is only now taking place, and there is still considerable discussion over how it should be done. Thus, if you are a student, you are probably coming into an upperclass course in Numerical Methods after having had the classical exposure to calculus and differential equations that has been pretty much standard since Newton (but without Newton's numerical experience!) or so I have assumed while writing this book. (If anybody asks me, I'll be glad to suggest a quite different curriculum that I feel would be far preferable for the final quarter of the twentieth century.)

Having disposed of the question "*Why* learn about numerical methods?" let me now get around to the equally important question of *how* to learn about them. I shall be brief.

Numerical equation solving is still largely an art, and like most arts it is learned by practice. Principles there are, but even they remain unreal until you actually apply them. To study numerical equation solving by watching somebody else do it is rather like studying portrait painting by the same method. It just won't work. The principal reason lies in the tremendous variety within the subject. By contrast, analytic solution methods work for very restricted classes of problems. Thus we know how to solve ordinary differential equations analytically, provided they are linear and with constant coefficients! Let them have variable coefficients, and we become quite unsure of ourselves. If any nonlinearity creeps in, you might as well give up. But all three kinds of ordinary differential equations can be solved by numerical methods—provided, of course, that a solution exists. Thus it would be quite surprising if one numerical method succeeded everywhere—and no single method does!

The art of solving problems numerically arises in two places: in choosing the proper method and in circumventing the main road-blocks that always seem to appear. So throughout the book I shall be urging you to go try the problems—mine or yours.

I have tried to make my explanations clear, but sad experience has shown that you will not really understand what I am talking about until you have made some of the same mistakes that I have made. I hesitate to close a preface with a ringing exhortation for you to go forth to make fruitful mistakes; somehow, it doesn't seem quite the right note to strike! Yet, the truth it contains is real. Guided, often laborious, experience is the best teacher for an art. If all you desire is a conversational knowledge of an art, you've chosen the wrong subject, the wrong author, and just possibly the wrong profession. It is one of the minor paradoxes of our language that, even in the 1970s, you learn how to solve real problems only by getting your hands dirty with rational numbers—although rational problems can frequently be solved only with real numbers. Good luck!

FORMAN S. ACTON

PART I

FUNDAMENTAL METHODS

BEFORE WE BEGIN—THE RAILROAD RAIL PROBLEM

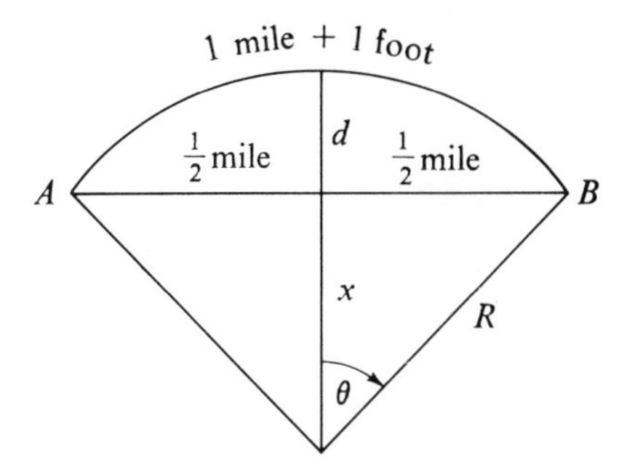

The following problem often appears as a puzzle in Sunday Supplements. The difficulties are numerical rather than formulative and hence it is an especially appropriate challenge to the aspiring numerical analyst. We strongly urge that the reader solve it in his own way before turning to the "official" solution.

A railroad rail 1 mile long is firmly fixed at both ends. During the night some prankster cuts the rail and welds in an additional foot, causing the rail to bow up in an arc of a circle.

The classical question concerns the maximum height that this rail now achieves over its former position. To put it more precisely: We are faced (see Figure) with a chord of a circle AB that is exactly 1 mile long and the corresponding arc AB that is 1 mile plus 1 foot and our question concerns the distance d between the chord and the arc at their midpoints.

The relationships available are the simple ones from trigonometry involving the subtended half-angle, θ, the standard functions of θ, and the Pythagorean relationship. The student at this point should attempt to solve the problem before turning to the solution given in Chapter 2. He should attempt to find the distance d to an accuracy of three significant figures. In his effort he will probably be faced with subtracting two large and nearly equal numbers, which will cause an horrendous loss of significant figures. He can live with this process by sheer brute force, but it will involve using eight-significant-figure trigonometric tables to preserve three figures in his answer. The point of the problem here is to find *another* method of calculating d, one that does not require such extreme measures. The three-figure answer can, indeed, be obtained rather easily using nothing more than pencil, paper, and a slide rule. The student should seek such a method.

CHAPTER 1

THE CALCULATION OF FUNCTIONS

Although one tends to think of engineering computation primarily in terms of integrating differential equations, finding the roots of polynomials, and solving transcendental equations, in actual fact a considerable part of the computer's efforts go into evaluating functions. A computer is built to add or multiply numbers—it is not built to find the cosine of an angle or evaluate a Bessel function. Neither are desk calculators, so human operators must resort to tables when these functions are encountered. But the stored program computer has too limited a memory to store extensive tables and hence it usually resorts to some algorithm that computes the required function directly from its argument. The commoner engineering functions, such as sines and cosines, logarithm, and exponentials, are so prevalent that most computer languages have machine-language subroutines already built into them to provide the customer with these standard functions upon request. The ease with which he may call them, however, should not blind the programmer to the fact that it takes perhaps a hundred machine operations to produce a single cosine, so he should not cause them to be computed unnecessarily. Specifically, if he wishes to evaluate an expression such as

$$y = a \cdot \cos^3 x + b \cdot \cos^2 x + c \cdot \cos x + d$$

he should write

$$T = COS(X)$$

$$Y = ((A*T + B)*T + C)*T + D$$

so that the cosine is only evaluated once. The expression

$$Y = A*(COS(X))**3 + B*(COS(X))**2 + C*COS(X) + D$$

will take at least three times longer.

Once we pass from the common functions we must supply our own algorithms and they are certain not to be short. Thus even though we think of our problem as the evaluation of the integral

$$\int_0^4 \frac{e^{-\sin^2 bx}}{1 + x^2} dx$$

by Simpson's rule, the computer actually spends well over seven eighths of its time evaluating the functions e^{-y} and $\sin x$. In the face of these facts we should attempt to understand the principles of functional evaluation, and this chapter is intended as an introduction.

The arctangent

We first consider the inverse tangent function because it exhibits most of the standard characteristics that we may expect in mathematical functions—even though most computer languages will already have an arctan subroutine built into them, and it is thus unlikely that the reader will have to write such a routine. We shall consider *four* algorithms for computing the inverse tangent of x:

1. Infinite (Taylor's) series.
2. Continued fraction.
3. Rational function approximation.
4. Gauss algorithm.

and we remark that most functions have the first three types of algorithms, while the last one comes under the heading of a special method. The reader should program at least two of these algorithms and test them on his computer for their relative efficiencies over a range of arguments.

The *series* for the arctangent

$$\arctan x = x - \frac{x^3}{3} + \frac{x^5}{5} - \frac{x^7}{7} + \cdots \qquad x \leq 1 \tag{1.1}$$

diverges for x greater than unity. This limitation is not crucial, for trigonometric identities allow us to take the reciprocal of x if it is greater than unity, finding the angle whose tangent is $1/x$, hence the angle whose cotangent is x, which is the complementary angle. Summarizing, for a given x greater than unity, θ may be computed from

$$\theta = \frac{\pi}{2} - \tan^{-1}\frac{1}{x} = \tan^{-1} x \tag{1.2}$$

and so we may always deal with x in the range $0 \leq x \leq 1$. Unfortunately this does *not* banish our troubles, for the series (1.1) converges very slowly for the larger values of x. (The reader should estimate how many terms of the series are needed to give five-figure accuracy for x equal to 0.4, 0.8, and 0.95.)

When x *equals* unity, the various powers of x help not at all in making the terms of the series smaller, and we have only the denominator factor. For five decimal places we need a denominator of 10,000 – or 5000 terms of the series! Clearly we wish to avoid the argument unity and, since the world is a continuous place, we better avoid arguments *close* to unity as well, for the behavior of the series there will not be much better. We may tolerate a 10-term approximation for a common function but not 5000 terms!

A *continued-fraction* representation for $\tan^{-1}x$ is

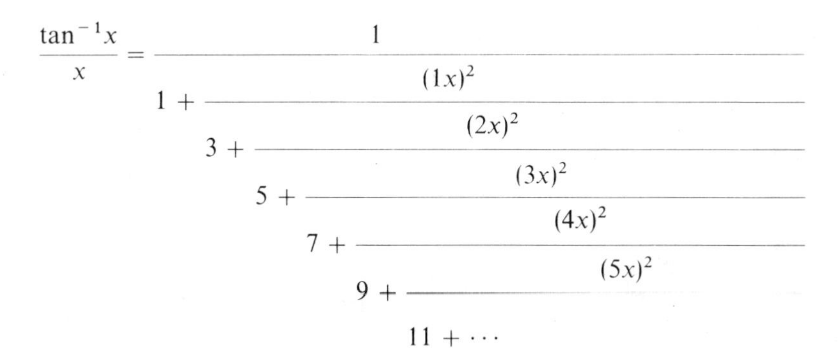

$$\frac{\tan^{-1}x}{x} = \cfrac{1}{1 + \cfrac{(1x)^2}{3 + \cfrac{(2x)^2}{5 + \cfrac{(3x)^2}{7 + \cfrac{(4x)^2}{9 + \cfrac{(5x)^2}{11 + \cdots}}}}}}$$

and we may truncate it where we will – or dare. Printers hate this form and you will usually find it expressed as

$$\frac{\tan^{-1}x}{x} = \frac{1|}{|1+} \frac{(1\cdot x)^2|}{|3+} \frac{(2\cdot x)^2|}{|5+} \frac{(3\cdot x)^2|}{|7+\cdots}$$

but your author always has to write it out in the first form before he is really sure of what he is doing.

One problem with the continued fraction is exactly this question of where to stop. With the series we merely keep calculating additional terms until the next is small enough so that we don't care any more. With the continued fraction we may decide that cutting after 7 is a good idea, but then to make sure that cutting at 9 is not significantly better we have to evaluate the entire truncated fraction a second time – our previous evaluation being quite useless, since the evaluation is from the bottom upward. (But see the rational function below.)

There is no sharp cutoff on values of x for which the continued fraction ceases to represent the arctangent. It merely requires more terms for a fixed accuracy as x increases. Thus values of x in the neighborhood of *unity* are not critical—although for reasons of efficiency we will probably still wish to throw large values of x back onto the 0–1 range by the reciprocal identity (1.2).

A *rational function* representation is given by two recursion relations, each of three terms and identical in form, although with different starting values. We define

$$\begin{array}{ll} \alpha_0 = 1 & \beta_0 = 1 \\ \alpha_1 = 1^2 & \beta_1 = 3 \\ \alpha_2 = 2^2 & \beta_2 = 5 \\ \vdots & \vdots \end{array}$$

Then

$$\begin{array}{ll} P_0 = \alpha_0 = 1 & Q_0 = \beta_0 = 1 \\ P_1 = \alpha_0\beta_1 = 3 & Q_1 = \beta_0\beta_1 + \alpha_1 x^2 = 3 + x^2 \end{array}$$

and thereafter

$$\left.\begin{array}{l} P_n = \beta_n P_{n-1} + \alpha_n \cdot x^2 \cdot P_{n-2} \\ Q_n = \beta_n Q_{n-1} + \alpha_n \cdot x^2 \cdot Q_{n-2} \end{array}\right\} \qquad n = 2, 3, \ldots$$

Finally

$$\frac{\tan^{-1} x}{x} \approx \frac{P_n}{Q_n} = \frac{\alpha_0}{\beta_0 +}\,\frac{\alpha_1 x^2}{\beta_1 +}\,\frac{\alpha_2 x^2}{\beta_2 + \cdots}$$

with the approximation improving as n increases. Again there is no sharp cutoff in x—larger values merely requiring more iterations, that is, larger n. Actually this algorithm is *algebraically* identical to evaluating the continued fraction to any number of steps, P_2/Q_2 being equivalent to the continued fraction truncated at 5, P_3/Q_3 being equivalent to the continued fraction truncated at 7, and so on. *Arithmetically*, the algorithms are quite different with different rounding problems, different programming difficulties. The principal advantage of this iterative evaluation of the rational function over the continued fraction lies in its ability to produce the next approximation from the preceding approximation with only an incremental calculational effort—precisely the property that the continued fraction lacks. We need only carry out the P and Q recursions one step further and form P_{n+1}/Q_{n+1} to get the next approximant. (The student should carry out the computation of $\tan^{-1} 0.8$ via the continued fraction and the rational recurrence to gain a feeling for their relative efficiencies.)

A difficulty peculiar to this rational function algorithm, however, is that the sizes of P_n and Q_n increase rapidly with n and there is serious danger of exceeding the largest number that the computer can represent. Since only the ratio P/Q

is of interest and since the recursions are linear in P and Q, we may choose, at any stage, to divide the four current values of P and Q by the same arbitrary constant without injuring the algorithm. But we must remember to test for excessively large P's and Q's and then carry out this normalization, or we will find ourselves with embarrassing machine overflows.

Finally, there is the algorithm attributed to *Gauss* that is based, ultimately, on the half-angle trigonometric identities. We define as before

$$\theta = \tan^{-1} x$$

then further define

$$a_0 = (1 + x^2)^{-1/2}$$

$$b_0 = 1$$

We continue with the recurrence

$$\left.\begin{cases} a_{i+1} = \frac{1}{2}(a_i + b_i) \\ b_{i+1} = \sqrt{a_{i+1} \cdot b_i} \end{cases}\right\} \qquad i = 0, 1, 2, \ldots$$

(Note that the subscripts under the radical are different.)

As we continue the recursion, a_i approaches b_i. When they agree to the precision in which we are interested we compute θ from

$$\tan^{-1} x \approx \theta = \frac{x}{(1 + x^2)^{1/2} \cdot L}$$

where

$$L = a_n = b_n$$

Like the continued fraction, this algorithm has no sharp limit on x but does exhibit precision difficulties as x gets large. It is relatively insensitive to x in the number of iterations it requires to achieve a prescribed accuracy. Again we advise the reader to try several values of x, comparing his computational labor with, say, the series for those same values.

The choice of a method

A person wishing to calculate the arctangent function has a wide choice of methods. Most functions of engineering interest are not so generously endowed, so perhaps with them the choice is easier. But whatever the function, the approximator must still decide a host of nagging details—details that annoy because they raise conflicts between aesthetics and expediency. Should one devise an approximation that gives only the five significant figures currently needed or the seven that the computer can handle? Should it hold only over the limited range of arguments at hand or be a general approximation for any argument that is reasonable in future problems? What kind of balance between memory storage requirements, accuracy of the approximation, and speed of evaluations is reasonable? These are questions on which even experts will often differ. We here can only offer some advice aimed principally at the beginner. If it seems obvious, so much the better, for we shall be in agreement.

The principal struggle, on which we feel strongly, concerns how much effort should go into making an approximation routine more efficient, more comprehensive than the current application justifies. Our answer: *When in doubt, make it more general!* Take advantage of all the accuracy that single-precision arithmetic can offer[1]; give back functional values for *all* reasonable arguments. Not only will you do mankind a service, but the life you save may be your own. The usual history of a so-called "one-shot" calculation in a computer center involves a resurrection a few days later, with different parameter values. Either the earlier ones turned out to have been wrong, or a new application of the process has just been discovered – but no matter, the one-shot calculation has returned to plague us in a slightly changed form. And you may bet your bottom dollar that those new parameter values will lie just outside any narrowly hand-tailored range of acceptable arguments in your old functional approximation routine. So beat the rap; make your approximation general *ab initio* rather than later when you've forgotten how the routine really works.

Finite algorithms

The preceding algorithms for the arctangent were all infinite. We truncated the series and the continued fraction at places where the remaining terms would contribute nothing new to the digits we needed in our answers. Similarly, we stopped the Gaussian iterations when we had sufficient accuracy in our result. In this sense we evaluated *approximations* to the arctangent, all infinite representations of a function requiring such truncation.

Some algorithms for functions, however, are *finite* and, except for rounding errors, exact. We may compute the cosine of 2θ via

$$\cos 2\theta = 2\cos^2\theta - 1$$

provided we have $\cos\theta$ readily available. Similarly, we can produce the sequence $\{\sin n\theta\}$ for all positive integers n from the three-term *recurrence relation*

$$\sin(n+1)\theta = 2\cdot\cos\theta\cdot\sin n\theta - \sin(n-1)\theta$$

with only the values of $\cos\theta$ and $\sin\theta$ being required to start. The number of

[1] Indeed, if your machine is irresponsible enough to provide less than 40 to 48 bits for its mantissa, you should preserve *double*-precision accuracy for any engineering functions.

arithmetic operations needed to produce any particular $\sin n\theta$ is very finite. The infinite part of the algorithm is buried in the starting values of $\cos\theta$ and $\sin\theta$.

For reasons of efficiency these finite recurrence algorithms are very desirable – when they work. Unfortunately, since earlier values of the sequence are used to produce the later values, the rounding errors in earlier values may be amplified by the recurrence to the point where they destroy its usefulness. On the other hand, another recurrence may act to suppress the rounding errors. Almost all the standard functions of mathematical physics and engineering possess three-term recurrence relations that are useful for computing the function over *some* range of their argument, but for other ranges the recurrences become unstable. This topic is treated in Chapter 16, so here we merely point to the possible difficulty by way of warning.

Evaluating series of orthogonal functions

The efficient technique for evaluating the finite Fourier series is not obvious – until one has seen it done. The conciseness of the standard mathematical representation obscures the proper approach. Suppose we wish to evaluate

$$g(\theta) = \sum_{k=0}^{8} b_k \cos k\theta \tag{1.3}$$

where the b_k are known numbers. At various times we will be supplied with a value for θ and told to find $g(\theta)$. If we are writing a computer program, we shall probably construct a small repetitive loop that multiplies each of the cosines by its coefficient and accumulates the products. Thereby we would waste enormous amounts of computer time! Even if we were evaluating our expression by hand, the notation suggests that we compute eight separate arguments, look up eight separate cosines in a table, multiply by the appropriate b_k's, and accumulate. Again a lot of our time would have been wasted.

The essential fact, obscured by our mathematical notation, is the ease with which $\cos k\theta$ may be computed from adjacent cosines in our series. Once θ is specified, the entire sequence of cosines is specified, and, in one sense, it is a waste of time to compute them separately from their series definitions or to look them up separately in tables. Once we know $\cos\theta$ all the rest of the information is contained in the b_k's. The important relation is the standard three-term recurrence relation

$$\cos(k-1)\theta - 2\cos\theta\cdot\cos k\theta + \cos(k+1)\theta = 0 \tag{1.4}$$

Using this relation we may express any $\cos k\theta$ in terms of $\cos\theta$ and some constants. In evaluating (1.3) we may do even better – *we may express the entire finite Fourier series in terms of* $\cos\theta$ *and two constants.*

The essential device is to define a new sequence of coefficients c_k recursively by

$$c_k = (2\cos\theta)\cdot c_{k+1} - c_{k+2} + b_k \qquad k = 8, 7, \ldots, 0 \tag{1.5}$$

and then begin evaluating at the high value of k, defining c_9 and c_{10} to be zero. We may then compute $c_8, c_7, \ldots, c_0$ entirely from the b_k and one evaluation of $2\cos\theta$. Further, the arithmetic is very finite, each c_k requiring only one multiplication and two additions.

It is instructive to derive our final result by formally substituting for the b_k in (1.3). We have

$$g(\theta) = [c_8 - 2\cos\theta\cdot c_9 + c_{10}]\cos 8\theta + [c_7 - 2\cos\theta\cdot c_8 + c_9]\cos 7\theta$$
$$+ \cdots + [c_1 - 2\cos\theta\cdot c_2 + c_3]\cos\theta + [c_0 - 2\cos\theta\cdot c_1 + c_2]\cdot 1$$

which rearranges into

$$g(\theta) = c_8[\cos 8\theta - 2\cos\theta\cdot\cos 7\theta + \cos 6\theta]$$
$$+ c_7[\cos 7\theta - 2\cos\theta\cdot\cos 6\theta + \cos 5\theta] + \cdots$$
$$+ c_2[\cos 2\theta - 2\cos^2\theta + 1] + c_1[\cos\theta - 2\cos\theta] + c_0$$

We note that all the terms except the last two now disappear because of (1.4). Thus our final evaluation of $g(\theta)$ is given by

$$g(\theta) = c_0 - c_1\cdot\cos\theta \tag{1.6}$$

There may be some accumulation of error in the recursive calculation of the c_k, but Clenshaw [1955] has shown that this is not crucial in the final result.

This method for evaluating a finite series of orthogonal functions that are connected by a linear recurrence is a technique generally available for all the standard "special functions" of mathematical physics. Thus Legendre polynomials, Chebyshev polynomials, Bessel functions, and others have two- or three-term recurrence relations that remove the need for evaluating any of the higher members of the series. A simple recursion in the coefficients followed by a still simpler use of c_0 and c_1 usually suffices for evaluating the several functions themselves. The device is very efficient but seems not to be sufficiently publicized. (The student should derive the corresponding expressions for evaluating a finite sine series and also a series of Chebyshev polynomials.)

Further finite algorithms

Interpolation in a table is another finite algorithm for evaluating functions. Because of the large number of functions and the rather small memories of early computers, this technique found little favor. With contemporary computers tending to have memories measured in the tens and even hundreds of thousands of words, however, condensed tables plus a sophisticated interpolation scheme may well give those economies of time that are essential in very long problems. The interpolation techniques are discussed in Chapter 3. Here we need only to point out that the table would have to be generated internally from some basic algorithm, since the alternative is to introduce it from punched cards—far too time consuming and conducive to error.

Finally, the simplest finite algorithm is a *transformation* of a function into another for which an adequate algorithm is already encoded. The integral

$$F(b) = \int_0^\infty \frac{e^{-bt}}{1+t} dt$$

by a simple change of the dummy variable becomes the product of two somewhat more familiar functions. On first replacing t by $(s - 1)$ and then replacing bs by t, we have

$$F(b) = \int_1^\infty \frac{e^{-b(s-1)}}{s} ds = e^b \cdot \int_1^\infty e^{-bs} \frac{ds}{s} = e^b \int_b^\infty e^{-t} \frac{dt}{t} = e^b \cdot Ei(b)$$

The exponential integral, $Ei(b)$, has the same problems at the origin that $F(b)$ does, but it is a fairly common function and may already be known to the local computer laboratory. Its approximation, in turn, includes the logarithm, $\ln b$, explicitly removing the trouble at the origin into one of the commonest functions. Presumably $\ln b$ has already been adequately encoded, even for small arguments. (This "removal of the singularity" is a recurring theme throughout our book; see Chapter 15.)

Rephrasing a function so that it actually is calculated as two or more other functions can, however, be dangerous. Once $\cos x$ is available, we may wish to save computational steps by producing $\sin x$ from

$$\sin x = \sqrt{1 - \cos^2 x}$$

but small arguments will give us precision troubles that do not arise when $\sin x$ is computed directly.

Power series from differential equations

The chief virtue of a power series for representing a function is its availability. The *Handbook of Mathematical Functions* [Abramowitz and Stegun, 1964] (hereafter called AMS 55) is full of them. Assuming we cannot find the series in standard references, it can usually be generated from a differential equation, and most functions satisfy some differential equation of first or second order. One must, however, be on one's guard against a few standard difficulties, chief of which is the presence of singularities. A power series is, after all, a rather long polynomial, and while polynomials can wiggle in a number of satisfactory ways, there *are* geometries they are notoriously poor at representing. No polynomial, for example, ever had an asymptote – vertical or horizontal – so if your function has one of these, beware a power-series representation. Either there will be none, or it will be grossly inefficient. (Rational functions, on the other hand, can become asymptotic to all sorts of lines, so they may do quite well.)

Before we apply the classical techniques for generating a power series from a differential equation, we would comment on a somewhat pedestrian but persistent difficulty that arises from power series with alternating signs. Let us consider for a moment that most wonderfully convergent of all common series:

$$e^{-x} = 1 - \frac{x}{1!} + \frac{x^2}{2!} - \frac{x^3}{3!} + \frac{x^4}{4!} - \cdots$$

Mathematically, the series is absolutely and uniformly convergent for all x – the highest blessing that can be bestowed. Let us consider, however, a value of x equal to 10. The term

$$+\frac{x^4}{4!} = +\frac{10{,}000}{24} \approx 416.6\ldots$$

What is worse, the term $x^{10}/10!$ is nearly 3100. Since e^{-10} is a number very much less than 1 (actually 0.000041), it is clear that other terms about as big as 417, but negative, must cancel all the figures in front of the decimal point and some behind as well. Since a computer can only hold a fixed number of significant figures, all those in front of the decimal point are not only useless – they are crowding out needed figures at the right end of the number. Unless we are very careful we will find ourselves adding up a series that finally consists entirely of rounding errors.

In this particular example we have a simple cure, for e^x is the reciprocal of e^{-x} and we may use the other series without the alternating signs, but such an easy alternative is not usually available. There is the additional point that as x increases we must take more terms for a fixed accuracy, but this is a less important factor. The cancellation of significant figures in the alternating series is an absolute limitation on the accuracy we may achieve. Further effort by taking more terms will not help. Only multiple-precision arithmetic will get us around these troubles, and that is a cure that is usually worse than the disease. We must seek elsewhere.

Let us now turn to one version of the common but not completely trivial error function

$$F(x) = \frac{2}{\sqrt{\pi}} \int_0^x e^{-t^2} \cdot dt = \operatorname{erf}(x) \tag{1.7}$$

We can find the Maclaurin series easily by replacing the exponential by its series and integrating term by term. We get the alternating series

$$F(x) = \frac{2}{\sqrt{\pi}} \cdot x \cdot \left[1 - \frac{x^2}{1!3} + \frac{x^4}{2!5} - \frac{x^6}{3!7} + \cdots\right] \tag{1.8}$$

that has already lost us one significant figure when x is 2, having a maximum term of about 1.4. Furthermore, it requires 19 terms before it yields seven significant figures at this argument. We shall probably be unhappy about using (1.8) for an x much larger than unity, where nine terms suffice for seven figures.

Fortunately, Taylor's and Maclaurin's series are not the only ones available. The strong influence of e^{-t^2} in our function suggests that we might seek a series of the form

$$F(x) = e^{-x^2} \cdot (\text{series})$$

which is most expediently found via the differential equation for $F(x)$. We differentiate (1.7) twice to find

$$F'(x) = Ce^{-x^2}$$

$$F''(x) = C(-2x)e^{-x^2}$$

so that

$$F''(x) + 2xF' = 0 \tag{1.9}$$

Letting

$$F(x) = e^{-x^2} \cdot G(x)$$

we obtain

$$G'' - 2xG' - 2G = 0 \tag{1.10}$$

for G. Since $F(0)$ is zero, so must $G(0)$ also be zero. Similarly,

$$G'(0) = \frac{2}{\sqrt{\pi}} = C$$

Thus we let

$$G = (b_0 + b_1 x + b_2 x^2 + \cdots) \tag{1.11}$$

with

$$b_0 = 0$$

$$b_1 = \frac{2}{\sqrt{\pi}}$$

Differentiating (1.11) twice and substituting it into the differential equation (1.10), we require the expression to be an identity for all values of x; hence the coefficients of every power of x must be zero. We obtain

$$-2b_0 + 2 \cdot 1 \cdot b_2 = 0$$
$$-2b_1 - 2 \cdot 1 \cdot b_1 + 3 \cdot 2 \cdot b_3 = 0$$
$$-2b_2 - 2 \cdot 2 \cdot b_2 + 4 \cdot 3 \cdot b_4 = 0$$
$$-2b_3 - 2 \cdot 3 \cdot b_3 + 5 \cdot 4 \cdot b_5 = 0$$

which gives all even b_j as zero and the odd b_j as

$$b_1 = \frac{2}{\sqrt{\pi}}$$
$$b_3 = \tfrac{2}{3} b_1$$
$$b_5 = \tfrac{2}{5} b_3$$
$$b_7 = \tfrac{2}{7} b_5$$

or the series

$$F(x) = \frac{2}{\sqrt{\pi}} e^{-x^2} \cdot x \cdot \left[1 + \frac{2x^2}{1 \cdot 3} + \frac{(2x^2)^2}{1 \cdot 3 \cdot 5} + \frac{(2x^2)^3}{1 \cdot 3 \cdot 5 \cdot 7} + \cdots \right] \tag{1.12}$$

This representation requires us to evaluate e^{-x^2} once, but the series does not alternate, so we do not lose precision from cancellation of leading digits. On the other hand, the series converges quite slowly once the argument x exceeds $1/\sqrt{2}$, so from the computational labor point of view we are less happy about (1.12) than its more carefree cousin (1.8).

Asymptotic series

Many functions have a second series representation which works best for large x—an *asymptotic* series. These are divergent series, but if truncated properly they are often very efficient representations of their function.

Suppose we wish to evaluate the complementary error function, erfc(x), which is defined as

$$\operatorname{erfc}(x) = 1 - \frac{2}{\sqrt{\pi}} \int_0^x e^{-t^2} \cdot dt = \frac{2}{\sqrt{\pi}} \int_x^\infty e^{-t^2} \cdot dt$$

The series

$$\operatorname{erfc}(x) = \frac{2}{\sqrt{\pi}} e^{-x^2} \cdot \frac{1}{2x} \cdot \left[1 - \frac{1}{(2x^2)} + \frac{1 \cdot 3}{(2x^2)^2} - \frac{1 \cdot 3 \cdot 5}{(2x^2)^3} + \cdots \right] \tag{1.13}$$

can be obtained by inserting t/t into the integrand and then repeatedly integrating by parts. It is not hard to show that the last term of such an integration, an integral, is of the same order of magnitude as the last term of the preceding series. Since we are obliged to discard the final integral, it is our error term. Thus our error is about the size of the last term that we retain in the series.

For any given value of x, however, this alternating series (1.13) diverges. More precisely, the terms start to decrease in absolute value, reach a minimum, and thereafter grow without end because the factorial in the numerator will eventually outweigh the corresponding power of any *fixed* x in the denominator. Putting it differently, we get each new term of (1.13) by multiplying the denominator of the current term by $2x^2$ and the numerator by $(2n + 1)$. For a fixed x the denominator factor is constant while the numerator factor grows without limit. This view might incline us toward pessimism.

On the other hand, if we truncate the series (1.13) at its *minimum* term, we know that it represents our function to within an error of the size of that term – which will be very small indeed if x is, for example, 10. (What size is that term?) We are thus faced with the curious phenomenon of a truncated divergent series providing a good representation of our function, at least for sufficiently large arguments x.

This behavior characterizes all *asymptotic* series. For any specified argument, there is an accuracy beyond which the series cannot hope to represent the function. The larger x is, the more accurate the asymptotic series can be, for the minimum term is smaller. This phenomenon has nothing to do with rounding errors or limited precision of arithmetic within some computer. The limitations are inherent in the form of a series which specifies the size of its minimum term. If, for example, we wish to evaluate (1.13) for an x equal to 3, the most precision we can hope to obtain from this series is

$$\frac{1 \cdot 3 \cdot 5 \cdots 17}{(18)^9} = 0.000174$$

which, since this term must be compared to unity, is somewhat less than four significant figures. If four figures is insufficient, then we cannot use this asymptotic series at this argument. We can, however, derive *other* computational forms from this series that may yield numerical approximations with greater accuracy for the same arguments. (See Chapters 11 and 12.)

Asymptotic series are frequently found, as we have here, by repeated integration by parts. The other principal technique is from differential equations, first having replaced the independent variable by its reciprocal. Thus, if we know that we should express $\text{erfc}(x)$ as

$$\text{erfc}(x) = \frac{2}{\sqrt{\pi}} \int_x^{\infty} e^{-t^2} \cdot dt = \frac{2}{\sqrt{\pi}} e^{-x^2} \cdot G(x) \tag{1.14}$$

where $G(x)$ will be the asymptotic series, we take the differential equation for the integral

$$F = \int_x^\infty e^{-t^2} \cdot dt$$

which is

$$F'' + 2xF' = 0$$

and generate the differential equation for $G(x)$, giving

$$G'' - 2xG' - 2G = 0 \tag{1.15}$$

Finally, we replace the independent variable by its reciprocal. Letting

$$y = \frac{1}{x}$$

we have

$$\frac{dG}{dx} = -y^2 \frac{dG}{dy} \qquad \text{and} \qquad \frac{d^2G}{dx^2} = y^4 \frac{d^2G}{dy^2} + 2y^3 \frac{dG}{dy}$$

so (1.15) becomes

$$y^4 G'' + (2y + 2y^3)G' - 2G = 0 \tag{1.16}$$

Making the usual series substitution

$$G(y) = b_0 + b_1 y + b_2 y^2 + \cdots$$

into (1.16) and equating coefficients, we find

$$\begin{aligned} 0 = b_0 &= b_2 = b_4 = \cdots \\ b_1 &= \text{arbitrary} \\ b_3 &= -\tfrac{1}{2} b_1 \\ b_5 &= -\tfrac{3}{2} b_3 \\ b_7 &= -\tfrac{5}{2} b_5 \\ &\vdots \end{aligned}$$

so

$$G(y) = b_1 y \left[1 - \frac{1}{2} y^2 + \frac{1 \cdot 3}{2^2} y^4 - \frac{1 \cdot 3 \cdot 5}{2^3} y^6 + \cdots \right]$$

and our final representation of (1.14) is

$$\operatorname{erfc}(x) = \frac{2}{\sqrt{\pi}} e^{-x^2} \cdot \frac{1}{2x}\left[1 - \frac{1}{(2x^2)} + \frac{1 \cdot 3}{(2x^2)^2} - \frac{1 \cdot 3 \cdot 5}{(2x^2)^3} + \frac{1 \cdot 3 \cdot 5 \cdot 7}{(2x^2)^4} - \cdots\right]$$

as before. Whether this technique is easier than integration by parts depends very much on the integral.

The gap

The asymptotic series (1.13) we investigated in the previous section has a rather stubborn lower limit. We cannot get more than four significant figures for any x less than 3.0. At 2.0 it only gives two significant figures – extremely unsatisfactory. On the other hand, the power series for $F(x)$ that are efficient for small x, (1.8) and (1.12), rapidly lose their attractiveness as x grows beyond unity, or even beyond $1/\sqrt{2}$. They lose efficiency by requiring more terms, while the alternating series loses precision as well.

We are thus left with an embarrassing gap

$$0.707 \leq x < 3.0$$

within which we have proposed no effective approximation. If we happen to discover the continued fraction for erfc(x), we may close the gap fairly well. Failing that expedient, a technique called economization comes to our rescue. (See Chapter 12.) If we economize the asymptotic series, not only can we move its useful lower limit down to 0.707, meeting the other series, but we can even achieve an accuracy of nine significant figures! Here lies the real power of economization. We are tempted to generalize and say that it makes automatic approximating of nasty functions by only two series a really practical process. Unfortunately for our ego, Hastings [1955] has made a small change of variable, letting $y = 1/(1 + 0.3275911x)$, and then has approximated the error integral to seven significant figures using a *single* series for the entire range $0 \leq x < \infty$. These flashes of inspiration are not granted to most ordinary mortals, however, much less to a machine.

The infinite product

Although series and continued fraction representations are by far the commonest useful forms for function approximation, one should not forget the infinite product. The sine function, for example, may be written

$$\frac{\sin z}{z} = \prod_{k=1}^{\infty}\left(1 - \frac{z^2}{k^2\pi^2}\right) \tag{1.17}$$

which converges quite efficiently when z is small. As the notation suggests, the form permits evaluation for complex values of z—at the price of complex multiplications. The series for $\sin z$ is so well known and tractable that it quite overshadows this product representation, but for some applications the ability to split out one or more factors explicitly may give the infinite product a distinct advantage. Thus, if we had to solve the equation

$$\sin x = x^n \left(1 - \frac{x^2}{\pi^2}\right)$$

we would strongly prefer the product form, permitting us to recast the problem as

$$\prod_{k=3}^{\infty} \left(1 - \frac{x^2}{k^2\pi^2}\right) = \frac{x^{n-1}}{1 - \dfrac{x^2}{4\pi^2}}$$

in which the left side is relatively insensitive to changes in x. We may thus set up a rapidly converging iteration to solve the equation. The infinite-product forms of many common functions may be found in AMS-55 [Abramowitz and Stegun, 1964].

Recurrence relations—continued

While most persons are quite familiar with infinite series as a tool for computing transcendental functions, recurrence algorithms seem to have received much less attention than they deserve. For they are often more efficient than the series and they may also be used in computational situations for which series are inappropriate. We here explore some of the uses, and limitations, of this tool.

The cosine, to return to a familiar function, satisfies the simple recurrence relation

$$\cos(n+1)\theta + \cos(n-1)\theta = (2 \cdot \cos\theta) \cdot \cos n\theta \tag{1.18}$$

which for a fixed θ has the form

$$C_{n+1} + C_{n-1} = b \cdot C_n \tag{1.19}$$

This is a linear recurrence on n that permits any of three consecutive cosines to be calculated if the other two are known. Of course b needs to be known, too, but it does not change and thus must be computed only once. An obvious use might be to calculate cos(0.4), cos(0.6), cos(0.8), . . . , starting with cos(0) and cos(0.2), which are easily obtained from tables or the series. We illustrate, b being 2 cos(0.2) or 1.960 134:

		Error
C_0	1.000 000	0
C_1	0.980 067	0.4
C_2	0.921 063	2
C_3	0.825 340	4
C_4	0.696 714	7
C_5	0.540 313	11

We see a disturbing error in our values, an error that was initially small but is now growing at an increasing rate. It has destroyed the last two significant figures and will soon become much worse. Clearly, recurrence relations have some problems! If we look back, we find our value of C_1 is off by nearly half a unit as the inevitable concomitant of rounding. The multiplier b, which in this problem is twice C_1, is off by a full unit and we might feel that this is the source of the trouble. Accordingly, we correct b and compute again:

		Error
C_0	1.000 000	0
C_1	0.980 067	0.4
C_2	0.921 062	1
C_3	0.825 337	1
C_4	0.696 708	1
C_5	0.540 303	1

This time the error in C_1 is doubled, but thereafter it remains constant at one unit in the last place.

In this application of our algorithm (1.19) the value of the multiplier b is approximately 2. If the errors in two successive C's are equal and of the same sign, then they will be maintained with neither diminution nor increase—since the more recent error gets doubled but then has the earlier one subtracted from it. If, however, the error in C_2 should be $+\varepsilon$ and the error in C_3 should be $-\varepsilon$, the error in C_4 will be -3ε and that of C_5 will be -5ε. Thus in our example we have been lucky, for our algorithm *is* a noise amplifier—though not an exponential one.

We shall now try another recurrence, this for the common Bessel function $J_n(x)$. It is

$$J_{n-1}(x) + J_{n+1}(x) = \frac{2n}{x} J_n(x) \tag{1.20}$$

and, for fixed x, is a recurrence for the sequence of J_n's. This time, however, the coefficient of J_n depends on n – giving the computation a rather different behavior. For our experiment we choose to compute $J_2(1)$, $J_3(1)$, ... from $J_0(1)$ and $J_1(1)$. We obtain

n			Error
	$J_0(1)$	0.765 198	from tables
0	$J_1(1)$	0.440 051	from tables
1	$J_2(1)$	0.114 904	+ 1
2	$J_3(1)$	0.019 565	+ 2
3	$J_4(1)$	0.002 486	+ 9
4	$J_5(1)$	0.000 323	+ 73
5	$J_6(1)$	0.000 744	+ 723

Here our computational process breaks down very quickly. The errors present in J_n get multiplied by $2n/x$ in the next step and only get reduced by the error in J_{n-1} through the subtraction. For our value of x, the coefficient $2n/x$ is always large, and the process is thus clearly unstable. Not that we can't use unstable processes in computation, far from it! But we clearly have to be careful about how far we push an unstable process, and we must take the precaution of starting with enough significant figures so that the instability has not yet progressed into the figures that interest us before we stop. In our current example the errors are damned further by the fact that the values of $J_n(x)$ are getting smaller, but this is not the important point. *Any* exponentially increasing error is a cause for serious concern unless the function is increasing at least as rapidly.

But wait a moment! The recurrence formula (1.20) can be used in either direction. Perhaps it would be good strategy to use it to generate the $J_n(1)$ in their direction of *increasing magnitude* – that is, of decreasing n. Then maybe the error would increase no faster than the function itself and be therefore tolerable. We experiment:

$J_7(1)$	0.000 002	from tables
$J_6(1)$	0.000 021	from tables
$J_5(1)$	0.000 250	
$J_4(1)$	0.002 479	+ 2
$J_3(1)$	0.019 582	+ 19
$J_2(1)$	0.115 013	+ 110
$J_1(1)$	0.440 470	+ 419
$J_0(1)$	0.765 927	+ 729

Better, not in the absolute sense, for the absolute error is very much the same. But the *relative* error is nearly constant. If we begin with values of J_7 and J_6 that carry four and five significant figures, respectively, we get

$J_7(1)$	0.000 001 502	
$J_6(1)$	0.000 020 938	
$J_5(1)$	0.000 249 754	+ 0
$J_4(1)$	0.002 476 602	+ 0
$J_3(1)$	0.019 563 06	+ 0
$J_2(1)$	0.114 902	+ 1
$J_1(1)$	0.440 044	− 6
$J_0(1)$	0.765 186	− 12

Our values here are quite good, though a buildup of errors seems finally to be eroding the digits that interest us.

We finally compute our Bessel functions once more, using the homogeneity of the recurrence to allow us to ignore scale factors; that is, kJ_n satisfies (1.20) just as well as J_n does. We also use the fact that $J_n(1)$ tends toward zero as n increases. It is not difficult to discover (perhaps from a table) that $J_8(1)$ is zero to nearly six decimal places. Accordingly, we shall use the value of zero and we shall take $J_7(1)$ to be unity. Of course, $J_7(1)$ is not 1, and so what we write down is really $kJ_7(1)$, where the value of k is not known to us – and our value of J_8 may also be considered to contain the same unknown scale factor since $kJ_8(1)$ is $k \cdot 0$, still *zero*. We now compute via the standard recurrence down to $kJ_0(1)$. (In our example, since x is an integer, we may use integer arithmetic – throwing away nothing.) We have

kJ_8	0	
kJ_7	1	
kJ_6	14	
kJ_5	167	
kJ_4	1,656	
kJ_3	13,081	→ 0.019 5634
kJ_2	76,830	
kJ_1	294,239	
kJ_0	511,648	→ 0.765 1979 (the error is less than 2 in the last figure)
	$k = 668{,}648$	

We then normalize our values by the relation, also linear,

$$J_0(x) + 2[J_2(x) + J_4(x) + \cdots] = 1$$

which, when multiplied through by k, gives

$$kJ_0(x) + 2[kJ_2(x) + kJ_4(x) + \cdots] = k$$

and we get k from our table, which is then used to divide each tabular entry to produce the actual $J_n(1)$ shown. The accuracy is excellent. In fact, this is the efficient and preferred method for computing J_n even when only the final J_0 is wanted. If we should need the entire sequence, they are all available at the price of one more division for each J_n.

It may happen that the argument b lies close to a zero of $J_0(x)$ so that the final application of our recurrence necessarily produces a large cancellation of significant figures. Under these circumstances any rounding noise that may have accumulated will necessarily appear as a large fraction of the final functional value. This warning does not constitute a criticism of the general method, however, since it applies with equal force to evaluations of $J_0(b)$ from the power series. When we know in advance that we are trying to evaluate a function near one of its zeros, we usually must seek special methods if we would preserve more than a small number of significant figures. This topic is too specialized for treatment in a general text. We content ourselves with referring the reader to AMS 55 for special series applicable near the zeros of the commoner Bessel functions and to an enlightening article by Olver [1952].

Approximation of *imprecisely* known functions

At the risk of making a lot of enemies, your author herewith states his conviction that a chapter on this subject does considerable harm—to the point of being downright dangerous if it makes *positive* recommendations. The topic is complicated by the great variety of information that may be known about the statistical structure of the errors in our functional values that we want to approximate. At the same time, people who wish to make these approximations are usually seeking a quick uncritical solution to a problem that can only be treated with considerable thought. To offer them even a hint of a panacea, usually with the words "least-squares fit," is to permit them to drown with that famous straw firmly grasped. The subject needs a book, and several are available [Acton, 1966]. Here, we limit ourselves to an imaginary conversation proffering largely negative advice.

Consider some hypothetical data from a wind tunnel that have been plotted in Figure 1.1. An analytical approximation is needed since the coefficient $C(m)$ enters into a differential equation that we wish to integrate numerically by a predictor-corrector method. We need an interpolation formula. We could fit a

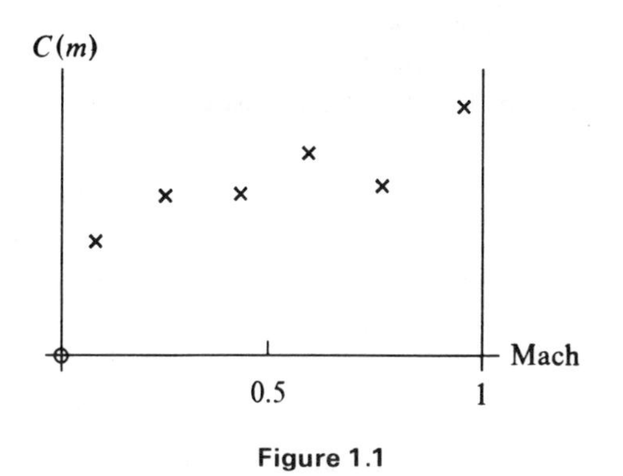

Figure 1.1

polynomial, or other functional form, exactly through all the points. "Absurd!" So says the aerodynamicist. "We know that the coefficient $C(m)$ is a smooth function of m." Well then, how about an average value – a horizontal straight line weighting all the points equally (one form of a least-squares fit!). "Of course not – can't you see there is an increase in C with m?" (This from the aerodynamicist.) O.K.! A straight line? "But the function curves, is concave downward!" Really? Statistical tests won't support you. That last point.... "Bother that last point! Everybody knows that measurements near Mach 1 are unreliable! Besides, I have a theory that says C depends on m in at least a quadratic manner." So now we are asked to fit our data by a parabola, ignoring the last point. But perhaps a cubic is needed to extract all the information from our data? "Yes, a cubic would be great! Err... how do we decide?" Well, that depends on what your functional model is and what information you have about the types and magnitudes of experimental errors that probably went into these points. A glazed look has been creeping into our aerodynamicist's eyes during this last sentence and he finally snaps out of it with an impatient, "What I really need is a Least-Squares Fit! Why don't you just do it and stop bothering me?"

Preliminary adjustment of the function

Approximation of functions is still a fine art, though most of this art occurs right at the beginning. It makes a tremendous difference whether or not the shape of the curve to be approximated bends in a way that permits efficient representation by the series or continued fraction we wish to use. We have considerable control of the shape of our function by our choice of independent variable and, to a lesser extent, of the dependent variable. Thus the curve $\sqrt{x}$ versus x has a vertical tangent at the origin which no polynomial can swallow. Even when we try to stay away from the origin by restricting our approximation to $1 \leq x \leq 10$, the infinite slope just offstage is still felt and prevents efficient polynomial representation. But if we plot $\sqrt{x}$ versus $10(2x-1)/(x+9)$ we find our curve has become almost a straight line (Figure 1.2), and approximating this new function by a polynomial is quite efficient. If we are willing

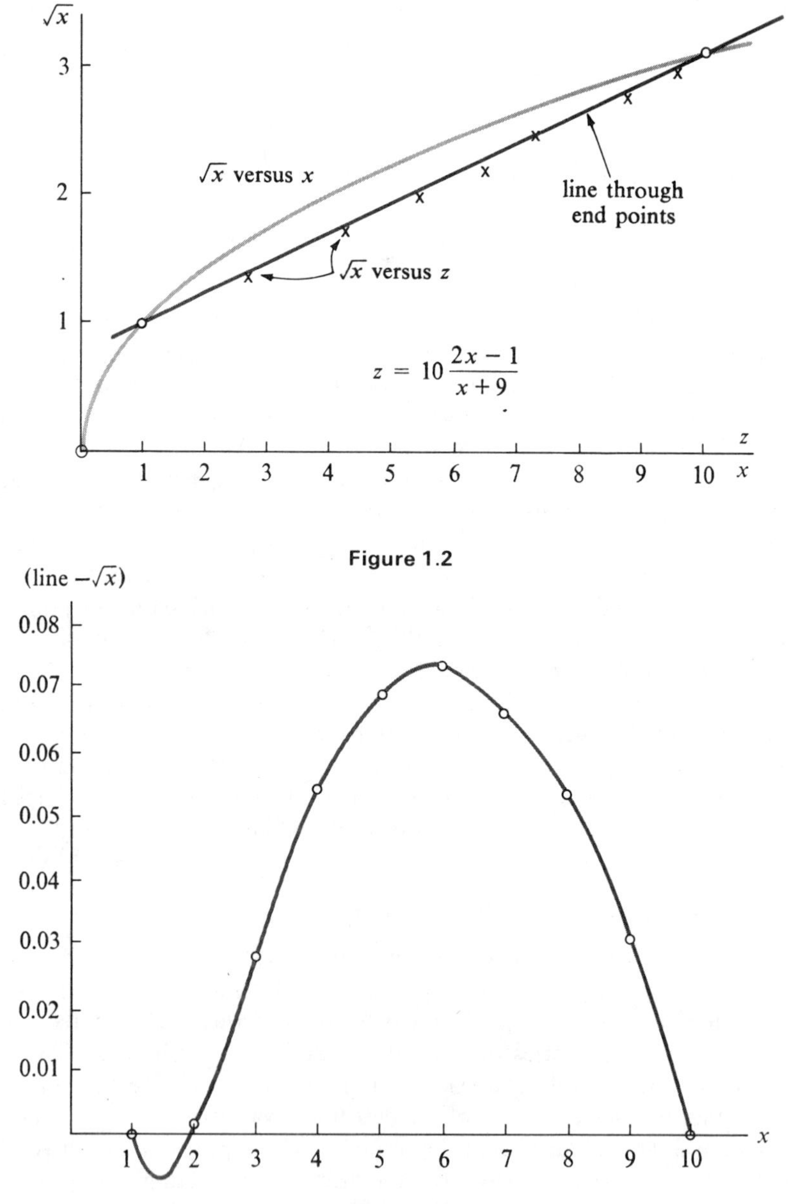

Figure 1.2

Figure 1.3

to use a parabola to represent the deviations from this line (Figure 1.3) then we obtain our square root to nearly three significant figures from

$$y = 10\frac{2x - 1}{x + 9}$$

$$\sqrt{x} = (0.760 + 0.2402y) + [0.0721 - 0.0041(x - 5.8)^2]$$

The change in the independent variable has made the curve bend the right way for polynomial approximation.

There are many such changes of variables that will produce similar improvement; some are undoubtedly better than this one (which was quickly chosen after a few minutes of slide-rule calculation). Finding the really good changes is very much a matter of patience and experience—it cannot be automated. Hastings, probably the greatest artist among the approximators [1955], in seeking values of $\log_2 x$ on the range $2^{-1/2} \leq x \leq 2^{1/2}$ changes the independent variable to $(x - 1)/(x + 1)$ and thereby gets 10 significant figures with only four terms of a series.

We have commented previously on the futility of computing e^{-x} from its series because of the accumulation of rounding errors from the differencing of large numbers that cannot contribute directly to the result. Fortunately, e^x has no such drawback and may be used instead. But e^x has its own minor problem—it has no symmetries and hence its power series possesses both even and odd terms. A symmetric function has a series that requires only half as much arithmetic for evaluation to a specified power. We shall usually gain a more efficient approximation if we can recast our general function into symmetric form. We note that the odd function

$$\tanh u = \frac{e^u - e^{-u}}{e^u + e^{-u}}$$

implies

$$e^x = \frac{1 + \tanh(x/2)}{1 - \tanh(x/2)}$$

To evaluate the hyperbolic tangent we have the continued fraction

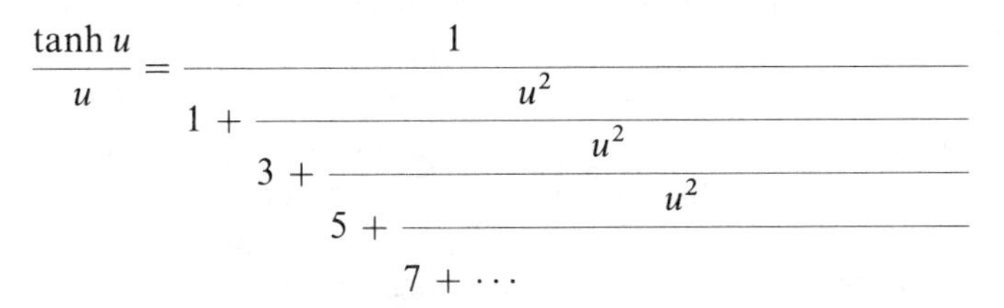

$$\frac{\tanh u}{u} = \cfrac{1}{1 + \cfrac{u^2}{3 + \cfrac{u^2}{5 + \cfrac{u^2}{7 + \cdots}}}}$$

or, if we prefer even series, we may write

$$\frac{\tanh(x/2)}{x/2} = 1 - \frac{z^2}{3} + \frac{2}{15}z^4 - \frac{17}{315}z^6 + \cdots$$

with

$$z = \frac{x}{2}$$

The price for the symmetry is a division by 2 (trivial on a binary machine) and the three arithmetic operations of the final fraction. There is another cost, however, for the hyperbolic tangent is not quite so tractable a curve to approximate as is e^x, so we do not really cut the series in half for fixed accuracy.

The approximation of functions for automatic computation

In creating a satisfactory functional approximation, three distinct but related decisions must be made:

1. What form of our function shall be approximated, and over what ranges?
2. What basic approximating tools shall be used?
3. Shall we optimize our approximation?

Probably the most difficult question is the first: the form of the function. Shall we approximate e^x and e^{-x}, or only e^x, or only $\tanh(x/2)$? If we choose e^x, shall we approximate it over (0, 1) or (0, 2) or (0, 0.1) – using the property that

$$e^{x+y} = e^x \cdot e^y$$

to reduce our original problem to the approximating range? Thus, for example, we could plan that $e^{4.3214}$ be calculated by being split into $e^4 \cdot e^{0.3} \cdot e^{0.0214}$ only the last factor being evaluated from our approximating algorithm, the others being supplied from a 20-entry table.

Similarly, for the periodic functions we must decide whether to evaluate both $\sin x$ *and* $\cos x$ for all arguments over one period directly from their series, or to use the half-angle formulas to reduce the argument to a smaller range that would permit the series to be shortened. The function thus found would then be parlayed back up by the double-angle formulas into the desired function. These are all questions of computational efficiency inside the complete subroutine. We are not proposing that each customer be required to reduce his arguments to the first quadrant or even the first period before invoking a cosine subroutine – no responsible programmer would do *that*. We *are* raising the point that it may save computational labor to put in some halving and doubling steps in order to gain the efficiencies of a shorter, simpler, approximation by confining this approximation to a smaller range of arguments.

Some functions, of course, do not permit easy condensation and we must approximate them over their entire range – probably using at least two different techniques. Thus

$$F(x) = \int_0^\infty \frac{e^{-t}}{x^2 + t^2} dt$$

has no periodic or other simplifying property that permits us to restrict x to some small standard range. We are stuck with all positive values of x and must approximate F by one technique for $(0, 3)$ and another for $(3, \infty)$.

Once a range of argument is chosen, we still have several approximating tools available. Shall we use a Taylor's series? A continued fraction? An infinite product? A recurrence relation (not based on any of these)? A small table with sophisticated interpolation? Again, questions of computational efficiency will decide.

A final function

To illustrate some of the points we have been making, we shall consider how we might reasonably approximate the function

$$F(b) = \int_0^\infty \frac{\tan^{-1} bx}{1 + x^2} dx \qquad 0 \le b < \infty \tag{1.21}$$

for use on automatic computers. We shall suppose that our accuracy requirements are not severe – an approximation that is good to three or four significant figures will suffice.

The function appears to be well behaved (Figure 1.4). We can easily evaluate it at unity, where it is $\pi^2/8$, and at both extremes. We wonder if we must seek two approximations, one for large b and one for small. Integration

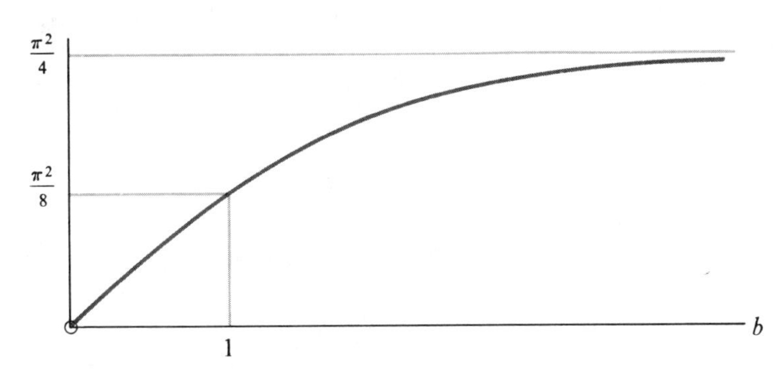

Figure 1.4. $F(b)$ plotted as a function of b

by parts answers this question:

$$
\begin{aligned}
F(b) &= \Big[\tan^{-1}bx \cdot \tan^{-1}x\Big]_0^\infty - \int_0^\infty \tan^{-1}x \frac{b\,dx}{1+b^2x^2} \\
&= \frac{\pi^2}{4} - \int_0^\infty \frac{\tan^{-1}(y/b)}{1+y^2}\,dy = \frac{\pi^2}{4} - F\left(\frac{1}{b}\right)
\end{aligned}
\tag{1.22}
$$

Thus we need consider only values of b that are less than unity—a considerable simplification.

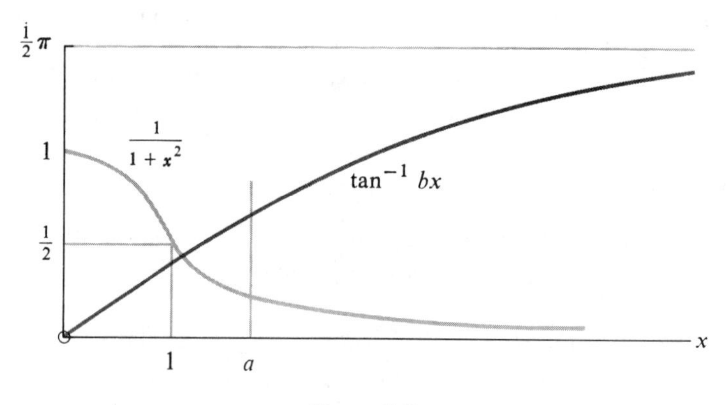

Figure 1.5

Looking at a plot (Figure 1.5) of the functions in the integrand of (1.21) we see that the principal contribution arises for small x. One strategy would be to approximate the arctangent rather well for smaller arguments in the hope that the integration could then be performed analytically. We have a choice between the power series and the continued fraction, the latter being much more efficient as the argument gets near one. We can also approximate the arctangent function rather well for large arguments with a series in reciprocal powers of the argument. Thus we might seek our approximation via

$$
F(b) = \int_0^a \frac{\tan^{-1}bx}{1+x^2}\,dx + \int_a^\infty \frac{\tan^{-1}bx}{1+x^2}\,dx \tag{1.23}
$$

$$
\begin{aligned}
F(b) &\approx \int_0^a \frac{(3bx + \frac{4}{5}b^3x^3)\,dx}{(3 + \frac{9}{5}b^2x^2)(1+x^2)} \\
&\quad + \int_a^\infty \left[\frac{\pi}{2} - \frac{1}{bx} + \frac{1}{3b^3x^3} - \cdots\right]\frac{dx}{1+x^2}
\end{aligned}
\tag{1.24}
$$

where the first integrand comes from replacing $\tan^{-1}bx$ by its continued fraction truncated after the $4b^2x^2/5$ term. Both the integrals in (1.24) can be expressed in closed form, yielding

$$F(b) \approx \frac{5B}{18b^2}\ln\left(1 + \frac{3}{5}b^2a^2\right) + \frac{D}{2}\ln(1 + a^2) + \frac{\pi}{2}\tan^{-1}\left(\frac{1}{a}\right)$$
$$- \frac{1}{2b}\ln\left[1 + \frac{1}{a^2}\right] + \frac{1}{6b^3}\left[\frac{1}{a^2} - \ln\left(1 + \frac{1}{a^2}\right)\right] \tag{1.25}$$

with

$$B = \frac{b^3}{\frac{3}{5}b^2 - 1}$$

$$D = \frac{b(\frac{4}{5}b^2 - 3)}{\frac{9}{5}b^2 - 3}$$

where a remains to be chosen so that the two approximations for $\tan^{-1}bx$ are about equally adequate. For the particular approximations used here this breakeven point is near where bx is equal to 1.5. Thus we might use

$$ba = 1.5$$

to determine a in (1.25).

Values of *b* near 1

The previous approximation has increasing accuracy problems as b moves toward unity, for this brings the joint in our two replacements for $\tan^{-1}bx$ closer to the origin (Figure 1.5). Since it is near this joint, at x equal to a, that the greatest errors occur, we suffer more when they are multiplied by larger values of $1/(1 + x^2)$. Accordingly, we should prefer to have some other approximation method for larger values of b—say $\frac{1}{2} \leq b \leq 1$.

At unity our function is analytically evaluable, so we try to express it in terms of that value and a difference therefrom. We split the integral at unity:

$$F(b) = \int_0^1 \frac{\tan^{-1}bx}{1 + x^2}\,dx + \int_1^\infty \frac{\tan^{-1}bx}{1 + x^2}\,dx$$
$$= \int_0^1 \frac{\tan^{-1}bx}{1 + x^2}\,dx + \int_0^1 \frac{\frac{1}{2}\pi - \tan^{-1}(y/b)}{1 + y^2}\,dy \tag{1.26}$$
$$= \frac{\pi^2}{8} - \int_0^1 \frac{\tan^{-1}(x/b) - \tan^{-1}bx}{1 + x^2}\,dx$$

Thus far there is no approximation, and the numerator of the integral in (1.26) goes to zero as b approaches unity. From here we may proceed in several directions according to our needs. Numerical quadratures of (1.26) will give the greatest accuracy, but analytic approximation can do quite well. In either case, one further transformation seems desirable to avoid the subtraction of the two arctangents. We have

$$F(b) = \frac{\pi^2}{8} - \int_0^1 \frac{\tan^{-1}\left[\dfrac{(1/b - b)x}{1 + x^2}\right]}{1 + x^2}\, dx \tag{1.27}$$

Since the argument of the arctangent in (1.27) is never going to be larger than $\frac{3}{4}$ (why?) we may use its truncated continued fraction replacement to good advantage. Retaining terms through $4y^2/5$ as before, we finally obtain

$$F(b) = \frac{\pi^2}{8} - \frac{c}{9} - \frac{5}{18}\frac{c}{R}\ln\left[\frac{(2 + a - R)(a + R)}{(2 + a + R)(a - R)}\right] \tag{1.28}$$

where

$$c = \frac{1}{b} - b$$

$$a = 2 + \tfrac{3}{5}c^2$$

$$R = c\sqrt{\tfrac{3}{5}(2 + a)}$$

The utility of our two approximations may be judged by looking at their errors in Table 1.1. The second approximation (1.28) is very good for b near unity, but, as expected, falls off when b is small. The first approximation (1.25) is best for small b and retains its limited utility rather well all the way up to unity.

Table 1.1. Value of $\displaystyle\int_0^\infty \frac{\tan^{-1}bx}{1 + x^2}\,dx$ and the errors of two approximations

b	$F(b)$	Error of (1.25)	Error of (1.28)
0	0	0	
0.1	0.33114	0.0010	
0.2	0.52719	0.0018	
0.3	0.67575	0.0027	
0.4	0.79574	0.0035	−0.0021
0.5	0.89608	0.0040	−0.00034
0.6	0.98192	0.0047	−0.00004
0.7	1.05661	0.0051	−0.000004
0.8	1.12244	0.0055	-1.4×10^{-7}
0.9	1.18105	0.0057	0
1.0	1.23370	0.0059	0

More sophisticated approximations

The example of the previous section is a typical direct approximation generated by your author in two hours upon receiving a request from a friend. His accuracy requirements were not severe; the urgency was relatively great. The resulting approximation, while satisfying the customer, left aesthetic gnawings in the stomach of its author and he later returned to the problem to see if it could not perhaps be solved more elegantly and, of course, more precisely. It can!

The major dissatisfactions with the previous approximation arise from the treatment of the second integral in the version for small b. We had split the integral

$$F(b) = \int_0^{a/b} \frac{\tan^{-1} bx}{1 + x^2}\, dx + \int_{a/b}^{\infty} \frac{\tan^{-1} bx}{1 + x^2}\, dx = S(b, a) + L(b, a)$$

and in the second integral replaced the arctangent by

$$\frac{\pi}{2} - \tan^{-1} \frac{1}{bx}$$

and then expanded the new arctangent in its series which, though valid for bx greater than unity, is notoriously inefficient unless $(bx)^{-1}$ is rather small—and this unfortunately is not where the major contribution of the second integral occurs (Figure 1.5). Further, this replaces a positive integral by the difference of two expressions that could be of comparable magnitudes. Finally, when the series replacement and the analytic integration are carried through, the resultant approximation is a series in $(1/b)^n$—terms that are not at all disposed to fade quietly away as b decreases!

In seeking a better approach, your author tried most of the manipulations that ought to occur to a frantic freshman who feels that analytic integration is still possible for anything that looks this simple. Thus letting

$$L(b, a) = \int_{a/b}^{\infty} \frac{\tan^{-1} bx}{1 + x^2}\, dx \qquad 0 \le b \le 1$$

we tried:

1. Replacing the fraction $1/(1 + x^2)$ by

$$\frac{1}{x^2(1 + 1/x^2)} = \frac{1}{x^2}\left(1 - \frac{1}{x^2} + \frac{1}{x^4} - \cdots\right)$$

and then integrating the resulting expression analytically—a maneuver that is valid, though not necessarily efficient, whenever x is greater than unity.

2. Changing the dummy variable almost trivially, letting

$$bx = y$$

whence

$$L(b, a) = b \int_a^{\infty} \frac{\tan^{-1} y}{b^2 + y^2}\, dy$$

3. Integrating by parts to produce

$$L(b, a) = \left[\tan^{-1} bx \cdot \tan^{-1} x \right]_{a/b}^{\infty} - b \int_{a/b}^{\infty} \frac{\tan^{-1} x}{1 + b^2 x^2}\, dx$$

4. Integrating by parts somewhat differently to produce

$$L(b, a) = \int_{a/b}^{\infty} \frac{\tan^{-1} bx}{2x} \cdot \frac{2x \cdot dx}{1 + x^2} = \left[\frac{\tan^{-1} bx}{2x} \cdot \ln(1 + x^2) \right]_{a/b}^{\infty}$$

$$- \int_{a/b}^{\infty} \ln(1 + x^2) \left[- \frac{\tan^{-1} bx}{2x^2} + \frac{b}{2x(1 + b^2 x^2)} \right] dx$$

5. Integrating by parts still differently to produce

$$L(b, a) = \int_{a/b}^{\infty} \tan^{-1} bx \cdot \frac{dx/x^2}{1 + 1/x^2} = \left[\tan^{-1} bx \left(- \tan^{-1} \frac{1}{x} \right) \right]_{a/b}^{\infty}$$

$$+ b \int_{a/b}^{\infty} \frac{\tan^{-1}(1/x)\, dx}{1 + b^2 x^2}$$

These operations may be permuted and combined to produce still other representations of $L(b, a)$, but the present collection will suffice. Commenting first upon step 2: The change in the dummy variable has left the numerator unchanged and has decreased the denominator, thereby *increasing* the value of the integral. This is not usually desirable. Instead of a larger integral we would prefer to get most of our integral out in some directly evaluable explicit form. Here we have extracted nothing except the factor b, and we have correspondingly inflated the integral. No gain.

In step 3 we at least extract something, but we also inflate the integral in a manner similar to step 2. Worse, we transform our evaluation into the difference between two quantities—itself a questionable maneuver. If the integral were noticeably small we might tolerate the subtraction, but currently some doubts nag and we prefer to look further.

In step 4 we have an even more obscure situation. Despite first appearances, the expression is not a difference. The factor $\ln(1 + x^2)$ is certainly positive, while the bracketed term in the integral is the *slope* of $\tan^{-1} bx/2x$—a decreasing function on the range we consider, hence negative. Thus this form behaves properly and might have received further exploration had it not been so messy.

With step 5 we become happy. The bracket in front is simply

$$\tan^{-1} a \cdot \tan^{-1} \frac{b}{a}$$

while the integral is clearly positive and analytically lucid. Additional transformations will be easy and possibly profitable. One approach is to trivially change the dummy variable of the integral in step 5 by letting bx become y to give

$$L(b, a) = \int_a^\infty \frac{\tan^{-1}(b/y)}{1 + y^2} dy = \int_a^\infty \left(\frac{1}{1 + y^2}\right)\left[\frac{b}{y} - \frac{1}{3}\left(\frac{b}{y}\right)^3 + \frac{1}{5}\left(\frac{b}{y}\right)^5 - \cdots\right] dy$$

and then to replace the arctangent by its series. This gives us a series of terms in b^{2n+1} which *do* fade away quickly as b becomes small. The only remaining worry is the old one about the inefficiency of the series for the arctangent when the argument is near unity. When a is 1.5 we suspect that quite a few terms might be needed for reasonable precision.

A second approach to the integral of step 5 is to make a reciprocal transformation of the dummy variable to give

$$L(b, a) = \tan^{-1} a \cdot \tan^{-1} \frac{b}{a} + \int_0^{1/a} \frac{\tan^{-1} by}{1 + y^2} dy$$

and now we discover our remaining integral to be identical in form to that we obtained previously for $S(b, a)$ (page 33). The only difference is the upper limit, which is $1/a$ instead of a/b. We thus obtain the economies of being able to use one technique for approximating two integrals. Further, if we choose a to be $\sqrt{b}$ the two integrals (of page 33) coalesce to give

$$F(b) = \left(\tan^{-1} \sqrt{b}\right)^2 + 2 \int_0^{1/\sqrt{b}} \frac{\tan^{-1} bx}{1 + x^2} dx \tag{1.29}$$

where we may now concentrate on efficient approximation of the one integral. Since the largest argument of the arctangent is never greater than unity we should do better than in our first attempts.

Our problem is now the finite quadrature

$$S(b, l) = \int_0^l \frac{\tan^{-1} bx}{1 + x^2} dx \qquad 0 \le bl < 1 \tag{1.30}$$

Expansion of $(1 + x^2)^{-1}$ is unpromising because the value of x will exceed unity, so we concentrate on replacing the arctangent over $(0, bl)$ in such a way as to permit analytic integration. The two obvious techniques are:

1. A continued fraction representation, or
2. A Chebyshev series.

With either approach we observe that the integrand is *odd*; hence a replacement of the dummy variable x^2 by z will cut the order of any polynomial or rational function replacement in half, considerably simplifying further manipulations.

Turning first to Chebyshev polynomial expansions, we find (Clenshaw [1955])

$$\tan^{-1}kx = 2\sum_{n=0}(-1)^n\frac{\tan^{2n+1}\alpha}{2n+1}T_{2n+1}(x) \qquad k = \tan 2\alpha \tag{1.31}$$

Since we shall want our argument of $T_{2n+1}(y)$ to be integrated over (0, 1) we make a change in the dummy variable of (1.30). Letting x become ly we get

$$S(b, l) = \int_0^1 \frac{\tan^{-1}bly}{1 + l^2y^2}\cdot l\cdot dy = \frac{1}{l}\int_0^1\left[\frac{\tan^{-1}bly}{y}\right]\frac{y\cdot dy}{(1/l)^2 + y^2}$$

Using (1.31), substituting d for $1/l^2$, and changing y^2 to z we have

$$S(d, \alpha) = \sqrt{d}\sum_{n=0}(-1)^n\frac{\tan^{2n+1}\alpha}{2n+1}\int_0^1\left[\frac{T_{2n+1}(y)}{y}\right]\cdot\frac{dz}{d+z} \tag{1.32}$$

where

$$bl = \tan 2\alpha \qquad \text{and} \qquad d = \frac{1}{l^2}$$

and we note that the expression in the brackets is a polynomial of only degree n in the new variable z. Analytic integration is easy. We exhibit the case for $n = 2$ as typical. Here

$$\frac{T_5(y)}{y} = 16y^4 - 20y^2 + 5 = 16z^2 - 20z + 5$$

Using synthetic division,

$$\begin{array}{ccc|l} 16 & -20 & 5 & -d \\ 0 & -16d & 16d^2 + 20d & \\ \hline 16 & -(16d + 20) & (16d^2 + 20d + 5) & \end{array}$$

we obtain the integrand

$$16z - (16d + 20) + \frac{16d^2 + 20d + 5}{z + d}$$

which immediately integrates to produce

$$\tfrac{16}{2} - (16d + 20) + (16d^2 + 20d + 5)\ln\left(1 + \frac{1}{d}\right)$$

for this term. The general rule is easily inferred and a complete subroutine constructed. We must, however, remember that we are still basically replacing

$\tan^{-1}bx$ by a polynomial representation on the range $(0, bl)$. If b becomes as large as 1 we can expect poor results, for the power series for $\tan^{-1}x$ diverges beyond this point. The use of Chebyshev polynomials extends the range of convergence, but even they have increasing difficulty in delivering accurate approximations as the argument increases. So we must not push l too far. Fortunately, in (1.29) we never have to go beyond an argument of $\sqrt{b}$—which is quite tractable on the range $0.1 \le b \le 0.9$. For *very* small b we may need another approach, but this is easy to devise from the original statement of our function (1.21) and we leave it as an exercise.

Our present approach produces the approximation

$$\begin{aligned} F(b) = {} & \left(\tan^{-1}\sqrt{b}\right)^2 + 2\sqrt{b}\left\{\tan\alpha \cdot \ln\left(1 + \frac{1}{b}\right)\right. \\ & - \frac{(\tan\alpha)^3}{3}\left[4 - (4b + 3)\ln\left(1 + \frac{1}{b}\right)\right] \\ & + \frac{(\tan\alpha)^5}{5}\left[\frac{16}{2} - \frac{16b + 20}{1} + (16b^2 + 20b + 5)\cdot\ln\left(1 + \frac{1}{b}\right)\right] \\ & - \frac{(\tan\alpha)^7}{7}\left[\frac{64}{3} - \frac{64b + 112}{2} + \frac{64b^2 + 112b + 56}{1}\right. \\ & \left.\left. - (64b^3 + 112b^2 + 56b + 7)\ln\left(1 + \frac{1}{b}\right)\right] + \cdots\right\} \end{aligned} \tag{1.33}$$

Taking terms through $(\tan\alpha)^7$ gives $F(b)$ correct to six or seven significant figures over the range $0.1 \le b \le 0.9$, being somewhat more precise for the smaller arguments.

The approximation for b near 1—again

Increased familiarity with Chebyshev manipulations encouraged us to return to our approximation for larger b to see if perhaps the use of the Chebyshev expansion there might not yield something interesting. We already had a sufficiently accurate representation of that integral via a continued fraction replacement, but idle curiosity nagged us to try the other technique, too. We have

$$F(b) = \frac{\pi^2}{8} - \int_0^1 \frac{\tan^{-1}\left(\dfrac{cx}{1 + x^2}\right)}{1 + x^2}\,dx$$

In replacing the arctangent we want the argument of the Chebyshev polynomials to cover the range (0, 1), necessitating a trivial rewriting of the integral as

$$I(c) - \int_0^1 \frac{\tan^{-1}\left(\frac{c}{2}\cdot\frac{2x}{1+x^2}\right)}{1+x^2}dx$$

thus giving

$$I(c) = 2\sum_{n=0}(-1)^n \frac{\tan^{2n+1}\alpha}{2n+1}\int_0^1 T_{2n+1}\left(\frac{2x}{1+x^2}\right)\cdot\frac{dx}{1+x^2} \qquad \tan 2\alpha = \frac{c}{2}$$

and we observe that the integrals no longer involve the parameter c, thus permitting evaluation once and for all. After considerable manipulation we find that they are merely

$$\frac{(-1)^n}{2(2n+1)}$$

and our approximation for large b takes on the simple form

$$F(b) = \frac{\pi^2}{8} - \sum_{n=0}\frac{(\tan\alpha)^{2n+1}}{(2n+1)^2} \tag{1.34}$$

where

$$\tan 2\alpha = \frac{c}{2} \qquad \text{and} \qquad c = \frac{1}{b} - b$$

This approximation works very well for all b greater than about 0.1, the number of terms of the series required for eight-decimal accuracy being shown here for several arguments. As one expects, the approximation succeeds most efficiently at the larger arguments. One wonders if this approximation ought not to have been discovered more directly, but discovery seldom follows the shortest path. Indeed, your author still sees no simpler route. [See Addendum, page 40]

b	0.1	0.2	0.3	0.4	0.5	0.6	0.7	0.8	0.9
n	32	18	13	10	8	7	6	5	4

A road map of an approximation

Figure 1.6 is a graphic summary of our several approximations for $F(b)$. We include some indication of the tactics taken, as well as the accuracies achieved. Not shown are the many false starts, expansions that led nowhere, and just plain bad results due to manipulative blunders. The total effort invested, which includes at least 10 baby computer programs to check the effectiveness of the results, is frighteningly large. At least three weekends disappeared into what "should have been" a simple straightforward procedure. Your author, however, suspects that this history, far from being unusual, is typical. He includes it here to correct the impressions given to students by all-too-slick presentations that

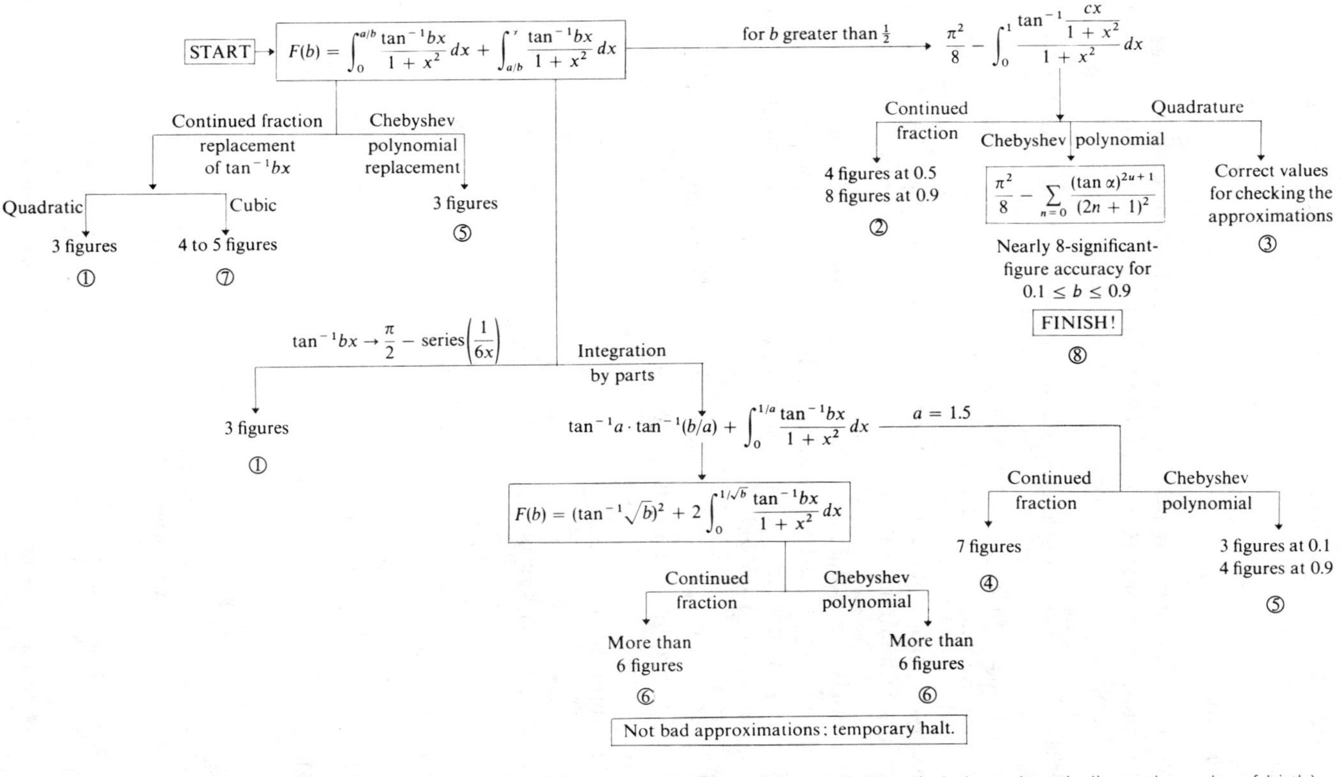

Figure 1.6. The genealogy of a family of approximations to a well-behaved function (the circled numbers indicate the order of birth)

they are somehow incompetent if they do not get excellent results with their first attempts!

Problems

1. Write a computer program to check the relative efficiency of three ways of computing arctan x. Run it for $x = 0(0.1)1.0$ and provide some glimpses of the progress of the iterations where appropriate.

2. Write a program to evaluate $J_0(x)$ or $\cos x$ to six decimals from its series:

   a. Starting with the first term and going out.
   b. Starting "sufficiently far out" and coming in.

   Run your program for all x's between 0 and 10 at steps of $\frac{1}{2}$. Compare your results with tabular values and explain any inconsistencies.

3. If you have to evaluate $x \cdot \ln x$ many times for x in the range $(10^{-5}, 10^{-4})$ and you do not trust the routine for $\ln x$ inside the computer system, how would you proceed? (*Hint*: Use the property $\ln by = \ln b + \ln y$ to scale your argument.)

4. Evaluate erfc(2.0) from the asymptotic series and compare with the tabular value. Is the discrepancy the proper size?

Addendum from page 38

And then, by the side of the Mosel in 1984, a bottle of wine supplied yet another way to evaluate that integral, $F(b)$, via its derivative:

$$F'(b) = \int_0^\infty \frac{x}{(1 + b^2x^2)(1 + x^2)}\,dx = \frac{\ln(b)}{b^2 - 1}$$

from which we get

$$F(b) = \int_0^b \frac{\ln(p)}{b^2 - 1}\,dp = \sum_{n\ \text{odd}} \frac{b^n}{n}\left(\frac{1}{n} - \ln(b)\right) \qquad (0 \le b \le 0.4)$$

and this simpler form of (1.34)

$$F(b) = \frac{\pi^2}{8} - t - \frac{t^3}{3^2} - \frac{t^5}{5^2} - \cdots \quad \text{with} \quad t = \frac{(1 - b)}{(1 + b)} \qquad (0.4 \le b \le 1)$$

that give 8 correct decimals with no more than 9 terms and 11 decimals with 14. Over the second range there is an even more efficient, tho less systematic, series

$$F(b) = \frac{\pi^2}{8} - q \cdot \left(1 - \frac{q^2}{18} + \frac{7x^4}{1800} - \frac{31x^6}{105840} + \cdots\right) \quad \text{with} \quad q = -\ln(b)$$

whose derivation we leave as a puzzle. (But it delivers at least 11 figures with no more than 9 terms!)

CHAPTER 2

ROOTS OF TRANSCENDENTAL EQUATIONS

Among all the topics of numerical analysis, none offers as much pleasure to the hand computor as transcendental equations. Equations such as

$$b \cdot x \cdot \cos x - \sin x = 0 \tag{2.1}$$

are fun to solve, the great variety they afford providing a constant challenge to one's ingenuity. The finding of the interesting roots is a pattern of trials and failures brightened with just enough successes to keep one going. Unfortunately, for the automatic computer this same pattern is more frustrating than encouraging, so that transcendental equations become a prickly and somewhat unsatisfactory field for the machine-oriented analyst. No general theory seems possible and the variety that stimulates the human still bewilders the machine. Some theory there certainly is, but it tends to make precise the areas that are already quite clear. The big gray areas where enlightenment is needed remain uncomfortably gloomy.

All transcendental equations must be solved iteratively. That is, we guess an approximate root and plug it into some formula derived from our equation to obtain a next approximation that we hope is closer to the root. This approximation in turn is used to get a third one . . . and the iteration ceases when it has either converged to the root with sufficient accuracy for our needs or has indicated that it is not going to converge. Or sometimes it converges so slowly that we just get tired and stop. The ideal iterative scheme will converge quickly to a root no matter what starting value is presented to it. The usual iteration will do almost everything but! The ideal scheme would give us both speed and certainty. We usually must settle for the one or the other. We can easily find iterations that converge fairly well once we are close to the root, but these same iterations will usually diverge wildly if started with a value that is at all far from the root we seek. We are thus back in the classic position of being able to find the answer efficiently once we know where it is. Our theories concern convergence rates in the neighborhood of the root, while our need is for a theory that will enable us to find these neighborhoods. For the present we get them as best we may—usually from a priori information not systematically deduced from the equation.

Since the topic lends itself to pictures and geometric arguments, we shall proceed with a few examples before worrying about much theory. Our equation (2.1) may be rewritten in the form

$$bx = \tan x$$

whose solutions occur at the intersections of the curves

$$y = bx \qquad \text{and} \qquad y = \tan x \tag{2.2}$$

In Figure 2.1 we picture the geometry when b is 0.4. After the obvious root at the origin, the first interesting solution occurs between π and $\frac{3}{2}\pi$ at about 4.0. Proceeding heuristically, it seems geometrically obvious that we can use a possibly bad guess about x, call it x_1, to get a fairly good value of the y coordinate for the point we wish to find simply by substituting x_1 into the straight-line equation. We have (Figure 2.1)

$$y_1 = (0.4) \cdot (4.0) = 1.6$$

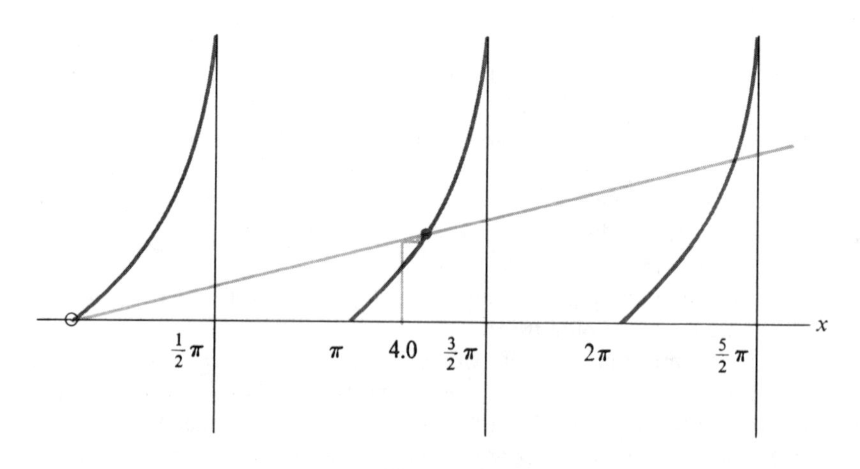

Figure 2.1

If we now use this value of y_1 in the tangent equation, we move horizontally in our figure, arriving on the tangent curve at a point which is quite close to the root, giving us a fairly good value for x_2. We have

$$1.6 = \tan x_2 \qquad \text{or} \qquad x_2 = 1.012 + 3.142 = 4.154$$

(but we must remember that we are on the *second* branch of the tangent curve, so π must be added to the argument found in the tables). Repeating our operations will improve our approximations, giving

x	y
4.0	1.6
4.154	1.662
4.171	1.668
4.173	1.669

Geometrically, we have been approaching the intersection in Figure 2.1 by a series of steps (Figure 2.2) that arise from our use of each equation in the direction in which it is *flattest*. Thus the straight line gives a good value of y even if x is rather poor because it is flat when viewed in the form

$$y = f(x)$$

while the tangent curve gives a good value of x for rather poor values of y

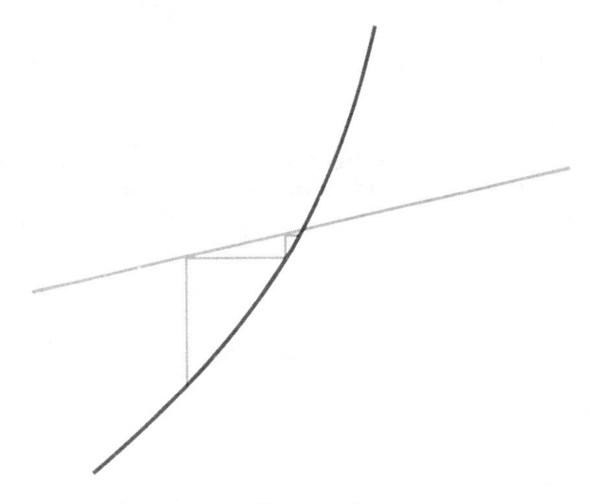

Figure 2.2

because it is flat when viewed as

$$x = g(y)$$

Analytically, our rapid convergence occurs because the two derivatives are small. Thus for the straight line

$$\frac{dy}{dx} = b = 0.40$$

while for the tangent curve

$$\frac{dx}{dy} = \frac{1}{dy/dx} = \frac{1}{\sec^2 x} = 0.27$$

and the convergence of the process is quite good.

This example is atypical in two ways: It converges quickly and it converges from all starting values within the region of that branch of the tangent curve. We do not even need this second property since the geometry allows us to pick rather good starting values. For all roots after the first, an x equal to $\pi(n + \frac{1}{2})$ will do quite nicely, since the root must lie close to the vertical asymptote of the tangent.

A spiral geometry instead of a staircase arises from the equation

$$x \sin x - b \cos x = 0$$

The student should try this one with $b = 0.4$ and 3.0 – with the special aim of discovering how to get good starting values, as well as a successful iteration.

In all these iterations, of course, we may accidentally attempt to use the equations backward. If we did not bother to draw Figure 2.1 we might have used our equations as

$$y = \tan x \qquad \text{and} \qquad x = \frac{y}{b}$$

and we would have gaily walked our steps *away* from the root – with disaster occurring in about three steps! (Try it on your slide rule.)

Equation (2.1) need not be broken into the two equations (2.2) that we

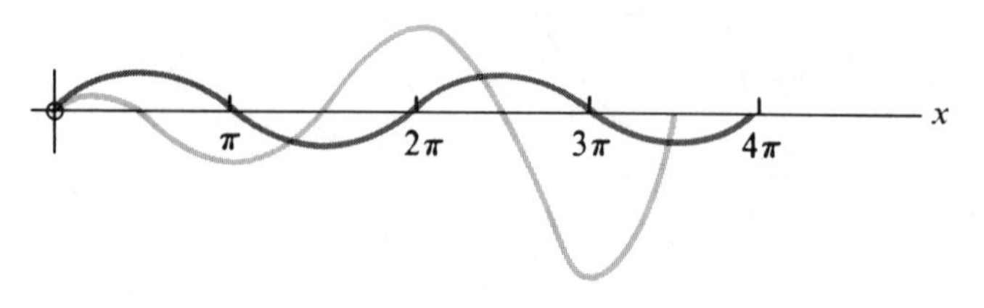

Figure 2.3

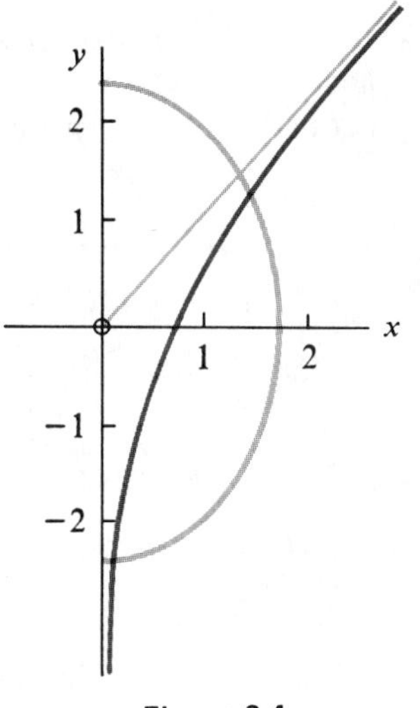

Figure 2.4

actually used. With less experience, we might have produced

$$y = bx \cdot \cos x \qquad \text{and} \qquad y = \sin x$$

which may also be plotted (Figure 2.3). They lead to an iteration, too, although it is less pleasant both in computational convenience and in its rapidity of convergence. Still another version might be

$$y = \cot x \qquad \text{and} \qquad y = \frac{1}{bx}$$

This one works quite well, being not seriously different from (2.2). Already we see that one problem can be phrased in several forms for which no prior guidance exists. After trying the several forms, we then know which is the efficient one, but there is no infallible way to find it.

Using geometric intuition

Let us find the roots of the equation pair

$$x^{10} + y^6 = 2^8 \tag{2.3}$$

$$e^x - e^y = 1 \tag{2.4}$$

A little thought will show that the graphs of these two equations have the rather simple forms shown in Figure 2.4. Clearly, there are two roots. We shall first deal with the one in the fourth quadrant, since it is easier.

The root near the y axis at -2.2 has a small positive value for x. Further, the exponential equation is nearly vertical, so it should provide a reasonably good value for x even if a rather bad guess for y were substituted into it. We try our guess (-2.2):

$$e^x = 1 + e^{-2.2} = 1.1108$$

$$x_1 = \ln(1.1108) = 0.10508$$

and, on substituting this into the algebraic equation (2.3), which is nearly horizontal and hence a good source for y, we obtain

$$-y = (256 - x_1^{10})^{1/6} = (256 - 0.000000000163)^{1/6} = 256^{1/6} = 2.5198$$

A second use of the exponential equation gives us

$$x_2 = \ln(1.080452) = 0.077003$$

and our iteration has converged to at least six significant figures, since the value of x is clearly irrelevant for the y coordinate of this root, the closed curve being so nearly horizontal here.

The first-quadrant root is not so cooperative. Straightforward iteration is possible but slow, as the 45° geometry of Figure 2.4 suggests. Since x must be slightly greater than y we would use the algebraic equation for x as

$$x = (256 - y^6)^{1/10} \tag{2.5}$$

and the exponential equation in the form

$$y = \ln(e^x - 1) \tag{2.6}$$

Starting with 2.0 for y, slide-rule computation gives us convergence in three iterations.

y	x
2.0	1.69
1.48	1.735
1.54	1.732
1.537	–

We shall perhaps do better if we observe that the exponential equation is sensitive primarily to the distance from the bisector of the first quadrant – that is, to the changes in the value of $(x - y)$ rather than to changes in either variable separately. It might be better if we could rearrange the equation so that it gives this quantity directly. Likewise, our algebraic equation in the vicinity of the sought root changes most rapidly with $(x + y)$ and again it might speed our iteration if that equation could be made to yield this quantity. The first change is easy. We merely rewrite (2.6) as

$$y = \ln[e^x(1 - e^{-x})] = x + \ln(1 - e^{-x})$$

or

$$u = y - x = \ln(1 - e^{-x}) \tag{2.7}$$

(note that u will be negative). We can also change the algebraic equation, but the resulting expression is so long that the increase in evaluation time defeats our computational purpose. Since one strongly convergent step per iteration is quite satisfactory provided the other steps are not actual noise amplifiers, we proceed with (2.7) and

$$\begin{aligned} y &= x + u \\ x &= (256 - y^6)^{1/10} \end{aligned} \tag{2.8}$$

Starting with our same value of y we use (2.7) and (2.8) to obtain

y	x	u
2.0	1.69	−0.205
1.485	1.735	−0.194
1.541	1.732	—

The accuracy of the computation for y is enhanced since the small quantity u is now produced directly, but the rate of convergence is the same as before. If we want faster convergence on this problem we shall have to use a different computational strategy.

The need for good starting values

An iterative process must somehow be started; a first approximation to the answer must be procured from somewhere. Furthermore, this first value may have to be a quite good one if the iteration is to converge. For automatic computation the problem of the initial value looms large and forbidding. It is at once the chief characteristic of iterative algorithms and their principal curse. But it can also be a blessing, for it permits efficient use of good a priori knowledge. If we are good enough guessers to start an iterative algorithm with a first approximation that is, in fact, correct, then it converges in one cycle—a handsome payoff to our intuition or experience. In general, however, one hears more moaning about the curse than rejoicing about the blessing—though this may merely reflect a correlation between incompetence and volubility.

One seldom discovers good starting values by gazing intently at the iterative algorithm itself, although the sketch of the equation that suggested the algorithm will frequently also suggest a reasonable way to begin it. Indeed, a sketch is the best source of good starting values and all experienced hand computers rely heavily on them. The human capacity for pattern recognition seems to provide enough help so that one is tempted to say that starting values are no problem to the hand computor who can sketch his problem—that if he can't, he probably has no business trying to compute it!

At first glance, the programmer of an iterative routine is in a much worse predicament. His problem is often bigger, his geometric grasp of it weaker. And automatic digital computers have notoriously poor geometrical intuitions! Fortunately, this statement of his difficulty is misleading. The problem that needs automatic computation is seldom a single one; it is a *family* of similar problems that gets programmed. By the time we have written and checked out a program for the roots of a transcendental equation, we could have computed the answers by hand for one or two of the parameter sets. No, it is not the isolated problem that reaches the computer. Since we now have admitted a family of equations that differ in the values of their several parameters, we may sort them into some sensible order on these parameters. Having then found the roots of the first equation, possibly by hand computation, we have quite good first approximations for the roots of "nearby" equations in the family. So the plight of the programmer is not so dismal after all! While your author hesitates to say that this strategy will always solve the starting-value problem, he does feel that it will serve for the usual system of transcendental equations that enters the computation laboratory.

One procedure that is not usually reasonable, however, is a systematic search for roots by a computer by evaluating the function over the space of the independent variables unguided by a priori information. It just takes too long. Transcendental functions are not cheap to evaluate, and to require such evaluation at typically 40 points in one dimension (or 1600 points in two dimensions) is a very wasteful procedure that can almost always be avoided at the price of a little thought. Polynomials are exceptions, but they are so exceptional that they should not be discussed along with other transcendental equations.

How to stop an iteration

With hand computation the question of when to stop an iteration seldom arises. The computor sees his iterations and usually has a quite good idea about how many of the figures are accurate. He simply stops when he has enough of them. When the same iteration is to be performed on an automatic computer, however, we must supply an adequate substitute for the human decision—and we must supply it *in advance* of the computation. While the process is not difficult, it does contain a few booby traps for the beginner that we would illustrate.

Suppose you were implementing a computer routine to find the square root of any number your computer could handle without resorting to double-precision arithmetic. Doubtless you would be using the effective divide-and-average process, probably starting with *unity* as your first approximation to the root. (There are better strategies for the first approximation, but they are not germane to our topic.) Thus your iteration might be summarized by

$$r_{i+1} = 0.5\left(r_i + \frac{N}{r_i}\right)$$

and your first idea would be to test the size of

$$|r_{i+1} - r_i|$$

against some ε to see if further iterations should be taken. Perhaps you decide that an agreement of 0.0000002 will be adequate. Then comes the unhappy thought that N (whose root we seek) might be quite small—say 0.000000000000035213789 on a machine that can carry about eight significant figures in a floating-point format. The *root* is smaller than our ε and hence the iteration would stop before obtaining even two significant digits! Something more sophisticated is needed.

We may then decide to use

$$\left|\frac{r_{i+1} - r_i}{r_{i+1}}\right|$$

as our critical quantity—the *relative* error instead of the error itself. This measure, if correctly implemented, will guarantee us a fixed number of significant figures instead of a number of accurate digits. We like this idea, but the implementation can be tricky for small roots lest we find ourselves computing the ratio of two small numbers that will be unstable enough to magnify their rounding error into a disturbingly large quantity. Specifically, we should *not* compute our test quantity in the form

$$\left|1 - \frac{r_i}{r_{i+1}}\right|$$

where the subtraction that follows the division will bring into prominence all the error that occurred in the division. More important, we must not set ε so small that, with the maximum possible rounding error, our test quantity cannot shrink below it. In an understandable eagerness to preserve all the precision that is present, we are inevitably tempted to set ε as small as we can. But exactly how small ε may be depends upon the most careful analysis of the detailed rounding properties of the particular computer. Sometimes they are really quite strange—and often incorrectly stated in the machine manuals!

One-dimensional iterations

The iterative schemes we have examined are attractive primarily because they permit a rather complicated expression to be broken down into a pair of often easily visualized functions whose revealed geometry then gives us guidance as we seek the roots. This geometric revelation is not lightly to be discarded, but the price we pay can be some rather inefficient bouncing back and forth between two curves when we obviously would prefer to go more directly toward the root. (See the arrow in Figure 2.5.) We can, of course, devise methods that speed up

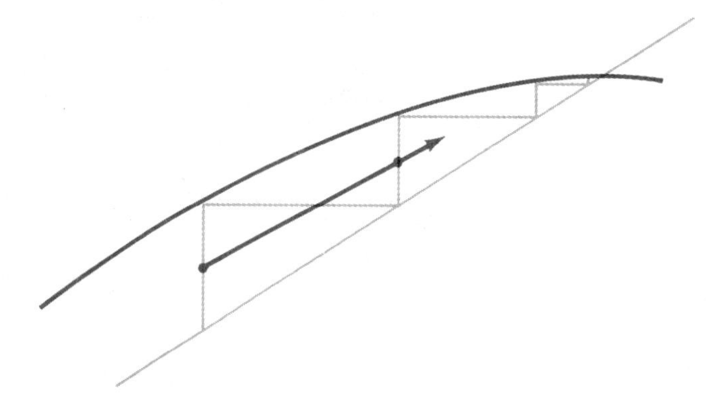

Figure 2.5

the convergence of our staircase processes, but these fall into the category of first aid—highly important after an accident but not a deliberately sought way of life. We feel that a different strategy might frequently serve us better.

If we seek the point, pictured in Figure 2.6, where

$$y = f(x)$$

crosses the x axis, we could follow the philosophy of sliding down a tangent line to get our next approximation to the root. With the exception of the first slide, the approach seems quite rapid. We spend our energy going in the right direction—at least in this example. On the other hand, we can easily produce geometries where tangent sliding can become quite dangerous (Figure 2.7), the danger usually being caused by a horizontal spot in our curve. Maxima and minima have the unpleasant property of being quite flat for some considerable

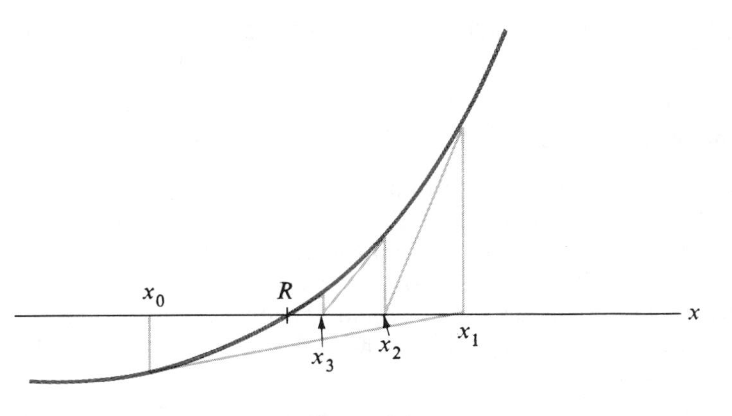

Figure 2.6

distance, so the chance that a blind stab for a root will land on a dangerous piece of the curve is actually fairly large. We do not have to hit the extremum in order to be sent on a wild ride; we need only land somewhere in its vicinity.

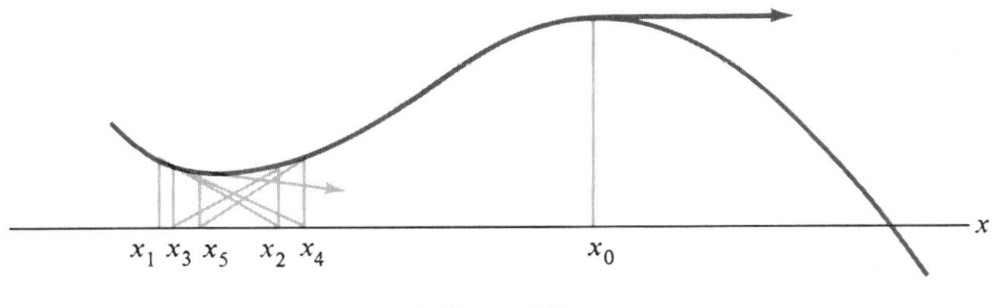

Figure 2.7

These geometric arguments, presented somewhat heuristically here, have their quite precise analytic counterparts. Newton's method, the official and familiar name of this tangent sliding philosophy, consists of the algorithm

$$x_{i+1} = x_i - \frac{f(x_i)}{f'(x_i)}$$

and we see that we are in trouble if $f'(x_i)$ is zero or even if it is small. Since f' is the slope, a small value gives the wild ride. In any automatic computer application of Newton's method we must be sure to check the size of the step we are about to take, rejecting it if it is improbably large. Such a rejection could call for an automatic switch to a less sensitive algorithm, or it might merely print out the current values of the function and its derivative—in effect, yelling for help. But it should never quietly take the giant step and thereby create a mess for somebody else to clean up.

Two-point iterations

The mercurial properties of Newton's method arise from its use of derivative information gathered at one point. If at that point the curve is headed toward a nearby root, then the method will get there very efficiently. But the *if* is a big one. We may prefer to approach more slowly, more surely. To achieve greater stability we may use *two* points on the curve and abandon any direct derivative computation. (Indeed, for many engineering functions their very complexity may preclude direct derivative computation.) If we insist that the two points should straddle a root (Figure 2.8) we shall produce a quite conservative algorithm

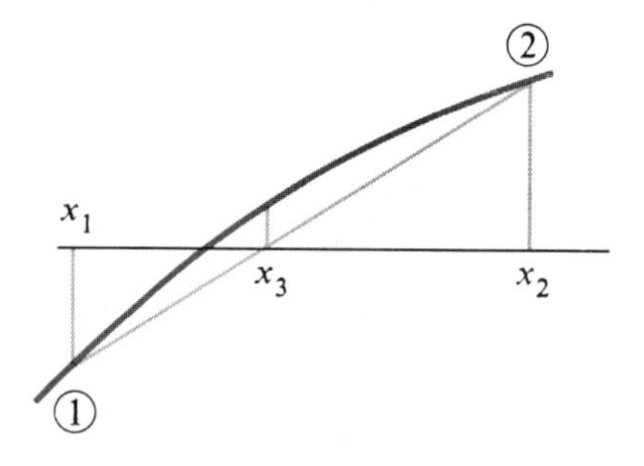

Figure 2.8

that always converges—though quite slowly. We simply take a straight line through our two points and use its crossing of the axis as our next approximation. We then throw away the point that lies on the same side of the root and repeat the process. The algorithm is called *False Position.* About the only thing that can go wrong with it, once a straddling pair of points has been found, is to discover that the points are straddling a discontinuity rather than a root (Figure 2.9). Such mistakes can occur rather easily in automatic stepping searches, where the test is apt to be a change of sign in the function. Approximate derivative information, even if quite crude, is useful in avoiding this mistake. It can be obtained by simple differencing as we step along. It would be better, however, to get the original pair of points by careful analysis, thus avoiding the blind inefficient stepping procedure altogether.

If we abandon the requirement that the two points straddle the root and merely keep the two most recently computed points at each iteration, we produce the *secant method.* The new point will sometimes be produced by an extrapolation (Figure 2.10) rather than only by an interpolation, as in False Position.

Convergence will be noticeably faster but will not be completely guaranteed. The process is somewhat more stable than Newton and is, predictably, somewhat slower. You pays your money and takes your choice.

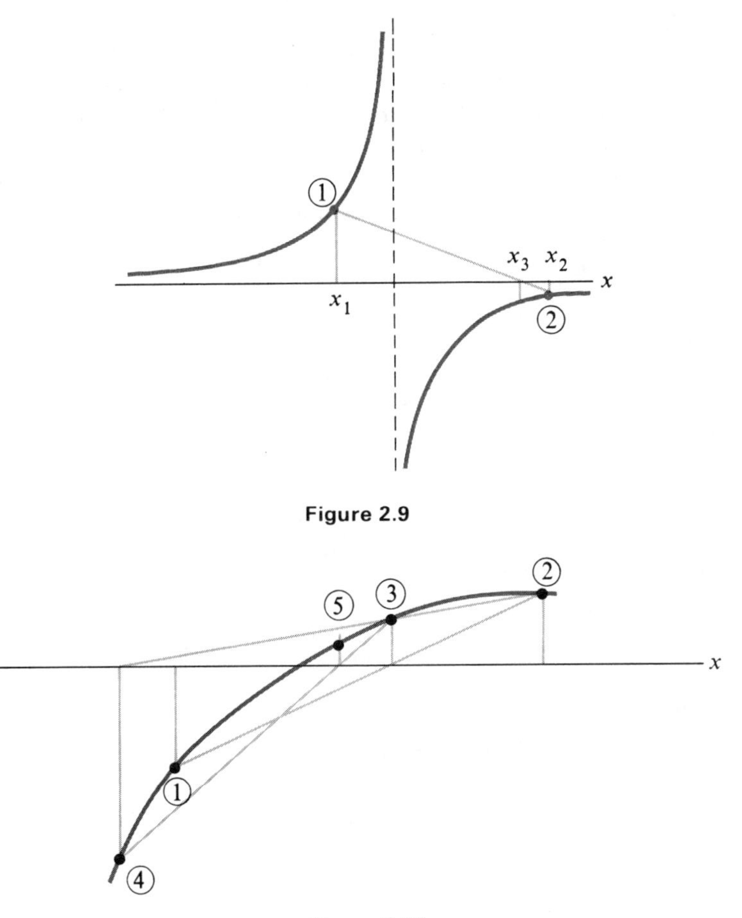

Figure 2.9

Figure 2.10

For both the secant and False-Position algorithms the formula is the same:

$$x_{i+1} = x_i \cdot \frac{y_{i-1}}{y_{i-1} - y_i} + x_{i-1} \cdot \frac{y_i}{y_i - y_{i-1}}$$

The only difference is the decision about which of the two old points is to be discarded.

The speed of convergence to a root

In spite of all our warnings about the importance of a good starting value in an iterative process, the *rate* of convergence is still of some interest. Once we are close enough to ensure convergence, we can increase the efficiency of our computation by changing to a more rapidly convergent algorithm. This change is totally unimportant if we seek only three-figure accuracy, but for automatic computers, where the subroutines are usually expected to supply seven or eight figures, the savings are marked. Accordingly, we shall discuss some of the elementary ideas about speed of convergence and computational labor for the iterative processes.

We first consider our familiar algorithm for the square root of N. We have

$$x_{i+1} = \tfrac{1}{2}\left(x_i + \frac{N}{x_i}\right) \tag{2.9}$$

If the root is b and we let

$$\varepsilon_i = x_i - b \tag{2.10}$$

we may rewrite (2.9) as

$$\varepsilon_{i+1} + b = \tfrac{1}{2}\left[\varepsilon_i + b + \frac{N}{b + \varepsilon_i}\right]$$

Noting that N is b^2, we have

$$\varepsilon_{i+1} = -b + \tfrac{1}{2}\left[\varepsilon_i + b + \frac{b}{1 + \varepsilon_i/b}\right]$$

If ε_i/b is small compared to 1, the fraction may be expanded in a series to give

$$\varepsilon_{i+1} = -b + \tfrac{1}{2}\left[\varepsilon_i + b + b\left(1 - \frac{\varepsilon_i}{b} + \frac{\varepsilon_i^2}{b^2} - \frac{\varepsilon_i^3}{b^3} + \cdots\right)\right]$$

and we obtain

$$\varepsilon_{i+1} = \frac{\varepsilon_i^2}{2b} - \frac{\varepsilon_i^3}{2b^2} + \cdots$$

Assuming we are close to a root, ε_i^3 is much smaller than ε_i^2, so we usually say that our process converges *quadratically*; that is,

$$\varepsilon_{i+1} = \frac{\varepsilon_i^2}{2b}$$

This squaring of the small error at each iteration has the effect of doubling the number of significant figures each time.

Turning to Newton's method in general (our square-root algorithm is a particularly simple example of it) we note that

$$x_{i+1} = x_i - \frac{f(x_i)}{f'(x_i)}$$

may be written

$$\varepsilon_{i+1} = \varepsilon_i - \frac{f_i}{f_i'} \tag{2.11}$$

and, if we expand $f(x)$ in a Taylor series around the root, b, we have

$$f(x_i) = \cancel{f(b)} + \varepsilon_i \cdot f'(b) + \varepsilon_i^2 \frac{f''(b)}{2} + \cdots$$

$$f'(x_i) = f'(b) + \varepsilon f''(b) + \cdots$$

Thus we may express (2.11) as

$$\varepsilon_{i+1} = \varepsilon_i - \frac{\varepsilon_i f' + \varepsilon_i^2 \dfrac{f''}{2} + \cdots}{f' + \varepsilon_i f'' + \cdots} = \frac{\varepsilon_i f' + \varepsilon_i^2 f'' + \cdots - \varepsilon_i f' - \varepsilon_i^2 \dfrac{f''}{2} - \cdots}{f' + \varepsilon_i f'' + \cdots}$$

so

$$\varepsilon_{i+1} \approx \varepsilon_i^2 \frac{f''}{2f'} \tag{2.12}$$

Hence we see that all Newton-method algorithms are quadratically convergent provided $f'(b)$ is not zero—that is, provided the root is isolated. (The student should repeat this analysis assuming the root is double. What is the rate of convergence? How may the algorithm be altered to restore the quadratic convergence?)

Two-point formulas

Turning to the interpolative algorithms of False Position and the secant method, we find the next approximation to the root given by

$$x_3 = x_2\left(\frac{y_1}{y_1 - y_2}\right) - x_1\left(\frac{y_2}{y_1 - y_2}\right) = \frac{x_2 y_1 - x_1 y_2}{y_1 - y_2} \tag{2.13}$$

Again expanding y in a Taylor series about the root R and letting

$$\varepsilon_i = x_i - R$$

we may write (2.13) as

$$\varepsilon_3 = \frac{(\varepsilon_2 + R)\left(\varepsilon_1 f' + \varepsilon_1^2 \frac{f''}{2} + \cdots\right) - (\varepsilon_1 + R)\left(\varepsilon_2 f' + \varepsilon_2^2 \frac{f''}{2} + \cdots\right)}{(\varepsilon_1 - \varepsilon_2) f' + (\varepsilon_1^2 - \varepsilon_2^2)\frac{f''}{2} + \cdots} - R$$

which becomes

$$\varepsilon_3 = \frac{\frac{f''}{2f'}\varepsilon_1\varepsilon_2 + \cdots}{1 + \frac{f''}{2f'}(\varepsilon_1 + \varepsilon_2) + \cdots} \approx \frac{f''}{2f'}\varepsilon_1\varepsilon_2 \tag{2.14}$$

In the method of False Position the location of x_1 is usually fixed throughout the several iterations by the requirement of keeping the root bracketed. Thus ε_1 is also fixed and the next error has the form

$$\varepsilon_{i+1} = C_0 \varepsilon_i \tag{2.15}$$

which is clearly *first* order. In the secant method, where x_{i-1} is always replaced by x_{i+1} without any bracketing requirements, the algorithm is algebraically the same and the error term is still given by (2.14) except that now ε_1 and ε_2 are both getting smaller at each iteration. We have

$$\varepsilon_{i+1} = C_1 \cdot \varepsilon_i \cdot \varepsilon_{i-1} \tag{2.16}$$

and only the first iterate has the form (2.15). This is not as fast as a second-order process, but it is closer to second order than it is to first. A better estimate can be obtained by considering the size of $C\varepsilon_{i+1}$ after several iterations. Taking

$$C\varepsilon_2 = C\varepsilon_1 = M$$

we have

$$\begin{aligned} C\varepsilon_3 &= C\varepsilon_2 \cdot C\varepsilon_1 \approx M^2 \\ C\varepsilon_4 &= C\varepsilon_3 \cdot C\varepsilon_2 \approx M^3 \\ C\varepsilon_5 &= C\varepsilon_4 \cdot C\varepsilon_3 \approx M^5 \\ C\varepsilon_6 &= C\varepsilon_5 \cdot C\varepsilon_4 \approx M^8 \end{aligned}$$

and generally

$$C\varepsilon_{i+1} = M^{\gamma}$$

where

$$\gamma_{i+1} = \gamma_i + \gamma_{i-1} \qquad \text{and} \qquad \gamma_1 = \gamma_0 = 1$$

The general expression for γ_i is

$$\gamma_i = \frac{1}{\sqrt{5}}\left[\left(\frac{1+\sqrt{5}}{2}\right)^{i+1} - \left(\frac{1-\sqrt{5}}{2}\right)^{i+1}\right]$$

with the second term rapidly becoming unimportant relative to the first. We thus feel that

$$|\varepsilon_{i+1}| \to |\varepsilon_i|^{(1+\sqrt{5})/2} \cdot K$$

and hence the process is of order $(1 + \sqrt{5})/2$, or 1.62, which lies between 1 and 2, as suspected.

The ideas of computational strategy

One of the more persistent messages of this book is an exhortation to suit the tool to the task. This exhortation, in turn, demands that both the tools and the task be understood. The essential shape of the problem to be solved must be discovered and then used as the driving force for the solution strategy. An elegant algorithm applied to the wrong problem is not apt to produce much enlightenment. Nor are large number of random trials to be justified simply because they are relatively easy. Seeking needles in haystacks, even with an electronic computer, is time consuming. It behooves the problem analyst to seek carefully for the important characteristics of his problem and then to pick a strategy that is appropriate. This art is difficult, for equations are often not what they at first appear nor are our numerical tools even close to universally applicable. A method that can find the isolated root of a cotangent-like curve may fail spectacularly when used on a polynomial with a triple root and a few bumps nearby. The separation of the essential from the merely incidental in an equation is still an art—and it is not advanced by books that try to develop numerical analysis as a systematic science. It is the *differences* between these numerical tools that are essential if they are to be used intelligently. Systematic development stresses the similarities, putting the spotlight on elegance of derivation at the cost of elegance in utilization. It may make the subject temporarily palatable to a few more pure mathematicians, but it does great harm to the student who wants to learn how to get answer. In this text we emphasize the variety inherent in numerical methods and urge a careful matching of the computational impedances.

Unimportant terms of an equation

Far too many persons make snap judgments without examining the evidence before them. Give the equation

$$x^2 - 10x + 1 = 0 \tag{2.17}$$

to almost anyone, asking him to find the smaller root, and the chances are overwhelming that he will try out the old schoolboy formula for quadratic equations. It's quadratic, isn't it? What else is there to do? Well, it is *not* a quadratic insofar as the *smaller* root is concerned. That root, being near 0.1, is largely determined by the *linear* terms of the equation and the x^2 is merely a minor perturbation. The correct strategy for solving (2.17) is to treat it as a linear equation, relegating the x^2 to an unimportant place on the right-hand side of the equation. It should never be permitted to escalate our operation into the solution of the quadratic—with its square roots and other complexities. The correct computation follows the algorithm

$$x_{i+1} = \frac{1 + x_i^2}{10} \qquad x_0 = 0$$

which gives

$$0$$

$$0.10$$

$$0.101$$

$$0.1010201$$

on the successive iterations. The computation is very efficient, as the solution of a linear equation should be. The quadratic formula, in its usual form, gives

$$x = \frac{10 - \sqrt{96}}{2} = \frac{-b \pm \sqrt{b^2 - 4ac}}{2a}$$

and we are immediately involved with not only a square root but a serious loss of significant figures in the subtraction. Of course, we may patch up this difficulty by using the plus sign to find the larger root and then get the smaller root by noting that their product is unity, or we could have used the nearly unknown form of the quadratic formula

$$x = \frac{-2}{-10 - \sqrt{96}} = \frac{-2c}{b \pm \sqrt{b^2 - 4ac}}$$

that gives small roots stably (with the minus sign)—but this is merely making partial amends for a wrong decision. Our equation simply was not a quadratic and we should never have got involved with the quadratic formula at all.

Turning to another polynomial,

$$x^3 - 2x - 5 = 0 \tag{2.18}$$

we soon see that there is a single real root near 2. In its present form this equation is most certainly a cubic, albeit a rather simple one. It will yield its root quite efficiently under Newton's method, which is probably the best approach to most cubics, including this one – the cubic "formula" of college algebra being both hard to remember and cumbersome to evaluate. If, however, we wish to apply our unimportant-term philosophy to (2.18) we may shift the root downward by two units, using repeated synthetic division, thus making it small enough for the equation with the shifted roots to become merely a linear equation with cubic perturbations. We obtain

$$\begin{array}{cccc|l}
1 & 0 & -2 & -5 & \underline{2} \\
x & 2 & 4 & 4 & \\
\hline
1 & 2 & 2 & -1 & \\
x & 2 & 8 & & \\
\hline
1 & 4 & 10 & & \\
x & 2 & & & \\
\hline
1 & 6 & & &
\end{array}$$

giving $x^3 + 6x^2 + 10x - 1 = 0$

or $x = \frac{1}{10}(1 - 6x^2 - x^3)$

which rapidly yields 0.0

0.100

0.0939

0.0946

so that our original root was 2.0946. By shifting we have *created* unimportant terms.

Now that we are sensitized to the importance of unimportant terms we might be tempted by this delicacy:

$$x^4 + 0.20x^3 + 3.14x^2 + 0.10x - 4.10 = 0 \tag{2.19}$$

A quartic? Maybe, but those odd powers seem to enter with quite small coefficients. Is this perhaps a biquadratic? We can try writing it as

$$y^2 + 3.14y - 4.10 = -0.20x^3 - 0.10x \tag{2.20}$$

with

$$y = x^2$$

and, for a first approximation, ignore the terms on the right. We obtain, rather crudely,

$$y = \begin{Bmatrix} 1 \\ -4 \end{Bmatrix} \quad \text{or} \quad x = \begin{Bmatrix} \pm 1 \\ \pm 2i \end{Bmatrix}$$

These are undoubtedly good values with which to enter some other algorithm, but the real roots can also be used directly in (2.20) to good effect. The successive values achieved by substituting on the right with ± 1 and then solving the resulting quadratic gives us

1.0	-1.0
0.964	-1.050
0.9705	-1.0571
0.9680	-1.0581

But again, Newton would be faster. We could also shift by unity and create a linear equation with quadratic, cubic, and quartic perturbation terms. This tactic gives us

$$x = \frac{1}{10.98}(-0.34 - 9.74x^2 - 4.2x^3 - x^4)$$

and

0.970

0.9682

0.96815

but the labor of shifting the quartic must be counted against the otherwise considerable efficiency of the process.

As a final example of equations that may have rather different characteristics than they at first glance imply, consider finding the root near $\frac{5}{2}\pi$ of

$$(0.4) \cdot x \cdot \cos x - \sin x = 0 \tag{2.21}$$

Since this root is near, and slightly less than, the point where $\cos x$ is zero and $\sin x$ is 1, we may do better to rewrite (2.21) in terms of the deviation from $\frac{5}{2}\pi$. Letting

$$y = \tfrac{5}{2}\pi - x \tag{2.22}$$

we have

$$(0.4) \cdot (\tfrac{5}{2}\pi - y) \cdot \sin y - \cos y = 0$$

Replacing the trigonometric functions by the first few terms of their series, we have

$$0.4(\tfrac{5}{2}\pi - y)y\left(1 - \frac{y^2}{6} + \frac{y^4}{120} - \cdots\right) = \left(1 - \frac{y^2}{2} + \frac{y^4}{24} - \cdots\right)$$

or

$$y = \frac{1 - \dfrac{y^2}{2} + \dfrac{y^4}{24} - \cdots}{(0.4)(7.8540 - y)\left(1 - \dfrac{y^2}{6} + \dfrac{y^4}{120} - \cdots\right)}$$

which iterates to give y as

$$0.318$$

$$0.32057$$

$$0.31529 \longrightarrow 7.53869 \text{ for } x$$

Thus our transcendental equation turns out to be not only algebraic but effectively a perturbed linear equation.

The choice of a method

Choosing a computational *strategy* is a difficult art for the beginner. It depends both on the specific problem and the tools to be used. Taking a simple example, we consider how to find the root of

$$\tan x = \frac{1}{x} \tag{2.23}$$

that lies just above 2π. A man working with a hand (desk) calculator and a good set of six-place tables might prefer to rewrite the equation as

$$\cot x = x \tag{2.24}$$

so he can scan the table to see where the cotangent becomes approximately equal to 2π (at 0.158), then mentally add this quantity to 2π (getting 6.44), and reenter the table to get an adjusted value (0.154). This technique, well suited to the tools as long as mental interpolation suffices, rapidly becomes laborious if more decimal places are needed. We soon have to shift to more elaborate tables, interpolating in some moderately sophisticated form, and we soon tire of the game – especially when it becomes clear that convergence is something less than rapid. The method will work, but the rate of return on our computational investment is unattractive if we want more than three decimals in x. For three-decimal answers, however, this strategy is the best one available to a human armed with tables.

Newton's method should always be considered when many figures are needed. If we write

$$f = \tan x - \frac{1}{x} \tag{2.25}$$

then

$$f' = \sec^2 x + \frac{1}{x^2} = 1 + \frac{1}{x^2} + \tan^2 x$$

Near the root, tan x will be nearly $1/x$ and will be a *small* contributor to f'. Thus we may eliminate $\tan^2 x$ from f' by writing it as

$$f' = 1 + \frac{2}{x^2} \tag{2.26}$$

To avoid interpolation we enter our tables with the closest tabular entry and get

$$x_1 = 0.154 + 2\pi \qquad f = -0.000118 \qquad f' = 1.04827$$

whence

$$\Delta = 0.000113$$

giving

$$x_2 = 0.154113 + 2\pi$$

All six decimals are probably good since we had three that were correct in x_1.

Let us now change our conditions by supposing that an automatic digital computer is to do the arithmetic. We can, of course, try to implement Newton's method directly, but the problem of a first approximation arises to nag us briefly. We soon see that 2π will do, at the cost of one more iteration. More serious, however, is the struggle the computer probably has to evaluate tan x. Lacking tables, it must compute this function from some continued fraction approximation at a considerable effort—perhaps 50 arithmetic operations before it has tan x in register. Is all this work really necessary? Perhaps our problem can be posed directly as an algebraic iteration with no explicit reference to trigonometric functions. We can write tan x as the ratio of sin x/cos x, then expand each function around 2π to give

$$\sin y = \frac{\cos y}{x} \tag{2.27}$$

where

$$x = 2\pi + y$$

then

$$y\left(1 - \frac{y^2}{3!} + \frac{y^4}{5!} - \cdots\right) = \left(1 - \frac{y^2}{2!} + \frac{y^4}{4!} - \cdots\right) \Big/ (2\pi + y)$$

and finally

$$y = \frac{1}{2\pi + y} \cdot \frac{1 - \dfrac{y^2}{2!} + \dfrac{y^4}{4!} - \cdots}{1 - \dfrac{y^2}{3!} + \dfrac{y^4}{5!} - \cdots} \tag{2.28}$$

In this form we may easily iterate, starting with y at 0. We obtain

0
0.159155
0.153910
0.154120
0.154113

The number of iterations is greater but the total computational labor is considerably less – for an automatic computer. (A man without adequate tables will obviously choose the same algorithm.) We could also employ Newton's method with (2.27), although its form does not seem particularly suited to that treatment.

A nastier problem

Turning to another problem we seek the real roots of

$$f = (x - 0.6)(x - 1.3)^2(x - 2.0)^3 + 0.01234 \ln x \tag{2.29}$$

A rough sketch (Figure 2.11) of the polynomial part and the logarithmic curve

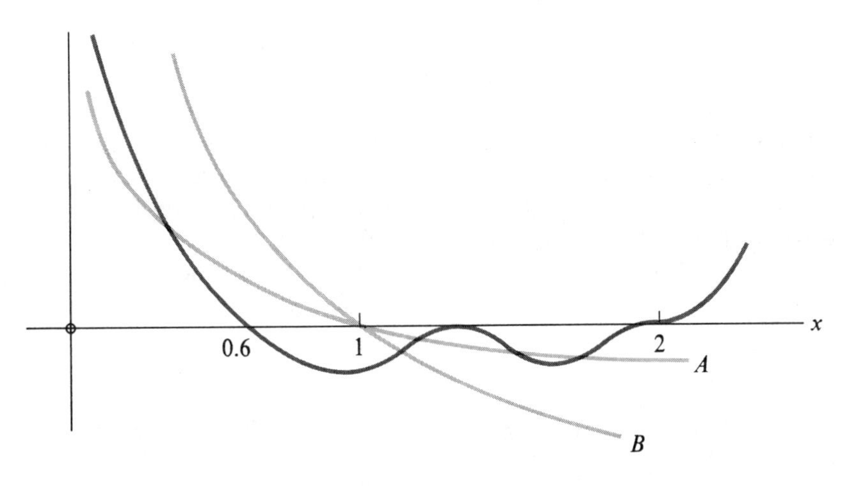

Figure 2.11

shows that there must be a root near 1.2 and that there may or may not be two more roots near 1.7, depending on whether the logarithmic curve looks like A or B. A preliminary evaluation at 1.7 is inconclusive, merely showing the curves to be close in this region but not settling the question about the possible intersection. The sketch also suggests two possible roots in the range (0, 0.6).

Geometrically this problem behaves like a polynomial. The roots may exist as close pairs, may be double, or may only be figments of the viewer's imagination. Careful exploration may be necessary. Unfortunately, the $\ln x$ term denies us many methods specifically designed for polynomials, so we will probably fall back on False Position and Newton. Since the function f has several flat places near the questionable roots, we would be wise to avoid Newton there at least temporarily. The chances of skittering all over the graph are far too large. Only when we go after the one clear isolated root is Newton definitely desirable.

In the questionable regions we should concentrate on finding the zero of the derivative, f', that would separate the pair of close roots, provided they exist. This extremum is isolated, so the search should be efficient via False Position and Newton. Once the extremum is located, a functional evaluation there for f will settle the question about the existence of the roots. If they really are there, a simple application of a Taylor series at the extremum will give good approximations to the two roots, which may then be considered to be well separated, permitting use of Newton's method on each separately should more significant figures be required. Thus we note that at the extremum we have

$$f(\Delta) = f_e + \Delta f'_e + \frac{\Delta^2}{2!} f''_e + \cdots$$

so the displacements Δ of the roots are given by

$$\Delta = \pm\sqrt{\frac{-2f_e}{f''_e}}$$

If the quantity Δ is imaginary, so are the roots. The second derivative information that we need will already have been available if we were homing in on the extremum by Newton's method. The student should try this problem by automatic computer—perhaps varying the coefficient of the $\ln x$ term by 1 per cent to see if his program can always handle the somewhat delicate questions that arise.

Computational tactics

We have been considering computational *strategy*: the identification of the essential difficulties in a problem and the mapping of a strategic approach well suited to the problem and the available tools. We now turn to items that could be called computational *tactics*: difficulties of detail that, although not central to the problem, still tend to occur across many problems and must be overcome if the grand campaign is to prosper. The loss of significant figures in the middle of an otherwise exemplary computation is usually a minor annoyance, but it can occasionally become a major disaster. In hand computation the danger can be assessed when it appears, remedial action being improvised as necessary. But in the bowels of a lengthy machine computation, such disappearances must be anticipated and prevented lest they vitiate the final results—with no one the wiser until too late. At the very least, a loss of figures must be tested for; at best, it will have been o'ercome! We say more on this subject below.

The somewhat allied topic of numerical indeterminacy also raises tactical difficulties. We may have to evaluate the function

$$f = \frac{\sin x}{x}(1 + x^2)$$

for various x. If zero is a possible value and if we are writing a program for an automatic computer, then trouble looms. A hand computor knows that $\sin x/x$ tends to unity at the origin, but an electronic machine must be told. Further, it is not enough simply to test for the unique troublesome x in order to provide a value for the function there; we must also beware when x is small. For if f is indeterminate when x is zero, it is *nearly* indeterminate when x is *nearly* zero. So we must provide an alternative computation for small x. How small an x and what alternative computation depend on the accuracy we seek. Clearly we may replace $\sin x$ by its power series, explicitly dividing out x to get

$$\frac{\sin x}{x} = 1 - \frac{x^2}{3!} + \frac{x^4}{5!} - \frac{x^6}{7!} + \cdots$$

Once we decide on the accuracy we want and the critical value, x_c, below which this series will be used, the sufficient number of terms becomes obvious. There are no real difficulties here; to recognize the problem is to suggest its solution. But sometimes it takes a most suspicious mind to anticipate the foci of indeterminacy. As an exercise, we consider some fairly obvious examples.

Singularities—computational and analytic

If $f(v)$ is well behaved, the integrals

$$\int_0^x \frac{f(v)\cdot(1-\cos v)}{\sin^2 v}dv \qquad \int_0^x \frac{f(v)\sin v}{v}dv \qquad \int_0^x f(v)\cdot v\cdot \ln v\cdot dv$$

$$\int_0^x \frac{e^{-v}}{\sqrt{v}}dv \qquad \int_0^x \frac{f(v)}{\sqrt{|1-v|}}dv$$

all have computational singularities at obvious places. Some of these are easily dealt with, some require considerable work—but all must be regarded with suspicion and positive action taken in every case. It is the height of computational folly to hope that a singularity may be avoided or that its influence may not be felt. Nor can one rely on the many-figured accuracy of a digital computer to save one from this little extra thought and work. Murphy's law guarantees that if a singularity is ignored, it will upset one of the first problems tried with the program. We may rewrite our first example as

$$\int_0^x \frac{f(v)}{1 + \cos v} dv$$

so that the computational singularity at zero is easily removed (but what about π?), whereas

$$\int_0^x \frac{f(v) \cdot \sin v}{v} dv$$

requires a series expansion and a test to see if v is small enough to require the algebraic evaluation rather than the trigonometric one. These computational singularities arose in the form of numerical indeterminacies that were easily handled. The term $v \ln v$ is also numerically indeterminate but is somewhat nastier. We know that its value at the origin is zero, but small values of v are computationally unpleasant. If the function $f(v)$ in our integral is sufficiently complicated to prevent integration by parts, the normal way to get rid of the embarrassing $\ln v$ term, then we might wish to go through the trouble of expanding f in a Maclaurin series. This maneuver turns our integral into a sum of integrals, each of which may be integrated by parts to give an expression that is well behaved for small v.

Some singularities are more than merely *computational.* When an integrand actually becomes infinite, we must determine whether the integral itself is infinite or not. If not, then we must find a way to remove the singularity into an integrable form and eliminate it by analytic integration. If, on the other hand, the *integral* is really infinite at some point, then we must live with it, expressing the singular part as an infinite function of the appropriate strength in order to remove it from the numerical quadrature, lest it bollix up the use of a quadrature formula that almost certainly assumes polynomic behavior of the integrand.

One should remember that an infinity in a derivative constitutes non-polynomic behavior, too, so that it is not sufficient merely to satisfy oneself that the integrand is finite. We could, for example, integrate our fourth example by parts to get

$$2\,e^{-x} \cdot \sqrt{x} + 2\int_0^x e^{-t}\sqrt{t} \cdot dt$$

but $\sqrt{t}$ has an infinite slope at zero and the use of Simpson's rule near zero will not give the best results. We shall be better advised to change the variable of integration by letting t become u^2, thereby producing

$$\int_0^x \frac{e^{-t}}{\sqrt{t}}\, dt = 2\int_0^{\sqrt{x}} e^{-u^2} \cdot du$$

which is well behaved near the origin. (Neither form is suitable for the standard polynomial-based numerical quadrature formulas when x is large, those values being best obtained from asymptotic series. But that is not a problem of singularities – merely one of inappropriate functional form – and will not be discussed further here.)

The final integral, which has square-rootish troubles in the middle of its range of integration, is left to the reader as an exercise!

The railroad rail problem—answer

(But why not have another try at it before reading further? Especially if your answer was not close to 45 feet.)

In solving any physical problem with numbers, it is usually convenient to normalize the dimensions so that they become human-sized numbers – that is, numbers between 0.1 and 10. In the railroad problem we shall take our unit to be the *half* mile. In Figure 2.12 we have shown the essential geometry of the problem and defined our several variables thereon. The quantity e is known and is equal to 1/5280. We could express the half-arc length as $1 + 0.000189$, but since it is precisely the *difference* between the chord and arc lengths that produces the crucial geometry, one suspects that it is this difference, e, that we need to use in our calculations rather than the arc length itself. Thus we preserve e as a separate entity. We easily find

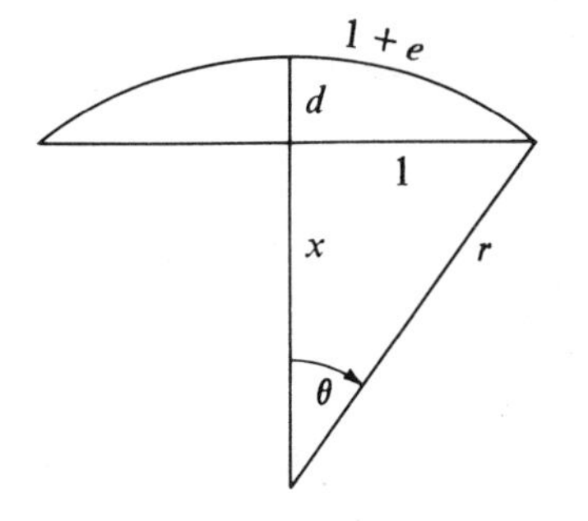

Figure 2.12

$$1 + x^2 = r^2 = (x + d)^2 \tag{2.30}$$

$$\sin\theta = \frac{1}{r} \tag{2.31}$$

$$r\theta = 1 + e \tag{2.32}$$

where $e = 0.000189394$.

By dividing (2.31) by θ we have

$$\frac{\sin\theta}{\theta} = \frac{1}{r\theta} \tag{2.33}$$

and then

$$\frac{\sin\theta}{\theta} = \frac{1}{1+e} = 1 - \frac{e}{1+e} \tag{2.34}$$

Using the series for $\sin\theta$ we have

$$1 - \frac{\theta^2}{3!} + \frac{\theta^4}{5!} - \frac{\theta^6}{7!} + \cdots = 1 - \frac{e}{1+e} \tag{2.35}$$

giving us an expression each side of which is a quantity very slightly less than unity. Having written our equation in this form it is trivial to cancel the 1's from both sides of the equation – this being the crucial step in getting an accurate computable form for our problem. Basically the information in our problem is contained in e, a small quantity, and from this we must compute θ or θ^2, likewise a small quantity. In the form in which the equation first stood [Equation (2.34)] this information was masked by being contained in expressions that were very close to unity, but the series expansion for $\sin\theta$ has permitted us to get rid of the masking large number. Factoring out $\theta^2/3!$ and solving for θ^2 we obtain

$$\theta^2 = \frac{6e}{(1+e)\left(1 - \dfrac{\theta^2}{20} + \cdots\right)} \tag{2.36}$$

Since it is clear that θ^2 will be of the order of 0.001, it is equally clear that terms higher than θ^2 need not be retained in the denominator factor on the right. Indeed, if only three significant figures is our aim we need not even retain the θ^2 term. We find immediately

$$\theta = 0.033708 \qquad \text{and} \qquad r = 29.672$$

Eliminating x from (2.30) we obtain

$$1 - 2rd + d^2 = 0$$

which looks like a quadratic in d but is really a linear equation with a slight quadratic perturbation. Accordingly we write it as

$$d = \frac{1 + d^2}{2r}$$

which we solve iteratively to obtain

$$d = 0.016856$$

(the d^2 term contributes something in the fifth significant figure). Remembering that our units are half-miles we must multiply by 2640 to find that d is 44.499 feet —a value that always surprises the readers of the Sunday papers. (A rough application of Pythagorus to the almost triangle with the sides $1, d$, and a hypotenuse $1 + e$ would yield a crude value for d of 0.0194, or approximately 51 feet. Thus the order of magnitude is easily ascertainable by quite unsophisticated techniques.)

This problem illustrates two important points in numerical computation. The first, already emphasized in the calculations, concerns the removal of large masking quantities when the crucial information is carried by small numbers. Sometimes considerable analytical ingenuity may be required. The second point is simpler but needs to be emphasized because it runs counter to the standard high school algebra training: that a polynomial equation with the unknown raised to the second power is not always a quadratic equation! Of course it has two roots, but the root of interest to us in two of our equations has been the one near zero. In an equation with a constant term the order of unity and the linear term with a coefficient much larger than unity, the quadratic term will carry little relative weight for such a root. Thus this equation is really a linear equation with a small quadratic perturbation and the root will be determined primarily by the linear and the constant terms. Accordingly, the quadratic term should be put over onto the right-hand side and temporarily ignored. A first approximation is obtained entirely from the linear part of the equation; this approximation is then substituted into the quadratic term and the equation re-solved as a linear equation. This is faster, more expedient, without the problem of square roots, and—most important—stable! If one treats the equation as quadratic according to the usual high school formula one quickly gets into the position of subtracting large nearly equal numbers to find a small difference—precisely the problem we have been trying to avoid. (There are other forms of the quadratic formula which avoid this difficulty but they involve more computational labor, as they still require square roots.)

Loss of significance in alternating series

A frequent significant-figure difficulty arises when we evaluate an alternating series. By way of example consider the series for $J_0(x)$, a series that is absolutely and uniformly convergent—and virtually useless for x greater than 4. This assertion may come as a bit of a shock to those who have never computed, for absolute uniform convergence is the highest mathematical accolade to which a series can aspire. It means that for *any* value of x the series will approximate the function that it represents to arbitrary accuracy provided enough terms are taken and provided all the arithmetic is performed with sufficient precision. This last proviso is the weak link, for in practical computation the precision of any one datum in the computer is sharply limited to some small standard number of significant figures.

If we examine our series

$$J_0(x) = 1 - \frac{x^2/4}{(1!)^2} + \frac{(x^2/4)^2}{(2!)^2} - \frac{(x^2/4)^3}{(3!)^2} + \cdots$$

while mentally substituting in the value 10.1 for x we see that the magnitudes of the terms increase up through x^{10} (why?) and then decrease steadily to insignificance. The largest absolute term is roughly 680, but it is largely offset by the negative term that follows it. We also notice that the final result, which is not large (-0.25), is reached by subtracting several large negative numbers from several large positive numbers. Since the final result has its first significant digit in the first place after the decimal point, all the figures in all the numbers to the left of that position must ultimately cancel by subtraction—and are thus irrelevant for our purposes. This means that the computed value of the largest term contains three figures that are irrelevant—three figures that occupy valuable space since they thereby displace the figures that will ultimately be needed to determine our answer. The big part of the numbers squeezes out the small, but only the small part survives in our answer. Thus for the argument 10.1 we have barely five significant figures in our final answer if we evaluate the series term by term on a computer capable of retaining eight significant digits. Since a loss of only two significant figures is usually considered serious, we see that even for x equal to 4 the series for $J_0(x)$ is barely adequate. This behavior is typical for alternating series.

More about significant-figure loss

The hidden loss of significant figures is probably the most persistent computational trouble to plague even the experienced user of automatic digital computers. He must continually be on guard lest he ask for a sequence of arithmetic operations that yields an answer of seriously diminished precision. Since the trouble arises in manifold guises and can be detected only by alert, suspicious programmers, your author takes every opportunity to warn—preferably by example. Except for the reader's patience, it would be difficult to overstress this subject.

The simple expression

$$f_1(k) = \sqrt{1 - k^2}$$

occurs frequently, and, for most values of k, is quite harmless. In many conformal mapping problems, however, the values of k get perilously close to unity, where the expression requires careful treatment. Let us suppose that k is an input parameter, known to us, having the value

$$k = 0.99987\,65432\,10$$

Since k occurs throughout our program in several contexts, we may be tempted merely to store it and compute all the functions of it that we need. But when we consider significant figures the function we need is

$$f_2(k) = \sqrt{1 - k} \cdot \sqrt{1 + k}$$

It is *not* really a function of k; it is a function of $1 - k$, which is quite another animal. If we are working on a computer that has eight-significant-figure floating-point arithmetic, then storage of k to eight significant figures only permits computation of $1 - k$ to five significant figures. Similarly, calculation of $f_1(k)$ by squaring k, subtracting from unity, and taking the square root will produce the sequence

0.9998 7654	k
0.9997 5340	k^2
0.0002 4690	$1 - k^2$
0.0157 3052	$\sqrt{1 - k^2}$
(0.0157 2999)	correct value

which, not too surprisingly, is wrong in the fifth significant figure. Even if we were slightly more clever and computed $f_2(k)$ *but started* with k, we would get the same wrong answer. Our loss is arising from starting with a five-figure value for $1 - k$. The proper technique, of course, is to store a new variable, k_2, equal to $(1 - k)$ and to compute directly with that wherever appropriate. We can always generate k to full accuracy from it, whereas the reverse is not true. Having started with $(1 - k)$ to eight figures, we must use $f_2(k)$ to assure eight significant figures in $\sqrt{1 - k^2}$, but at least by avoiding $f_1(k)$ we *can* preserve all eight of them – provided our machine rounds properly! The general rules are clear: Store the small quantities; compute the larger ones; never subtract nearly equal quantities.

Around a computer laboratory we hear "Look what the machine did to me this time" as a programmer discovers the wrong result. In our present example, however, we must emphasize the machine did not "do this to the programmer." The programmer did it to himself. With very little thought he could have used an eight-significant-figure tool to produce a nearly eight-significant-figure result. The fact that he got less than six must be blamed primarily on his lack of foresight. The machine contributed to his troubles by allowing him only eight figures, which is disturbingly close to the minimum that one should aim to preserve in order to be reasonably sure of useful results in many engineering calculations. Since even with care, thought, and effort, three or four digits may be eroded away in a computation of moderate size, we do not wish to throw away three at one fell swoop. The student should evaluate the same function both ways via a six-figure machine and compare its results with ours. (It is a sad commentary on our business ethics that some computer companies recommend the six-figure machines for scientific and engineering computation. They know better!)

Turning to a less obvious example, we consider evaluating the expression

$$F(b) = \int_0^b \sin^4\theta \cdot d\theta \qquad 0 < b < 1.5 \tag{2.37}$$

The function may be integrated analytically, giving

$$F(b) = \tfrac{3}{8}b - \tfrac{1}{4}\sin 2b + \tfrac{1}{32}\sin 4b \tag{2.38}$$

For larger values of b we have no problems with (2.38), but if b is small we have a loss of significant digits. For example, b equal to 0.16 gives

$$F(0.16) = 0.060 - 0.07864\,16401 + 0.01866\,23575$$

$$= 0.00002\,07174 = 2.07174 \times 10^{-5}$$

—a loss of three significant digits.

We could have foreseen this problem by noting that, for small values, $\sin\theta$ is close to θ and hence

$$F(b) = \int_0^b \theta^4 \cdot d\theta = \frac{b^5}{5} = 0.00002\,09715$$

Since the first term of (2.38) is 0.06, three figures must disappear through subtraction. We may cure our problem by expanding the sine functions in series, in either (2.37) or (2.38) to produce directly useful contributions to $F(b)$. If many figures are necessary, neither of these series manipulations is pleasant, being fraught with algebraic dangers. Typically one must perform them by hand three times before feeling reasonably secure. Algebraic manipulation of series by the computer itself will help greatly here, if one has it available. If not, your author recommends the series replacement in (2.37), combined with judicious grouping. We illustrate: Assuming that terms through θ^{10} are wanted,

$$\sin^4\theta = \theta^4\left[1 - \frac{\theta^2}{3!} + \frac{\theta^4}{5!} - \frac{\theta^6}{7!} + \cdots\right]^4 = \theta^4\left[1 - \frac{\theta^2}{3!}\left(1 - \frac{\theta^2}{20} + \frac{\theta^4}{840}\right)\right]^4$$

$$= \theta^4\left[1 - 4\frac{\theta^2}{3!}\left(1 - \frac{\theta^2}{20} + \frac{\theta^4}{840}\right) + 6\left(\frac{\theta^2}{3!}\right)^2 \cdot \left(1 - \frac{\theta^2}{20} + \cdots\right)^2\right.$$

$$\left. - 4\left(\frac{\theta^2}{3!}\right)^3 \cdot (1 - \cdots)^3 + \cdots\right]$$

$$= \theta^4\left[1 - \frac{2}{3}\theta^2\left(1 - \frac{\theta^2}{20} + \frac{\theta^4}{840}\right) + \frac{\theta^4}{6}\left(1 - \frac{2\theta^2}{20} + \cdots\right) - \frac{4\theta^6}{6.36}\right]$$

Here we have grouped the terms so that we can repeatedly use the *bi*nomial expansion (who remembers the multinomial expansion?) and proceed only as far as terms that will be needed. The final result is

$$\sin^4\theta = \theta^4\left(1 - \frac{2}{3}\theta^2 + \frac{1}{5}\theta^4 - \frac{34}{27 \cdot 35}\theta^6 + \cdots\right)$$

so the integral becomes

$$F(b) = \frac{1}{5}b^5 - \frac{2}{21}b^7 + \frac{1}{45}b^9 - \frac{34}{35 \cdot 27 \cdot 11}b^{11} \tag{2.39}$$

This expression, evaluated for 0.16, gives

$$\begin{array}{rl} & +\ 0.00002\ 0971520\ 00 \\ & -\ 0.00000\ 0255652\ 82 \\ & +\ 0.\qquad\quad 0001527\ 10 \\ & -\ 0.\qquad\quad 0000005\ 75 \\ \hline F(0.16) = & \ 0.00002\ 0717388\ 53 \end{array}$$

and we see that no cancellation occurs. Indeed, our first expression, (2.38), becomes progressively worse as b decreases while (2.39) improves. On the other hand, we cannot push (2.39) much beyond 0.2 for b, so both forms will normally be required in any reasonably complete program that aims for eight significant figures. This bifurcation of the computation is the rule rather than the exception. So common is it that we are tempted to warn the programmer that if he has provided only one route for the evaluation of any private function, he is probably courting trouble.

Reducing the problem to small increments

We have many examples of numerical difficulties caused by the subtraction of nearly equal quantities. The cure is always some analytic transformation that permits the "exact" subtraction of the large quantities, leaving an equation in terms of a new unknown that is small. Sometimes the quantities subtracted are, indeed, exact—as in the railroad rail problem—but sometimes they are merely sufficiently precise, so that their inexactitude is of no concern to us. The devices for the expedient transformation of the problem are at least as varied as the types of problem themselves. Here we can only give an example or two.

Consider the maximum values of $x \cos x$, which are given by the roots of

$$\cos x - x \sin x = 0 \qquad \text{or} \qquad \cot x = x \tag{2.40}$$

We shall restrict the rest of our discussion to the first root, lying near 0.860 (as a casual glance down the table of cotangents will quickly reveal), the others being easier to find. The root lies close to 0.8603—we used a six-figure table but want a more precise determination. We could, of course, use Newton's method on either form of (2.40), but this immediately involves us in finding our trigonometric functions to many figures, possibly 8 or 10, and this is awkward, especially if we have to do it more than once. Most of us will not have tables that will permit linear interpolation to so many figures. We probably will have tables that give $\sin x$ and $\cos x$ to many figures at some interval, possibly 0.01, and our purpose here is to use such tables to remove the bulk of the subtractions in (2.40), thus leaving us with more tractable quantities to manipulate. We define

$$0.86 + y = x$$

and rewrite (2.40) as

$$\cos(0.86 + y) - (0.86 + y)\sin(0.86 + y) = 0$$

so that

$$(\cos 0.86)(\cos y) - (\sin 0.86)(\sin y) - [0.86 + y] \cdot [(\sin 0.86)\cos y + (\cos 0.86)\sin y] = 0$$

Calling

$$\cos 0.86 = c \qquad \text{and} \qquad \sin 0.86 = s$$

and noting y to be small, we have

$$c\left(1 - \frac{y^2}{2}\right) - [0.86 + y] \cdot \left[s\left(1 - \frac{y^2}{2}\right) + cy\left(1 - \frac{y^2}{6}\right)\right] - sy\left(1 - \frac{y^2}{6}\right) = 0 \tag{2.41}$$

which becomes

$$[c - 0.86s] = y[2s + 0.86c] + y^2\left[\tfrac{3}{2}c - \frac{0.86}{2}s\right] + y^3\left[-\tfrac{2}{3}s - \frac{0.86}{6}c\right] - \cdots$$

Finally we write

$$[2s + 0.86c]y = [c - 0.86s] - y^2[\tfrac{3}{2}c - 0.43s] + y^3\left[\tfrac{2}{3}s + \frac{0.86}{6}c\right] - \cdots \quad (2.42)$$

where all terms in brackets may be evaluated once and for all, directly from tabulated quantities (that is, no tabular interpolation). Since y is small we may reasonably hope that the first term on the right of term (2.42) gives the major contribution. We then solve (2.42) iteratively. For our problem we obtain

$$2.07678\,13499y = 0.00069\,28666 - y^2(0.65278\,39015) + y^3(0.59874\,44)$$

and we quickly find

$$y_0 = 0.00033\,3625$$

so that the y^3 term is negligible and the quadratic term is $-0.00000\,00726$, giving a final value of

$$y = 0.00033\,3552$$

Hence our root is

$$0.86033\,3552$$

The major subtraction took place in the term $(c - 0.86s)$ and the accuracy with which we may evaluate y depends on the number of places to which c and s are known. The other bracketed terms have no critical cancellation of significant digits.

Most transcendental equations may be solved precisely by a similar device. It is for this reason that many-figured tables even at a fairly coarse interval are quite useful. They permit us to reduce the problem to a small unknown and use simple approximations for the function containing it. We commend to the reader's attention the AMS 55: Handbook of Mathematical Functions (U.S. Government Printing Office) as a very sophisticated compendium of tables and formulas. Its low cost leaves the technical man no excuse for ever being without an adequate set of tables.

Multiple roots

Root-seeking algorithms are widely known to "have trouble" with multiple roots and nearly multiple, that is, close, roots. The precise sort of trouble, however, seems not to be very clear in the public mind. We here attempt to relieve some of this confusion. First, let us consider a functional equation

$$f(x) = 0 \tag{2.43}$$

with a *multiple* root of order k at b. Here two sorts of difficulties occur: The *rate* of convergence to b is distinctly slower for most algorithms unless the algorithm is altered to include information about the order of the root, and the *precision* with which $f(x)$ may be computed deteriorates as the root is approached—and deteriorates faster than it would if the root were single and isolated. A supposed difficulty that, in fact, does not occur is divergence of the iteration to some wildly erroneous root. Such instability as the iterations exhibit is confined to small bouncings around within a region called [Wilkinson, 1959] a circle of indeterminacy that includes the root. Just where within this region the root is located is impossible to determine except by increasing the precision of the functional evaluation arithmetic. If we assume that the algorithm has been programmed sensibly (dangerous!) so as to lose no information unnecessarily, then the root will be located more precisely only by going to double-precision arithmetic. By our choice of single-precision arithmetic, every root of (2.43) has acquired its own region of indeterminacy to which it is confined and to which any root-seeking algorithm will converge once it is close enough. The size of the region is a function of the condition of that particular root, one factor of which is its multiplicity, but the arithmetic nature of the equation also exerts considerable influence. Multiplicity generally increases the size of the region of indeterminacy but does not decrease the stability of the iteration.

We may express $f(x)$ as

$$f(x) = (x - b)^k \cdot g(x) \qquad \text{with } g(b) \neq 0 \tag{2.44}$$

so that

$$f'(x) = k(x - b)^{k-1} \cdot g(x) + (x - b)^k \cdot g'(x)$$

and

$$w = \frac{f}{f'} = \frac{(x - b)^k \cdot g(x)}{(x - b)^{k-1} \cdot [k \cdot g(x) + (x - b) \cdot g'(x)]} = \frac{(x - b) \cdot g(x)}{k \cdot g(x) + (x - b)g'(x)} \tag{2.45}$$

We see that w, the usual Newton correction term, approaches indeterminacy as x approaches b—and nothing can be done about it because we cannot reach the useful form at the right of (2.45) by cancelling the troublesome $(x - b)$ factors without knowing b. This formulation of the problem shows why people fear numerical instability. But clearly, in any reasonable algorithm, no correction term is accepted if it is larger then the preceding correction, once a limit is being approached. Thus the indeterminacy in w need never knock us away from the region in which the root lies. The region of indeterminacy of the root arises from the precision with which $f(b)$ may be evaluated, concerning which point the equations of this paragraph say nothing.

If we analyze the rate of convergence of Newton's method for (2.44) we find that the error ε_2 after one iteration is

$$\varepsilon_2 = \left(1 - \frac{1}{k}\right)\varepsilon_1 + C\varepsilon_1^2 \tag{2.46}$$

where ε_1 is the error before the iteration. Thus for single roots convergence is quadratic, but for multiple roots it is linear unless the method is altered to read

$$x_{i+1} = x_i - \frac{kf}{f'}$$

which effectively removes the first term on the right of (2.46), restoring quadratic convergence. Since we do not usually know the multiplicity of our root, this modification is little used. In theory one might estimate k from the successive iterates, but the labor seems of questionable value.

A more attractive device for speeding up the rate of convergence at a multiple root is the use of w as our basic function, rather than f. As is clear from (2.45), w has a single root at b whatever be the multiplicity of the root of f. The price we pay for faster convergence is a higher level of numerical indeterminacy in our function. Since w' is given by

$$w' = \frac{f'^2 - f''f}{f'^2}$$

we have

$$\frac{w}{w'} = \frac{ff'}{f'^2 - f''f}$$

$$= \frac{(x-b)^{2k-1}g[kg + (x-b)g']}{(x-b)^{2k-2}\{[kg + (x-b)g']^2 - g[k(k-1)g + 2k(x-b)g' + (x-b)^2 g'']\}}$$

$$= \frac{(x-b)\left[1 + (x-b)\dfrac{g'}{kg}\right]}{1 + \dfrac{(x-b)^2}{k}\left[\left(\dfrac{g'}{g}\right)^2 - \dfrac{g''}{g}\right]}$$

and we see that considerable cancellation occurs in the denominator—cancellation that can only be accomplished by subtracting nearly equal quantities during the evaluation. Thus we pay for our simple root in w through a more nervous ill-conditioned computation. The increased rate of convergence may still be worth this price.

The device of using

$$u(x) = \frac{f(x)}{x - b}$$

to suppress a root already found will still work even if that root happens to be multiple. In particular, if the root of f was double, then u has a simple root and we get a good rate of convergence with some computational nervousness because of the first-order indeterminacy at the root. If the root was triple, presumably our original convergence to b was slow, then the convergence to b the second time (using u) will be somewhat faster but still linear, while the third convergence [using $u_1 = f/(x - b)^2$] will finally be quadratic, as the root is now effectively simple.

Nearly multiple roots

When do two roots cease to be two single roots and become a double root? Is there an intermediate condition that is worth separate discussion under the name *nearly double roots*? The answer to the first question is clear: Two roots coalesce into a double root when their circles of indeterminacy overlap. If we cannot tell which of two roots we have computed because of the precision limitations with which we can evaluate our function, then the roots have the same computational value and are double by any operational meaning of that word. If we decide to recompute our roots in double-precision arithmetic, the double root may separate into two single roots when their circles of indetermin-acy have been reduced.

The second question is harder, for it really is asking if the presence of a nearby second root seriously interferes with the convergence properties of our algorithm—and thus can be discussed only within the context of a specific algorithm. Newton's method, for example, is very unhappy if applied to an *extremum* of a function. Since two close roots of a continuous function necessarily have a horizontal tangent about halfway between them, there is a very real chance that casual attempts to apply Newton's method will end disastrously. On the other hand, convergence from one side can be guaranteed once we are there—so our only fear is that of starting the first iteration, accidentally, between the roots. A simple test that weeds out enormous increments will take care of that problem. The method of False Position, in contrast to Newton, is not the least disturbed by *extrema*. It only needs to bracket the root to find it. Thus two close roots are troublesome to False Position only to the extent that they prevent one from finding a point *between* them in order to apply False Position in the first place. An *odd* number of roots that are nearly multiple will cause False Position no trouble.

The equation $(px + q)e^x = (rx + s)$

The preliminary analysis necessary for even an innocent-looking equation can often prove complicated. In the hope that it may prove instructional we present a treatment of the problem

$$\frac{rx + s}{px + q} = e^x \tag{2.47}$$

clearly the intersection of a general rectangular hyperbola with the standard exponential curve. One parameter is superfluous, but for manipulative convenience we prefer to leave it in, since various questions about the geometries of the hyperbola lead to different normalizations. The determinant

$$\begin{vmatrix} r & s \\ p & q \end{vmatrix} = rq - ps = D$$

is central to our discussion. In particular, if $D < 0$, our hyperbola lies in the first and third quadrants formed by the asymptotes and the roots are either

1. Two well-separated roots, one on each branch, or
2. One root, on the first quadrant branch, the horizontal asymptote being negative.

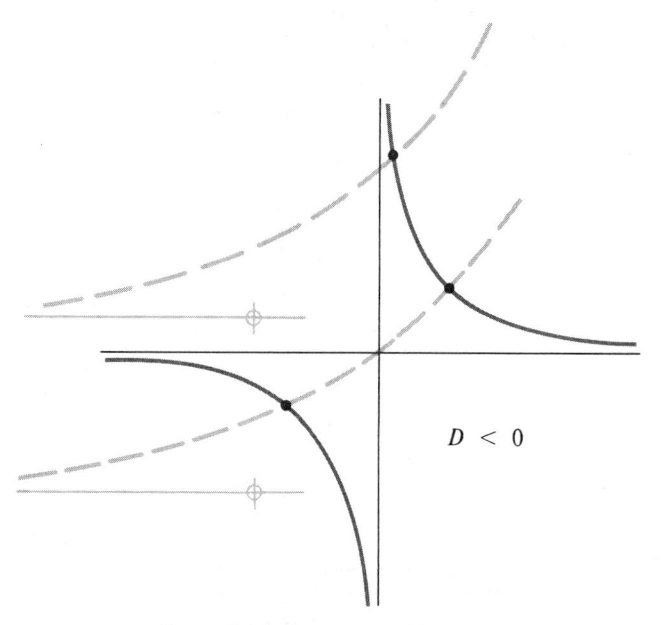

Figure 2.13. Two relative placements of a hyperbola and the exponential e^x

For these geometries there is little to be said, all standard techniques working quite well and good first approximations being available by inspection (Figure 2.13).

For $D > 0$ our hyperbola lies in the second and fourth quadrants formed by the asymptotes, and the geometry is thereby complicated. Because both the hyperbola and the exponential have horizontal asymptotes to the left, their intersections will involve only *one* branch of the hyperbola, but it can be either one (see Figure 2.14). Pictorial clarity in Figure 2.14 requires that we show the

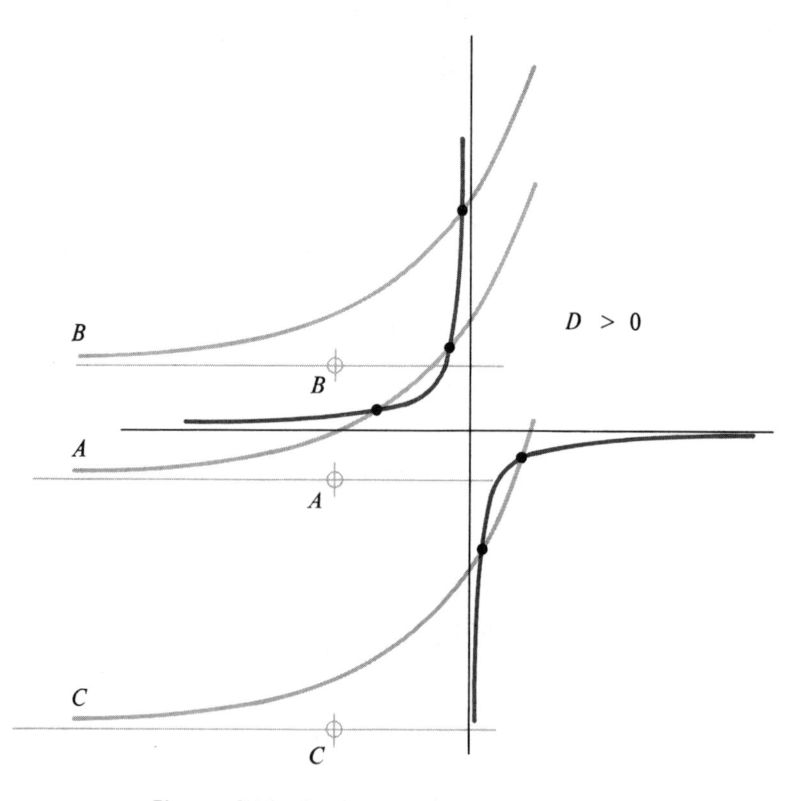

Figure 2.14. Several possible positions of the exponential curve relative to a rectangular hyperbola

hyperbola only once, while moving the exponential curve. The reader must remember, however, that the exponential, having no parameter, is fixed in the coordinate system, so the origin moves with it. Thus for curve B the origin is O_B, for curve C it is O_C. We note that if the horizontal asymptote of the hyperbola is positive (A) and we are considering the second quadrant intersections, we see that only two roots can occur – possibly double. If the horizontal asymptote is strongly negative, only one root will occur (B). If, however, the horizontal asymptote is slightly negative, the possibility of three roots arises, either adjacent pair of which may be double (D in Figure 2.15). This is the most complicated geometry. And it is even possible to find one hyperbola in which all three roots coalesce into a triple root. Intersection of the exponential with the fourth quadrant branch of the hyperbola can give only two roots, possibly double (C).

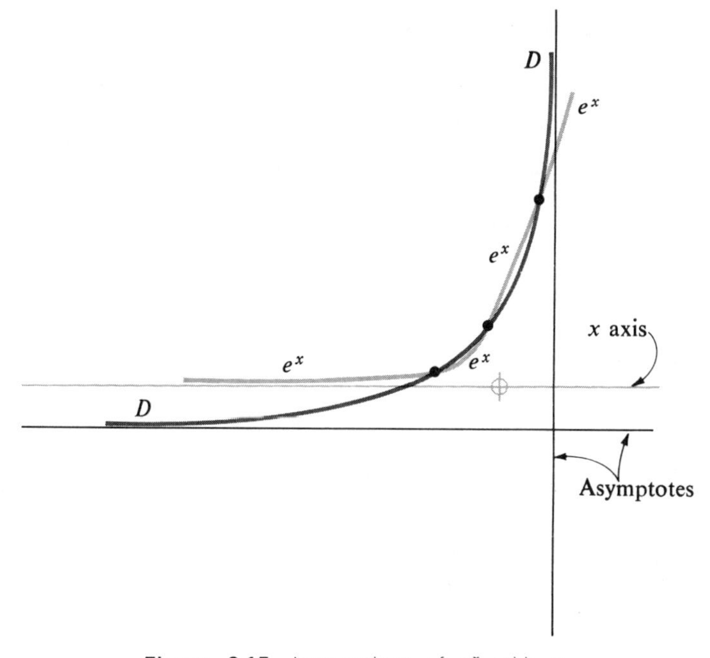

Figure 2.15. Intersections of e^x with a rectangular hyperbola in three distinct places

For all the single (isolated) root geometries there are easily discovered iterative algorithms, although Newton's method will usually be more efficient. These iterative strategies depend on the fact that the roots usually lie near an asymptote – itself easily evaluated from the parameters (p, q, r, s) – and thus one form of the equation exists in which the deviation from that asymptote is expressible in terms of factors that are, relatively, quite stable. Near a vertical asymptote we may thus write our equation in the form

$$(\text{asymp.} - x) = f(\text{stable factors in } x)$$

and solve for the x on the left.

Manipulations

We may express our hyperbola as

$$w = \frac{rx + s}{px + q} \qquad \text{or} \qquad x = \frac{s - qw}{pw - r} \tag{2.48}$$

and also as

$$(pw - r)(px + q) = -D$$

In this latter form we see that the asymptotes occur at

$$w_a = \frac{r}{p} \qquad \text{and} \qquad x_a = -\frac{q}{p}$$

while $|D|$ controls the bending geometry or curvature. A large absolute value of D gives a gentle curve and small values imply sharp curves. (In the limit, D equal to zero gives the asymptotes.) The slope of the hyperbola is given by

$$\frac{dw}{dx} = \frac{D}{(px + q)^2}$$

although other forms may also be used.

Double roots

Concentrating our attention morbidly upon *tangency* between the exponential and a branch of our hyperbola, we note that this requirement becomes one of $w = w'$ at a point x_d. Since the exponential has this property everywhere, our hyperbola must share it at tangency. Formal differentiation of (2.48) and equating to (2.48) gives us the tangency condition to determine x_d. We have

$$(rx_d + s)(px_d + q) = D \tag{2.49}$$

and there may be two such points. If we solve we obtain

$$x_d = -\frac{q}{p} + \frac{D}{2rp}\left(1 \pm \sqrt{1 + \frac{4rp}{D}}\right) \tag{2.50}$$

at which point

$$w_d = w'_d = \frac{r}{p}\left[1 + \frac{D}{2rp}\left(1 \pm \sqrt{1 + \frac{4rp}{D}}\right)\right] \tag{2.51}$$

Note that

$$\frac{D}{rp} = \frac{q}{p} - \frac{s}{r}$$

We may now expand the vertical scale of our hyperbola

$$w = \frac{r}{p}\left(\frac{x + s/r}{x + q/p}\right)$$

by increasing r and, provided we also increase s proportionally, we will not affect D/rp or q/p or hence x_d. Thus we may expand our hyperbola until it is tangent to e^x at x_d.

For real values of x_d we need

$$D(D + 4rp) \geq 0$$

and a triple root presumably requires D equal to $-4rp$, since then the two places, x_d, a double root *could* have occurred have coalesced. The geometry of Figure (2.15) requires a slightly negative horizontal asymptote, implying

$$\frac{r}{p} < 0$$

so r and p must have opposite signs. We thus normalize with

$$p = -1 \qquad \text{and} \qquad r > 0$$

Our sketch suggests that the vertical asymptote is positive, hence

$$-\frac{q}{p} > 0$$

so we take q positive. We may now construct typical examples for the several geometries of curve A in Figure (2.14). Thus the double roots occur with

$$\begin{bmatrix} 1 & 1 \\ -1 & 4 \end{bmatrix} = 5 = D \qquad \text{at } x_d = \begin{cases} 0.382 \\ 2.618 \end{cases}$$

the equations being

$$3.836\left(\frac{x + 1}{4 - x}\right) = e^x \tag{2.52}$$

and

$$5.236\left(\frac{x + 1}{4 - x}\right) = e^x \tag{2.53}$$

With (2.52) there is also an isolated root at 3.462, obtainable iteratively from

$$x = 4 - 3.836e^{-x}(1 + x)$$

while (2.53) has an isolated root at -0.456 obtainable from

$$x = -1 + \frac{5e^x}{5.236 + e^x}$$

Special case:

$$\begin{bmatrix} e & e \\ -1 & 3 \end{bmatrix} = 4e = D$$

gives $x_d = 1$ with a *triple* root there.

Our discussion of double roots gets into trouble if x_d is zero. Then the condition from (2.49) is

$$qs = D$$

while from (2.48) we have

$$\frac{s}{q} = 1 \qquad \text{so} \qquad D = q^2 > 0$$

and

$$rq - pq = q^2 \qquad \text{so} \qquad r - p = q = s$$

As a simple example we take

$$\frac{2x + 1}{x + 1} = e^x \qquad (D = 1)$$

which has a double root at $x_d = 0$. The geometry is shown in Figure (2.16).

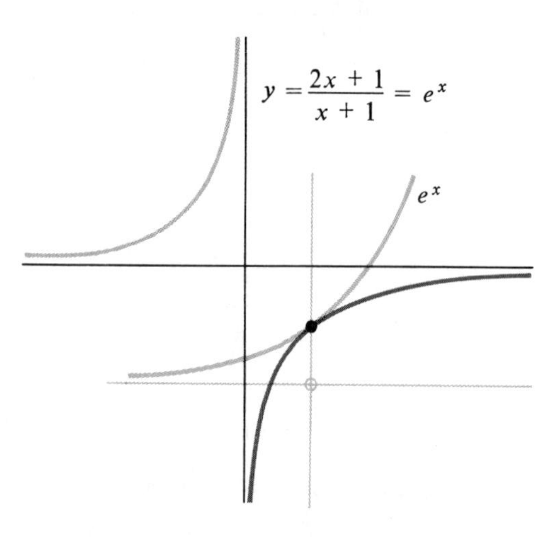

Figure 2.16. A double root configuration

There are no other roots. The student should find iterations that converge to the two roots when they are well separated.

No roots

Between A and C of Figure 2.14 there lies a region in which no roots can occur (Figure 2.17). The limiting configuration of A is the triple root, while that of C

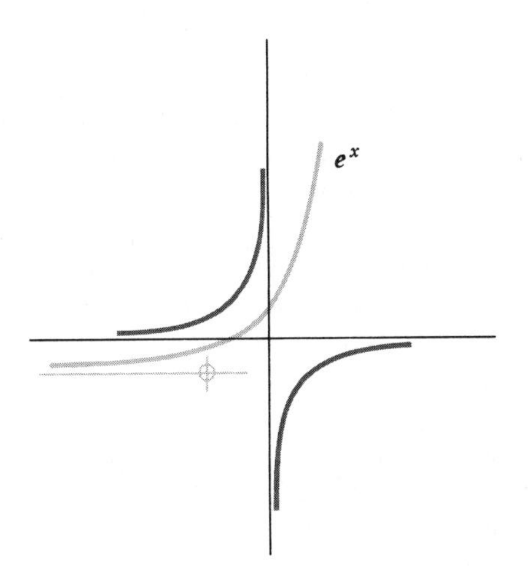

Figure 2.17. A configuration with no roots

is the double root. Looking at (2.50) and (2.51) and remembering that $4rp$ is negative, see that no multiple roots are possible if

$$0 < D < -4rp$$

for *any positions* of the asymptotes. *Thus no roots are possible* if D is in this range. Finally, after this amount of analysis, we are ready to write that computer subroutine, reasonably confident that it won't bomb out on its first use because of our inadequate understanding of the several facets of the problem.

Learning the art of computing—a word to the student

It seems obvious that the person who would learn the art of computing must compute. Unfortunately, some numerical methods lend themselves rather badly to student exercises that can be accomplished with reasonable expenditures of time and effort. Equations in one variable, however, are nearly ideal. Results–

or evidences of trouble—usually come quickly, and analysis of observed behavior is not impossible and definitely rewarding. Accordingly, your author urges you to put considerable effort into working the exercises of this chapter. The skills learned here, the realism gained, will stand you in good stead when you must undertake problems where the complications are greater and you are forced to rely on hunches and intuition.

Most of the problems of this chapter may be done either by hand methods with the help of a slide rule, a desk calculator, and a good set of tables (*Chambers's Shorter Six-Figure Mathematical Tables* [Comrie, 1966] are easily the best known to your author), or they may be done on an automatic digital computer. If the computer is available and you already have a little experience with writing simple programs in FORTRAN (ALGOL, BASIC, and other higher-level languages are equally good), then computer solution is preferable. You will almost certainly have to work out at least one sample calculation per problem by hand, first to gain some understanding of the problem and to provide a check for your programming, but your method of solving the problem will be influenced by what is easily done on the computer rather than by what can be done easily on a slide rule. Thus your basic solution strategies will often be different from those of the student who uses hand-calculation techniques throughout.

If you have never used an automatic computer, never even written a FORTRAN program to sum the squares of the integers from 1 to 10, then you better go learn as soon as possible—but don't learn your programming by trying to solve *these* problems on a computer! If you do you will get into the uncomfortable position of being unsure whether your troubles are arising from a mistake in your FORTRAN grammar or a misunderstanding of the computational algorithm—and people in this position almost always look in the wrong place! Keep your troubles separated and do these problems by hand-calculational methods.

Whichever route you follow, your main object is to acquire a geometric feeling for how these functions twist and turn and from this feeling to develop techniques for root finding that have a reasonable chance of succeeding. Newton's method is neither a universal nor a complete answer to one-dimensional root finding. Even Newton needs to be given good starting values, and for some functions their special properties suggest other approaches.

One final piece of advice: If an algorithm is behaving in an unexpected way, *find out why*! Don't blindly assume your expectations were at fault. If they were, you should aim to improve them by critical examination. If they were not, you are not carrying out the algorithm you thought you were. Passing on to the next problem at this point is a big waste of invested effort. Stick to the problem until you understand it. Good luck!

Problems

1. Solve for the small root of

$$x^2 - 10{,}000x + 1 = 0$$

by the quadratic formula and also by the method of the text.

2. Write a computer program that finds the first three positive roots of

$$\sin x - \frac{b}{x}\cos x = 0$$

by Newton's method. Try it on

$$\tan x - \frac{b}{x} = 0$$

Explain any differences in behavior.

3. Repeat Problem 2 using an iteration scheme for each function.

4. Find the roots of

$$\cos x - 0.80 + px^2 = 0$$

by a computer program. Begin with $p = 0.10$. What happens as p is increased?

5. Find a positive root of

$$e^{-x^2} = \frac{1}{1 + b^2x^2}$$

when $b = 2.0$; when $b = 4.0$. Are there values of b for which no nonzero roots exist?

6. What are the difficulties in writing a computer program to find *all* the real roots of

$$\sin x - bx = 0$$

where b can lie anywhere in the range (0.5, 0.90)?

7. Find the real roots of

$$x^6 - 12.1x^5 + 59.5x^4 - 151.85x^3 + 212.6625x^2 - 156.6x + 48.5625 = 0$$

8. Find the real roots of

$$x^4 - 3.0x^3 + 3.37x^2 - 1.680x + 0.3136 = 0$$

9. If you had to produce a computer program that would find all the roots greater than 0.02 for

$$bx - \sin\frac{1}{x} \qquad 0.1 < b < 0.5$$

how would you proceed?

10. Find the real roots of

$$x^3 + 1.5x^2 - 5.75x + 4.37 = 0$$

via your own computer program, printing out each iteration.

11. Find the real roots of

$$x^4 - 1.73x^2 + 0.46x + 1.275 = 0$$

12. Where are the maxima and minima of $\sin(x - \sin x)$? Where are the roots? Give as good a quantitative discussion as possible without actually resorting to extensive computation.

13. Where are the roots of $\sin(x + \pi \cos x)$?

14. Construct some polynomials with isolated, multiple, and close real roots, then try to find them by a program of your devising.

CHAPTER

3

INTERPOLATION—AND ALL THAT

The derivation of higher-order interpolation formulas is an esthetically pleasing exercise in the expedient manipulation of symbols. It also survives as a heavily emphasized cornerstone of most classical numerical analysis texts. We certainly cannot dispense with interpolation theory: Most of our standard techniques for the solution of differential equations and even the solution of transcendental equations, not to mention quadrature, numerical differentiation, and interpolation itself, rest directly on the replacement of intractable functions by interpolating polynomials. On the other hand, the amount of interpolation theory that one requires for these purposes can probably be absorbed from a one-hour lecture or five pages of a book – rather than the customary lengthy chapter that always seems to turn up in works carrying "Numerical Analysis" in their titles. Aside from its esthetic appeal, interpolation theory once had a prominent practical role in connection with obtaining accurate numerical values from tables of the less common mathematical functions. In this age of electronic computation, however, the use of tables has declined in favor of recomputation of functional values – so that even this justification of interpolation theory has waned and we are forced to give short shrift to the subject. Unless, that is, one happens to be exceptionally enamored of it and decides to give the full-blown treatment for the sheer beauty of it. Since your author does not belong to this school, his exposition will be correspondingly brief.

In this chapter higher interpolation formulas are devices for occasional interpolation in numerical tables—their original function, incidentally—when economic limitations demand that the table be too compressed for linear interpolation. Thus we shall give the expedient formulas, illustrating their practical use. These formulas, useful in different contexts, themselves differ sharply in appearance, but they are *all representations of the same interpolating polynomial.*

Many persons are unaware of the considerable compression in a table that even the use of quadratic interpolation permits. A table of $\sin x$ covering the first quadrant, for example, requires 541 pages if it is to be linearly interpolable to eight decimal places. If quadratic interpolation is used, the same table takes only *one* page having entries at one-degree intervals with functions of the first and second differences being recorded together with the sine itself. And the additional labor of quadratic interpolation is not all that bad. We simply evaluate

$$\sin(x + p) = (ap + b)p + c$$

where a, b, and c are read directly from the table without any modification whatsoever. If a person requires one or two trigonometric functions per month and has a desk calculator, he will probably be quite content to perform this slight additional work in return for being able to condense his shelf space for the common functions by a factor of 500.

Whenever we must find an accurate (say six- or eight-figure) value for a tabled function and must use only desk calculators, by far the most useful interpolation formulas are Bessel's and Everett's. We shall skip their derivation and merely illustrate their use. Most of their virtues are self-evident. If we let p denote the *fractional* distance from the next smaller tabled argument to our value of x, then Bessel's interpolation formula may be written

$$f(p) = (1 - p)f_0 + pf_1 + (\delta_0'' + \delta_1'')B_p'' + \delta_{1/2}''' \cdot B_p''' + (\delta_0^{(iv)} + \delta_1^{(iv)})B_p^{(iv)} + \cdots \quad \textbf{(3.1)}$$

where

$$B_p'' = \frac{p(p-1)}{2!2} \qquad B_p''' = \frac{p(p-1)(p-\frac{1}{2})}{3!}$$

$$B_p^{(iv)} = \frac{(p+1)(p-2)}{3\cdot 4} \cdot B_p'' = \frac{(p+1)p(p-1)(p-2)}{4!2} \qquad B_p^{(v)} = \frac{(p+1)(p-2)}{4\cdot 5} B_p'''$$

$$B_p^{(vi)} = \frac{(p+2)(p-3)}{5\cdot 6} B_p^{(iv)} \quad \text{etc.} \qquad = \frac{(p+1)p(p-1)(p-2)(p-\frac{1}{2})}{5!}$$

The B's may be calculated from their definitions, though hand calculators will find a small table of the B's to be extremely convenient. (One of the best is found in Chambers's Six-Figure Mathematical Tables [1949], pp. 532–539.) They do not, of course, depend on the function being interpolated.

The f's and δ's are the standard entries in the difference table

$$\begin{array}{ccccc} f_{-1} & & \delta''_{-1} & & \delta^{(iv)}_{-1} \\ & \delta'_{-1/2} & & \delta'''_{-1/2} & \\ f_0 & & \delta''_0 & & \delta^{(iv)}_0 \\ & \delta'_{1/2} & & \delta'''_{1/2} & \\ f_1 & & \delta''_1 & & \delta^{(iv)}_1 \end{array} \quad \text{where} \quad \begin{aligned} \delta''_0 &= \delta'_{1/2} - \delta'_{-1/2} \\ &= f_1 - 2f_0 + f_{-1} \end{aligned}$$

Note that the first two terms of Bessel's formula constitute the standard linear interpolant, although not in its usual hand-calculational form. Each additional term represents a correction which gets smaller both because the B's themselves diminish and because of the usual shrinkage in differences of nice functions as the order increases. A table that gives only functional values and second differences—a frequent and typographically convenient arrangement—may be easily interpolated through third differences with no struggle worse than a single subtraction to get δ'''. Additional terms require somewhat more work. In many tables most of the effect of the fourth differences has been thrown back into the tabulated second differences by a device made possible by the similarities between the B'' and $B^{(iv)}$ coefficients, as well as by the smallness of the fourth differences themselves. For further details see Comrie [1966]. Suffice it here to say that tables with "thrown-back" fourth differences are to be interpolated at most through B''' terms.

Example: Suppose we have available a table of sin x given every 2° to 10 decimals, together with the even differences δ'', $\delta^{(iv)}$, and $\delta^{(vi)}$, as shown in Figure 3.1. We wish to find sin 32.2345° so

$$p = \frac{0.2345}{2} = 0.11725$$

From p we may calculate the B's directly, obtaining

	0.5333515984	linear interpolant
$B'' = -0.025875609$	+343347	
	0.5333859331	quadratic interpolant
$B''' = 0.00660259299$	−2355	
	0.5333856976	cubic interpolant
$B^{(iv)} = 0.004535785$	+73	
	0.5333857049	quartic interpolant
$B^{(v)} = -0.000694427$	0	
	(same)	quintic interpolant

When we multiply each B by its proper δ as indicated in (3.1), we obtain the corrections shown above, together with the successive interpolants they yield.

We can shorten these B's somewhat if we note that our δ'' data are seven figures, and hence the correction they produce when multiplied by B'' will be no more

x	$\sin x$	$-\delta^2$	$+\delta^4$	$-\delta^6$
26	0.43837 11467 891			
28	0.46947 15627 859	57 19787 827		
30	0.50000 00000 000	60 91729 809	7421 834	
32	0.52991 92642 332	64 56249 957	7865 948	9 591
34	0.55919 29034 707	68 12904 157	8300 471	10 103
36	0.58778 52522 925	71 61257 886	8724 891	10 634
38	0.61566 14753 257	75 00886 724	9138 677	
40	0.64278 76096 865	78 31376 884		
42	0.66913 06063 589			

Figure 3.1

than six figures. Hence six-figure accuracy for B'' would have sufficed. Similar arguments will permit us to shorten B''' to four significant figures and $B^{(iv)}$ to 3. $B^{(v)}$ in this example can be entirely ignored.

The commonly available tables of B's given in *Chambers's* supply only three figures for B'', two figures for B''', and one for $B^{(iv)}$ – making them suitable for interpolation in our table of Figure 3.1 only to five or six figures. On the other hand, our sample table is given at an extremely coarse interval simply for didactic purposes and does not represent good tabular practice if 10 figures were regularly to be required. In a practical table, the interval would be smaller and the δ's correspondingly reduced. For many tables, the Besselian coefficients given in *Chambers's* are adequate for full tabular accuracy.

Everett's formula uses only the functional values and the *even* differences, but it uses all the items appearing on *two* tabular lines. We use

$$f(p) = \begin{cases} (1-p)f_0 + E_0'' \cdot \delta_0'' + E_0^{(iv)} \cdot \delta_0^{(iv)} + \cdots \\ +pf_1 + E_1'' \cdot \delta_1'' + E_1^{(iv)} \cdot \delta_1^{(iv)} + \cdots \end{cases}$$

where

$$E_0'' = -\frac{p(p-1)(p-2)}{3!}$$

$$E_0^{(iv)} = -\frac{(p+1)p(p-1)(p-2)(p-3)}{5!} = \frac{(p+1)(p-3)}{4 \cdot 5} E_0''$$

and

$$E_1'' = \frac{(p+1)p(p-1)}{3!}$$

$$E_1^{(iv)} = \frac{(p+2)(p+1)p(p-1)(p-2)}{5!} = \frac{p^2-4}{4 \cdot 5} \cdot E_1''$$

and again the two left terms constitute linear interpolation. Everett's and Bessel's coefficients are simply related, the two formulas being merely slight rearrangement of each other. We have

$$E_0'' = B'' - B''' \qquad E_0^{(iv)} = B^{(iv)} - B^{(v)} \qquad B'' = \tfrac{1}{2}(E_1'' + E_0'')$$

$$E_1'' = B'' + B''' \qquad E_1^{(iv)} = B^{(iv)} + B^{v} \qquad B''' = \tfrac{1}{2}(E_1'' - E_0'')$$

Mere rearrangements are not to be sneered at if much interpolation is to be performed, but for the occasional value the relative merits of Bessel and Everett are obscure. The reader is hereby urged to use one or the other of these formulas several times until he feels secure in its manipulation and then to stick with it. The throwback feature is equally as practical for Everett as it is for Bessel.

Example of Everett interpolation

Using the same data of Figure 3.1 and seeking the same value, sin 32.2345°, we again have $p = 0.11725$. We may then calculate Everett's coefficients only from p to find

$E_0'' = -0.0324782024$	0.5333515984	linear interpolant
	→ +340993	
$E_1'' = -0.0192730164$	0.5333856977	cubic interpolant
$E_0^{(iv)} = +0.00523021$		
	→ +73	
$E_1^{(iv)} = +0.00384135$	0.5333857050	quintic interpolant

Again, each E is multiplied by its corresponding *even* difference. With Everett's formula we get a succession of *odd* interpolants, linear, cubic, quintic, and so on. While the amount of labor is slightly less than that of Bessel's, the difference is unimportant. If an *even* interpolant is specifically desired, the last pair of Everett coefficients may be replaced by their Besselian counterparts and that part of the formula turned into the corresponding piece of Bessel's. Note that, except for rounding vagaries, the two formulas give identical results for the odd interpolants. This is true for *all* the common interpolation schemes since they all pass the same polynomial through the function being interpolated.

Interpolation without differences: Aitken

If one is so unfortunate as to be forced into hand interpolation in a table which inconsiderately has failed to supply the higher differences, then Bessel and Everett are not directly useful. We may either repair the omission by differencing the table, which is dangerous, or resort to interpolation formulas, such as

Aitken's, which require no differences. Lagrange is still another possibility, but only if the maker of the table has been careful to specify the order of the interpolation that is necessary—for Lagrange carries no built-in measure of the accuracy he supplies. Reckless use of Lagrangian interpolation will yield many digits and with great efficiency, but one will have little idea about their validity.

Aitken's scheme of interpolation has a number of minor variations that are all mathematically equivalent but vary considerably in the ease with which a human can carry them out. We present our favorite:

1. Write down several tabular arguments x_i in a column (1) (see Figures 3.2 and 3.3), the top item x_0 being chosen to be the argument *closest* to the x for which we seek $f(x)$. The subsequent arguments should be arranged so that each one is the closest tabular argument not yet chosen. In an evenly spaced table this rule will usually cause the arguments to alternate around x (the exceptions occur near the ends of tables when the interpolation process may have to approach mostly from one side).

$$
\begin{array}{c}
(1) \\
x_0 \\
x_1 \\
x_2 \\
x_3 \\
x_4
\end{array}
\quad
\begin{array}{c}
(3) \qquad\qquad\qquad\qquad\qquad\qquad (2) \\
\left[\begin{array}{lllll}
f_0 & & & & x_0 - x \\
f_1 & d_{01} & & & x_1 - x \\
f_2 & d_{02} & d_{012} & & x_2 - x \\
f_3 & d_{03} & d_{013} & d_{0123} & x_3 - x \\
f_4 & d_{04} & d_{014} & d_{0124} & x_4 - x
\end{array}\right]
\end{array}
$$

Figure 3.2

To find sin 32.2345°:

x_i	$\sin x_i$						$x_i - x_\varepsilon$
32	0.52991 92642						−0.2345
34	0.55919 29035	0.53335 15984					1.7655
30	0.50000 00000	0.53342 72979	[0.53338 50103]				[−2.2345]
36	0.58778 52523	0.53331 16578	0.53338 68560	0.53338 56977			3.7655
28	0.46947 15628	0.53346 30107	[0.53338 43814]	[57129]	57049		[−4.2345]
38	0.61566 14753	0.53327 03556	0.53338 74569	56937	57052	050	+5.7655

Figure 3.3

2. Write down the values $(x_i - x)$ over at the right in column (2) and the corresponding tabular entries f_i in column (3), next to column (1).
3. Compute each of the d's by evaluating a 2×2 determinant comprising (a) the item immediately to the left of the d being computed; (b) the item at the top of the column containing (a), and finally the two $(x_i - x)$ items that sit on the same rows as items (a) and (b). The determinant is then divided by the difference of the two x_i's—or of the two $(x_i - x)$ factors if you prefer—subtracting the upper one from the lower.

Thus in our example (Figure 3.3) the term enclosed in the box in dashed lines, 57129, was computed from the four items enclosed in solid boxes, according to the scheme

$$\begin{vmatrix} 0.5333850103 & -2.2345 \\ 0.5333843814 & -4.2345 \end{vmatrix} \Bigg/ (-2) = 0.5333857129$$

the denominator being $(28 - 30)$ from the corresponding rows in the leftmost column. In this calculation, those leading figures in the first column that are identical, 0.53338, may be dropped provided they are supplied in the result. If this scheme is carried out by rows, then the d entries usually converge smoothly toward the final interpolated value and one may easily decide when to stop by noting the degree of agreement between the d items that sit directly below one another at the right ends of the rows. A blunder quickly appears as a worsening of agreement rather than the expected improvement. Finally, since each d is just a linear interpolation between two earlier d's, only those digits need be carried in which they differ—thereby somewhat reducing transcription problems.

This Aitken interpolation scheme is quite general. Neither the spacing nor the order of the tabular arguments is mathematically restricted. Computationally, however, a different order will produce different round-offs and hence slightly different final results. Psychologically (and this is most important in hand computation) the physical nearness on the page of the numbers that most closely agree is reassuring, while the use of the *top* item in a column is far more efficacious in preventing blunders than are those rearrangements that choose other elements—even nearby ones.

Lagrangian interpolation

If Lagrangian interpolation is particularly appropriate, then we would demonstrate how it should be carried out, for most of the standard treatments are incomplete or restricted to the theoretical aspects rather than the computational. We follow Comrie [1949]. The basic Lagrangian formula is straightforward. For four points it reads

$$f(x) = \frac{(x - x_0)(x - x_1)(x - x_2)}{(x_{-1} - x_0)(x_{-1} - x_1)(x_{-1} - x_2)} f_{-1}$$
$$+ \frac{(x - x_1)(x - x_2)(x - x_{-1})}{(x_0 - x_1)(x_0 - x_2)(x_0 - x_{-1})} f_0$$
$$+ \frac{(x - x_2)(x - x_{-1})(x - x_0)}{(x_1 - x_2)(x_1 - x_{-1})(x_1 - x_0)} f_1 \tag{3.2}$$
$$+ \frac{(x - x_{-1})(x - x_0)(x - x_1)}{(x_2 - x_{-1})(x_2 - x_0)(x_2 - x_1)} f_2$$

and generalization to n points is direct. The formula is highly systematic, so it is not difficult to program its evaluation on an automatic computer. Further, its generality is attractive, there being no requirement that the x_i be equally spaced – merely distinct. The disadvantages, however, are threefold:

1. Having evaluated the formula for a set of x_i and an x one obtains an interpolant, $f(x)$, with no indication of its accuracy.
2. The process does not lend itself readily to inverse interpolation.
3. As the hand calculational process it is psychologically incompatible with most people. (Try it!)

The accuracy problem may be resolved *a priori* by studying the function and recording the degree of Lagrangian interpolation that suffices for full tabular accuracy or else internal evidence may be sought by performing the interpolation *twice*: once on the set of points used in (3.2) and once on a shifted set (say x_{-2}, x_{-1}, x_0, x_1). The second solution to our difficulty is, of course, expensive. Your author far prefers Aitken when the spacing of the x's is uneven and Everett when uniformly spaced tables are involved. He observes, in particular, that his personal error rate is much higher for Lagrangian interpolation in many-figured tables than when he uses the other methods. For automatic computers, Lagrangian interpolation is occasionally the efficient procedure. Were it not for such uses, we would not bother to discuss the process. (On the other hand, for derivational and other theoretical purposes it is indispensible.)

When our table has uniform spacing the Lagrangian formula simplifies and it becomes practical to table the *Lagrangian coefficients*, L_i, so that formula (3.2) becomes

$$f(x) = L_{-1}(p) \cdot f_{-1} + L_0(p) \cdot f_0 + L_1(p) \cdot f_1 + L_2(p) \cdot f_2 \tag{3.3}$$

and the amount of calculation labor drops to a most reasonable level if the argument p is a tabular entry of the *Lagrangian coefficient table*. There are quite complete tables of Lagrangian coefficients, though they tend to be bulky. AMS 55 gives a set that is about right for semiannual use by the hand computor, and they are at an argument of 0.01 – meaning that if one starts from a table of (say) cosines given at intervals of 2°, then interpolation to 0.02° would be possible by using (3.3) while reading the $L_i(p)$ directly from AMS 55. Cumulation of products on most calculators is straightforward, although machines with incomplete carry may get into trouble if intermediate cumulants become negative – as they usually do. [The $L_i(p)$ are not all positive.]

If we wish to interpolate to an argument that is not in our table of Lagrangian coefficients, we have two choices: We may subtabulate in the crucial region, using one application of (3.3) for each intermediate value we need to construct, and then apply (3.3) once more on the local table of smaller interval that we have just constructed; or we can use the device illustrated below.

To find sin 32.2345° from the data of Figure 3.3 we see that we need $p = 0.11725$ (the tabular interval is 2°). Using the six-point Lagrangian coefficients from AMS 55, we see that the closest argument is 0.12, and we must produce a correction for m equal to -0.00275. The obvious correction will involve the first and possibly higher differences. Defining n as the closest argument in our table of Lagrangian coefficients (0.12 in our example), then

$$p = n + m$$

and we define

$$A = \frac{m(2n - 1)}{4} = 0.0005225$$

The arithmetic to be carried out may be written

$$\begin{aligned} f(p) = f_{-2} \cdot L_{-2} & \\ + f_{-1} \cdot (L_{-1} + A) & \\ + f_0 \cdot (L_0 - m - A) & \\ + f_1 \cdot (L_1 + m - A) & \\ + f_2 \cdot (L_2 + A) & \\ + f_3 \cdot L_3 & \end{aligned}$$

but below we have carried out the $\sum_i f_i \cdot L_i$ first to give the six-point Lagrangian interpolant for 32.24, using the tabulated coefficients from a 10-place table to get

$$0.53346\ 69004$$

Then we computed the m correction and the A correction, giving

$$\begin{array}{rl} 0.5334669004 & \\ -805025 & m \text{ correction} \\ \hline 0.5333863979 & \\ -6933 & A \text{ correction} \\ \hline 0.5333857046 & \end{array}$$

The result is off by 4 in the last place. These corrections can be applied after the value for sin 32.24° has been found, or they may be combined into the Lagrangian coefficients. With most desk calculators it is simpler to apply them serially right after the corresponding $L_i(p)$ while the f_i is still on the machine's keyboard.

Tables of Lagrangian coefficients may be used to simplify hand interpolation by Everett's formula, for Everett's coefficients are merely Lagrangian coefficients of various orders. Thus we may avoid the deviltry of Lagrangian interpolation even while using the tools devised to lead us to it. The connections between the two sets of coefficients are

$$E_0'' = L_{-1}^4 \qquad E_0^{(iv)} = L_{-2}^6 \qquad E_0^{(vi)} = L_{-3}^8$$

$$E_1' = L_2^4 \qquad E_1^{(iv)} = L_3^6 \qquad E_1^{(vi)} = L_4^8$$

The virtues of Everett arise from the decreasing size of the terms. The first two terms give the linear interpolant, the next two a third order, the next a fifth, and so on. Thus each set of terms produces a smaller quantity to be added to the previous result and we can stop whenever it becomes clear that the next correction would be to digits that are of no particular interest. Further, the more complicated calculations occur on the less important later corrections, minimizing the probability of a blunder. Lagrangian interpolation places equal emphasis on all terms—making the reasonableness of the result undeterminable until the calculation is complete. Psychologically it is so bad that your author cannot recommend it for hand calculation.

Special methods for special functions

Most of the well-known tabled functions have specific properties that can usefully be employed in interpolation. Since the better tables give examples, there is no special need to memorize these tricks. In programming a function for a computer, however, one can sometimes use these devices to save tabular entries, and so we mention a few here by way of reminder.

The *derivatives* of functions may be easily available as tabular entries—so then a Taylor series around the nearest tabular entry is an adequate interpolating formula. Sines, cosines, and exponentials leap immediately to mind.

If the function of the *sum* of two arguments is easily expressed in terms of functions of the separate arguments, we may usually avoid the interpolation problem by turning it into the combination of two or more tabular entries. Thus

$$\sin 30.142^\circ = (\sin 30^\circ)(\cos 0.142^\circ) + (\cos 30^\circ)(\sin 0.142^\circ)$$

and the sine and the cosine of small angles are available from one or two terms of their series – but don't forget to express the argument in radians! Similarly,

$$e^{8.742} = e^8 \cdot e^{0.7} \cdot e^{0.042}$$

with a 20-entry table sufficing to cover both the *integer* arguments and the *tenths*, while the series for e^x will quickly supply the final factor. Thus a simple routine combined with a 20-entry table is a feasible machine method for producing e^x – though there are better [Kogbetliantz, 1960].

In a high-speed digital system tangents are normally computed as the ratio of a sine and cosine in order to avoid a separate routine. But if tangents only are needed, a continued fraction is available that is good for small arguments and usable up to at least $\frac{3}{8}\pi$. The Maclaurin series is even more restricted but is usable. Arguments larger than $\frac{1}{4}\pi$ may be handled by computing the cotangent of the complementary argument, a function that in turn has a reciprocal which is the tangent of this complementary argument. Thus we need never enter our series or fraction with arguments larger than $\frac{1}{4}\pi$ or, if we prefer, one application of the half-angle identity will enable us to work with angles less than $\frac{1}{8}\pi$. Whether the extra arithmetic required to reduce the argument and to obtain the final function is balanced by the savings in evaluating a shorter continued fraction is a delicate question that we shall leave to the specialists.

Problems

1. Using the table of Figure 3.1, calculate sin 34.5678° via Bessel's or Everett's formula. Do it again using Besselian coefficients from a table (*Chambers's* or AMS 55).
2. From sin 34° calculate cos 34°, then use Taylor's series to get sin 34.5678°.
3. Calculate sin 34.5678° from sin 34° and sin 0.5678°, together with their cosines, first obtaining the functions of the small angles from their series.
4. Use Aitken's method on the data of Figure 3.1 to calculate sin 34.5678°.
5. Care to try Lagrange's four-point formula *without* any table of coefficients?

CHAPTER 4

QUADRATURE

Just as the carriage maker was rendered unemployed by the modern automobile, so seems the modern computer to have doomed the deriver of quadrature formulas. Not that the evaluation of well-posed integrals was ever very difficult – far from it! The incessant repetition to generations of engineers about integrals merely being areas under curves has led countless numbers of those good people to draw the curves and carefully count squares to find the integrals they seek. The method is simple, direct, and laborious. Now that the digital computer is available to do the labor, the major difficulty is removed. The simplicity and the directness of the method are virtues against which most alternative techniques cannot prevail. At this we must shed a not entirely crocodilian tear, for the systematic cultivation of quadrature formulas has been one of the crowning glories of classical numerical analysis, espoused of Newton, Euler, and Gauss. Since numerical analysis receives its life from its utilitarian ends, the derivation has now been reduced to a desiccated, precious, and somewhat primitive art form. If your author feels that the loss is not too great, it is because the eager systematizers of derivations seldom bother to give criteria of excellence for their product. Thus generations of would-be users have been left floundering in the sea of competing formulas whose relative merits are far from clear. In desperation they usually have seized upon the easily remembered Simpson's rule and plunged bravely ahead. With this intuitive approach their instincts served them well by getting them through expediently to a reasonable answer that could have been done better, but not by very much.

The criteria of goodness in quadrature formulas

Let us admit at the outset that there is no simple criterion of goodness for a quadrature formula. Depending upon the problem and the equipment there are several *desiderata* and they tend to conflict. This, in part, explains the persistence of several classes of quadrature formulas in the literature. Some of the *desiderata* will become apparent as our discussion progresses.

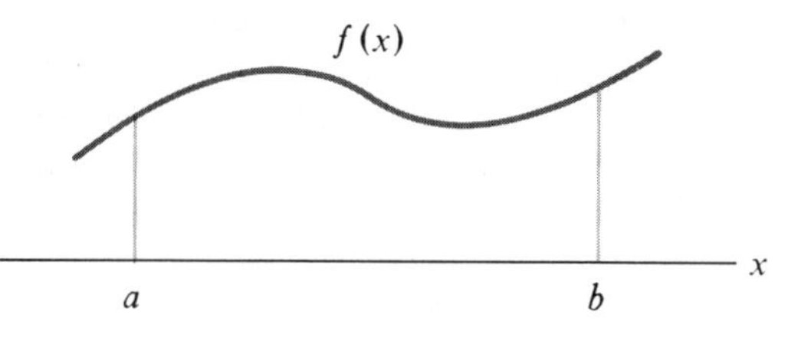

Figure 4.1

Consider the problem of finding the area under the curve in Figure 4.1. For the present we shall assume $f(x)$ to be given by an explicit formula which may be evaluated at any x in the range (a, b). If we have no special knowledge about $f(x)$ other than its apparent "smoothness," we will probably be tempted to sample it at regular intervals with a number of uniformly spaced ordinates (Figure 4.2). The number of the ordinates will depend on the accuracy with

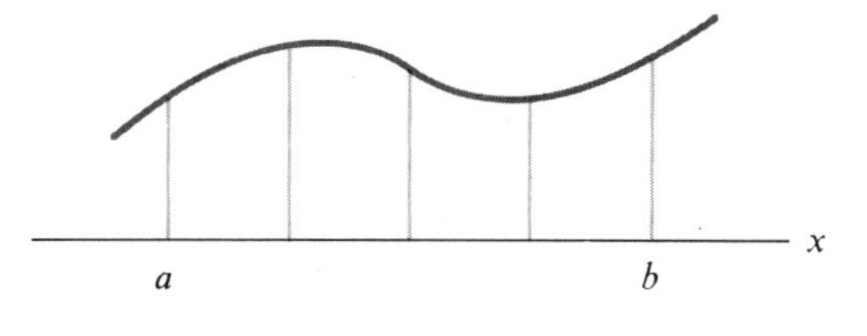

Figure 4.2

which we wish to find our area. Since we lack precise information about the number of ordinates required, a reasonable strategy would be to evaluate the area for some number, n, ordinates, then cut the interval in half by inserting $n - 1$ additional ordinates and evaluate again. If the two evaluations agree sufficiently for our purposes, we stop. Otherwise we repeat the subdivision until we achieve the desired accuracy. The fact that one can construct pathologic functions which will defeat this strategy does not invalidate it. The important question is whether a different strategy will yield the required area with less work—always assuming the $f(x)$ belongs to some vaguely specified but usefully broad class of well-behaved functions. It is this assumption that frustrates those who would make quadrature a precise science. Historically, numerical analysts have not produced a useful definition of the class to which $f(x)$ must belong and, indeed, most have not even tried. Analytic specification is difficult (one suspects it is impossible in any practical sense) and the alternative, statistical definition in terms of those functions that come into the computer lab to be integrated, is esthetically distressing.

By far the commonest assumption is that $f(x)$ may be represented adequately by a polynomial of some degree, provided we only go up far enough. In point of computational fact this assumption is seldom satisfied. It is equivalent to saying that $f(x)$ has a good Taylor's series representation almost everywhere – and in the world of computation there aren't very many of those functions! As a practical restriction, one does well to remember that no polynomial ever had an asymptote – vertical or horizontal. When faced with such integrands, stop! Don't automatically reach for a Simpson.

Once we assume adequate representation by a polynomial it is easy to compute the error term (p. 111) for most quadrature formulas, thus giving a measure of adequacy. As the formula becomes more complicated the accuracy increases, but so does the difficulty of using it. When hand calculation was the rule, one could afford to toy around with complicated formulas, piecing several of them together to fit a stubborn piece of geometry. As computation has become more automated a premium on simplicity has emerged. As a consequence, the simplest possible formula, repeatedly used, seems to be the current favorite. Why not count squares – provided we have an automatic computer to do it?

A few formulas

The Newton–Cotes quadrature rules are commonest for use in hand computation. Their equally spaced ordinates that include the bounding ones seem reasonable, and the lower-order formulas possess rather simple coefficients. We list the first four, together with their error terms:

$$A_2 = \frac{h}{2}(y_0 + y_1) \qquad - \frac{1}{12}y''h^3 \text{ (trapezoidal)}$$

$$A_3 = \frac{h}{3}(y_{-1} + 4y_0 + y_1) \qquad - \frac{1}{90}y^{(\mathrm{iv})}h^5 \text{ (Simpson)}$$

$$A_4 = \frac{3h}{8}(y_{-1} + 3y_0 + 3y_1 + y_2) \qquad - \frac{3}{80}y^{(\mathrm{iv})}h^5$$

$$A_5 = \frac{4h}{90}(7y_{-2} + 32y_{-1} + 12y_0 + 32y_1 + 7y_2) - \frac{8}{945}y^{(\mathrm{vi})}h^7$$

The decrease in the error terms is not steady, giving many persons a preference for the formulas with odd numbers of ordinates. It seems likely that this is an artifact of some unreasonable assumptions about the polynomic behavior of our function y and is not the important point. Other users worry about whether accuracy should be sought by repeated application of a few-point formula or the less frequent application of a many-point formula. Thus, in Figure 4.2 we could apply the trapezoidal rule four times, Simpson's rule twice, or the five-point formula just once. The greatly disparate weights that appear on the various ordinates in the higher-order formulas go against one's intuition—especially *negative* weights—so most persons settle for repetition of a simpler formula. Simpson's rule is a frequent choice, though some stick by the trapezoidal.

Gaussian quadrature gives up equal spacing of the ordinates and permits their locations to be chosen so as to optimize the estimation of the area—still assuming that our integrand is a polynomial. Somewhat surprisingly, the critical locations for the ordinates turn out to be the roots of the Legendre polynomial of appropriate degree. (The elegant proof uses only simple tools and is recommended to the unhurried student. Unfortunately, it does not fit conveniently in our exposition. See Ralston [1945] or Milne [1949]). When our region of integration is normalized to span $(-1, +1)$ the simpler of these formulas are

$$G_2 = y_{-1} + y_1$$

with $x_{\pm 1} = \pm 0.57735$

$$G_3 = 0.88889y_0 + 0.55556(y_{-1} + y_1)$$

with $x_0 = 0 \qquad x_{\pm 1} = \pm 0.77460$

$$G_4 = 0.65215(y_{-1} + y_1) + 0.03485(y_{-2} + y_2)$$

with $x_{\pm 1} = \pm 0.33998 \qquad x_{\pm 2} = \pm 0.86114$

$$G_5 = 0.56889y_0 + 0.47863(y_{-1} + y_1) + 0.23693(y_{-2} + y_2)$$

with $x_0 = 0 \qquad x_{\pm 1} = \pm 0.53847 \qquad x_{\pm 2} = \pm 0.90618$

where x_i is the abscissa and y_i the corresponding ordinate. If our assumption about the polynomial character of the integrand is correct and if both the abscissas and the weights are accurately evaluated (difficult, since they are irrational), then the Gaussian quadrature formula represents areas accurately up to *twice* the degree that the corresponding Newton–Cotes formula does. In the real world where none of these conditions is apt to be met, the advantage of the Gaussian formula over the Newton–Cotes is still real, but one suspects that it has suffered more degradation than has its less ambitious counterpart. Some careful experiments are needed.

Quadrature with increasing precision

A sensible machine quadrature strategy over a finite region begins by applying an even-spaced rule at an interval that is too coarse, then adds intermediate points until successive stages give the area to the required accuracy. With the trapezoidal rule this strategy is trivially implemented, since one need only supply the ordinates halfway between the ordinates already included and cut the interval size in half. Thus for five points we have

$$A_1 = H(\tfrac{1}{2}y_0 + y_2 + y_4 + y_6 + \tfrac{1}{2}y_8) = H \cdot T_1$$

while for nine points

$$\begin{aligned} A_2 &= h(\tfrac{1}{2}y_0 + y_1 + y_2 + y_3 + y_4 + y_5 + y_6 + y_7 + \tfrac{1}{2}y_8) \\ &= h(T_1 + y_1 + y_3 + y_5 + y_7) \end{aligned}$$

with

$$h = \frac{H}{2}$$

With Simpson's rule the implementation is slightly trickier. By writing down a sequence of Simpson formulas applied on successively halved intervals one soon perceives that the ordinates added *this time* enter with weight 4, while those that were added *last time* have their weight reduced to 2 and do not thereafter change. The interval h is, of course, halved each cycle. Thus

$$\frac{3}{h}A_n = (\text{first term}) + 4(\textstyle\sum \text{new ordinates})$$

On the next cycle we take the second term of our equation, divide it by 2, then add it to the first term to give the new first term. The new second term is four times the sum of the new ordinates, and h is halved. The iteration begins with the first term equal to the sum of the two end ordinates, and the one "new" ordinate is taken at the midpoint. As an example, consider the area under $\sin x$ over $(0, \pi/2)$.

h	[first term]	$4 \cdot$ [new]	Area
$\frac{1}{4}\pi$	1.0	2.82843	1.00228
$\frac{1}{8}\pi$	2.41421	5.22625	1.00014
$\frac{1}{16}\pi$	5.02734	10.25166	1.00001

Similar tactics can be devised for the higher-order Newton–Cotes formulas, though plainly your author feels that one should not use them for most integrands.

Adaptive quadrature schemes for computer use usually stay with the trapezoidal rule, halving any *sub*region that did not agree adequately with its previous stage. The saving in computational labor is large, for no further work is expended on the parts where the area is already adequately estimated. (The increasing precision philosophy described above for Simpson could equally well be modified to apply only to those subregions that still needed refinement, though the control would be far trickier.)

Richardson, Romberg, and plain John Doe

In seeking a sufficiently precise value for a quadrature, most persons apply some standard formula at an interval they hope is adequate, then cut the interval in half and apply the same formula twice as often. If the agreement is not good enough they halve again. In this way they generate a sequence of estimates

$$A_0(h)$$

$$A_1\left(\frac{h}{2}\right)$$

$$A_2\left(\frac{h}{4}\right)$$

$$A_3\left(\frac{h}{8}\right)$$

The question is then apt to arise about whether this sequence converges to the true area, A, and if so, if there is any sensible way of extrapolating to the limit from the few members already at hand. Such a strategy could save much labor, for each succeeding member of the sequence requires twice the computation of its predecessor. For calculations of low and moderate precision, say up to six significant figures, this observation is not particularly important, but when very high precision is needed, such as in long orbital satellite trajectory calculations, the savings could be tremendous.

One simple form of this extrapolation is based on the assumption that the derivative in the principal error term remains the same as we progress along our sequence of areas. Thus, if we are using the trapezoidal rule, we have

$$A = \frac{H}{2}(f_0 + f_2) + CH^3 = A_0 + 8Ch^3$$

and then

$$A = h(\tfrac{1}{2}f_0 + f_1 + \tfrac{1}{2}f_2) + 2Ch^3 = A_1 + 2Ch^3$$

where the A_0 and A_1 are the area estimates actually calculated and H is $2h$. The constant C contains the second derivative of the integrand. The assumption, often untrue, that C is the same in both equations will hold if the second derivative is roughly constant over the integral. Multiplying the second equation by 4 and subtracting, we hopefully eliminate the error term of order h^3 – leaving us with an error term in h^5. (Why?) We obtain

$$3A = 4A_1 - A_0 \qquad \text{or} \qquad A = \frac{4A_1 - A_0}{3}$$

as our extrapolation formula.

This philosophy of *Richardson extrapolation*, as it is called, may be applied to any of the Newton–Cotes quadrature formulas or, indeed, to any general computation that is based on a fixed grid h with an error describable in powers of h. Further, it may be applied more than once, for we now have one estimate (A) with an error term beginning with h^5 and we could get a second such estimate by halving h and doubling our number of ordinates, then using A_2 and A_1 instead of A_1 and A_0. Then we could combine these two h^5 estimates to generate an h^7 estimate – subject to the same sort of assumption about the universality of the higher derivative values. One tends, however, to become suspicious of assumption piled upon dubious assumption, and so these more sophisticated extrapolations were not historically popular.

One particular adaptation of this idea, however, has been proved to converge even without all the stringent assumptions. Only boundedness of the second derivative seems required. This *Romberg* integration starts with a sequence of trapezoidal-rule estimates for the area, each estimate halving the h of the preceding one. We may arrange them in a column, this time with two subcripts.

$$A_{00}$$
$$A_{01}$$
$$A_{02}$$
$$A_{03}$$

where

$$A_{0,k} = \frac{b-a}{2^k}(\tfrac{1}{2}f_0 + f_1 + \cdots + f_{2^k-1} + \tfrac{1}{2}f_{2^k})$$

Then *from these* we construct a triangular array of convergents by the rule

$$A_{n,k} = \frac{1}{4^n - 1}(4^n A_{n-1,k+1} - A_{n-1,k})$$

where the subscripts are defined by Table 4.1. The principal error expression for $A_{n,k}$ increases its powers of h by two each time we go one column to the right, while h itself halves each row downward. In practice, one stops when two successive entries at the right of Table 4.1 agree. The method is quite easy to program and seems to be one of the most satisfactory machine methods for doing mechanical quadratures when high precision is required. Its general philosophy also underlies one of the high-precision schemes for integrating ordinary differential equations [Bulirsch and Stoer, 1966] (see page 136).

Table 4.1. Romberg integration convergents

A_{00}			
A_{01}	A_{10}		
A_{02}	A_{11}	A_{20}	
A_{03}	A_{12}	A_{21}	A_{30}

Specialized quadrature formulas

In addition to some of the simple standard formulas quoted in the previous sections, one occasionally wishes to use a formula specifically derived to meet a special situation. Heretofore, all our formulas have been designed to give areas under curves as linear combinations of ordinates that fully spanned the area. Thus, in Figure 4.2 we used $f(a)$ and $f(b)$, as well as a number of ordinates between, in order to represent the area bounded by a and b. There are occasions when the bounding ordinate is inconvenient to evaluate, and so we might wish to express the area of Figure 4.3 in terms of the *interior* ordinates. Clearly we may do this, since our quadrature formulas are simple analytic integrations of

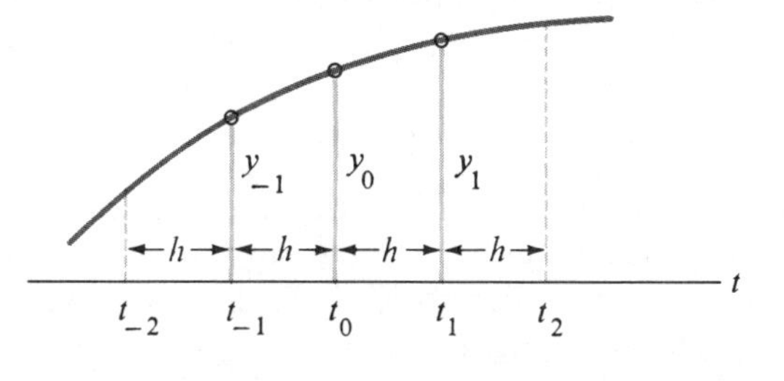

Figure 4.3

the corresponding interpolation formulas—themselves being polynomials. If we have fitted a polynomial through three interior points we may extrapolate it on one or both sides to any distance we wish and then integrate under the extrapolated parabola. Prudence, however, suggests that we not extrapolate too far because, in Figure 4.3 for example, our parabola is only tied to $f(x)$ at three points. Near those three points we may assume it approximates $f(x)$ rather well, but who can say what happens as we quietly drift away from the well-controlled region. For this reason, the usual extension is only one unit, h, on one or both sides. These formulas find their principal use for integrating ordinary differential equations from initial conditions, and we derive one of them below to illustrate a very general and powerful technique.

The method of undetermined coefficients

Although the digital computer has made a large collection of interpolation and quadrature formulas unnecessary, nevertheless we must occasionally derive special formulas for special geometries. The general technique known as the *method of undetermined coefficients* is by far the most useful one. We strongly urge the student to master it. In particular we derive the quadrature formula that gives the area under an extrapolated parabola that is determined by three equally spaced ordinates (Figure 4.3). Our troubles are purely manipulative, there being no theoretical difficulties. The three points determine a unique parabola and Lagrange's interpolation formula shows that it may be expressed as

$$y = \frac{(t - t_0)(t - t_1)}{(t_{-1} - t_0)(t_{-1} - t_1)} y_{-1} + \frac{(t - t_1)(t - t_{-1})}{(t_0 - t_1)(t_0 - t_{-1})} y_0 + \frac{(t - t_{-1})(t - t_0)}{(t_1 - t_{-1})(t_1 - t_0)} y_1 \qquad (4.1)$$

where everything with a subscript is a constant. Equation (4.1) has the general form

$$y = c_{-1}(t) \cdot y_{-1} + c_0(t) \cdot y_0 + c_1(t) \cdot y_1 \qquad (4.2)$$

where each $c_j(t)$ is a quadratic in t, hence y is a quadratic—the unique parabola. When we integrate (4.2) with respect to *t between definite limits*, the $c_j(t)$ functions become constants and our formula must therefore have the final form

$$A = b_{-1}y_{-1} + b_0 y_0 + b_1 y_1 \qquad (4.3)$$

where the b_i are known constants that are universal for all parabolas. This general form will appear no matter what the range of our integration, although the numerical values of the coefficients b_i will be different if we change the range. Our problem is to find the numerical values of these b_i coefficients for the geometry of Figure 4.3, where the range of integration extends a distance h beyond the three ordinates on both sides. We could, of course, merely integrate Lagrange's equation, but the manipulation is tedious and tends to produce errors. Ours is a considerably easier technique. It is also quite generally useful. The student should master it.

We observe that (4.3) must hold exactly for *all* parabolas, which have the general form

$$y = c + gt + kt^2 \tag{4.4}$$

Our strategy is to pick a particular parabola, compute the required area from our knowledge of calculus, and substitute in (4.3) for the area and for the three ordinates—leaving us an equation in the three as-yet-undetermined coefficients b_i. This then is a linear equation for the three b's. If we now choose a second parabola that is linearly independent of the first and perform the same set of operations, we will produce a second equation for the same three undetermined coefficients. Similarly, a third independent parabola will produce a third equation for the three b's and now these three linear equations may be solved to give the universal values for the coefficients b_i.

A simple parabola (we might as well make life easy for ourselves) is

$$y = t^2$$

which we show in Figure 4.4. The shaded area is clearly $2{\cdot}(2h)^3/3$, while y_{-1} and

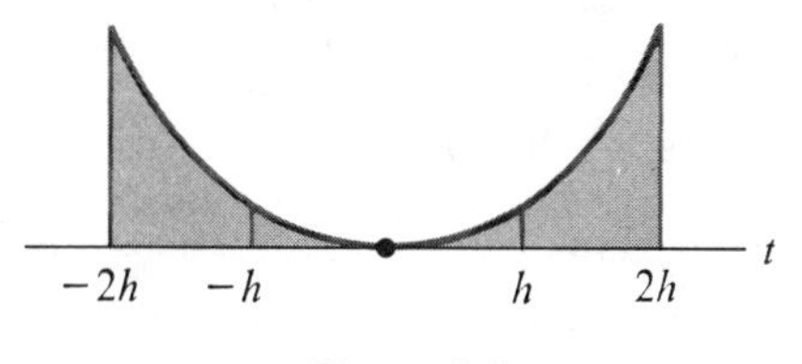

Figure 4.4

y_1 are both equal to h^2 and y_0 is zero. Substituting in (4.3) we have

$$\frac{16h^3}{3} = b_{-1} \cdot h^2 + b_0 \cdot 0 + b_1 \cdot h^2 \tag{4.5}$$

At this point we wish to make a minor but aesthetically pleasing change in our derivation. It is clear from dimensional considerations that (4.3) might be better expressed as

$$A = h(b_{-1}y_{-1} + b_0y_0 + b_1y_1) \tag{4.6}$$

in which the dependence on the horizontal dimension is explicitly displayed in

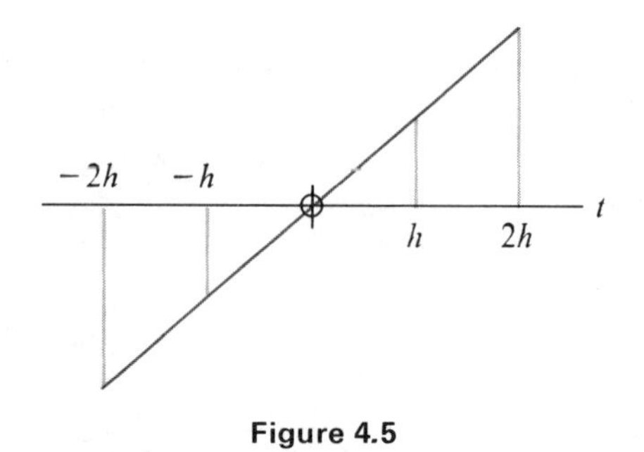

Figure 4.5

the factor h. This change is not essential and had we not made it, it would merely have produced a factor of h in all our subsequent algebra—slightly annoying but scarcely fatal. With this redefinition, however, the coefficients b_i are dimensionless and hence our formulas will involve only numbers. Our new version of (4.5) now reads

$$\tfrac{16}{3} = b_{-1} + b_1 \tag{4.7}$$

A second parabola, albeit a somewhat degenerate one, is the equation

$$y = t \tag{4.8}$$

pictured in Figure 4.5. This leads immediately upon integration to

$$0 = h(-h \cdot b_{-1} + 0 \cdot b_0 + h \cdot b_1)$$

which may be simplified to read

$$0 = b_{-1} - b_1 \tag{4.9}$$

A final very degenerate parabola is the horizontal straight line, which gives

$$4 = 1 \cdot b_{-1} + 1 \cdot b_0 + 1 \cdot b_1$$

Solving these three equations simultaneously we obtain

$$b_{-1} = b_1 = \tfrac{8}{3}$$

$$b_0 = -\tfrac{4}{3}$$

and on substituting the values back in (4.6) we find our quadrature formula to be

$$A = \frac{4h}{3}(2y_{-1} - y_0 + 2y_1) \tag{4.10}$$

This formula holds exactly for all parabolas—and hence holds approximately for all curves for which parabolas are good approximations. The only requirement is the equal spacing of the ordinates and the area is the one that is bounded by the ordinates at $\pm 2h$.

The error term

Having derived a quadrature formula in the previous section we might wonder what sort of accuracy it will give if used on curves that are not parabolas. This legitimate speculation unfortunately has no precise answer, but some insight may be obtained by computing a formal error term under the assumption that our nonparabolic function has a Taylor expansion whose early terms represent it well. We want the error

$$\text{error} = \int_{-2h}^{2h} y(t) \cdot dt - \frac{4h}{3}(2y_{-1} - y_0 + 2y_1) \tag{4.11}$$

We expand y in a Taylor's series about y_0,

$$y(t) = y_0 + ty_0' + \frac{t^2}{2!}y_0'' + \cdots \tag{4.12}$$

and substitute throughout (4.11), remembering that y_1 is $y(h)$ and y_{-1} is $y(-h)$. All terms up through y_0'' contribute nothing, as they must, for the first three terms of (4.12) constitute a parabola – for which (4.10) is exact. We might, however, expect the term in h^3 to contribute to our error. Substituting $t^3 y_0'''/3!$ into (4.11), we get

$$\text{error}_3 = \frac{y_0'''}{3!}\int_{-2h}^{2h} t^3 \cdot dt - \frac{y_0'''}{3!} \cdot \frac{4h}{3}[2(-h)^3 + 2(h)^3]$$

and both terms disappear by virtue of the odd symmetry. Thus even the h^3 term of the Taylor expansion is adequately represented by our quadrature formula.

When we substitute t^4 in (4.11) we obtain

$$E_4 = \int_{-2h}^{2h} t^4 \cdot dt - \frac{4h}{3}[2(-h)^4 + 2(h)^4] = \tfrac{2}{5}(2h)^5 - \tfrac{16}{3}h^5 = \tfrac{112}{5}h^5$$

Appending the constant coefficient of the t^4 term in the Taylor's series we have

$$\text{error} = \tfrac{112}{5}h^5 \frac{y^{(\text{iv})}}{4!} + \cdots = \tfrac{14}{45}h^5 \cdot y_0^{(\text{iv})} + \cdots \tag{4.13}$$

where further terms will arise from higher-order terms in the Taylor's series. If these terms are sharply decreasing in magnitude, then the error committed by using our quadrature formula is given approximately by the one term of (4.13). This is a serious qualification, for many functions do not have Taylor's series that are so obliging. Further, if absolute error bounds are wanted, we must be able to evaluate the fourth derivative of the integrand – no mean feat if it is at all complicated. About the best that can be said for these error terms is that they indicate the *relative* accuracy of various quadrature formulas, always assuming the adequacy of the Taylor expansion in eliminating the higher terms. Perhaps more important, they indicate, by the power of h, how much improvement one may expect if one halves the interval and applies the formula twice.

Infinite regions

Functions having tails that stretch out toward infinity (Figure 4.6) do not lend themselves immediately to mechanical quadrature. If the area under them is indeed finite, then $f(x)$ probably subsides as e^{-x}, though it may only decline slowly as $1/x^2$, or worse. Whatever the analytic behavior, it is certainly not polynomic, so the standard quadrature formulas are not appropriate. Although

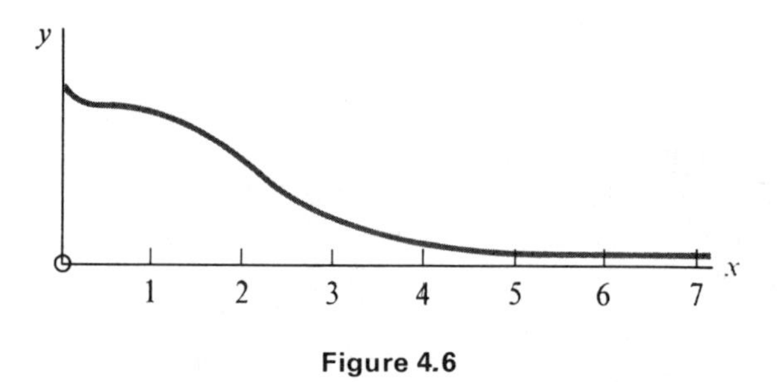

Figure 4.6

a variety of specialized formulas may easily be devised, the expedient course usually is to change the independent variable in a way that transforms our infinite integral onto a finite range. We could, for example, cut our area in Figure 4.6 at 2, evaluating the part to the left by Simpson's (or anybody else's) rule. Then the infinite tail could be mapped onto (0, 1) in a new variable y_1 by the substitution

$$y_1 = \frac{2}{x}$$

for the dummy variable of the integration. We must be somewhat cautious lest we introduce some nasty singularity in the integrand by our transformation, but with moderate care no trouble usually arises that can compare with the problems we had before the transformation. The form of the integrand may suggest a different transformation, such as

$$y = e^{-x} \qquad \text{or} \qquad y = \frac{1}{x^2}$$

but the essential idea remains.

Quite often, especially with particularly simple integrands, some approximation may permit analytic integration for the area out in a tail. Thus once b is large enough in

$$\int_b^\infty \frac{dx}{1+e^{-x}+x^2} < \int_b^\infty \frac{dx}{1+x^2} = \left[\tan^{-1}x\right]_b^\infty = \left(\frac{\pi}{2} - \tan^{-1}b\right) = \tan^{-1}\frac{1}{b}$$

the offending e^{-x} can just be discarded as unimportant, giving the second integral for a close upper bound. On the other hand, it may be "frozen" at the value e^{-b}, thereby yielding the lower bound

$$\int_b^\infty \frac{dx}{(1+e^{-b})+x^2} = \frac{1}{\sqrt{1+e^{-b}}}\left(\frac{\pi}{2} - \tan^{-1}\frac{b}{\sqrt{1+e^{-b}}}\right)$$

For b equal to 5, for example, these expressions equal

$$0.197395$$

and

$$0.197379$$

thereby defining the area of the tail to at least five significant figures. Such devices, when available, are usually preferable to the transformation gambit, although occasionally a very long analytic formula full of transcendental functions will require more computational labor than the alternative transformation followed by quadrature.

Estimating the tail—more examples

When performing numerical quadratures with infinite limits most of the contribution usually occurs quite early. We therefore seek to stop our use of Simpson's rule at a conveniently small value of the dummy variable – estimating the area remaining under the tail by some explicit formula. As we have already seen, by this time some terms in the integrand will have become disproportionately small compared to others, so we may frequently neglect enough of the complications to permit analytic integration – or at least to permit efficient expansion in some series. Consider the quadrature

$$I(b,c) = \int_c^\infty \frac{dx}{1+bx^2+x^4} \qquad c \geq 2 \tag{4.14}$$

This integral actually has a complicated analytic closed form (which we shall let the curious student seek!), but in the sequel we behave as if it did not. Thus we shall attempt to estimate its value, sometimes directly but sometimes by finding sufficiently close upper and lower bounds to serve as estimates. [The original problem may be presumed to be the evaluation of the integral from *zero* to infinity—the part from 0 to c having been accomplished via Simpson's rule. The total area we seek, for c equal to 2 and b equal to 1, happens to be $\pi/(2\sqrt{3})$, or 0.906900, so the tail to be estimated must be compared to approximately 0.91 when its accuracy is being judged.]

1. *Discarding the small terms.* Since all terms in the denominator are positive, discarding any of them will yield an upper bound. A drastic discard of all but the x^4 term gives

$$I < \int_c^\infty \frac{dx}{x^4} = \frac{1}{3c^3} = 0.041667 = I_{11}$$

A less drastic discard, that of unity, also yields an integrable quantity. We have

$$I < \int_c^\infty \frac{dx}{x^2(x^2+b)} = \frac{1}{b}\left(\frac{1}{c} - \frac{1}{\sqrt{b}}\tan^{-1}\sqrt{\frac{b}{c}}\right) = 0.036352 = I_{12}$$

This result, while a better bound than the previous one, suffers here by being expressed as the difference of two larger quantities. With our values (1, 2) for (b, c) we lose one significant figure and hence would probably prefer to replace the arctangent function by its series, thereby effecting the cancellation of the $1/c$ terms analytically. (But beware the series for the arctangent if its argument gets much larger than 0.5; see Chapter 1.)

2. *Freezing the small terms.* By replacing some of the denominator terms by their values at the lower limit we can often reduce the complexity of the integral sufficiently to permit analytic integration. By writing our integral as

$$\int_c^\infty \frac{dx}{x^4(1 + b/x^2 + 1/x^4)} > \frac{1}{1 + b/c^2 + 1/c^4}\int_c^\infty \frac{dx}{x^4}$$

$$= \frac{1}{1 + b/c^2 + 1/c^4}\cdot\frac{1}{3c^3} = 0.031746$$

we obtain a lower bound that now traps our tail between 0.0317 and 0.0417 (to use the cruder upper bound). We can also difference these analytic bounds to get an expression

$$\frac{b}{3c^5}\cdot\frac{1 + 1/bc^2}{1 + b/c^2 + 1/c^4}$$

that could be used to determine how large c must be to guarantee any desired degree of closeness of these bounds. But these bounds are rather crude. We can do better.

By writing our integral as

$$\int_c^\infty \frac{dx}{(x^4+1)\left(1+\dfrac{bx^2}{1+x^4}\right)} > \frac{1}{1+\dfrac{bc^2}{1+c^4}} \int_c^\infty \frac{dx}{x^4+1}$$

$$= \tfrac{17}{21}(0.040689) = 0.032939$$

and freezing the second factor at c we keep the denominator too large, thereby creating the lower bound shown. On the other hand, by simply freezing the bx^2 term at bc^2 we can create an upper bound. We get

$$I < \int_c^\infty \frac{dx}{x^4+bc^2+1} = \int_2^\infty \frac{dx}{x^4+5} = 0.037051$$

Both these bounds depend on our ability to integrate $(x^4+a^4)^{-1}$ analytically.

3. *Expansion in a series.* We observe that the second factor in

$$\int_c^\infty \frac{1}{x^4} \cdot \frac{1}{1+\dfrac{1+bx^2}{x^4}} dx$$

is of the form $(1+\varepsilon)^{-1}$, which, ε being small, may be expanded by the binomial theorem to give

$$\int_c^\infty \frac{1}{x^4}\left[1 - \frac{1+bx^2}{x^4} + \left(\frac{1+bx^2}{x^4}\right)^2 - \cdots\right] dx$$

$$= \left(\frac{1}{3c^3} - \frac{1}{7c^7} + \frac{1}{11c^{11}} - \cdots\right) - b\left(\frac{1}{5c^5} - \frac{2}{9c^9} + \frac{3}{13c^{13}} - \cdots\right)$$

$$+ b^2\left(\frac{1}{7c^7} - \frac{3}{11c^{11}} + \cdots\right) - b^3\left(\frac{1}{9c^9} - \cdots\right)$$

where the coefficients are highly regular and easily discovered. *There is no approximation here*, so this technique may be used to give our integral to any precision we may have the patience to pursue. (The correct value for our integral, by the way, is 0.035591.)

4. *More sophisticated bounds.* We can write our integral as

$$\int_c^\infty \frac{dx}{(x^2+b/2)^2 + (1-b^2/4)} \tag{4.15}$$

and then simplify it by substituting

$$x = \sqrt{\frac{b}{2}} \tan\theta$$

to give

$$\sqrt{\frac{b}{2}} \int_{\theta_c}^{\pi/2} \frac{\sec^2\theta \cdot d\theta}{\frac{b^2}{4}\sec^4\theta + 1 - \frac{b^2}{4}} = \left(\frac{2}{b}\right)^{3/2} \int_{\theta_c}^{\pi/2} \frac{\cos^2\theta \cdot d\theta}{1 + a\cos^4\theta} \tag{4.16}$$

Since

$$b = 1$$

we have

$$a = \frac{4}{b^2} - 1 = 3.0 \qquad \text{and} \qquad \theta_c = \tan^{-1}\frac{c}{\sqrt{b/2}} = \tan^{-1} c\sqrt{2}$$

For c larger than 2, θ_c is up near $\frac{1}{2}\pi$, the $\cos^4\theta$ term is small and we may decide how cavalierly we wish to treat it. If we neglect it entirely we get the upper bound

$$I_4 < \left(\frac{2}{b}\right)^{3/2} \int_{\theta_c}^{\pi/2} \cos^2\theta \cdot d\theta = \left(\frac{2}{b}\right)^{3/2} \frac{1}{2}\left(\tan^{-1}\frac{\sqrt{b/2}}{c} - \frac{c\sqrt{b/2}}{c^2 + b/2}\right) = 0.36159$$

which is a little better than I_{12}. (But again there is a subtraction problem.)

On the other hand, we may prefer expansion of (4.16) to yield

$$I = \left(\frac{2}{b}\right)^{3/2} \int_{\theta_c}^{\pi/2} (\cos^2\theta - a\cos^6\theta + a^2\cos^{10}\theta - \cdots)\, d\theta$$

which, by changing to the complementary angle for the dummy variable, may be expressed as

$$I = \left(\frac{2}{b}\right)^{3/2} \int_{0}^{\varphi_c} (\sin^2\varphi - a\sin^6\theta + a^2\sin^{10}\varphi - \cdots)\, d\varphi$$

with

$$\varphi_c = \tan^{-1}\frac{\sqrt{b/2}}{c} \approx 0.34$$

and the first term may usefully be evaluated analytically. Again nothing has been discarded. The correction terms are best computed by replacing $\sin\varphi$ by one or two terms of the series and integrating analytically. (Obviously $\sin^6\varphi$ integrates to something beginning with $\varphi^7/7$, which is quickly estimated to be of the order of 0.0007. From such crude estimates one can decide how many terms must be included in the evaluation program for the expected ranges of b and c. In particular, it is *not* useful to integrate $\sin^6\varphi$ analytically, since this produces the result as the very small difference of large quantities—one of these being the integral of $\sin^2\varphi$.)

Other bounds for our integral (4.14) are available if we observe that it can be integrated analytically for b equal to *zero* or *two*. We thus obtain for

$$2 \geq b \geq 0$$

$$0.0318 = \tfrac{1}{2}\left(\tan^{-1}\frac{1}{c} - \frac{c}{1+c^2}\right) \leq I(b,c) \leq \frac{1}{4\sqrt{2}}\ln\frac{1-c\sqrt{2}+c^2}{1+c\sqrt{2}+c^2}$$
$$+ \frac{1}{2\sqrt{2}}\left(\tan^{-1}\frac{1}{c\sqrt{2}+1} + \tan^{-1}\frac{1}{c\sqrt{2}-1}\right) = 0.0407$$

the numerical values being for c equal to 2. These expressions could be used to decide how large c should be taken to guarantee that the estimated area under the tail will be sufficiently accurate. But this last example is somewhat artificial in the sense that if one can obtain these analytic bounds, one can also probably obtain an analytic form for the original problem.

Tail estimation for oscillatory infinite integrals

If we must evaluate

$$F(b) = \int_0^\infty \frac{dx}{x^2 + b\cos x} \qquad 0 < b \leq 1 \tag{4.17}$$

we have two types of difficulties:

1. $F(b)$ has a singularity as b approaches *zero*.
2. We would prefer not to integrate numerically out to ∞ in x.

Here we shall look at the second difficulty, estimating the tail of the quadrature, even though it is usually the less serious of the two. Our problem is to evaluate

$$\int_c^\infty \frac{dx}{x^2 + b\cos x}$$

for b not too small, say *unity*. If we look at a graph of the integrand (Figure 4.7) we see that it oscillates around the curve x^{-2} and that the oscillations decrease monotonically in importance. We see immediately that the integral

$$\int_c^\infty \frac{dx}{x^2} = \frac{1}{c}$$

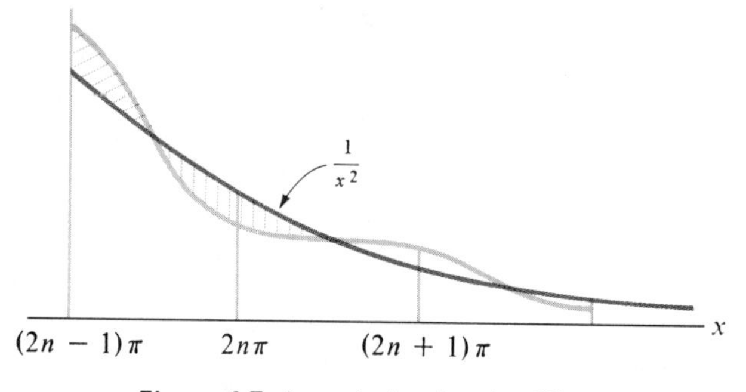

Figure 4.7. A graph showing the difference between x^{-2} and $(x^2 + b\cos x)^{-1}$

will be a good approximation *provided* we begin at one of the points where $\cos c$ is ± 1, that is, at $m\pi$, for then the two successive lobes of ignored area that occur over the next half-period will be of opposite sign (see the two shaded areas) and must partly cancel. Further, if we begin at x equal to $(2n - 1)\pi$, use of the curve x^{-2} will result in an *under*estimate, whereas beginning at $2n\pi$ will produce an *over*estimate.

We wish to compute the approximate size of this underestimation—the area between the actual curve and x^{-2} between $(2n - 1)\pi$ and $2n\pi$. We thus wish to find

$$D_n = \int_{(2n-1)\pi}^{2n\pi} \left(\frac{1}{x^2 + b\cos x} - \frac{1}{x^2}\right) dx = -b \int \frac{\cos x \cdot dx}{x^2(x^2 + b\cos x)}$$

$$= -b \int_{-\pi/2}^{\pi/2} \frac{\sin y \cdot dy}{(C + y)^2[(C + y)^2 + b\sin y]} \qquad C = (2n - \tfrac{1}{2})\pi$$

where the last integral is formed by taking the origin of the dummy variable at $(2n - \frac{1}{2})\pi$. Splitting into two integrals and redefining y to be $-y$ in the first, we have

$$D_n = b\int_0^{\pi/2} \frac{\sin y \cdot dy}{(C - y)^2[(C - y)^2 - b\sin y]} - b\int_0^{\pi/2} \frac{\sin y \cdot dy}{(C + y)^2[(C + y)^2 + b\sin y]}$$

$$= b\int_0^{\pi/2} \sin y \frac{[(C + y)^4 - (C - y)^4] + b\sin y[(C + y)^2 + (C - y)^2]}{(C^2 - y^2)^2[(C + y)^2 + b\sin y][(C - t)^2 - b\sin y]} dy$$

$$= b\int_0^{\pi/2} \frac{\sin y \cdot (C^2 + y^2)(8Cy + 2b\sin y)}{(C^2 - y^2)^2[(C + y)^2 + b\sin y][(C - y)^2 - b\sin y]} dy \tag{4.18}$$

At this point we need to get a feeling for the size of D_n. If it is so small that we can neglect it, we shall be very happy—but we are not apt to be so lucky. In any event, D_n becomes smaller as n (and hence C) increases, so we have the choice of using our expression for D_n to find out how big n must be before we *can* neglect D_n, or we may attempt to evaluate D_n and use it as a correction term to the area of the tail under $1/x^2$. Under the first of these choices we shall probably seek a strict *upper bound* for D_n; under the second, we shall attempt to *estimate D_n as closely as possible,* trying not to bias it either up or down. Under either strategy we shall first remove all factors of C to give

$$D_n = \frac{8bC^3}{C^5}\int_0^{\pi/2} \frac{\dfrac{\sin y}{y}(1+z^2)(y^2)\left(1+\dfrac{b\sin y}{4Cy}\right)\cdot dy}{(1-z^2)^2\left[(1+z)^2+b\dfrac{\sin y}{C^2}\right]\left[(1-y)^2-b\dfrac{\sin y}{C^2}\right]}, \quad z=\frac{y}{C} \tag{4.19}$$

All the factors under the integral sign have been arranged to be either powers of y or else close to unity for all values of y. For example, $\sin y/y$ remains between 1.0 and 0.64; if n is 2, z lies in the range (0, 0.143). Thus, for b equal to 1, a crude estimate for D_2 is $(8b/3C^5)(\pi/2)^3$, or 0.00064, while the total area is of the order of 1.5. If we want six significant figures, then one or two significant figures in our estimate of D_2 will suffice. We may replace $\sin y/y$ by its approximation

$$\frac{\sin y}{y} \approx 1 - 0.1661y^2$$

The maximum value of z^2 is 0.02, which can therefore be neglected when compared to unity, as can $(b\cdot\sin y)/C^2$ terms. We now have

$$D_n \approx \frac{8b}{C^5}\int_0^{\pi/2} \frac{(1-0.1661y^2)\left[1+\dfrac{b}{4C}(1-0.1661y^2)\right]y^2}{(1+z)^2(1-z)^2}dy \tag{4.20}$$

Multiplication of the denominator factors gives $(1-z^2)^2$, permitting us to replace it by unity–and we now have an easily integrable analytic approximation to D_n:

$$D_2 \approx \frac{8b}{C^5}0.957198 = 0.00004764$$

of which at least the first two figures are significant.

If we now decide that we really want a strict upper bound for D_2, we return to (4.19) and proceed somewhat differently. This time we must replace the denominator factors by expressions that are the same size or *smaller* numerator factors that are the same or *larger*. We get

$$B_n = \frac{8b}{C^5}\int_0^{\pi/2} \frac{(1-0.1661y^2)y^2\left[1+\frac{b}{4C}(1-0.1661y^2)\right]}{\left(1-\frac{\pi^2}{4C^2}\right)^2\left[\left(1-\frac{\pi^2}{4C^2}\right)^2-\frac{2\pi b}{C^3}-\frac{b^2}{C^4}\right]}dy$$

$$= \frac{D_n}{\left(1-\frac{\pi^2}{4C^2}\right)^2\left[\left(1-\frac{\pi^2}{4C^2}\right)^2-\frac{2\pi b}{C^3}-\frac{b^2}{C^4}\right]} = \frac{D_n}{0.9162} = 0.000052$$

As we remarked earlier, it is clear from Figure 4.7 that the net shaded area in the *next* half-period will be smaller in magnitude and opposite in sign. The value of D_3 can easily be calculated, since the integral part is unchanged to the accuracy we are using and we need only substitute the value $(2n + \frac{1}{2})\pi$ for C. We get

$$D_{n+1} = -\frac{8b}{[(2n+1)\pi]^5}(0.957) = -0.0000255$$

To get the final value of the integral we take the total area under x^{-2} and add 0.000048 to get an upper bound, then from this we subtract 0.000026 to give a lower bound. If these values are not sufficiently close, we may always compute D_n for one or two more half-periods, applying them as further corrections—remembering, of course, that their precision is only about 2 in the second significant figure because of the approximations made in going from (4.19) to (4.20).

Singularities in the integrand

When the infinity occurs in the integrand (Figure 4.8) we face quite another problem. Simpson and his brethren are still inapplicable, but the cure for this trouble is not so simple as those of infinite limits. We can of course, interchange dependent and independent variables, thereby transforming our difficulty into the previous "long-tailed" type—but usually this transformation causes new problems of its own by seriously complicating the integrand and is only occasionally profitable.

As a typical example of the numerical problems we are discussing here consider the integral

$$\int_{0.01}^{1.00} \frac{dx}{e^x - 1} = \left[\ln(e^x - 1) - x\right]_{0.01}^{1.00} = 4.15149087$$

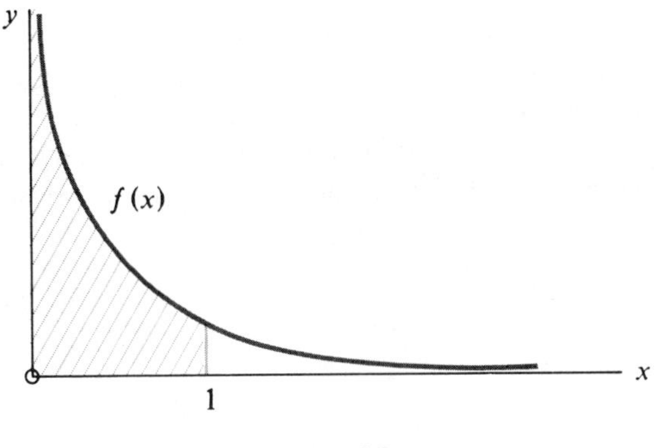

Figure 4.8

which has a closed analytic form through which we may check any approximation we use. The troubles, of course, arise because our integrand becomes infinite at zero – and zero is not far from 0.01, where most of the area under the integrand necessarily occurs. Thus the parabolas of Simpson's rule do not fit our integrand, which is trying to become vertical.

If we take a stubborn unsympathetic attitude and use Simpson's rule anyway, merely using it on finer and finer subdivisions of the interval (0.01, 1.0) until two successive applications agree, we ultimately get an answer. But we pay the price of inefficiency. To get our nine-figure answer the computer had to subdivide that interval into 8192 sections, applying Simpson's rule 4096 times – and that is considerable work even for an electronic computer, especially if it isn't necessary! To obtain a reasonable alternative method we note that this integrand behaves like $1/x$ near the lower limit. Thus we rewrite it as

$$\int_{0.01}^{1.00} \left(\frac{1}{e^x - 1} - \frac{1}{x} \right) dx + \int_{0.01}^{1.00} \frac{dx}{x}$$

perform the second integration analytically, and evaluate the first integral by Simpson's rule again. Since the integrand is well behaved at zero this time (it equals -0.5 there) and has acquired no new difficulties by our alterations, we expect that fewer applications of Simpson will be needed. This time it takes only 16 to deliver the nine-figure accuracy that required 4096 before. Most persons would consider price reduction by a factor of 256 worth going out of their way to obtain if they were buying an automobile. It is unfortunate that they sometimes don't feel the same way about their computing bills. (Oh yes – just in case you are wondering if the effect here is due primarily to the high accuracy we sought in our quadrature, we can tell you that for six figures the two methods require 1024 and 2 Simpson's, respectively, while for four-figure accuracy they used 128 versus 1. It's a bargain!)

Basically we are faced with a singularity that must be analytically "removed"—that is, we must ascertain just how quickly $f(x)$ goes to infinity at the singularity in order to subtract off or divide out a simple function with the same extravagant property—but one that can be integrated analytically. The *difference* between $f(x)$ and our matching function is necessarily well behaved at the

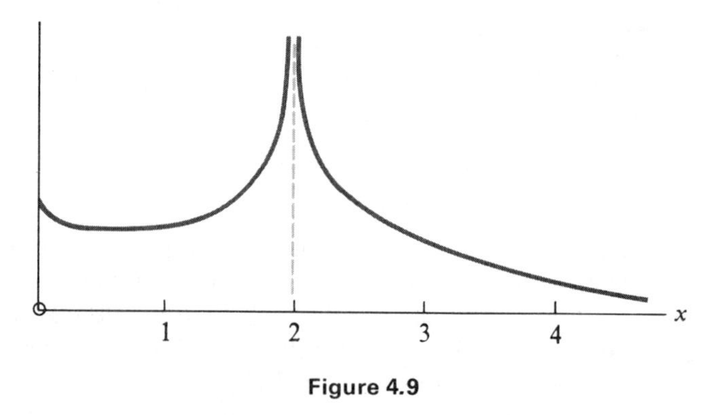

Figure 4.9

crucial point. Thus we may integrate it by the usual quadrature formula. We illustrate some techniques for the example of Figure 4.9. We wish to evaluate

$$\int_0^\infty \frac{e^{-x}g(x)}{\sqrt{|x-2|}}dx = \int_0^2 \frac{e^{-x}g(x)}{\sqrt{2-x}}dx + \int_2^\infty \frac{e^{-x}g(x)}{\sqrt{x-2}}dx \tag{4.21}$$

where $g(x)$ is presumed well behaved. Dividing the integral as shown in (4.21) we then transform to new dummy variables of integration, giving

$$e^{-2}\int_0^2 \frac{e^t g(2-t)}{\sqrt{t}}dt + e^{-2}\int_0^\infty \frac{e^{-t}g(2+t)}{\sqrt{t}}dt \tag{4.22}$$

Our troubles now consist of two removable singularities at the origin. We may either substitute further, letting t become s^2 and then perform numerical quadrature on

$$2e^{-2}\left[\int_0^{\sqrt{2}} e^{s^2}\cdot g(2-s^2)\cdot ds + \int_0^\infty e^{-s^2}\cdot g(2+s^2)\cdot ds\right] \tag{4.23}$$

or we may expand the $g(x)$ functions in the second integral of (4.22) around 2, giving a series of gamma functions, while still transforming the first integral in order to evaluate it by quadrature. As another alternative for the finite first integral in (4.22), expansion of both e^t and $g(2 - t)$ will permit explicit analytic integration into a series that, considering the small sizes of the limits, ought to converge rather quickly. The practicality of these expansion techniques hinges on the ease with which higher derivatives of g may be obtained.

As an example of "subtracting off the singularity" we may write the first integral in (4.22) as

$$\int_0^2 \frac{e^t g(2-t)}{\sqrt{t}}\,dt - \int_0^2 \frac{g(2) + t[g(2) - g'(2)]}{\sqrt{t}}\,dt + \int_0^2 \frac{g(2) + t[g(2) - g'(2)]}{\sqrt{t}}\,dt$$

Note that g and its derivative are evaluated at 2 and hence are constants in the last two integrals. The last integral may be integrated analytically while the first two combine to give a numerator that grows as t^2 near the origin. Thus this new integrand and its first derivative are zero at the origin, conferring a geometry that is hospitable to use of standard quadrature formulas. The second derivative is still infinite there, but that is a much more tolerable infinity than those in the first derivative or in the integrand itself. There are better techniques for this integral, but the subject of handling singularities is a large one; we devote Chapter 15 to it.

Evaluation of another integral with a singularity

As a final example, we consider the integral

$$I(b, x, h) = \int_{b-h}^{b} \frac{\ln|x - \xi|}{\sqrt{b^2 - \xi^2}}\,d\xi \qquad b - h \le x \le b$$

by several related techniques. There are two singularities: one at the upper limit b, which may be removed by a cosine substitution, and one at x, where the logarithm becomes infinite. Defining θ by

$$\xi = b\cos\theta$$

and θ_2 and θ_x by

$$b - h = b\cos\theta_2 \qquad \text{and} \qquad x = b\cos\theta_x$$

we substitute to obtain

$$I(b, h, x) = \int_0^{\theta_x} \ln(\xi - x)\cdot d\theta + \int_{\theta_x}^{\theta_2} \ln(x - \xi)\cdot d\theta$$

with the logarithmic singularities still present at θ_x. Integration by parts allows suppression of these infinities, yielding

$$\left[(\theta - \theta_x) \cdot \ln(\xi - x)\right]_0^{\theta_x} - \int_0^{\theta_x} \frac{(\theta - \theta_x)\sin\theta}{\dfrac{x}{b} - \cos\theta} d\theta$$

$$+ \left[(\theta - \theta_x) \cdot \ln(x - \xi)\right]_{\theta_x}^{\theta_2} - \int_{\theta_x}^{\theta_2} \frac{(\theta - \theta_x)\sin\theta}{\dfrac{x}{b} - \cos\theta} d\theta$$

and both terms in brackets disappear at θ_x. Furthermore, the integrands are identical and well behaved at θ_x, permitting consolidation of the two integrals into a single integral if we wish.

For the physical problem in which this integral actually occurred h was rather small compared to b. Thus the entire range of integration was confined to the neighborhood of the two singularities. We had

$$(\theta_2 - \theta_x)\ln(x - b + h) + \theta_x \cdot \ln(b - x) + \int_0^{\theta_x} \frac{(\theta - \theta_x)\sin\theta}{\cos\theta - \dfrac{x}{b}} d\theta + \int_{\theta_x}^{\theta_2} \frac{(\theta - \theta_x)\sin\theta}{\cos\theta - \dfrac{x}{b}} d\theta$$

-0.29618	-0.90677	-0.22579(1S)	-0.08806	$= -1.51680$
		-0.22847(2S)		$= -1.520$ (probably correct)

where the numbers come from the parameter values

$$b = 1.20 \qquad x = 1.12 \qquad h = 0.12$$

which imply

$$\theta_2 = 0.45103 \qquad \theta_x = 0.36721$$

and the integrals were evaluated separately, first by one application of Simpson's rule to each, then by two applications. The only approximations are the integral evaluations, and these may be indefinitely improved by quadrature on a finer mesh. Note that at θ_x both integrands take on the value -1.0. (If we combine the two integrals and apply Simpson only *once*, we get -1.510 for our value, an obviously cruder result.)

An alternative approach, pointed out by J. B. Rosser (who was unaware of the numerical parameter values), is to substitute $b\eta$ for ξ and bz for x to give

$$\int_{1-h/b}^{1} \frac{\ln b + \ln|z - \eta|}{\sqrt{1 - \eta^2}} d\eta = \ln b \cdot \int_{1-h/b}^{1} \frac{d\eta}{\sqrt{1 - \eta^2}} + \int_{-1}^{1} \frac{\ln|z - \eta|}{\sqrt{1 - \eta^2}} d\eta - \int_{-1}^{1-h/b} \frac{\ln(z - \eta)}{\sqrt{1 - \eta^2}} d\eta \tag{4.24}$$

The object of this maneuver is to create the central integral of (4.24), which, surprisingly, is not a function of z and has the value $-\pi \ln 2$. The first integral integrates analytically, while the last one has no *logarithmic* singularity on its range. We obtain

$$\begin{array}{ccccc} \theta_2 \cdot \ln b & - \ \pi \cdot \ln 2 & - \displaystyle\int_0^{\pi - \theta_2} \ln(z + \cos\theta) \cdot d\theta & & \eta = -\cos\theta \\ 0.08 & - \ 2.18 & + \ 0.97\,(1\mathrm{S}) & = -1.13\,(\text{bad!}) & \end{array} \qquad (4.25)$$

where the singularity at η equal to -1 has been banished by a cosine transformation. We still must evaluate the integral, which, for our values of θ_2 and z, is not well-behaved. Since θ_2 is rather small, $\cos\theta$ gets negatively large at the upper limit and nearly cancels z. Thus a direct application of Simpson's rule to this integral is asking for trouble unless we are willing to apply it on a quite fine mesh, and our purpose is thereby defeated. (We were seeking a reasonably efficient analytic formula.)

The singularity just offstage in the logarithm may be suppressed by an integration by parts. Our integral in (4.25) becomes, on defining θ_z to be the value of θ *at the singularity*,

$$\Big[\{\theta - (\pi - \theta_z)\} \ln(z + \cos\theta)\Big]_0^{\pi - \theta_2} + \int_0^{\pi - \theta_2} \frac{(\theta - \pi + \theta_z)\sin\theta}{z + \cos\theta}\, d\theta$$

$$2.114 - 2.651 = -0.537$$

and finally we have a reasonable integral to which Simpson's rule may be applied. We are not very happy about the subtraction, which loses us one significant figure, nor is the value of the integral necessarily well represented by a single application of Simpson's rule even here. A finer subdivision would give better results. This one, however, yields a final value of -1.56, which is a lot closer to -1.52 than the -1.13 we got from (4.25).

The moral of our story is clear. In our attempt to get rid of the logarithmic singularity via the middle integral in (4.24), we forced the computation of our integral into a small difference of large quantities, and one of these quantities still had the logarithmic singularity in it in the sense that it was just outside its range of integration. Hence the singularity contributed strong nonpolynomic behavior that had to be suppressed by the device we used previously (integration by parts). For a different set of parameters this method of evaluating the original integral could be useful, but it is quite poor here.

Problems

The first three problems are essential if the student would appreciate the principal points of this chapter.

1. Evaluate

$$\int_0^{\pi/2} \sin x \cdot dx$$

by the trapezoidal rule, by Simpson's rule. Use three and five ordinates with each rule.

2. Evaluate

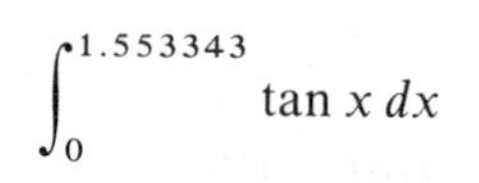

$$\int_0^{1.553343} \tan x \, dx$$

by the same techniques as in Problem 1. (The upper limit is 89° expressed in radian measure.) What differences do you observe? Explain them. Compare with the analytic answer.

3. Evaluate the integral of Problem 2 having first subtracted $(\pi/2 - x)^{-1}$ from $\tan x$ and then compensate by adding the analytic integral of this function to your result. Use three ordinates with Simpson's rule.

4. Write a program to evaluate

$$F(b) = \int_0^5 e^{-(b \cos x)^2} \cdot dx \qquad 0 \leq b \leq 5$$

Compute values for $b = 0(1)5$ and compare with those of your classmates. Who is right? How do you decide?

5. Evaluate

$$F(b) = \int_0^5 e^{-b^2 \cos x} \cdot dx \qquad 0 \leq b \leq 5$$

with a general quadrature program. Comment on its efficiency; accuracy. What difficulties occur?

6. Evaluate

$$\int_0^3 \frac{e^{-b/x}}{x} dx$$

for $b = 0.01$ both in this form and also after transforming the dummy variable to improve the geometry of the integrand. Compare with tabulated values of the exponential integral.

7. Develop a strategy for evaluating

$$F(b) = \int_0^\infty e^{-(x^2 + b/x)} \frac{dx}{x}$$

for $b \geq 1$ and, separately, for $b \approx 0.02$.

8. Evaluate

$$\int_5^\infty \frac{dx}{1 + e^{-x} + x^2}$$

by some suitable quadrature technique, first transforming the dummy variable of integration to a finite range. Check your results against the values on page 113.

9. Evaluate

$$\int_0^5 \frac{dx}{\sqrt{|1.5 - x|}}$$

by some quadrature technique.

10. Evaluate

$$\int_0^5 \frac{dx}{|2 - \sqrt{x}|}$$

by some technique.

11. Evaluate

$$\int_0^\infty \frac{dx}{1 + x^2 + x^4}$$

by numerical quadratures. Break the range at 2 and transform the tail suitably onto a finite range.

12. Devise a procedure for evaluating

$$\int_0^\infty \frac{dx}{x^2 + b\cos x} \qquad 0 < b \le 0.1$$

Five significant figures are desired.

13. Evaluate

$$\int_{0.001}^{1.0} (1 + e^{-at}) \ln t \cdot dt$$

for

$$a = 0.2(0.1)\,2.0.$$

14. Evaluate

$$\int_{0.8}^{1.0} \frac{\ln(1 - x)}{\sqrt{1 - x^2}}\, dx$$

15. Evaluate

$$\int_{0}^{x} \left(\frac{1 - e^{-t}}{t}\right) \sin t \cdot dt$$

for $x = 0.2(0.1)\,2.0$.

CHAPTER 5

THE INTEGRATION OF ORDINARY DIFFERENTIAL EQUATIONS

The integration of ordinary differential equations from *initial conditions* is primarily an exercise in judicious extrapolation. The principal worries are, or should be, ones of efficiency, stability, and accuracy. We have no trouble in finding techniques that will, with sufficient labor, produce an integral curve from reasonable differential equations and initial conditions. The literature is full of them. Our problem is to choose wisely, to recognize when the tool is suitable, and to avoid making a mountain out of a molehill. Conversely, we should learn to spot those unpleasantnesses that can constitute a real problem and to isolate them, lest we ask an integration algorithm to attempt a job for which it was never designed. In this chapter we examine the philosophy underlying the commonest class of reasonable integration schemes with the purpose of exposing their strengths and weaknesses. We shall not, in general, attempt to give systematic derivations, or to settle parochial arguments about the relative merits within the predictor-corrector class (for it is with these schemes we shall principally concern ourselves).

If we are given a first-order differential equation

$$\frac{dy}{dt} = f(t, y) \tag{5.1}$$

we may represent it graphically as a collection of slopes (Figure 5.1), for at each value of the independent variable, t, and the dependent variable, y, our equation defines a dy/dt which we may presume to be unique. If we are now given a point, A, through which our solution is asserted to pass, we may sketch that solution quite easily by drawing a curve smoothly through the slopes—probably with the aid of a French curve (Figure 5.2). We have *integrated* our differential equation and, if graphical accuracy suffices, this is often the expedient method.

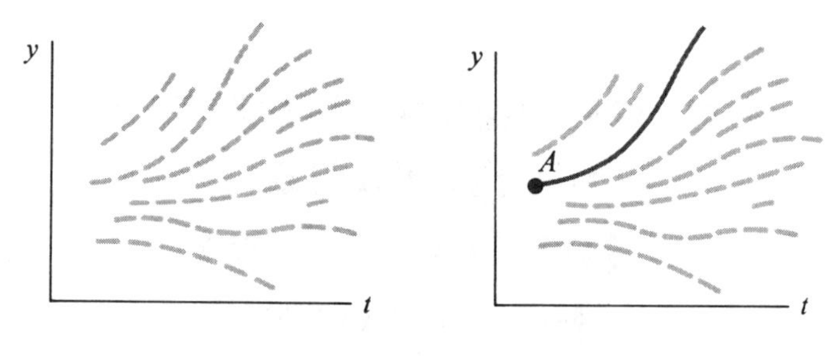

Figure 5.1

Figure 5.2

We resort to numerical methods and the automatic digital computer only when we need more accuracy or when the plotting labor threatens to overwhelm us. The numerical techniques themselves are direct elaborations of the graphical one, refined to produce error measures and to give other assurances that the points they yield do indeed constitute an approximate solution to the differential equation.

The basic extrapolative techniques

We here describe in graphical terms the basic predictor-corrector extrapolative technique. For expository clarity we suppress until later some details that actually are essential to the efficient performance of the method—trusting in the patience of the reader. In the same spirit we postpone discussion of how the integration is started, preferring to concentrate on the technique of keeping it going, which really contains the important concepts. Accordingly, we draw two pictures, the upper showing the derivative dy/dt plotted against the independent variable t, the lower showing the dependent variable y plotted against t (Figure 5.3). The basic extrapolatory device is the fitting of a polynomial to the three most recent points on the derivative (upper illustration), then *extrapolating the fitted polynomial*, and integrating under it to give an increment that yields y_1 at the next point. Since

$$y_1 - y_a = \int_{t_a}^{t_1} \frac{dy}{dt} \cdot dt = \text{the area } A \tag{5.2}$$

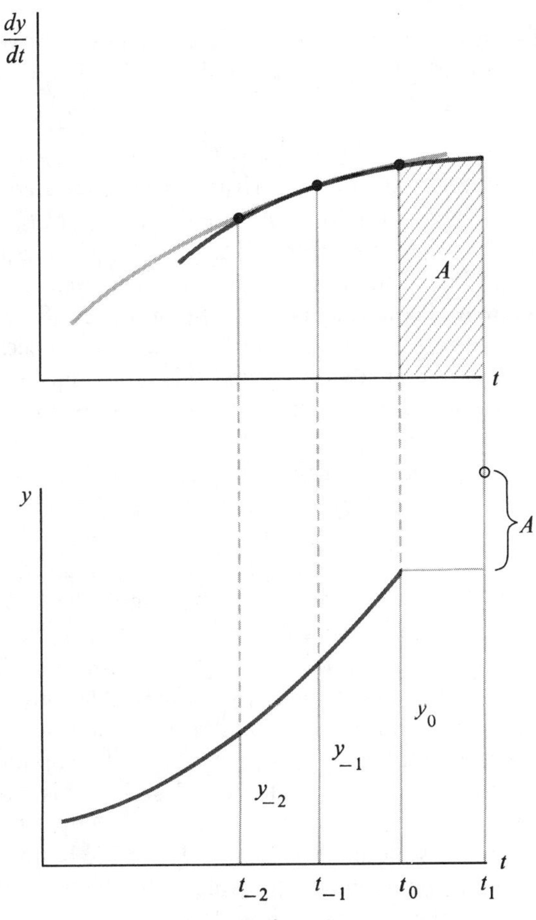

Figure 5.3. A predictor step

we may actually integrate from *any* previous point, t_a, to t_1, provided we then add the area, A, to the appropriate ordinate, y_a. In Figure 5.3 we have used t_0 and, correspondingly, y_0. The order of the polynomial to be fitted is not crucial, but a variety of esthetic and practical considerations usually conspire to make it a parabola—or at most a quartic.

This whole operation is called the *predictor step* and the ordinate that it produces is called y_{1_p} or y_1 *predicted*. We could continue onward, using this value of y_{1_p} as the Truth, but we might become somewhat apprehensive. After all, the foundations on which we built are somewhat shaky—the extrapolated curve was a parabola but the solution to our differential equation certainly is not—or we would not be resorting to numerical integration! So we seek to get the differential equation back into our computational cycle by pumping this ordinate y_{1p} through it to give the corresponding value of the derivative at t_1 via

$$y_1' = f(t_1, y_{1p}) \tag{5.3}$$

Now we may use this value of the derivative at t_1 (note that heretofore we have not had one available), together with the derivatives at t_0 and t_{-1}, to fit another parabola and to integrate under *it* to give a second estimate of y_1 – called y_1 *corrected* or y_{1c}. This value may be expected to be closer to the correct solution of the differential equation, not only because the differential equation has been invoked rather recently in its computation but also because no extrapolation of the fitted polynomial was needed this time. Since extrapolation is always dangerous, we feel much safer with the results of the corrector step. More important, the degree of agreement between y_{1p} and y_{1c} is a measure of the error still remaining in y_{1c} and hence provides us with an internal measure for deciding whether our procedure is precise enough for our purposes. This characteristic is the basic reason that predictor-corrector processes are almost universally recommended by contemporary numerical analysts. The fact that they are not always available in computer centers is simply due to human laziness – for most alternative integration schemes are easier to program.

We now wish to restate the basic computational steps, showing where the data enter the process and in which order. (See color plate on inside back cover.) This will, hopefully, throw light on later discussions about the stability of the process. We use pictures, appending the actual computational formulas that correspond to each step. Note that the description is more ponderous than the actual calculation. The particular predictor-corrector method set forth here is the famous Milne method [1953]. There are better ones, but this is not bad and it is somewhat simpler to explain than most. It differs slightly from the details in the previous paragraph. Step 5 is new, but reasonable. It might be awkward to march along with functional and derivative values that did not satisfy the differential equation!

A measure of accuracy

At each step forward in our marching process we have produced an internal measure of accuracy, a difference between y_{1b} and y_{1c}. For that loosely defined class of problems commonly called "well behaved," we shall usually be correct if we assume that the error remaining in y_{1c} (y_1 *corrected*) is less than

$$\frac{|y_{1c} - y_{1p}|}{20}$$

although conservative analysts may prefer a smaller number in the denominator, while more optimistic persons will use the number 29. Thus we may compute a plausible bound on our error and decide whether or not we can afford to live with it. If it is too large, we must throw away our most recent marching step, cut the interval in half, and start forward again. Since the corrector equation has an error term containing h^5, reducing the interval by a factor of 2 will reduce our error by a factor of 2^5, or 32 – which should be enough to bring it back under control. Any reduction in interval other than by a factor of 2 is administratively messy since it precludes the use of previous points and essentially involves us in restarting the problem. Cutting the interval in half permits us to use most of our previous points and only requires us to interpolate in one or two places. (See Figure 5.4.)

An unusually small error permits us to double our interval, which involves no additional calculation provided we have kept enough of our previous points available to the computer. (See Figure 5.5.)

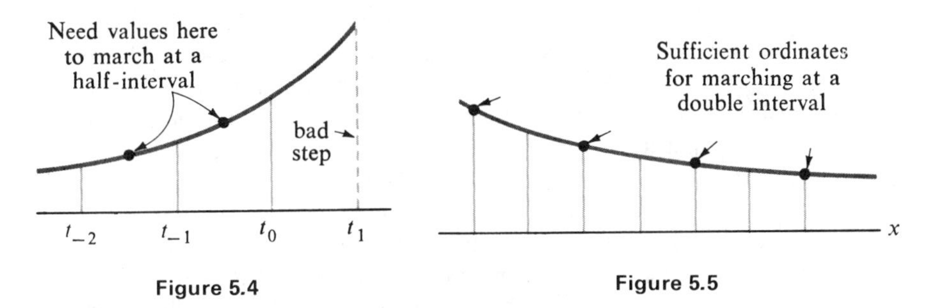

Figure 5.4

Figure 5.5

The iteration heresy

It is clear that y_{1c} in the predictor-corrector process just described receives its major contribution from the corrector equation. The predictor has not been entirely eliminated, but its influence is small in the numbers we finally record as our numerical solution. Among those who carry out this process on a desk calculator there is a strong psychological desire to continue iterating with the corrector equation inside each marching step – thereby completely eliminating the effect of the predictor. These people wish to take y_{1c} and pump it through the differential equation to get a better y', a supercorrected y – possibly even keeping this iteration up until the values change no more. By this time, of course, the predictor has been completely eliminated, but, and this is important, we have *not* thereby achieved a solution to our differential equation!

There are several quantities here and the relationships between them need clarification. Perhaps a diagram will help. We have at least y_{1p} and y_{1c} from our marching process. There is the desired but unknown solution, y, to the differential equation at this point. And finally there could be a whole sequence of supercorrected y's that would be produced by an infinite iteration of our corrector

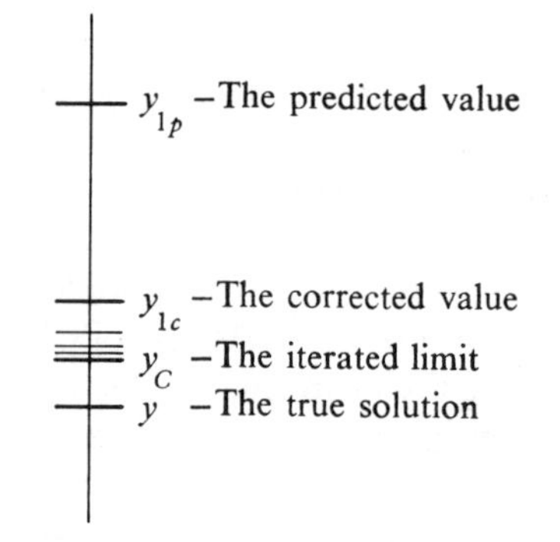

Figure 5.6

step which gradually converge on a value that we may call y_C. In Figure 5.6 these four quantities are arranged on an exaggerated vertical scale to show their probable interrelations. The move from y_{1p} to y_{1c} brings very satisfactory progress toward the unknown true solution, y, that we seek. Subsequent iterations of the corrector equations, however, yield rapidly diminishing returns as they converge toward y_C – the completely correct solution to our corrector equation. This y_C may be closer to our true solution than is y_{1c}, but the improvement is not worth the effort. If y_{1c} is not close enough, then it is more efficient to improve the value by cutting our marching interval in half than it is to make another circuit around our computational cycle. The iteration heresy places us in the uncomfortable position of working harder to get a wrong answer!

One must remember that we have replaced our differential equation by a *difference* equation whose solution, while close to that of the differential equation, is not identically equal to it. We can make it as close to the differential equation as we wish, *but only by reducing the step size, h,* of our marching process. Solving an unsatisfactory approximation more accurately does not really help!

We might note in passing that most of the error studies on predictor-corrector process appearing in the literature were not made on the practical process at all, but were made on the quite impractical process, just castigated above, that carries each step out until it produces y_C. Fortunately, we do not have to depend upon such error studies since the major problems with predictor-corrector integration schemes concern stability rather than accuracy. We shall have considerably more to say about stability both below and in Chapter 16.

Integrators that extrapolate on *h*

Since 1964 a third philosophy of integration has been gaining favor among solvers of ordinary differential equations. We outline only the central idea; the details are still under active investigation. In so doing, it is worth reflecting for a moment on one aspect of earlier methods. In both the Runge–Kutta (p. 139) and predictor-corrector schemes a basic step size h is picked and used for as long as it gives satisfactory accuracy. It may be adjusted from time to time in response to internal or external requests, but for relatively long runs h is maintained constant. The numbers produced by these schemes are solutions to somewhat complicated difference equations that, as h goes to zero, converge to the solution of the differential equation. *But h is never permitted to go to zero* or even partway to zero. It is simply held small enough so that the agreement between the difference equation solution and the differential equation solution is sufficiently good for our purposes. Because of this philosophy we are constantly building forward on slightly erroneous data. We would prefer at each step to get a *good* estimate of the solution to the differential equation rather than a good solution to a difference equation that approximates the differential equations.

The new philosophy is based upon the idea of Richardsonian "extrapolation to the limit" on h at each step. Under this philosophy several estimates for the dependent variable at the next time point, t_1, are generated using successively smaller substep sizes. In effect we do a quadrature on dy/dt between t_0 and t_1 (a distance of h) first using a one-step quadrature to get $y_{1,1}$. Then we do this quadrature again using the rule twice, having subdivided h into two equal subregions, to produce a second estimate $y_{1,2}$. Then still a third estimate $y_{1,3}$ is found by using the quadrature four times at a substep size of $h/4$. In this way we create a sequence of estimates for y_1 using successively smaller values of h. Then, when enough estimates are available, we use an extrapolation rule to estimate what $y_{1,\infty}$ would have been had h been permitted to go all the way to zero—and this extrapolated value is accepted as our final "true" y_1. We have completed one step forward in time and are now ready to begin the process all over again at t_1 to get y_2 at t_2.

This description is far from complete. Which quadrature formula should be used? Should we use the same one repeatedly as the number of substeps is increased or should we proceed up some systematic sequence of increasingly complex formulas? Not all quadrature rules give results that can be extrapolated on h in the same way, or even extrapolated at all! And where do we get those intermediate values of dy/dt that are the ordinates for the next finer application of the quadrature rule? On the answers to these questions hangs the effectiveness of this Richardsonian philosophy. Qualitatively, however, the central idea appeals. It seems right to make a serious effort to get *the* solution to the differential equation at each point rather than to pass on, even reluctantly, knowing that we have settled for a solution to a difference equation that can never be correct.

The utility of these Richardsonian methods depends upon their computational cost. Each forward Richardsonian step of size h is much more complicated than any of the classical Runge–Kutta or standard predictor-corrector steps. Furthermore, it requires notably more evaluations of the differential equation—usually the major component of the computational labor in any realistic differential equation problem. This greater computational cost per step can only be justified if we can take significantly fewer steps to produce the desired solution curves. Experiments have shown that Richardsonian methods do, indeed, permit large integration steps—like two steps per period of the cosine wave! This is a tremendous improvement over the earlier methods and if we are only interested in the end result we shall clearly opt for the new technique. If, however, we are interested in having a fair number of intermediate values for our solution curve, the advantages are less clear. The information would seem to be present but in a concealed form that requires an extrapolation for each desired output. If the extrapolative computation cost is not high, Richardsonian methods will still win. In 1970 there is still insufficient experience to yield definite answers, though the experimenters with Richardsonian techniques remain notably enthusiastic and your author suspects that their enthusiasm is justified. For the student who would try one of these methods we refer him to a published ALGOL program by Bulirsch and Stoer [1966]. The fundamental idea of this integration step is displayed in the next section, but we refer the reader to the current literature for the subtler questions concerning the relative advantages of the several competing methods for extrapolating these quadratures to zero with h.

The fundamental integration step of Bulirsch and Stoer

A Richardsonian integration step consists of two parts:

1. The assembling of a sequence of estimates of y_1 at t_1 using first h, then $h/2$, and then still smaller subdivisions of the basic step size.
2. Extrapolating from the sequence of estimates to that y_1 we might have obtained if the subdivision of h went all the way to zero.

We are deliberately vague here about the suitable subdivisions of h, for there is an interaction between those actually used and the appropriate rule for extrapolation. This level of detail is beyond our immediate purpose, so we shall confine ourselves to describing how Bulirsch and Stoer obtain a typical estimate for y_1 at t_1. They always use an *even* subdivision of h which we shall call k. Thus

$$k = \frac{h}{2j} \qquad j \text{ integral}$$

The basic substep is the simple *linear* extrapolation of y from t to $t + 2k$, an interval of $2k$, using the slope at the midpoint $t + k$. In terms of Figure 5.7,

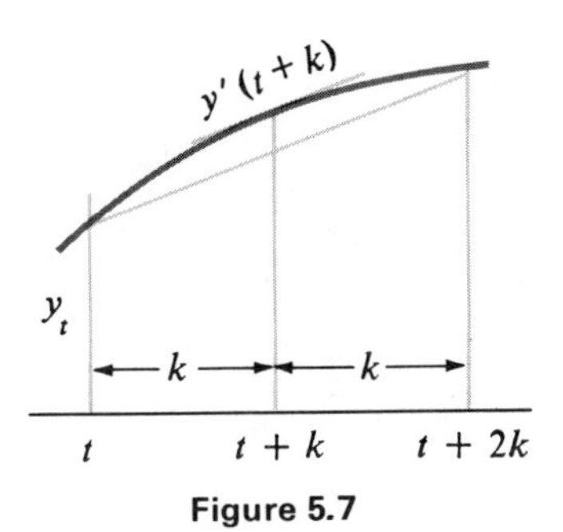

Figure 5.7

the estimate $y(t + 2k)$ is given by

$$y(t + 2k) = y(t) + 2k \cdot y'(t + k)$$

These basic steps are interleaved, each one producing a next value of y that can then be pumped through the differential equation

$$y'(t) = f[t, y(t)]$$

to give a slope. This slope may then be used with the preceding value of y to give the next y, and so on. To start this process, one step must be taken via the cruder formula

$$y(k) = y_0 + k \cdot y'_0$$

Likewise a final special step of length k is taken via the formula

$$y_n(k) = y_{n-1} + ky'_n$$

A complete Bulirsch and Stoer step for the subinterval $h/4$ is given in Figure 5.8. We compute

$$
\begin{array}{lll}
(0) & y_1 = y_0 + ky'_0 \Rightarrow y'_1 & \text{(the special starting step)} \\
(1) & y_2 = y_0 + 2ky'_1 \Rightarrow y'_2 & \\
(2) & y_3 = y_1 + 2ky'_2 \Rightarrow y'_3 & \text{general steps} \\
(3) & y_4 = y_2 + 2ky'_3 \Rightarrow y'_4 & \\
(4) & y_{4f} = y_3 + ky'_4 & \text{(the special ending step)}
\end{array}
$$

Finally, the two estimates (here y_4 and y_{4f}) for the final ordinate are *averaged* to give the final value of y at $t_0 + h$,

$$y(t_0 + h) = \frac{y_4 + y_{4f}}{2}$$

The smallest Bulirsch and Stoer step subdivides h into two parts, using the

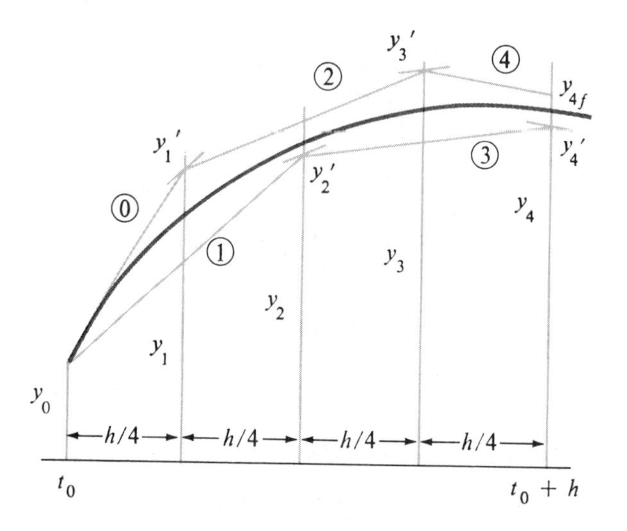

Figure 5.8. Basic Bulirsch and Stoer integration. Each line bears the number of the abscissa from which its slope was derived. The lines also are constructed in the order of their numbering. The final value of y at $t_0 + h$ is the average of the two values there

three steps (Figure 5.9)

(0) $$y_1 = y_0 + ky_0' \Rightarrow y_1'$$

(1) $$y_2 = y_0 + 2ky_1' \Rightarrow y_2'$$

(2) $$y_{2f} = y_1 + ky_2'$$

Finally an average of the two y_2's is taken to be the value of y at $t_0 + h$.

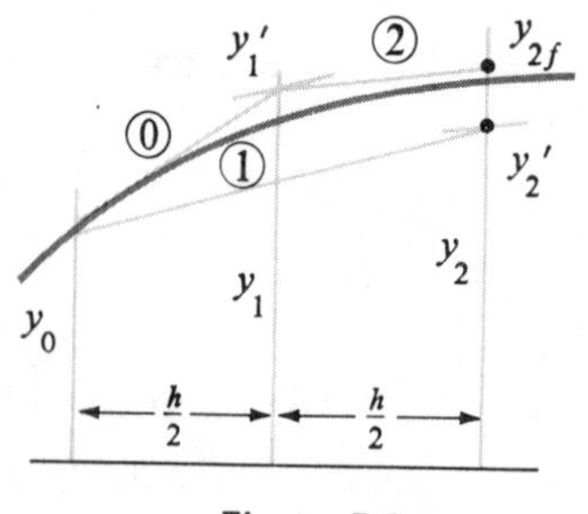

Figure 5.9

Starting the integration

The predictor-corrector marching process requires three or four accurate points on the solution curve in order to get started. Occasionally, we may be lucky enough to have these values from some previous computation, but normally we have to produce them ourselves. The method is unimportant as long as the points are accurate. The whole process of integration from initial conditions is one continuous extrapolation, interspersed with judicious checks to see that no intolerable new errors are being introduced. But if the starting values are wrong, the integration will produce the wrong solution, with no one the wiser. Special care is needed when starting.

Conceptually simplest, the calculation of starting values from Taylor's series is only occasionally useful. We have one value, y_0 at t_0 and the differential equation gives us y'_0. Differentiating the differential equation we now have

$$y'' = g(t, y, y')$$

which may be evaluated to give y''_0 – and in theory the process may be continued to give as many coefficients of Taylor's series as we want. Then we evaluate y at $(t_0 + h)$, $(t_0 + 2h)$, etc., to produce any small number of starting points. The difficulties arise when we try to differentiate a differential equation five or six times. Unless the equation is exceedingly simple, the derivatives grow rapidly in complexity and we despair at ever getting correct derivative expressions, let alone evaluating them accurately. The equations that engineers wish to integrate seldom have the form

$$y' = y + t$$

that usually grace numerical analysis texts as examples. They are much more apt to extend three times across the page, full of square roots, various transcendental functions, and absolute-value signs. Differentiating such equations even once is a major undertaking, so Taylor's series usually remain unused. (Equations of moderate complexity may, however, be differentiated symbolically by the computer itself if symbol-manipulation routines are available.)

A lesser trouble is a fundamental mismatch between the h that can be used in evaluating several points from the series and the step size h that we can afford to use in the predictor-corrector marching process for the same accuracy. The latter is much larger. Thus a series of four points from Taylor's series will typically be too close together for efficient marching, so we spend the next few cycles marching and doubling our step size to get up to a reasonable marching h.

Probably the best, and certainly the commonest, starting technique depends upon another marching philosophy that uses no past information – hence is applicable directly at the initial point. This philosophy, rather like the isoclines of Figure 5.1, probes out in front a half-step or a whole step to ascertain the derivative values at several points, then forms a weighted average slope and uses it to make one final irrevocable leap from t_0 to t_1. A whole family of methods embodies this philosophy. They are collectively known as Runge–Kutta methods and are well described in many sophomore calculus texts, as well as most other books on numerical methods, to which your author refers you [Lance, 1960; Ralston and Wilf, 1960]. Their lack of need for past information is a virtue when starting is the problem, but unfortunately these methods are often recommended and used for the entire marching process, even after we have a history of our function sitting available. Surely one should not throw away past information when extrapolating. We only handicap ourselves and are thus forced to do more work to compensate for the information we have ignored.

As a concrete measure of the price paid by Runge–Kutta marchers versus those using predictor-corrector methods of comparable accuracy, we observe that one predictor-corrector cycle requires two evaluations of the differential equation (steps 2 and 5 of the color plate), whereas the fourth-order Runge–Kutta process requires four. Since nearly all the computational labor in both methods is concentrated in the differential equation evaluation, assuming that it contains at least one transcendental function or is otherwise complicated or lengthy, we usually summarize by saying that Runge–Kutta requires *twice* as much work as predictor-corrector for the same accuracy.

More important than work efficiency is the fact that Runge–Kutta has no built-in accuracy check. Thus the cautious will fix the marching step at too conservative a size, or will do the problem twice, once at h and once at $h/2$ – paying a terrible price for assurance that the answers are correct. The irresponsible computor will merely step merrily along at a value of h that he hopes is adequate – and will get the numbers he deserves! In either case the arguments against Runge–Kutta are formidable. It should be used only to get starting values, which it does very well.

Equations of higher order

The predictor-corrector philosophy extends directly to higher-order differential equations with initial conditions. Consider

$$y'' = f(t, y, y')$$

$$y(a) = c$$

$$y'(a) = d$$

again ignoring the starting problem. In a typical marching cycle we find ourselves at t_0 with some reasonable amount of past history in the form of values of y, y', and y'' on a regularly spaced set of points (Figure 5.10) so that we now have three pictures instead of two. The prediction, or extrapolation, step normally takes place in the highest derivative. Thus we fit a parabola to the latest three points

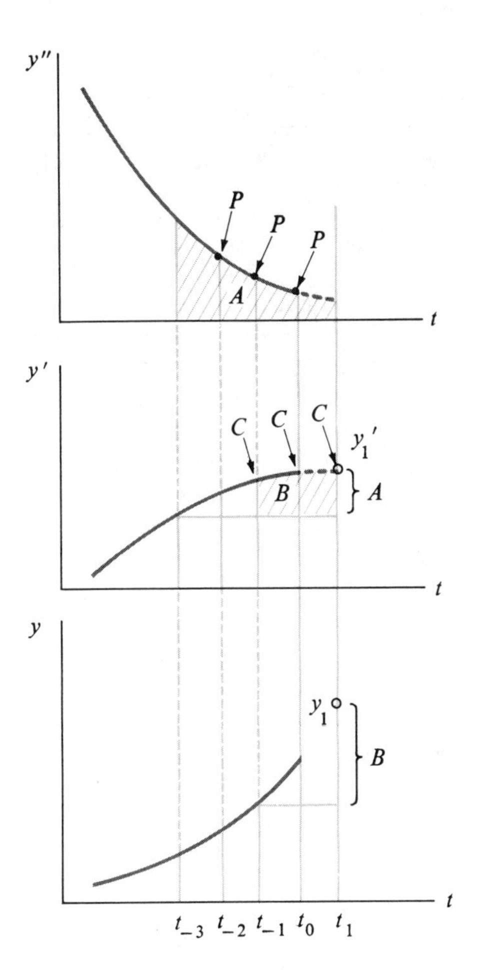

Figure 5.10. A typical predictor step for a second-order differential equation

(labeled P), extrapolate it, integrate under it, and add the area A to the appropriate ordinate in the next lower picture (y'_{-3}) to get y'_{1p}. The Milne formula is

$$y'_{1p} = y'_{-3} + \frac{4h}{3}(2y''_{-2} - y''_{-1} + 2y''_0)$$

Having one value out at the advanced abscissa t_1, we now have no further need for the predictor equation in this cycle. We go down from the y' picture to the y picture by the *corrector* philosophy that avoids extrapolation by fitting a parabola to the latest three points (labeled C), integrating under it, adding the area B to get y_1.

Now that we have values of y_1 and y'_1, as well as t_1, we can substitute into the differential equation to get y''_1 for the first time. Then we use the corrector philosophy in the y'' picture to give y'_{1c} (these last two steps are not shown in Figure 5.10), thus providing us with the essential check of our precision via the test

$$\frac{|y'_{1p} - y'_{1c}|}{20}$$

If the test is satisfactory we continue with the corrector philosophy to produce a new value of y, because the old one really was based on y'_{1p} rather than y'_{1c}, which we are about to accept as correct. For the same reason we pump y'_{1c} and y_1 through the differential equation to give y''_1, but we always stop just short of producing a third value of y'_1.

Higher-order equations add more pictures but introduce no new ideas. Note that the prediction equation is used only once per marching step. It is the least precise calculation in the cycle, so we want to avoid using it except when absolutely necessary. Since it usually gets used in the highest derivative, the subsequent integrations serve to damp out most of its imprecision by multiplying the errors with a factor of $h/3$ each time we descend to a lower picture via a corrector step. Finally, when all points at the new t_1 are available except the highest derivative, we employ the differential equation to supply that missing value.

Systems of higher order

Often higher-order differential equations are presented as systems of first-order equations. While in theory one could combine n first-order equations into a single nth-order equation, in practice this is usually difficult. Going in the other direction, splitting an nth-order differential equation into n first-order differential equations, is trivial, so we often implement the predictor-corrector strategy for the first-order system of n equations and, having it available, use it instead of the philosophy of the previous section. Some inefficiency is involved.

We now have n picture pairs, a derivative and a function picture for each of the n dependent variables. Figure 5.11 depicts a third-order system. We apply the predictor to each of the n derivative pictures to produce n predicted values, one for each $u_i(t_1)$. Then the equations

$$\begin{aligned}
u'_1 &= f_1(t, u_1, u_2, \ldots, u_n) \\
u'_2 &= f_2(t, u_1, u_2, \ldots, u_n) \\
&\vdots \\
u'_n &= f_n(t, u_1, u_2, \ldots, u_n)
\end{aligned}$$

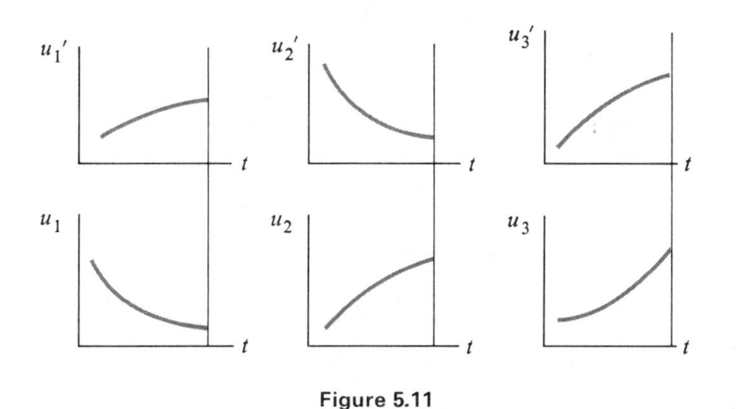

Figure 5.11

may be evaluated at t_1 to give the u_i' values, and finally a corrector cycle can be performed in each picture pair. The test is now n-fold, checking that all n predicted values agree adequately with their n corrected values, but no other modification of the simple predictor-corrector philosophy is needed.

Stability of the Milne process

Your author would be remiss if he left you with the impression that all virtue was on the side of the predictor-corrector marching processes, for even they sometimes have their troubles. We shall analyze a process we have spoken against: that of the *iterated* Milne corrector that produces y_C before moving on (see color plate on inside back cover). The recommended computational process of the color plate is simpler to execute but harder to analyze, and has very similar stability characteristics. We here opt for the greater expositional simplicity in our stability analysis by getting the predictor equation out of the computational cycle. Thus our solution can be assumed always to satisfy

$$y_1 = y_{-1} + \frac{h}{3}(y'_{-1} + 4y'_0 + y'_1)$$

Let us now apply this corrector equation to a simple differential equation

$$y' = ay \tag{5.4}$$

with a known answer

$$y = e^{at}$$

We have

$$y_1 = y_{-1} + \frac{h}{3}(ay_{-1} + 4ay_0 + ay_1) \tag{5.5}$$

or

$$\left(1 - \frac{ah}{3}\right)y_1 = \left(1 + \frac{ah}{3}\right)y_{-1} + \frac{4ah}{3}y_0$$

Setting

$$b = \frac{ah}{3} \tag{5.6}$$

we have the homogeneous *difference* equation

$$(1 - b)y_1 - 4by_0 - (1 + b)y_{-1} = 0 \tag{5.7}$$

which may be solved by assuming a solution of the form

$$y_n = \rho^n$$

Substituting in (5.7) and multiplying through by ρ we have

$$(1 - b)\rho^2 - 4b\rho - (1 + b) = 0$$

a quadratic with two roots

$$\rho_{\pm} = \frac{2b \pm \sqrt{1 + 3b^2}}{1 - b} \approx \frac{2b \pm (1 + \frac{3}{2}b^2)}{1 - b}$$

where the last term is an approximation that assumes b to be small relative to unity. Under the same assumption we may express

$$\frac{1}{1 - b} = 1 + b + b^2 + \cdots$$

and then our two roots are

$$\rho_+ = 1 + 3b + \cdots \qquad \text{and} \qquad \rho_- = -1 + b + \cdots$$

so the general solution to (5.7) is the linear combination of these two independent solutions. We have

$$y_n = A(1 + 3b)^n + C(-1 + b)^n \tag{5.8}$$

Returning to our differential equation (5.4): If a is greater than zero the solution is e^{at}, the growing exponential. We see that b also is positive (5.6) and hence the first term of (5.8) also grows, as it should if it is to represent the differential equation solution. The second term represents an extraneous solution introduced by us when we employed a second-order difference equation to represent the first-order differential equation. We naturally want this extraneous solution to fade quietly away—which it obligingly does if b is somewhat greater than zero, since the value of the term $(-1 + b)$ is then less than unity in absolute value and is raised to higher and higher powers. Everything is fine.

When a is *negative*, however, the story is sadder. Our solution is now the negative or descending exponential, e^{at}, and again the first term of (5.8) approximates it by steadily shrinking as it is raised to continually higher powers. The second term in (5.8), however, is obstreperous. Being negative and larger than unity in magnitude, it grows in size while steadily alternating in sign. Thus, even though its coefficient C may be small, it will soon grow to become the dominant term in the sum – obscuring its meekly declining companion that is dutifully following the solution to the differential equation. Graphically, we observe such phenomena as are pictured in Figure 5.12. The "average" of the points, in some strange sense, is correct – but that does us little good since we do not know how to average them.

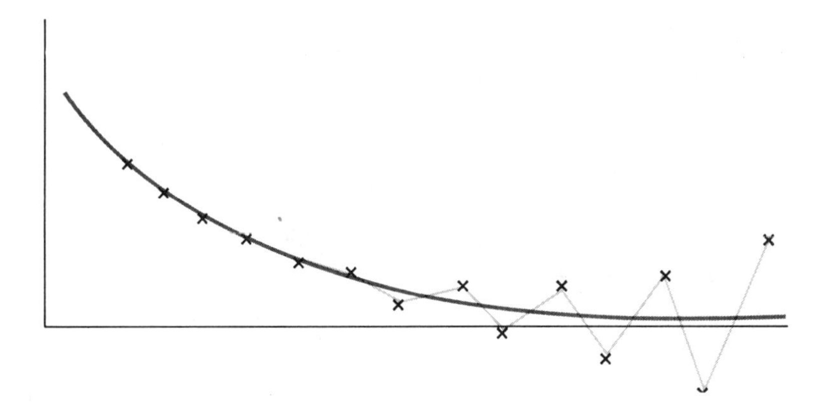

Figure 5.12. An unstable integration

Note that the trouble occurs because b is negative – not a. Thus for the geometry of Figure 5.12 we may safely carry out our marching process from right to left, that is, in the uphill direction. With this stratagem not only is a negative, as before, *but h is also negative.* Thus b is positive. Our solution grows, step by step, and so does the first term of (5.8), while the second term obligingly disappears. As one might suspect, the right-to-left integration of e^{at} for positive a, which goes downhill, leads to trouble.

Geometrically put, the argument is simple: Our difference equation has one solution that grows and one that shrinks as we step along. If the growing one is the desired solution, we have no troubles. If the shrinking solution is the one we want, its growing brother is a damned nuisance. Many predictor-corrector systems have extraneous solutions that quietly disappear under all reasonable circumstances. They tend to be a little more complicated than the Milne process but are safer to use. One of the commonest is the Adams–Bashforth–Moulton system, whose essential equations are

$$y_{1p} = y_0 + \frac{h}{24}(55y'_0 - 59y'_{-1} + 37y'_{-2} - 9y'_{-3})$$

$$y_{1c} = y_0 + \frac{h}{24}(9y'_1 + 19y'_0 - 5y'_{-1} + y'_{-2})$$

Still another process, due to Hamming, is

$$p_1 = \frac{1}{3}(2y_{-1} + y_{-2}) + \frac{h}{72}(191y'_0 - 107y'_{-1} + 109y'_{-2} - 25y'_{-3})$$

$$m_1 = p_1 - \frac{707}{750}(p_0 - c_0)$$

$$c_1 = \frac{1}{3}(2y_{-1} + y_{-2}) + \frac{h}{72}(25m'_1 + 91y'_0 + 43y'_{-1} + 9y'_{-2})$$

$$y_1 = c_1 + \frac{43}{750}(p_1 - c_1)$$

For a complete account of its quite pleasant sophistications, we refer you to Hamming's book [1962, p. 206].

Common difficulties

Initial-value problems often start at a numerically awkward place. The differential equation may be changing more rapidly there or, even more likely, may have an infinite extraneous solution there. Bessel's equation is typical. We have

$$y'' + \frac{1}{x}y' + y = 0 \tag{5.9}$$

with the initial conditions at the origin

$$y(0) = 1 \qquad y'(0) = 0$$

The singularity in the equation at the origin is kept under control by the zero value of y' there – but it does produce a numerically indeterminant form that cannot be ignored when seeking a numerical solution. The cure is simple: At the origin the second term of (5.9) becomes equal, by L'Hospital's rule, to y'', so that – at the origin only – our differential equation is

$$2y'' + y = 0 \tag{5.10}$$

We thus see that $J_0(x)$ begins like $\cos(x/\sqrt{2})$, and this fact is the simplest way to get starting values if we are carrying out the numerical integration by hand, but in an automatic program we must merely put in both forms (5.9) and (5.10) with a test on x to decide which will be used in computing y''.

A potentially more troublesome difficulty is the equation that changes its nature in midstream. Consider

$$y'' - \left(1 - \frac{13}{x} + \frac{12}{x^2}\right)y = 0$$

with the initial conditions at the origin

$$y(0) = 0 \qquad y'(0) = b$$

For large values of x the expression in parentheses is positive and essentially unity – so the solution behaves like the double exponential $Ae^x + Be^{-x}$. This can be nasty enough, for we are usually interested in the solution that declines as x becomes large, while its growing companion keeps intruding when least wanted because of rounding error. We have more to say about this difficulty in Chapter 6. But our troubles are only beginning! Very near the origin the bracket is large and positive, with the $12/x^2$ term dominating it. If only that term were in the parentheses, the solutions to differential equation would be x^4 and x^{-3}. Again the small solution is wanted, but his infinite brother hovers around ominously ready to pop up whenever we do anything wrong. In between small x's and large x's there lies a region where the term $13/x$ may dominate, thereby reversing the sign of the parentheses and turning our equation into a very pleasant well-behaved oscillatory type that gives no troubles. It is safe to say that any blind use of a numerical integration package on such an equation will usually end in disaster. Detailed discussion of the proper strategy is beyond the scope of this book. Our purpose here is strictly minatory – prophesying doom for the unwary.

As a final horrible example, consider the equation

$$(1 - x)y'' + fy' + y = 0$$

which briefly descends from second to first order at x equal to unity and, for good measure, changes from oscillatory to a double exponential at the same time. Insight can be gained by assuming f constant and moving the trouble to the origin, where, after several changes of the variables, the solution may be recognized to be basically $uZ_2(2\sqrt{u})$. Here Z_2 stands for the oscillatory Bessel function J_2 on one side and I_2, the exponential Bessel function, on the other. Good luck!

"Stiff" equations

We now turn to a somewhat rarer but no less troublesome class of equations

$$y'' - L(t)y = 0 \qquad L(t) \gg 0 \tag{5.11}$$

Nontrivial second-order differential equations can really only wiggle in two ways. They can oscillate like a cosine or behave like an exponential. Some equations may occasionally change their minds (at an inflection point) about which prototype they wish to emulate, but at most points they are opting for the one or the other. In our present example, we have a most decided exponential behavior, for $L(t)$ is large and positive. If L were constant at, say, 400, then y would be a linear combination of e^{20t} and e^{-20t} – with the negative exponential fading out of the solution extremely quickly as t grows.

In most physical systems where this exponential equation arises, however, the desired solution is the *negative* exponential. Such equations have come to be known as "stiff" differential equations. We know that our system behaves like Figure 5.13. Numerical integration by conventional methods starting at the origin is thus immediately in trouble. Any rounding error has the effect of introducing a small amount of the unwanted, or increasing, exponential solution and in the linear combination

$$Ae^{-20t} + \varepsilon\, e^{20t}$$

the contribution of the second term does not remain small very long no matter what ε may be. If ε is not *zero*, the second term will soon dominate and our numerical solution will have the behavior shown in Figure 5.14. The "extraneous" solution refuses to be ignored.

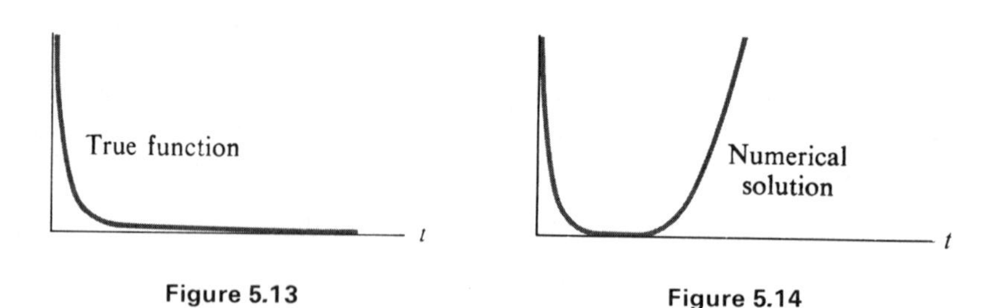

Figure 5.13

Figure 5.14

If we should apply standard extrapolative numerical integration techniques to (5.11) in its present form, we see that every evaluation of y'' from the differential equation is a powerful *noise amplifier*, in the sense that any error in y is multiplied by an L which is presumed large. The subsequent integration step imposes a damping on this error that is proportional to h or perhaps h^2, so that by taking small enough steps we can perhaps live with this noisy equation – but this accommodation is troublesome and expensive of computer time. Further, it does nothing to suppress the unwanted solution. Thus a brute-force integration that cuts the marching step down to 0.001 merely postpones the time when unsatisfactory computer output will force the customer to rethink his whole approach to the equation.

Here the Riccati transformation is useful. We let

$$y' = \eta y$$

so that

$$y'' = \eta' y + \eta^2 y$$

and (5.11) becomes

$$\eta' + \eta^2 = L(t) \tag{5.12}$$

which is mildly nonlinear but of first order. We have banished the extraneous solution though we have not eliminated it. But it is far enough away so that it cannot interfere with the numerical integration processes. (Where did it go?) Note that if L approaches a constant value, η becomes $\pm\sqrt{L}$ and η' becomes 0. Thus the solution curve for (5.12) has a geometry (Figure 5.15) without the sharp

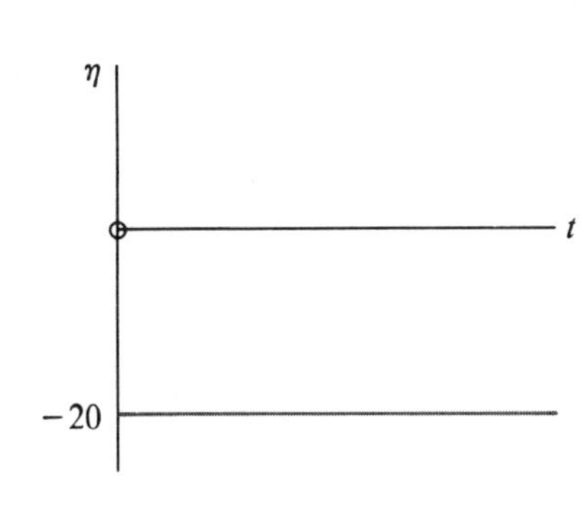

Figure 5.15

corner that plagues us in Figure 5.13 and without the small numbers that arise in an asymptotic approach to the t axis. Our transformation has given us a much more tractable geometry at the price of solving a nonlinear equation. It is a bargain that we should eagerly accept. Finally, we compute the logarithm of the original variable, y, from a quadrature.

The presence of a y' term in (5.11) would not seriously complicate our difficulties, if it enters linearly. The equation

$$y'' + by' - Ly = 0$$

becomes

$$\eta' + \eta^2 + b\eta = L$$

under the Riccati transformation. With minor modifications our previous remarks still apply.

These double-exponential problems in which the declining solution is desired are basically boundary-value problems rather than initial-condition problems. Unfortunately the right-hand boundary condition is at infinity and is essentially a rate-of-growth condition. At times the problem can best be handled by a transformation of the independent variable onto a finite range, followed by an algebraic replacement of the differential equation on a grid of points by a set of algebraic equations (Chapter 6). Unfortunately, unless the transformation is carefully chosen and the equation is particularly tractable, the resulting set of algebraic equations will not be linear – leading to perhaps more difficulties in their solution than we face in treating the original equation by the somewhat inappropriate initial value methods discussed above. Neither approach is particularly attractive, and specific problems must be decided on their individual merits.

Reversing the predictor-corrector cycle

If the form of a "stiff" equation precludes application of the Riccati transformation, we have a struggle on our hands, but there are still several useful strategems. Suppose that L is a slowly varying but complicated function of y as well as t in

$$y'' = L(y, t) \cdot y \qquad L(y, t) \gg 0$$

We can no longer easily eliminate y, but we can carry the calculation of y along with that of η, permitting concurrent evaluation of $L(t, y)$.

More serious are the difficulties raised by the nonlinear

$$y'' = \text{sgn}(y) \cdot L(y, t)y^2$$

where such devices are not effective. We still have our two separate problems:

1. The need to get a decreasing solution when an increasing one will also satisfy the equation.
2. The noise amplification in calculating y'' from y caused by L being large.

The first difficulty can be outflanked by reversing the direction of integration – going from large t toward small t – so that the solution we want is now the growing one and the declining solution is extraneous to our purpose. We will have a trial-and-error operation on our hands to hit the boundary value at the left, for we really are applying initial-value tactics to a boundary-value problem

—the requirement that we want the declining solution constituting the right-hand boundary condition. We have already commented on the difficulty of making an algebraic replacement for the exponential on an infinite range, so we need not here apologize for the initial-value treatment. It's grim, but what's better?

The second difficulty, noisy derivatives because of large L, can be overcome by reversing the predictor-corrector cycle. If we use the differential equation to calculate y from y'', then L appears in the denominator, where its largeness acts to *diminish* errors in y'' rather than to amplify errors in y. If we are now using the differential equation to go downward in our predictor-corrector picture (p. 141) we must go upward by some other device. Previously we integrated numerically to go downward, so it is clear that *numerical differentiation* will carry *upward*. Our new predictor-corrector cycle, then, is to extrapolate y with a parabola, go from y to y' and then on to y'' by numerical differentiation, and finally to back down to y through the differential equation. Numerical differentiation is not an attractive operation because it, too, is apt to be noisy, but the cycle now contains a powerful noise suppressor L strategically maneuvered into the denominator where it will restore good behavior to our numbers. Thus, by reversing both our direction of integration and our computational cycle, we can hope to integrate stiff equations that resist the Riccati treatment.

Another sticky equation

Consider the ordinary differential equation system

$$w' + 10w = 11e^t \tag{5.13}$$

$$w(0) = 20$$

Its analytic solution

$$w = e^t + 19e^{-10t}$$

is plotted in Figure 5.16.

Clearly we have two exponentials that have very little to do with one another, a brief crossover taking place around $t = 0.5$. (The actual minimum occurs at 0.477, where w is 1.77, but even by this time the contribution of the declining exponential is under 10 per cent of the solution. Between 0 and 0.1 the curve drops from 20 to 8.1.)

Geometrically, we have a quite sharp corner with a nearly vertical left side—a most unpolynomic geometry. Predictor-corrector schemes based on polynomial fitting are going to have trouble. Of course, they will sneak thru via the usual device of cutting the step size until they can assimilate the geometry, but efficiency will suffer. If we have many problems with this geometry we might better devise a less blunt approach.

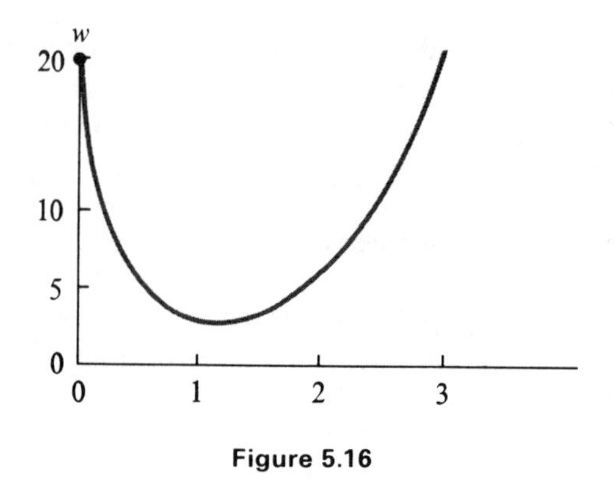

Figure 5.16

The cure is to separate, insofar as possible, the two solutions. The device: Define a new dependent variable by

$$w = e^t + v \tag{5.14}$$

whence

$$v' + 10v = 0 \qquad v(0) = 19 \tag{5.15}$$

The integration of this auxiliary system will offer no trouble provided one uses a predictor-corrector system that can handle declining exponentials (Milne cannot, but Adams–Bashforth can).

Of course a real problem is not so neatly altered, but the principle is applicable. We might have

$$w' + 10f \cdot w = 11g \cdot e^t$$

where f and g are nuisance functions of t (and perhaps w) that are nearly equal to unity but prevent analytic solution. Thus our geometry is the same—and so is the cure. We make the same definition (5.14) as before, obtaining

$$v' + 10 \cdot fv = e^t(11g - 10f - 1)$$

which has the same geometry as (5.15)—and hence is easy to integrate by standard predictor-corrector systems. If the subtractions that nearly cancel bother us, we may expand g and f in a Maclaurin series for small t—though it seems unnecessary in this example.

Highly oscillatory equations

In this chapter most of the discussion has centered on standard methods for integrating equations and perhaps not enough has been spent upon suitable transformation of the equation before any integration is undertaken. In this day of prepackaged integration systems the analyst must beware lest he drop his critical guard and blindly apply those standard techniques where they are inefficient or possibly even disastrous. By way of example, suppose we need the solution from initial values for an equation of the type

$$w'' + Lw = 0 \qquad L \gg 0 \tag{5.16}$$

where L is a large positive function that varies slowly with t, though not w. If L were constant at, say, 400, then we may solve (5.16) analytically, finding

$$w = A \cos 20(t + \theta)$$

determining A and θ from the values of w and w' at the origin. Since L is presumably nasty enough to prevent this analytic solution, we merely know that w will look rather like Figure 5.17—a nearly sinusoidal curve with rapid oscillations that have slowly varying amplitude and phase. If we need the solution out to some value of t such as 30, the standard initial-value stepping methods must take us over approximately $20 \cdot 30/2\pi$ or about 100 complete oscillations.

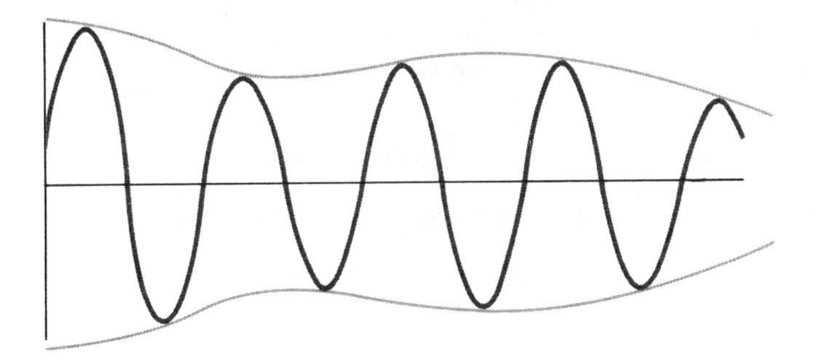

Figure 5.17

Since at least eight or ten points will be needed to follow each oscillation (and possibly many more if we insist on high accuracy) we may expect an expensive integration. If the problem is to be solved only once, we may sigh and pay, for the standard techniques will work, but if we face a whole set of these problems we better invest some of our time in devising a method that makes the oscillatory nature of the solutions work for us instead of against us. The payoff will more than justify the effort.

If the inefficiency of standard methods comes from the rapid variations of w, then we must seek to transform w into something that varies slowly. The obvious answer has already been mentioned: amplitude and phase. Thus we try

$$w(t) = y(t) \cdot \cos(z(t))$$

Substituting into (5.16) and separating into two differential equations we have

$$y'' - (z')^2 y + Ly = 0 \tag{5.17}$$

$$yz'' + 2y'z' = 0 \tag{5.18}$$

Equation (5.18) is separable and integrates once immediately to give

$$z' = \frac{B}{y^2} \tag{5.19}$$

so that (5.17) may now be written

$$y'' + Ly - \frac{B^2}{y^3} = 0 \tag{5.20}$$

and we see that it has become an equation containing only the amplitude. Once we integrate it, then the phase, z, may be obtained from (5.19) by a quadrature. The arbitrary constant B must be obtained from a boundary condition.

This Madelung transformation, as it is called, is especially useful when L is asymptotically constant, giving a simple harmonic solution to (5.16) for large t. Since (5.16) is homogeneous we may choose the value of y at large t to be unity, being prepared to scale our final solution later. Using (5.20) we immediately determine B from

$$B = y^2\sqrt{L} = \sqrt{L}$$

and we may integrate (5.20) in toward the origin without even troubling to

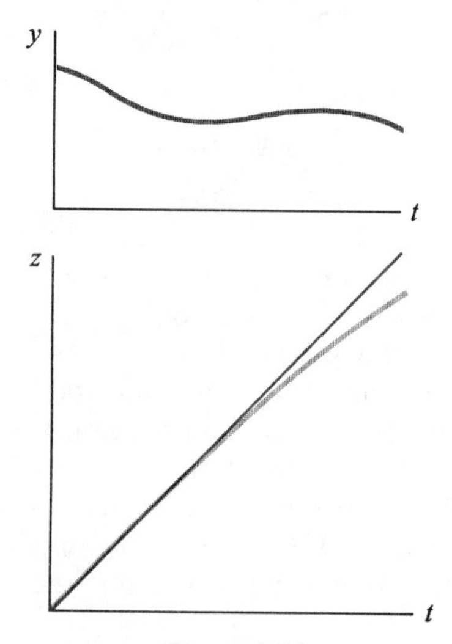

Figure 5.18

obtain starting values for y. The quadrature for z may be carried along in parallel, an unknown additive constant of integration, z_0, to be supplied from the initial conditions when we reach the origin.

In this approach to an highly oscillatory equation we have undertaken to integrate two equations instead of one in order to work with slowly changing dependent variables. Since the graphs of y and z may be expected to look like Figure 5.18, we shall probably be able to march along with giant steps, the decrease in the total number of steps more than compensating for their increased complexity.

Removing nonpolynomic behavior

If our equation defines a mildly perturbed negative exponential

$$\frac{dy}{dt} = -f(t, y) \cdot y \qquad f(t, y) \approx A \gg 0 \tag{5.21}$$

where $f(t, y)$ is a large and positive but slowly changing function of t and y, we may expect double trouble. Not only is our solution nonpolynomic, thereby rendering the standard integration algorithms inefficient, but also the largeness of f makes the differential equation (5.21) a noise amplifier by multiplying directly any errors that exist in y. Both of these difficulties can be circumvented if we approximate $f(t, y)$ by the constant A, locally, and let

$$y = v \cdot e^{-At}$$

so that our equation becomes

$$v' = (A - f)v \qquad (A - f) \approx 0$$

and there are no longer any large noise-amplifying factors in the differential equation that must be numerically integrated.

Replacing y by $v(1 + \gamma t)e^{-At}$ will yield the differential equation

$$v' = \left[\left(A - \frac{\gamma}{1 + \gamma t}\right) - f\right] \cdot v$$

which clearly approximates f by a slightly more complicated function than a mere constant. Other approximants may easily be found and the choice must be dictated by the geometry of f.

Initial-value problems—a prediction

Before the advent of the modern computers, integration of ordinary differential equations from initial conditions was just too laborious to invite much experimentation. Until the early 1950s such work as was done was apt to employ

Runge–Kutta algorithms. The popularization of predictor-corrector methods in 1949 by Milne [1949] coincided with the arrival of the stored-program computer to make reasonably efficient integration both possible and attractive. In 1951 Gill improved the accuracy of Runge–Kutta, and Rosser [1967] has recently remedied some of its efficiency problems by supplying internal accuracy measures, thereby restoring Runge–Kutta techniques to relative respectability. But the efforts of Bulirsch and Storer [1966] since 1964 seem most likely to produce the principal change in current initial-value integration habits. Their invocation of Richardsonian extrapolation to zero on h has produced algorithms that surpass predictor-corrector and Runge–Kutta methods in accuracy and efficiency for enough different examples to suggest that this class of methods will soon become standard for many problems. By 1970, however, only the more adventurous of computer centers are learning how to employ the available algorithms while the rough edges are still being polished. Thus, although we cannot at this writing confidently display dependable Richardsonian methods for the student, we feel that they should soon be available.

CHAPTER

6

ORDINARY DIFFERENTIAL EQUATIONS—BOUNDARY CONDITIONS

In Chapter 5 we discussed the numerical integration of systems of ordinary differential equations when all the necessary conditions occur at one value of the independent variable. We now turn to a rather different type of problem that at times looks deceptively similar—the boundary-value problem—where the conditions are given at two (or more) different values of the independent variable. We limit our discussion to second-order systems. They are quite sufficiently complicated to illustrate our points.

Perhaps the most familiar equation of second order is

$$y'' + y = 0 \tag{6.1}$$

With the boundary conditions

$$y(0) = 1 \qquad \text{and} \qquad y(1) = 2$$

the analytic solution is

$$y(t) = 1.7347 \sin t + \cos t$$

but we shall suppose that we have forgotten our analytic methods, forcing us to seek the numerical solution. If we try to invoke the methods of Chapter 5, we are immediately in trouble, since we cannot get started. This is most easily demonstrated by considering the Taylor's series where, at the origin, we have y and y'' and, by differentiating (6.1), all the higher *even* derivatives—but no method for finding y' and the other odd derivatives. We are stymied. The additional fact, the known value of y at x equal to unity, undoubtedly suffices to define the unique solution, but it is not immediately usable. The student should consider trying to start this numerical integration by other techniques such as Runge–Kutta, to see where they break down.

One way around our difficulty is to guess a value of $y'(0)$, thus enabling us to carry out a standard initial-value predictor-corrector integration. We shall not usually come out at the other given boundary point, so a second guess and a second complete integration will be needed. Again we shall not be correct, but now we have two starting values of y' with their corresponding solutions everywhere, including at the far end. We may now perform a linear interpolation to give our next trial value of $y'(0)$.

This iterative process of guessing $y'(0)$ and marching is obviously laborious, but a good bit of the labor would be required even if our methods were more sophisticated. The plain fact is that the boundary-value problem is inherently more complicated than the initial-value problem.

We remark in passing that if our nth-order equation is linear, n iterations are enough. Since any solution is a linear combination of n independent solutions, we can produce the final solution simply by forming the same linear combination, point by point, as suffices to give us the second boundary point. As usual, it is the nonlinear equation that gives us our major troubles.

Our treatment above of the two-point boundary-value problem should trouble the sensitive reader. The problem is symmetric in the independent variable, but our method of solution is not. Why, after all, begin by assuming a value for $y'(0)$ when $y'(1)$ would do as well? Perhaps we should seek a different approach to the boundary-value problem.

A symmetric process

A symmetric problem suggests that we seek a symmetric process. Our grand strategy will be to replace the differential equation by a set of algebraic equations on a grid of regularly spaced points. We need only get rid of the derivatives. Considering three equally spaced ordinates (Figure 6.1), we see that the fundamental definition of a derivative as the limit of the ratio $\Delta y/\Delta x$ suggests that

$$\left.\frac{dy}{dx}\right|_{1/2} \approx \frac{y_1 - y_0}{h}$$
$$\left.\frac{dy}{dx}\right|_{3/2} \approx \frac{y_2 - y_1}{h} \tag{6.2}$$

If we should now let h go to zero, these expressions for our derivatives would of course be exact – but no numerical analyst ever let an h go to zero in his life, so we must be content to accept the expressions (6.2) as mere approximations. Since we are faced with second derivatives, we repeat our philosophy by differencing the first derivatives. Thus

$$\left.\frac{d^2y}{dx^2}\right|_1 \approx \frac{\left.\dfrac{dy}{dx}\right|_{3/2} - \left.\dfrac{dy}{dx}\right|_{1/2}}{h} = \frac{y_2 - 2y_1 + y_0}{h^2} \tag{6.3}$$

will give a simple approximation for y'' at the ordinate labeled with subscript 1.

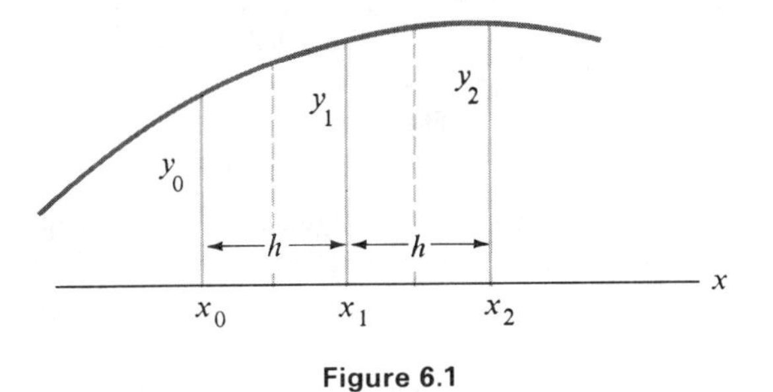

Figure 6.1

We now divide the entire range of our problem into five equal intervals, as shown in Figure 6.2. The precise number is not important provided we have enough to permit the solution curve to be replaced locally by parabolas. We have some reason to believe that our sketch of $y(x)$ in Figure 6.2 is qualitatively correct, hence six ordinates will probably suffice. If we expected the solution to look like Figure 6.3, however, this small number of ordinates would give a

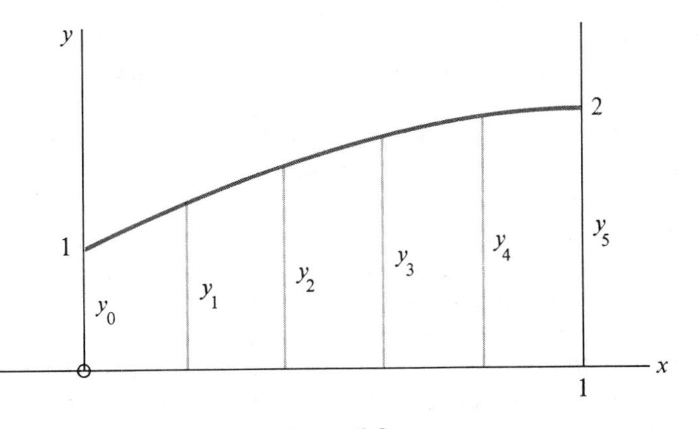

Figure 6.2

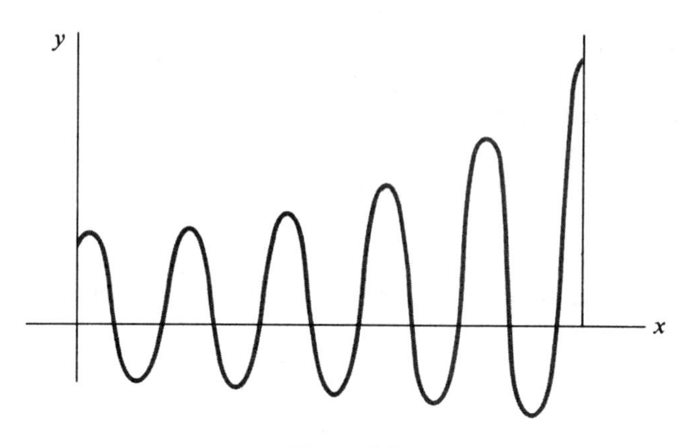

Figure 6.3

totally inadequate sampling of all the wiggles and we would be courting a numerical disaster. Fortunately, we usually have a sufficiently good qualitative idea about the shape of the solution we seek, so that this requirement is not onerous.

Using the four unknown y_i's we have defined in Figure 6.2, we may replace y'' at, say, x equal to 0.4 to express our differential equation there as

$$\frac{y_3 - 2y_2 + y_1}{h^2} + y_2 = 0 \tag{6.4}$$

or, clearing h^2 out of the denominator, as

$$y_1 - (2 - h^2)y_2 + y_3 = 0 \tag{6.5}$$

We can get one such equation at each interior point on the range, and since the boundary values for y give the end ordinates, we have one algebraic equation for each unknown y_j that we have introduced. We have replaced *one* ordinary *differential* equation by a set of n simultaneous *algebraic* equations. If the differential equation was linear, then the algebraic equations will also be linear and, unfortunately, the negative is equally true—a nonlinear differential equation produces a nonlinear algebraic equation system. Our problem has thus ceased to be that of integrating an ordinary differential equation with boundary conditions and has become one of solving four simultaneous linear algebraic equations. If we do so, we obtain the values shown in Table 6.1. As may be seen by comparing these numbers with the correct solution, the use of only four unknowns was remarkably effective, partially because a parabola is so very similar to one arch of a sine wave that our replacement of y'' was a quite good one. With other equations we may not be so fortunate. We must, for example, beware of the curve with vertical or horizontal asymptotes, for no parabola ever had them. But this geometry is, at the moment, a side issue, since we may suppose any nonpolynomic behavior to have been removed before we ever reach these numerical operations.

Table 6.1

Algebraic	1.3252	1.5973	1.8056	1.9416
Correct	1.3247	1.5966	1.8048	1.9411

If we need more accuracy than four ordinates can give, we must decrease h—the minimum sensible decrease being to halve its current value. Unfortunately this raises the number of unknowns to 9 and another halving will saddle us with 19. True, we shall have the same number of equations as unknowns, but it is one thing to solve a set of four linear equations and quite another to solve a set of 19! The general methods of high school are not really adequate. Fortunately, we do not have to use them, since our equations possess a quite simple and comforting structure that suggests several special methods.

Solution of the algebraic equations

Our set of equations that replaced the original problem (6.1) is

$$\begin{aligned}
-(2-h^2)y_1 + y_2 \qquad\qquad\qquad\qquad &= -1 \\
y_1 - (2-h^2)y_2 + y_3 \qquad\qquad &= 0 \\
y_2 - (2-h^2)y_3 + y_4 &= 0 \\
y_3 - (2-h^2)y_4 &= -2
\end{aligned} \tag{6.6}$$

with h equal to 0.2. Using matrix conventions we may rewrite (6.6) as

$$\begin{bmatrix} -1.96 & 1 & & \\ 1 & -1.96 & 1 & \\ & 1 & -1.96 & 1 \\ & & 1 & -1.96 \end{bmatrix} \begin{bmatrix} y_1 \\ y_2 \\ y_3 \\ y_4 \end{bmatrix} = \begin{bmatrix} -1 \\ 0 \\ 0 \\ -2 \end{bmatrix} \tag{6.7}$$

or, still more compactly, as

$$(A - h^2 I)y = b \tag{6.8}$$

All three forms have their uses in exposition and analytic manipulations, but when numbers are finally needed we must unpack our compressed notations and return to (6.6).

The important properties of this equation set are:

1. No more than the three unknowns are involved in any one equation, and these are adjacent – that is, the matrix is tridiagonal.
2. The largest coefficient in any equation lies on the principal diagonal of the matrix.

These two properties follow from our replacement of the second derivative by its standard approximation and may well appear even if the differential equation is more complicated. Some other regularities of our matrix, such as the remarkable constancy of the several diagonals, arise because this differential equation is particularly simple. In fact, this equation set is so regular that one may easily discover a general inverse for any order – but such facts are too specialized to concern us here.

The first property, that only three variables have nonzero coefficients in any one equation, means that the high school technique of eliminating one variable at a time is feasible as far as the arithmetic workload is concerned. The total number of arithmetic operations in solving such a *tridiagonal* set of linear equations is proportional to n, the number of equations, rather than to n^3, as is true of an arbitrary set of linear equations. For sets of size 9, 19, or larger, this fact is very important. If we choose this approach we may still have to do our eliminations circumspectly to avoid loss of significant figures, but at least the approach is not obviously silly, even for the hand calculator.

The second property, that the coefficients in each equation are dominated in absolute value by the principal diagonal, suggests that we consider the Jacobi algorithm in which the equation set is rewritten

$$\begin{aligned} y_1 &= \frac{1 + y_2}{1.96} \\ y_2 &= \frac{y_1 + y_3}{1.96} \\ y_3 &= \frac{y_2 + y_4}{1.96} \\ y_4 &= \frac{y_3 + 2}{1.96} \end{aligned} \tag{6.9}$$

and an approximate solution is plugged into the right-hand side. This allows us to calculate a new guess which we, in turn, plug into the right-hand side – iterating away secure in the comfortable knowledge that the process will converge. Comfortable, that is, if an automatic computer is doing the iterating and somebody else is paying the bill – for the convergence is apt to be disappointingly slow! But the program is trivial to write, computers are tolerant of slave drivers, and maybe the bill payer won't notice. The method is guaranteed to work even if the initial guess is as bad as all zeros!

Perhaps it is unfair to castigate so strongly the users of Jacobi here, for the method enjoys an important property, common to all iterative processes, that it rewards good insight with a quick and efficient solution. If we should happen to start the iteration with the correct answer, the process will stop in one iteration. Similarly, if we are close to the correct solution, then the total amount of calculation is noticeably less than by any elimination technique. Such possibilities for economy are not available in noniterative solution methods. Your author, however, has a somewhat jaundiced view of the initial approximations supplied by several generations of students, so he prefers to recommend the other methods for this tridiagonal problem. In particular, elimination with interchanging so as to always divide by the larger of the two possible pivots is probably his favorite. An example is given at the end of Chapter 13 (p. 358).

Boundary condition at infinity

When one of the boundary conditions is at infinity the numerical analyst must take special care lest he find he has chosen an inefficient or even futile strategy. The solution to his differential equation certainly has a tail, possibly even a fairly fat tail, as it goes offstage (Figure 6.4), but it eventually declines monotonically and it is surely not this part of the solution that is his principal concern.

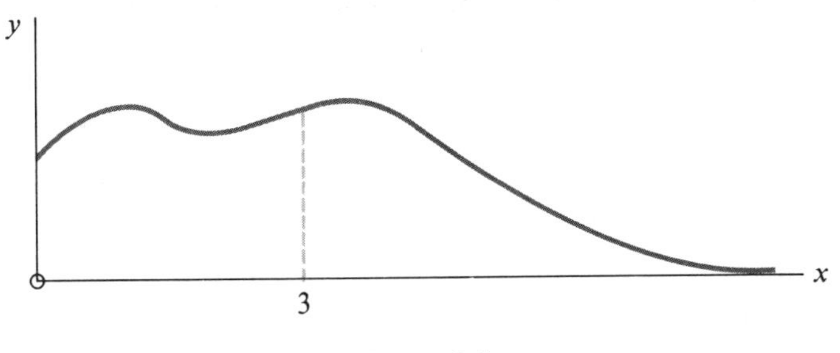

Figure 6.4

He may cut the tail off when it is effectively zero, but even then a direct replacement of the differential equation by a set of algebraic equations on a regular grid will necessarily put most of the ordinates out in the part of the tail that remains, thereby spending a lot of effort to gather relatively uninteresting information. We would prefer a better allocation of our resources. The standard direct replacement technique can, of course, be patched up by using *two* grids, one at a fine spacing over, say (0, 3), where the action is expected to be and a coarse grid out over the tail. We thereby incur some nastiness at the joint where we must find a representation for the second derivative in terms of unequally spaced ordinates – not an insuperable problem, for the method of undetermined coefficients (page 108) will provide the required formula – but the idea of a joint in a continuous problem is, of itself, unattractive. And then infinity is still a long way out!

A better strategy usually is to change the independent variable, bringing infinity into some more accessible place – like the origin. Whether we replace x by e^{-x}, x^{-1}, $(1 + x)^{-1}$, or some more exotic variable will depend on the specific differential equation. Very often such a change will introduce its own difficulties into the problem, so some circumspection is appropriate. If the first change seems to be less than ideal, by all means take time to explore several others before getting involved with the details of programming around the difficulties.

By way of being more specific, we now integrate a particular equation by several techniques, without any guarantee that we have included the best one. Consider the system

$$w'' - \left(1 - \frac{8}{1 + 2x}\right)w = 0 \tag{6.10}$$

with

$$w(0) = 1 \qquad \text{and} \qquad w(\infty) = 0$$

which we suspect a priori might have a solution that qualitatively resembles Figure 6.5. This sketch was achieved by noting that at the origin our equation is

$$w'' + 7w = 0$$

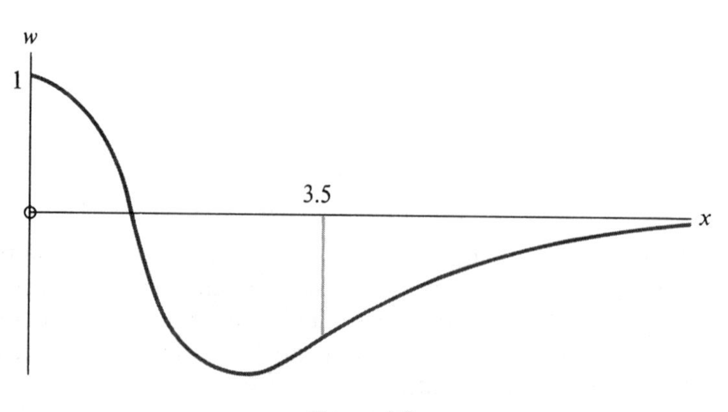

Figure 6.5

which has the solution

$$w = \frac{\cos[\sqrt{7}(x - \theta)]}{\cos[\sqrt{7}\theta]}$$

while at 3.5 it becomes a straight line (why?), and thereafter, if it is to satisfy the right-hand boundary condition, it must ultimately become the double-exponential equation with the decreasing solution Be^{-x}. Since we may have a crossing of the axis, the Riccati transformation is not useful over the whole range—and partitioning a boundary-value problem is superficially unattractive because of difficulties at the joint. Partitioning is, however, possible, so we place Riccati low on our priority list without altogether eliminating him.

We first try a simple transformation to remove the e^{-x} behavior explicitly. Substituting

$$v = e^{-x} \cdot w$$

into (6.10) we find our new differential equation to be

$$v'' - 2v' + \frac{8}{2x + 1} v = 0 \qquad \text{(6.11)}$$

with

$$v(0) = 1 \qquad \text{and} \qquad v(\infty) < e^x$$

Note that the boundary condition at infinity is a growth-rate condition.

Again, resorting to crude mental approximation, we temporarily suppose that for large x the final term of (6.11) may be neglected. The truncated equation has solutions

$$v = \text{constant}$$

$$v = e^{2x}$$

While the exponential solution immediately violates our supposition, the other solution fits in rather well. The standard techniques for following a "small" solution to a differential equation in the unwanted presence of a "large" one is to pursue the small solution in the direction in which the large solution is dying out. Thus our mental estimation concerning the nature of the two solutions suggests that we try integration from large x in toward the origin. Any component of the e^{2x} solution should then quickly die out, leaving the small solution quite uncontaminated by it—even supposing we make a poor choice of initial conditions at the right. Thus we try an integration using

$$v(10{,}000) = 1.0$$

$$v'(10{,}000) = 0$$

which assumes that v is a constant out to the right. We are relying on the homogeneity of the equation to permit us to rescale the solution so as to fit the boundary condition at the origin. In fact, the solution to (6.11) is a parabola, which we can obtain numerically quite easily via standard predictor-corrector methods once we adopt the strategy of integrating from right to left. Attempts to go from zero toward infinity will quickly blow up because bits of e^{2x} get into the solution, and there is no such thing as "a small amount" of e^{2x} when x starts to grow. The parabola we seek is soon quite submerged by the exponential under such tactics.

An alternative strategy uses the Riccati transformation, again integrating in from the right but switching to the original equation when the solution shows signs of crossing the axis. We let

$$\eta = \frac{w'}{w} \tag{6.12}$$

so that our equation becomes

$$\eta' + \eta^2 = 1 - \frac{8}{2x + 1}$$

with

$$\eta(\infty) = -1 \tag{6.13}$$

Taking 10,000 as an approximation to infinity, we march by standard predictor-corrector methods in to 3.5, accumulating the integral of η as we go in order to be able to evaluate w from

$$w = A\, e^{\int \eta \cdot dx}$$

(Remember that η is negative over the Riccati range.) The value 3.5 is not crucial, but since (6.10) changes from exponential to oscillatory there, this choice guarantees that w will not cross the axis while we are still integrating the Riccati equation.

Breaking the range—the joint

When the geometry of a boundary problem suggests breaking the range into two or more pieces we must decide how to handle the joint. We illustrate with our current example, an equation that is oscillatory on (0, 3.5) and exponentially declining on (3.5, ∞). Here the value 3.5 is an obvious place to break the system, although with many equations we have no such convenient point so that our choice will usually depend on an equitable distribution of our computational resources between the "main" part of the solution and the presumably less interesting "tail."

On (0, 3.5) we use the original equation

$$w'' - \left(1 - \frac{8}{2x + 1}\right)w = 0 \qquad w(0) = 1$$

replacing it on a uniform grid in the standard way. Calling the grid spacing h we obtain

$$w_{i-1} - \left[2 + h^2\left(1 - \frac{4}{x_i + 0.5}\right)\right]w_i + w_{i+1} = 0 \qquad i = 1, 2, \ldots, M - 1 \quad \textbf{(6.14)}$$

This gives us a set of $M - 1$ equations for the M unknown values $\{w_i\}$.

On $(3.5, \infty)$ we should prefer to transform our infinite interval onto (0, 1). Two obvious choices for the new independent variable are

$$u = e^{-(x-3.5)}$$

and

$$u = 4(x + 0.5)^{-1} \qquad \textbf{(6.15)}$$

For this problem they are about equally effective. We continue our example using the reciprocal (6.15) transformation. Our differential equation now becomes

$$\frac{u^4}{16}w'' + \frac{u^3}{8}w' - (1 - u)w = 0$$

or

$$u^2w'' + 2uw' - \frac{16}{u^2}(1 - u)w = 0 \qquad \textbf{(6.16)}$$

Replacing (6.16) on a grid of N points regularly spaced in u we have

$$(j^2 + j)w_{j+1} - \left[2j^2 + \frac{16}{u_j^2}(1 - u_j)\right]w_j + (j^2 - j)w_{j-1} = 0 \qquad \textbf{(6.17)}$$

$$j = 1, 2, \ldots, N - 1 \qquad w_0 = 0$$

being a set of $N - 1$ equations for N unknowns. Note, however, that the numbering of our unknowns $\{w_j\}$ runs in the opposite direction to the $\{w_i\}$ of (6.14) and that w_N is w_M. Thus we are now possessed of $N + M - 2$ equations for $N + M - 1$ unknowns. We need an equation at the joint.

If we diagram the points at which we have replaced the differential equation using only one scale (say, x) as in Figure 6.6, we note that the spacings of the points to the left of the joint are all equal but that the spacings to the right are not. (They are equally spaced in u.) If we express w'' at the joint in terms of the ordinate w_M and the two adjacent ordinates w_{M-1} and w_{N-1}, we may then write an algebraic replacement for the differential equation at the joint. For our example, it is

$$\frac{2}{h+k}\left[\frac{w_{M-1}}{h} - w_M\left(\frac{1}{h}+\frac{1}{k}\right) + \frac{w_{N-1}}{k}\right] = 0$$

or

$$w_{M-1} - (1+\beta)w_M + \beta w_{M-1} = 0 \qquad \beta = \frac{h}{k} \tag{6.18}$$

where k is the spacing between the joint and the next ordinate to the right *expressed in units of* x. Again, for our transformation, the value of k is $4/(N-1)$ while h is merely $3.5/M$.

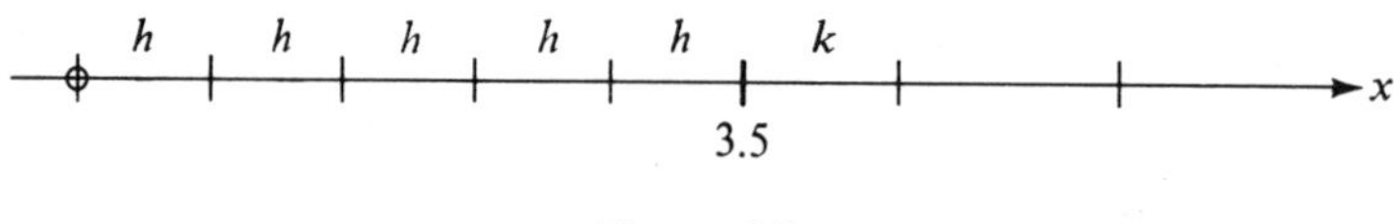

Figure 6.6

An alternative to using the differential equation at the joint is to demand that the solution and its first derivative be continuous across the joint. We may estimate w'_M from the last three ordinates on the left (including the one at the joint), and we can do the same on the right. Equating them, being sure to use the relation

$$\frac{dw}{dx} = \frac{dw}{du}\cdot\frac{du}{dx}$$

will give an equation at the joint that can be used instead of the differential equation replacement. For our example

$$\frac{1}{h}\left(\tfrac{1}{2}w_{M-2} - 2w_{M-1} + \tfrac{3}{2}w_M\right) = \frac{1}{k}\left(\tfrac{3}{2}w_N - 2w_{N-1} + \tfrac{1}{2}w_{N-2}\right)\cdot\left(-\tfrac{1}{4}\right)$$

Here h is $3.5/M$ as before, while k is the uniform spacing of the points in u, that is, $1/N$. Thus we finally have the equation

$$\frac{M}{3.5}(0.5w_{M-2} - 2w_{M-1} + 1.5w_M) + \frac{N}{4}(1.5w_N - 2w_{N-1}) = 0 \tag{6.19}$$

and w_M equal w_N.

For our problem the two approaches to the equation at the joint, using 12 points in each of the two regions, give nearly identical results. The error is about 5 percent. Intuitively, however, your author feels that the use of differential equation at the joint should generally prove superior, since it contributes one additional piece of information about this particular function – information that is lacking in a simple matching of first derivative. (Would it be a good idea to remove the e^{-x} behavior and then treat the remaining differential equation by this algebraic replacement technique? Why?)

The location of the joint, at 3.5, was picked somewhat arbitrarily. A second solving of the problem placing the joint at 2.0 – a round figure that is close to the extreme value of the solution rather than at the inflection point – gives greater accuracy. Also the use of the exponential transformation on the tail gives better results now than does a reciprocal. The error is about 2 percent, again using 12 points in each of the two ranges.

Exponential equations—second order

Second-order differential equations can really only do two things. They can oscillate like a cosine or they can behave like an exponential. To be sure, damping terms occur and are felt, but in any one region a second-order equation is still exponential or oscillatory. The exponential members of this family that arise from physical problems, moreover, tend to be *negative* exponentials, at least as their arguments become large, because this is the only way they can retain values small enough to have physical significance. Thus in the typical boundary-value problem over the range $(0, \infty)$ the equation is usually

$$w'' - L^2 w = 0 \tag{6.20}$$

at least after the independent variable has grown large enough. Furthermore, the right-hand boundary condition is zero – implying that

$$w \sim A\,e^{-Lx} \tag{6.21}$$

if L is essentially constant for large x. The functional form of L for smaller x may force us to numerical integration methods and we would here review the several alternatives and their difficulties.

Fundamentally we face a boundary-value problem. If the differential equation is not linear, the natural strategy of replacing it on a grid by a set of algebraic equations is not attractive, both because of the large range and because the algebraic set will, itself, be nonlinear. In spite of these difficulties this approach will usually be the best one, with our principal efforts going to linearize the algebraic equations to permit iterative solutions. We exhibit this technique on page 172.

The initial-value approach to (6.20) enables us to forget about the nonlinearities but exposes us to the unpleasant fact that the differential equation also has a positive exponential solution for large x. Thus any stepping method that will succeed over long distances must go from right to left – that is, must go in the direction in which the desired solution is increasing and the unwanted solution is decreasing. And, of course, any initial-value approach is inherently iterative, since we are trying to hit the distant boundary condition having guessed a second initial condition at the other end.

Both of these general strategies also suffer from the common difficulty that the solution is somewhat nonpolynomic, e^{-Lx} being very badly fitted by a parabola. We would be wise therefore to remove this known asymptotic behavior explicitly, leaving a hopefully more attractive function to integrate. The Riccati transformation, in regions where w does not cross the axis, is probably the simplest way to deal with the exponential behavior in homogeneous equations. We have

$$u = \frac{w'}{w} \tag{6.22}$$

and (6.20) becomes

$$u' + u^2 = L^2 \tag{6.23}$$

The solution of (6.23) that corresponds to the negative exponential solution of (6.20) now lies near $-L$. The other exponential solution is effectively banished by the Riccati transformation to a distant region, u equal to $+L$, where it cannot interfere. To the extent that the solution of (6.20) is e^{-Lx}, the solution to (6.23) is a constant. Parabolas and stepping integration methods generally are quite good at approximating constants.

Nonhomogeneous second-order exponential equations

When we have a linear second-order exponential equation with a nonhomogeneous term that tends to zero as x becomes large, Hartree [1958] and others recommend a stepping technique that succeeds in separating the two exponentials, although it leaves us following one of them, presumably with a process that assumes parabolas. We have

$$w'' - L^2 w = g(x) \tag{6.24}$$

with both g and w approaching zero as x becomes large. The central idea is to define another intermediate variable, v, by the equation

$$w' + Lw = v \tag{6.25}$$

Since, in this application, L is constant, the additional equation

$$v' - Lv = g \tag{6.26}$$

implies (6.24) and we see that v is a negative exponential. Accordingly, v may be integrated in from infinity, from right to left, and this integration will be stable. When the left-hand boundary is reached, the value of v may be used together with the value of w there to produce the initial condition for integrating (6.25). This second integration proceeds from left to right in the direction of increasing x and, since the unwanted complementary function is e^{-Lx}, no serious integration problems arise. The strategem is another example of balancing out the unwanted exponential while allowing the desired solution to come through. Since the problem is, however, linear, the device seems inferior to an algebraic replacement of the entire equation on a suitably transformed scale.

Linear versus nonlinear systems

The illustrative equation in the previous sections was linear, and thus we had a linear tridiagonal system of algebraic equations to solve. For such systems our difficulties consist largely in trying to choose the efficient method from among a large number of rather well-behaved methods. Not so when the differential equation is nonlinear! Then the algebraic system also consists of nonlinear equations, and we plunge from a wealth of feasible methods to a veritable pauper's smidgin. Differential equations can be linear in only one way, but they can be nonlinear in many. Since their nonlinearity transmits itself directly into the algebraic system, the large variety prevents any comprehensive treatment. We can only suggest approaches – including a return to an initial-value formulation, where nonlinearity was of no particular import. Let us, pray, consider an example.

The equation system

$$\begin{aligned} &y'' - y^2(1 - 0.2 \sin 2x) = 0 \\ &y(0) = 1 \\ &y(3) = 0 \end{aligned} \tag{6.27}$$

is, in your author's terminology, *mildly* nonlinear. Mildly, because the boundary conditions and the differential equation conspire to keep the dependent variable y on the range (0, 1), where squaring it will not cause any serious change in its behavior – just a mild warping. Thus we expect the solution to (6.27) to resemble qualitatively the much simpler linear system

$$\begin{aligned} &y'' - y = 0 \\ &y(0) = 1 \\ &y(3) = 0 \end{aligned}$$

which has the shape shown in Figure 6.7, made up of a linear combination of the positive and negative exponentials, but being primarily e^{-x}. We don't really expect the squaring of y to alter the behavior of the solution very much, nor can the term (0.2 sin $2x$) have serious effects, coming only as a perturbation on unity

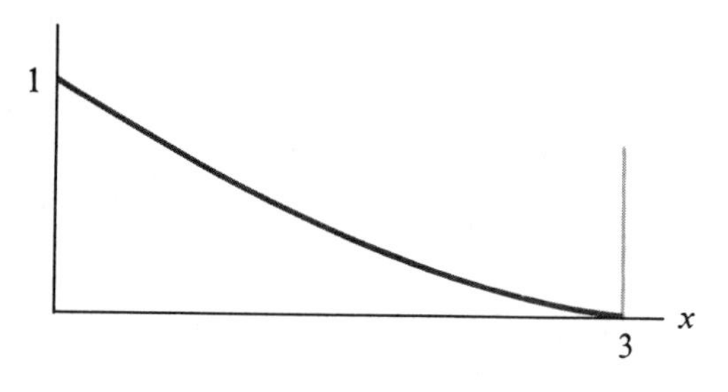

Figure 6.7

with a maximum possible size of 0.2. Thus we have a good feel for the shape of our solution and are prepared to deal boldly with nonlinearities that may block our path. Inserting five interior ordinates, our typical algebraic system is

$$-y_{n-1} + [2 + h^2(1 - 0.2 \sin 2x_n)y_n] \cdot y_n = y_{n+1} = 0 \qquad \textbf{(6.28)}$$

and we see that the $y_n{}^2$ term appears only with h^2, keeping its effect small. We shall solve this system in two ways, each reducing it to a linear system. The cruder method is to substitute an estimated solution, y_{n0}, for the innermost y_n, then solve the resulting linear equations to get the estimated solution—which in turn is substituted for the innermost y_n for the system to be solved again. The iteration, starting with e^{-x} as our initial guesses, gives

x	$y_0 = e^{-x}$	y_1	y_2	y_3	y_4	y_5	y_6
0.5	0.6065	0.71775	0.69171	0.69784	0.69636	0.69672	0.69663
1.0	0.3679	0.52602	0.48665	0.49604	0.49375	0.49431	0.49417
1.5	0.2231	0.37387	0.33394	0.34362	0.34124	0.34182	0.34168
2.0	0.1353	0.24199	0.21157	0.21907	0.21722	0.21767	0.21756
2.5	0.0821	0.11953	0.10393	0.10798	0.10689	0.10713	0.10707

The more sophisticated linearization depends upon replacing $y_n{}^2$ by a linear function of y_n—by the first two terms of its Taylor series expanded about our estimated value y_{n0}. Thus we substitute

$$y_n{}^2 \leftarrow y_{n0}^2 + (y_n - y_{n0}) \cdot 2y_{n0} = 2y_{n0}y_n - y_{n0}^2$$

into (6.27) to give

$$-y_{n-1} + [2 + 2y_{n0}h^2(1 - 0.2 \sin 2x_n)] \cdot y_n - y_{n+1} = y_{n0}^2 h^2 \cdot (1 - 0.2 \sin 2x_n) \tag{6.29}$$

We now have small elements on the right-hand side of our equation. Again using e^{-x} as our first estimate to the solution of (6.3), we get

y_1	y_2	y_3 (converged)
0.70049	0.69666	0.69664
0.50117	0.49422	0.49420
0.34960	0.34174	0.34171
0.22383	0.21761	0.21758
0.11022	0.10710	0.10708

and the convergence is gratifyingly faster. This replacement of nonlinear terms by their linear Taylor's series representation expanded about an estimated solution is the most effective *general* technique with systems of nonlinear equations. It should always be tried first unless some striking special feature of the system cries for a corresponding special method.

A much cruder method (not recommended) is simply to replace y_n^2 by y_{n0}^2, thereby moving the entire nonlinear term to the right side as a constant. We get no compensatory advantages for the crudity of our linearization, and the convergence will be slower. (For the hand computor, this remark is not entirely true. The matrix of these cruder equations, for this particular example, has a very simple form whose inverse is well known a priori. Hence no inversion of the matrix is needed. But arguments based on the characteristics of a particular example are at least suspect.)

Intelligent initial-value methods

Although we initially deprecated treating a boundary-value problem by initial-value methods, the relative advantage of the algebraic replacement technique for *strongly* nonlinear problems is far from clear. It is easy enough to replace any differential equation on a grid of points by a set of algebraic equations, but a strongly nonlinear set of algebraic equations causes no great joy to leap in the heart of a numerical analyst. The number of effective methods for their solution is distressingly small. The example in the preceding paragraph had mild enough nonlinearity to permit linearization of the equations—leading immediately to their effective solution. If no such strategem will work, then we shall be strongly tempted to fall back to our initial-value position.

For the simplest boundary-value problem, the second-order equation, we are given a boundary condition at each end of an interval. To treat it as an initial-value problem we need only assume a second boundary condition at one end, integrate via standard predictor-corrector techniques until we get to the other end, and there compare the value that we obtained with that which we desire. We can then make an adjustment in our assumed initial value, integrate a second time, and compare a second time. Our third choice for the unknown initial condition can be guided by an interpolative philosophy based on the two values we have already found at the far end of the range. The process is strictly brute force, contains no particular booby traps, and will ultimately succeed. On the negative side, it is a lot of work.

We integrated our equation (6.27) seven times with the values of $y'(0)$ in the table. The second column shows the corresponding values of $y(3)$ that we thereby obtained. The first two runs were made with guessed values for $y'(0)$. Thereafter linear interpolation on the two most recent runs was used to get the next initial condition. The last two runs, which bracket the correct boundary condition at 3 and are fairly close to it, gave the ordinates 0.34100 and 0.3374, respectively, for $y(1.5)$, showing that our algebraic replacement techniques above gave values somewhat too large there. [When run at half the interval the algebraic solution gave 0.3386 for $y(1.5)$.]

$y'(0)$	$y(3)$
−1.00	−0.78736
−0.92126	−0.58662
−0.69117	1.14802
−0.84345	−0.24776
−0.81642	−0.08492
−0.80232	0.01146
−0.80400	−0.00044

In its crudest form, just outlined, our repeated marching has been guided by an interpolation philosophy that is essentially False Position. The final value we obtain each time, $y(b)$, is a continuous function of the initial value we have assumed, $y'(a)$. In Figure 6.8 we show a plot of two final values coming from two initial-value integrations from two assumed values for the derivative. The False-Position philosophy clearly points to $y_2'(a)$ for our next trial. Since this trial will

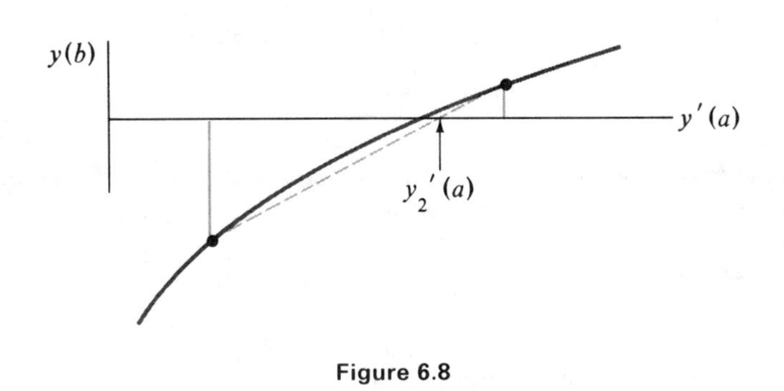

Figure 6.8

undoubtedly not be correct, we are then in a position to use higher-order interpolation in picking our next starting value—but our iterations would still seem to be linearly convergent. We might do better if we were able to bring Newton's method to bear on the problem of choosing the next trial initial value. To do so we obviously need the derivative of $y(b)$ with respect to $y'(a)$.

Let us consider our problem in the form

$$\begin{aligned} \dot{x} &= f(x, y, t) \qquad x(a) = A \\ \dot{y} &= g(x, y, t) \qquad y(b) = B \end{aligned} \tag{6.30}$$

In principle, we know that the solutions of the differential equations (6.30) are determined by (x_a, y_a)—the values of x and y at t equal to a—but only x_a is known to us. If we alter x_a or y_a, we change x and y everywhere, thus

$$x = x(x_a, y_a, t) \qquad \text{and} \qquad y = y(x_a, y_a, t)$$

Now define two new variables

$$\xi = \frac{\partial x}{\partial y_a}$$

$$\eta = \frac{\partial y}{\partial y_a}$$

and examine our differential equations as we alter one of the boundary conditions, y_a. Differentiating (6.30) with respect to y_a we obtain

$$\begin{aligned} \dot{\xi} &= \frac{\partial f}{\partial x} \cdot \frac{\partial x}{\partial y_a} + \frac{\partial f}{\partial y} \cdot \frac{\partial y}{\partial y_a} = f_x \cdot \xi + f_y \cdot \eta \\ \dot{\eta} &= \frac{\partial g}{\partial x} \cdot \frac{\partial x}{\partial y_a} + \frac{\partial g}{\partial y} \cdot \frac{\partial y}{\partial y_a} = g_x \cdot \xi + g_y \cdot \eta \end{aligned} \tag{6.31}$$

and we observe that if we change y_a but not x_a, then, at t equal to a, we have

$$\xi = 0 \qquad \text{and} \qquad \eta = 1$$

Consider one march: We assume a y_a at t equal to a and march to some value of y at b which we may call $Y(y_a)$. Presumably we are not lucky enough for it to be the value B that we desire. Newton's method says that a new better guess for y_a is given by

$$y_a - \frac{Y(y_a) - B}{Y'(y_a)} \tag{6.32}$$

both Y and Y' being evaluated at b. But

$$Y'(y_a) = \frac{\partial y}{\partial y_a} = \eta(b)$$

Since equations (6.31) are a pair of ordinary differential equations for ξ and η with known initial conditions, they can be integrated jointly with (6.30) to give $\eta(b)$, while the principal equations (6.30) give $Y(b)$. We thus have the essential numbers available to apply Newton's method (6.32) in choosing our next trial value of y_a.

When we run our standard problem by this technique we get the values for $y'(0)$ and $y(3)$ shown in the table. This time, the value for $y(1.5)$ was 0.33756, which may be presumed accurate. The early oscillations were wilder than with simple linear interpolation, but once convergence set in it was quadratic. This behavior is, of course, characteristic of Newton's method.

$y'(0)$	$y(3)$
−1.00000	−0.78736
−0.57761	3.48919
−0.69745	1.06114
−0.77567	0.21820
−0.80163	0.01646
−0.80392	0.00012
−0.80394	0.000000068

Problems

The differential equation

$$w'' - (1 + e^{-x})w = 0$$

with the boundary conditions

$$w(0) = 1 \qquad \text{and} \qquad w(\infty) = 0$$

has the analytic solution

$$w = 1.4514874\, I_2(2e^{-x/2})$$

where I_2 is one of the Bessel functions. Solve this boundary-value problem numerically in several ways, comparing your results with the analytic solution.

1. Direct replacement on $(0, 1)$ and $(1, \infty)$ by algebraic systems.

2. Direct replacement of the corresponding Riccati equation

$$\eta' + \eta^2 = 1 + e^{-x}$$

taking the *negative* value of η. Then use

$$\ln w = \int \eta \cdot dx$$

to get the original variable.

3. Initial-value integration of the Riccati equation in from infinity, where η is -1.

4. Use the energy equation in two different forms:

a. $$w' = -\left\{(1 - e^{-x})w - \int_x^\infty w^2 e^{-x} \cdot dx\right\}^{1/2}.$$

b. $$w^2 = \frac{1}{1 + e^{-x}}\left\{(w')^2 + \int_x^\infty w^2 e^{-x} \cdot dx\right\}.$$

In part *a* the strategy should be (1) assume $w(x)$; (2) compute in from ∞: w', w; and (3) normalize w and go to step (2). In part *b* the strategy is similar, though w' will be found by differentiation. One of the two methods should have stability problems.

CHAPTER 7

STRATEGY VERSUS TACTICS

When we first approach a numerical problem we must take care lest we become immersed too soon in the detail without having thought through clearly the general strategy of our attack. There is, of course, the alternative danger that we will concern ourselves exclusively with the grand design while ignoring inconvenient details that turn out to be decisive—thereby wrecking our plans with a thoroughness that borders on the catastrophic. This second danger, however, seems less severe than the first, against which we must accordingly warn the more loudly.

In this chapter we try to find *roots of polynomials*, and our concern is to select from among the multitude of published techniques a reasonable *strategy* when an automatic computer is to be used. Clearly a balance must be struck: We cannot decide which techniques fit well enough together to provide a reasonable chance for finding the more elusive roots of polynomials if we are not already familiar with the techniques themselves. At the same time, a fascination with the neatness and beauty of a particular method must not blind us to its weaknesses and thereby cause us to leave our flank exposed by a total commitment to the one beautiful, but inadequate, method. In making our principal points we spend rather more time on polynomial roots than the subject deserves, but our excuse is the expository one. Eigenvalues of matrices or linear simultaneous algebraic equations could have been used equally well, but we need only one subject in which to indicate the wealth of detail available in the literature, thereby emphasizing the necessity for keeping clearly in mind the objective of the computation so as to choose wisely from among the uncritical offerings of generations of numerical analysts.

Real roots come first

When we are given the polynomial $P_n(x)$ and told to seek its roots, our first reaction is to plot it. If it crosses the x axis at some points, those points are the real roots. We may find them to the desired precision, divide them out of the polynomial, and then tackle the smaller problem of finding the complex root pairs of the reduced polynomial. This general strategy of *finding and removing the real roots first* is valid both because real roots are easier to find than are complex roots and because it is easier to find roots of a lower-degree polynomial than to find them for one of higher degree. Roots tend to interfere with one another, so that as we remove them the remaining roots become more tractable. Naturally we want to remove the easiest ones first.

Another reason for plotting a polynomial is that its shape near a root frequently suggests the expedient method for extracting it. The graph also shows how many complex roots there are and often gives a rough idea about where they lie. But we must be careful here, for these insights are *geometric*; they arise from patterns in two dimensions, and machines are very poor at perceiving patterns. Wc must cxamine our human-oriented hand techniques carefully to see those parts of them that are useful to a machine.

Plotting a polynomial can easily be done by a machine. It need only evaluate $P_n(x)$ at regularly spaced values of x. If the machine can afford to wait while a human looks at the plot and then directs it how to proceed, we have a very powerful method for finding roots – especially if the intervener is an experienced person. But human intervention is seldom practical, so we assume that, as a matter of policy, we are going to do without it. (If someone wishes to build an interventionist option into his program, that is his business and could, under some circumstances, be very useful.)

The binary chop

Even without humans, some sort of systematic search is still very valuable. If the machine should evaluate a polynomial at a regular sequence of points along the x axis and if it should find that the value of the polynomial changes sign between two such points, then at least one real root has been trapped between these two values of x. The interval may then be cut in half by evaluating the polynomial at the obvious place, and a test of signs will locate a root in one of the two half-intervals. The process may be continued to any desired degree of precision, although other processes will often be more efficient, if less safe. This method of dividing in half an interval known to contain a root is called a *binary chop*, since it increases the precision to which the root is known by a factor of 2, or one binary digit, at each iteration (and because it "chops" the interval in half each time). It would seldom be tolerated by a human using hand techniques because of its tedious slowness, but it does lead to a machine program whose simplicity often tempts the programmer to stop here and let the machine churn away to an answer. He thereby shirks his responsibilities for efficient machine usage, but who will find out? (Anybody who bothers to examine the program will find out—but these people are few.)

Newton's method for an isolated real root

If we draw the graph of a polynomial with an isolated real root, the picture suggests a better approach (Figure 7.1). We can evaluate the polynomial *and its derivative* at a point x_0 near the root and then slide down the tangent line

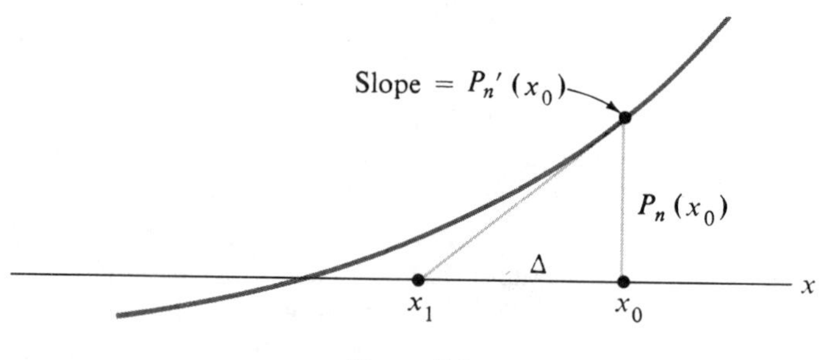

Figure 7.1

to the x axis to get a new point x_1 which is close to the root and hence is a better approximation to it. This simple idea, already known to us as Newton's method, is certainly one of the most effective techniques for finding isolated roots of polynomials, as it is for finding any isolated root of any transcendental equation whose derivative is expediently calculable. Starting with an x_0, we need only compute

$$x_1 = x_0 - \frac{P_n(x_0)}{P_n'(x_0)} = x_0 - \Delta \tag{7.1}$$

then repeat the calculation using x_1 in place of x_0. If the process converges, convergence is quadratic—that is, the number of correct significant figures roughly doubles each time. The only difficulties are the standard ones: Is the root isolated and are we near enough to it with our x_0? If we have managed to plot $P_n(x)$ against x, then we usually have a clear enough idea of the geometry to permit direct application of Newton's method to good effect. Automatic computers, however, never have that plot before them and, unless we take considerable precautions, we shall frequently find that they are trying to apply Newton's method where he is most unreliable indeed. Considering Figure 7.2, if we are so

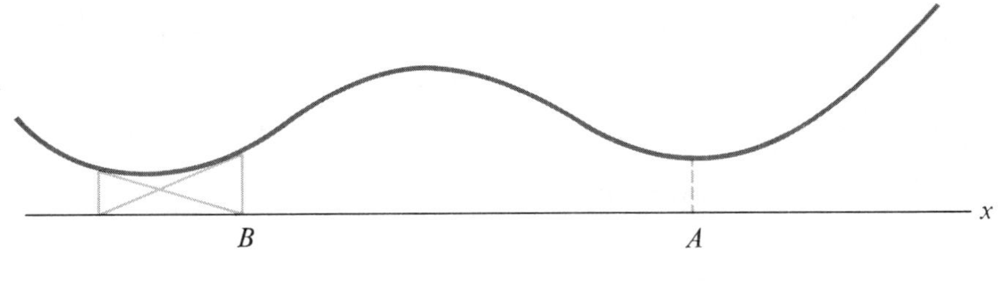

Figure 7.2

unfortunate as to choose point A, we will promptly divide by the zero slope, and point B will only lead to an endless cycle of operations that can waste valuable minutes of machine time before somebody notices that answers don't seem to be coming out. It is possible to build safeguards against these behaviors, but to do so increases the complexity of our program to the point where we begin to wonder if slower but less nervous methods might not be preferable. The argument that we are not *likely* to land exactly on A is true but is of no help, for a turning point is necessarily quite flat for some distance, and to land *nearly* on A is *almost* as bad as *actually* landing on it. If there are extrema around, we have a very high probability of landing near one unless considerable precautions are taken.

Polynomial evaluation techniques—synthetic division

Turning to computational detail for awhile, we consider how best to evaluate a polynomial and probably a derivative or two as well. Here the machine technique coincides with the hand technique that is already familiar from school. We rewrite our polynomial

$$a_4x^4 + a_3x^3 + a_2x^2 + a_1x + a_0$$

as

$$\{[(a_4x + a_3)x + a_2]x + a_1\}x + a_0 \tag{7.2}$$

and, for a particular x, evaluate outward starting from the inside. For the quartic, we need four multiplications and four additions. The computation that first forms the several powers of x, then multiplies by the four coefficients and adds, is appreciably longer—*and is less systematic*. Thus, on both counts, the inside-out evaluation wins.

This inside-out evaluation may be conveniently arranged in a form that is usually taught under the name *synthetic division*. The coefficients are written in descending order, supplying zeros if necessary, and the value of x is written somewhere—often at the right or left end. We continue the explanation with a specific example. Evaluate, for x equal to 3, the polynomial

$$2x^4 - 3x^3 + x - 2 \tag{7.3}$$

We write

$$\begin{array}{c|ccccc} & 2 & -3 & 0 & 1 & -2 \\ 3 & z & \nearrow 6 & \nearrow 9 & \nearrow 27 & \nearrow 84 \\ \hline & 2 & 3 & 9 & 28 & \boxed{82} \end{array} \tag{7.4}$$

and assert that the value is 82. The number at the bottom of each column is the sum of the numbers above. The diagonal arrows indicate multiplication of the bottom number by x to give the number in the second row, one column farther to the right. (We may consider the element z of this second row to be given as zero if we wish our operations to be completely systematic.) Comparing these observations with our polynomial, written

$$\{[(2 \cdot x - 3)x + 0]x + 1\}x - 2 \tag{7.5}$$

we see that the leftmost arrow is the innermost multiplication $(2 \cdot x)$, that -3 is then added to this product, that this sum is next multiplied by x, and so on. Thus the numerical arrangement (7.4) is simply a convenient hand-computational device for carrying out the arithmetic that is implied by the algebraic form (7.5).

The name synthetic division comes from another fact, not yet mentioned. If we divide a polynomial, $P_n(x)$, by a linear factor, $(x - b)$, we will generate a quotient polynomial $Q_{n-1}(x)$ of degree $n - 1$, and a constant remainder, R. This statement may be rewritten more compactly as

$$P_n(x) = (x - b) \cdot Q_{n-1}(x) + R \tag{7.6}$$

and it is true for all values of b and x. In particular (7.6) remains true when x equals b, and, on substituting, we find

$$P_n(b) = R$$

That is, the value of $P_n(x)$ when x is set equal to b is exactly the value of the *remainder* when $P_n(x)$ is divided by $(x - b)$. [The student should carry out the long division for the example if he has uneasy feelings about this proof. He will see that exactly the same arithmetic is performed as in (7.4) but that he carries along a lot of powers of x which serve only to keep the coefficients lined up properly.] In addition to evaluating the remainder, the bottom line of (7.4) contains the coefficients of $Q_{n-1}(x)$. Thus $2x^4 - 3x^3 + x - 2$ divided by $(x - 3)$ gives a quotient $2x^3 + 3x^2 + 9x + 28$, as well as a remainder of 82.

We can also find the derivative of $P_n(x)$, evaluated at b, by a continuation of the process. We merely repeat our synthetic division, using the quotient polynomial, the remainder of that division being the desired derivative. The proof is straightforward. We start with (7.6) and observe that we may write

$$Q_{n-1}(x) = (x - b) \cdot S_{n-2}(x) + T \tag{7.7}$$

Then, on substituting in (7.6) we get

$$P_n(x) = (x - b)^2 \cdot S_{n-2}(x) + (x - b) \cdot T + R \tag{7.8}$$

If we set x equal to b in (7.8), we again get the familiar fact that $P_n(b)$ is R, but if we first differentiate (7.8) with respect to x we have

$$P_n'(x) = 2(x - b) \cdot S_{n-2}(x) + (x - b)^2 \cdot S_{n-2}'(x) + T$$

and on setting x equal to b we find

$$P_n'(b) = T$$

Thus the value of the derivative is the remainder after dividing $Q_{n-1}(x)$ by $x - b$. The complete computation would appear as

	2	−3	0	1	−2
3	0	6	9	27	84
	2	3	9	28	82
3	0	6	27	108	
	2	9	36	136	

so $P_n'(3)$ is 136 and $S_2(x)$ is $2x^2 + 9x + 36$. [Check $P_n'(3)$ directly by formal differentiation and substitution.] The values of $P''(b)$ and higher derivatives are available by continuation of this process, but numerical factors begin to appear. The curious student should investigate to find out what factors appear and why, verifying his conclusions by specific example.

Searching along the real axis

By this time the student should have realized that Newton's method by itself will not suffice. It is a nervous method, full of idiosyncrasies and quirks that make it undependable except near an isolated root – where it becomes a very good method indeed. It is rather like the little girl of the nursery rhyme who had a curl in the middle of her forehead and

> "When she was good she was very very good
> But when she was bad she was horrid!"

We must have some method for finding, at least roughly, the location of these roots before calling on Newton to help. The most direct idea that keeps recurring is to step in methodical, pedestrian fashion along the real axis at some regular interval, evaluating $P_n(x)$ at each point while looking for a change in sign, which must indicate the presence of a root. But what size step? Consider Figure 7.3,

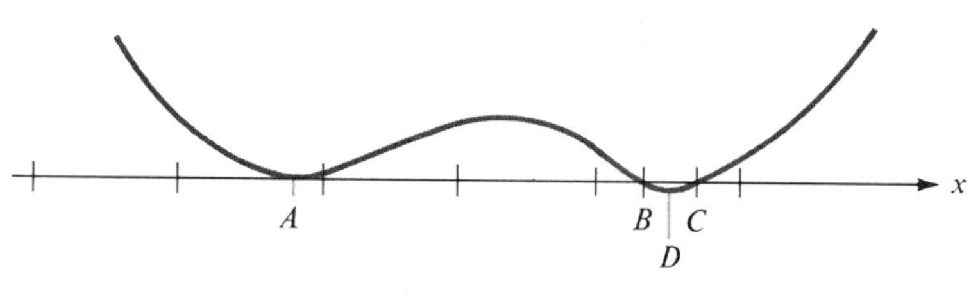

Figure 7.3

in which none of the four real roots would be detected by stepping along at the interval shown. Of course, if P_n' were also being calculated (not unreasonable, since it will be needed as soon as Newton is involved), then the search routine could become suspicious as it notes the passage of a minimum, and it might refine its search in this region to see if the minimum actually is low enough to cut the axis. But already our search is ceasing to be simple.

Another question concerns the range of the real axis to be searched. How far should we step to make sure that we have passed all the real roots? Do we march from -25 to $+25$ or -1000 to $+1000$? (That last search, at steps of 0.1 or even at steps of 1.0, is going to take even a computer a noticeable time.) Surely no such range is necessary for our sample quartic. Its coefficients are all human-sized numbers, roughly between ± 10, and we instinctively feel that the roots can't get too far away. Feelings, however, are not sufficient! We need something more precise. We know a lot about polynomials. We know, for example, that $-a_{n-1}/a_n$ is the sum of the roots, and that $(-1)^n \cdot a_0/a_n$ is their product. Thus if both these numbers are of reasonable size we suspect that the roots are also. But not quite. A cubic with roots of -1000, 0, $+1000$ will have both of these coefficients equal to zero—although the size of the roots will show up in another coefficient. The polynomial is

$$x^3 - 1{,}000{,}000x = 0$$

There are procedures, due to Sturm, which will give the exact number of roots between any two points on the real axis but they are cumbersome and seldom used.

While there are no simple useful bounds on the roots of a polynomial that will strictly delimit the region that has to be searched, one may frequently get an approximation to the largest root by taking the two highest order terms and solving them, on the theory that for large x they generally dominate—a theory that is true only if the largest root is greater than unity. Thus for

$$2x^4 - 3x^3 + x - 2 = 0$$

we would try solving

$$2x^4 - 3x^3 = 0$$

to get

$$x = \tfrac{3}{2} = 1.5$$

for our approximate largest root. (The correct root is 1.56.)

An alternative searching procedure, however, is more attractive. We replace x by $1/y$ in our polynomial

$$2x^4 - 3x^3 + x - 2 = 0$$

which then becomes

$$2\left(\frac{1}{y}\right)^4 - 3\left(\frac{1}{y}\right)^3 + \left(\frac{1}{y}\right) - 2 = 0$$

or

$$2 - 3y + y^3 - 2y^4 = 0$$

and we obtain another polynomial *whose roots are the reciprocals* of the roots of the original polynomial. Thus, if our original polynomial had roots at -4, $-\frac{3}{4}$, 1, 3 (it doesn't, but no matter), then the new polynomial would have roots at $-\frac{1}{4}$, $-\frac{4}{3}$, 1, $\frac{1}{3}$. In particular, all the roots of the polynomial in x that lie *outside* the range ± 1 now lie *inside* it for the polynomial in y, while those that were inside now fall outside. Thus we need only search for real roots from -1 to $+1$ for the original polynomial, then form the reciprocal polynomial and search for its roots in the same region. Further, the reciprocal polynomial is easy to form, as it has exactly the same coefficients as the original, but in the reverse order. If we consider a computer program we see that the same search routine can be used for both polynomials and that only a little administrative machinery need be added to reverse the coefficients and call the search for a second time. (But don't reverse them again and call the search a third time!) In terms of both computational efficiency and computational aesthetics we prefer this procedure.

Double roots—and higher forms of trouble

When the graph of our polynomial looks like Figure 7.3 we may expect trouble. There is a double real root at A and the roots at B and C are close together so that they interfere with the use of Newton's method, unless we are careful. If we *know* that Figure 7.3 describes our problem then the problem shrinks from a big to a little one. We can easily devise special methods for finding these roots—methods that turn the geometric facts from liabilities into assets. We only have to *know* that it is a nearly double root that we seek.

Suppose we do know. We may then seek to find the minima of the curve, that is, the roots of $P_n'(x) = 0$, which are isolated and hence easy to find by Newton's method. [Draw the graph of $P_n'(x)$.] If we expand $P_n(x)$ about one of these minima, say D, we have

$$P_n(x) = P_n(D) + \frac{\Delta^2}{2} \cdot P_n''(D) + \text{higher terms}$$

where the term containing Δ to the first power disappears because $P_n'(D)$ is zero, by definition of D. We now ask that $P_n(x)$ be zero and find

$$0 \simeq P_n(D) + \frac{\Delta^2}{2} \cdot P_n''$$

or

$$\Delta = \sqrt{\frac{-2 \cdot P_n(D)}{P_n''(D)}}$$

and we obtain the approximations for the roots B and C as $D - \Delta$ and $D + \Delta$, respectively. These approximations may be good enough or they may each be further refined by Newton's method—since we are already close to the root in comparison to the distance, 2Δ, to the next root. Thus, on this scale, each root is now "isolated" and the ordinary Newton's method will work quite well. The double root at A is trivially obtained, being discovered when we evaluate $P_n(D)$.

Our difficulties arise from two places: We seldom know that we are faced with a double root, or else we have higher multiplicities whose degree is unknown. To know the multiplicity of a root is to suggest the method for extraction, for we can use the Taylor series expansion to the proper number of terms as we did above for double roots. *Nearly* multiple roots are harder, but the technique above can be generalized to give a polynomial of the appropriate degree containing only the nearly multiple roots, which may then be easily solved to give approximate values. These values may then be considered to be separated and hence are refinable by Newton's method.

The serious problem is how to gain the insight so often gleaned from the graph—and there is no really satisfactory answer. The number of configurations is too great for an exhaustive catalogue, and to search by machine for all these possibilities would take more time than it is worth. We must therefore content ourselves with a reasonable analysis designed to reveal the more probable configurations and ask our routine to cry for help when it finds that it needs it.

Laguerre's method

A method due to Laguerre has attracted favorable attention and we set it forth briefly here, warning the reader that we are going to reject it as not sufficiently compatible with our other algorithms to warrant inclusion in a general polynomial system. It is, however, not typical of methods that can be found by leafing idly through the more applied mathematical journals of the last 50 years because it is quite efficient. These other methods are usually advanced by their authors with enthusiasm, but the examples—if any—almost always show the method in its best light. A balanced judgement usually comes only after a large amount of experimentation.

We first note a few useful facts. Writing

$$P_n(x) = (x - x_1)(x - x_2)\cdots(x - x_n) \tag{7.9}$$

then

$$\ln|P_n| = \ln|x - x_1| + \ln|x - x_2| + \cdots \ln|x - x_n|$$

and

$$\frac{d\ln|P_n|}{dx} = \frac{1}{x - x_1} + \frac{1}{x - x_2} + \cdots + \frac{1}{x - x_n} = G = \frac{P'_n}{P_n} \tag{7.10}$$

where we note that the absolute-value signs have disappeared and G may be positive or negative. Further,

$$-\frac{d^2\ln|P_n|}{dx^2} = \frac{1}{(x - x_1)^2} + \frac{1}{(x - x_2)^2} + \cdots + \frac{1}{(x - x_n)^2}$$

$$= H = \left(\frac{P'_n}{P_n}\right)^2 - \frac{P''_n}{P_n} \tag{7.11}$$

Thus H is always positive, if the x_i are real, and may be written

$$H = G^2 - R \qquad \text{where } R = \frac{P''_n}{P_n}$$

Now we make a rather drastic set of assumptions: that the root we seek, x_1, is isolated while the others are bunched at some distance from x_1. Denoting the distances by

$$a = x - x_1$$

$$b = x - x_i \qquad i = 2, 3, \ldots, n$$

we may rewrite (7.10) and (7.11) as

$$\frac{1}{a} + \frac{n-1}{b} = G$$

$$\frac{1}{a^2} + \frac{n-1}{b^2} = H$$

and then we may solve for a as

$$a = \frac{n}{G \pm \sqrt{(n-1) \cdot (nH - G^2)}} \tag{7.12}$$

where the sign is chosen to give the denominator that is larger in absolute value. The iteration then is

1. Choose a trial root, x_0, and evaluate G, P, H, and a.
2. Choose the next trial root from $x = x_0 - a$.
3. Go to step 1 until a is small enough.

Laguerre's method has the possibility of converging to a complex root, since the argument under the square-root sign may be negative. This property can be counted as an asset or as a liability depending on your intent and mode of operation. On an automatic computer, the need to switch all arithmetic to complex numbers on the next cycle of an iteration is a blasted nuisance, though certainly possible. In hand computation, the aggravation is possibly less. For your author, this property of Laguerre is a mild weakness – rather in the category of Newton's predilection for wandering off to East Limbo on encountering a minimum. It makes the inclusion of the algorithm in a general package rather difficult unless we circumscribe it with considerable additional machinery. The wide utility of Newton combines with his simplicity to make us willing to take the extra care. With Laguerre, we are less enchanted, in spite of his undeniable virtues, not the least of which is guaranteed convergence to a real root when all roots are real.

Complex roots

The finding of a complex root involves the simultaneous determination of *two* real numbers. This is harder than finding two real numbers separately, so we must expect to work considerably harder when seeking complex roots than we did when real roots were our quarry. To reduce the work as much as possible, we always try to remove all the real roots first, reducing the polynomial to an even degree, before attempting the complex roots.

Complex roots always occur in conjugate pairs (if the coefficients of our polynomial are real) and thus our polynomial has *real* quadratic factors. If $a \pm bi$ is a root pair, then the factors are

$$[x - (a - bi)] \cdot [x - (a + bi)] = x^2 - 2ax + (a^2 + b^2)$$

and we see that the second form has only real coefficients. We are at once confronted with a major strategic decision: Shall we seek the linear factor

$$x - (a + bi)$$

or the quadratic factor

$$x^2 - 2ax + (a^2 + b^2)?$$

The first choice commits us to arithmetic with complex numbers, while the second choice retains real numbers at the price of seeking a more complicated kind of factor. There is no clear economy in one choice over the other, but human psychology seems to prefer the second—avoiding complex numbers in normal computation as long as possible. Since computers do not care, we are left with no decision and, accordingly, shall describe methods for both strategies.

Newton's method was derived from the Taylor expansion by retaining linear terms. There is nothing in that derivation which demands the variable to be real, and it is quite valid for complex numbers. Thus we may choose our first approximation to an isolated root of the polynomial $P_n(z)$ to be complex

$$z_0 = a_0 + ib_0$$

and then

$$z_1 = z_0 - \frac{P_n(z_0)}{P_n'(z_0)}$$

where P and P' will now also generally be complex. The same difficulties with zero derivatives still plague here and the evaluation of P and P' will be about four times as laborious as before because of the complex arithmetic (although there are some tricks to reduce this a bit). The basic concepts are simple and the nature of the difficulties only becomes apparent if we draw pictures of the polynomial over the complex plane. (It takes two pictures now: one for the real part and one for the imaginary – or, better, one for the amplitude and one for the phase, the latter often being ignorable.) The difficulties show up as the rather large number of points where the surfaces are level, giving zero derivatives to divide by. Please also note that initial approximations for Newton *must* be complex, since a real value for x_0 can only lead to real values for P, P' and x_1.

Division of one polynomial by another

A general technique for dividing a polynomial by another polynomial of lesser degree is clearly one of the necessary tools when seeking quadratic factors. At the very least we must be able to try out a proposed quadratic factor to see what the remainder looks like. The algorithm is quite straightforward, though its description is a bit cumbersome. We describe it as it is carried out in hand calculation, leaving its transferral to machine to the reader.

We write the coefficients of the polynomial horizontally, in descending order, just as for synthetic division, and we put the coefficients of the divisor down the side, with the coefficient of the highest power of x normalized to unity, *but with the signs of all the coefficients reversed.* Thus to divide

$$2x^4 - 4x^3 + x^2 + 3$$

by $x^3 - 2x^2 - x + 1$ we would write

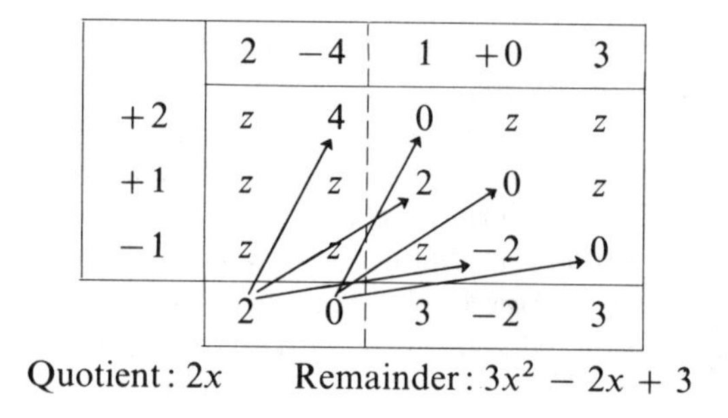

	2	−4	1	+0	3
+2	z	4	0	z	z
+1	z	z	2	0	z
−1	z	z	z	−2	0
	2	0	3	−2	3

Quotient: $2x$ Remainder: $3x^2 - 2x + 3$

and would obtain the figures shown on the left and across the top of our diagram. The figures at the bottom are merely the *sums* of columns. The diagonal arrows point to figures obtained by multiplying the figures at the bottom by the figures on the left – each row yielding a product that is placed one column farther to the right than that from the row above.

The rules are

1. Add the column.
2. Multiply this sum by each of the numbers on the left, starting with the uppermost, placing the products in successive columns to the right on the corresponding rows.

3. Go to 1 – but omit step 2 after the quotient has been computed (it will be of degree *one* in our example, hence will consist of two numbers). The dashed vertical line shows the limit of step 2.

There will be two triangles of blanks, as shown by the z's.

Consider a second example; the same quartic as before, but the divisor is the quadratic $(x^2 - 3x + 4)$.

	2	−4	1	0	3
3	—	6	6	−3	—
−4	—	—	−8	−8	4
	2	2	−1	−11	7

Quotient: $2x^2 + 2x - 1$ Remainder: $-11x + 7$

When our divisor is linear, the algorithm reduces to the familiar synthetic division.

Backward division

The polynomial $x^3 - 2x^2 - x + 1$ may be divided into $2x^4 - 9x^3 + x^2 + 3$ in a second way: we may start by dividing the *constant terms*, instead of the highest coefficients. Our remainder will then be a polynomial of degree equal to that of the dividend, and we probably have never had any use for this kind of long division – a sort of *backward division* – in our previous experience. It turns out to be useful later in this chapter, however, so we demonstrate it here. We arrange our computation as before, except that the divisor with reversed signs is now written down the right side, the constant term is normalized to unity, and the work proceeds from right to left. We obtain

x^4	x^3	x^2	x^1	x^0		
2	−4	1	0	3		
—	—	3	3	—	1	x^1
—	6	6	—	—	2	x^2
−3	−3	—	—	—	−1	x^3
−1	−1	10	3	3		

Thus

$$\frac{3 + x^2 - 4x^3 + 2x^4}{1 - x - 2x^2 + x^3} = 3 + 3x$$

with a remainder of

$$10x^2 - x^3 - x^4$$

(The power of x associated with each column remains as before and appears at the top of the diagram as a reminder. The powers of the divisor are in the opposite order from the forward division and are written at the right.)

Evaluation of a polynomial for complex arguments

The amount of arithmetic necessary to evaluate $P_n(z)$ for complex z may be reduced by constructing a real quadratic factor and noting that we may write

$$P_n(z) = (z - r) \cdot (z - \bar{r}) \cdot Q_{n-2}(z) + R_1(z)$$

where R_1 is linear in z. ($\bar{r}$ is the conjugate of r.) Letting z equal r we have

$$P_n(r) = R_1(r)$$

so that we need only divide P_n by the real quadratic factor, obtain the linear remainder, and substitute r—which is complex—into this R_1 instead of into P_n. A similar, though more complicated, algorithm is available for $P_n'(r)$, which we leave for the reader to discover. The division by the quadratic factor may be carried out in the standard way, using the formal division algorithm: To evaluate $x^4 + 4z^2 - 8z + 1$ at $z = 1 + 2i$ we first form

$$[z - (1 + 2i)] \cdot [z - (1 - 2i)] = (z^2 - 2z + 5)$$

and divide

	1	0	4	−8	1
2	—	2	4	6	—
−5	—	—	−5	−10	−15
	1	2	3	−12	−14

so that

$$R(1 + 2i) = -12 \cdot (1 + 2i) - 14 = \boxed{-26 - 24i}$$

The same result can be obtained from straight synthetic division by the value $(1 + 2i)$ as follows:

	1	0	4	-8	1
$1 + 2i$	z	$1 + 2i$	$-3 + 4i$	$-7 + 6i$	$-27 - 24i$
	1	$1 + 2i$	$1 + 4i$	$-15 + 6i$	$-26 - 24i$

Quadratic factors by remainder minimization

If we decide to seek quadratic factors one obvious tactic is to try a reasonable one and then look at the remainder to see if perhaps it is zero. If we were not so lucky, the next question is how to alter our trial divisor so as to reduce the size of the remainder. The philosophy of Newton may be employed here, in two dimensions this time, to make the two numbers in the remainder shrink toward zero, but other techniques are also available. Suppose, for example, our quadratic divisor is

$$x^2 + d_1 x + d_0$$

and we pick a pair of (d_0, d_1) values for a first trial. We obtain a remainder R that is not zero. Next we might increase d_1 slightly and try again. We could also try decreasing d_1 slightly. Then, with the original d_1, we could vary d_0 in both directions—again observing the size of R. By this time we have evaluated R at a star of five points in the d_0, d_1 plane and it will be obvious in which direction we must move to decrease $|R|$ (Figure 7.4). When the central point of our star

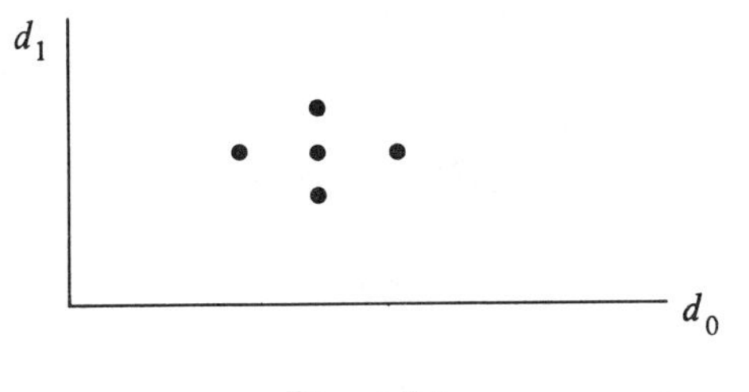

Figure 7.4

has the smallest R, we shrink the size of the star and repeat the calculations. This stepping search in two dimensions is simply the binary chop all over again, and it suffers from the same limitations. It is fairly stable, but it can get trapped in minima that do not involve a root. Worse, it does not improve its efficiency as the root is neared. Thus, in spite of its simplemindedness, we cannot recommend it.

Newton's method for quadratic factors of polynomials

If we divide a polynomial $P(x)$ by a quadratic $(x^2 + bx + c)$ we will usually obtain a linear remainder $(rx + s)$:

$$P(x) = (x^2 + bx + c) \cdot q(x) + rx + s \tag{7.13}$$

In order for our quadratic to be an exact factor of P, we must make r and s become zero. Thus we wish to know how r and s vary with b and c. If we are near a root ($r = s = 0$), then we can slide down a gradient to the bottom. Note that r and s are *not* functions of x but only of b and c. A two-dimensional Taylor's series gives

$$r = r(b_0, c_0) + \Delta b \cdot \left(\frac{\partial r}{\partial b}\right)_0 + \Delta c \cdot \left(\frac{\partial r}{\partial c}\right)_0 + \text{higher terms}$$

$$s = s(b_0, c_0) + \Delta b \cdot \left(\frac{\partial s}{\partial b}\right)_0 + \Delta c \cdot \left(\frac{\partial s}{\partial c}\right)_0 + \text{higher terms}$$

Setting r and s to zero we obtain two equations in the two increments to b and c,

$$\begin{aligned} 0 &= r_0 + \Delta b \cdot \left(\frac{\partial r}{\partial b}\right)_0 + \Delta c \cdot \left(\frac{\partial r}{\partial c}\right)_0 \\ 0 &= s_0 + \Delta b \cdot \left(\frac{\partial s}{\partial b}\right)_0 + \Delta c \cdot \left(\frac{\partial s}{\partial c}\right)_0 \end{aligned} \tag{7.14}$$

Note that x is not involved. As (7.13) implies, we can easily find r_0 and s_0 by a single division of P by $(x^2 + bx + c)$. The four partial derivatives may also be obtained in a similar fashion: Differentiate (7.13) with respect to c:

$$0 = \frac{\partial q}{\partial c} \cdot (x^2 + bx + c) + q(x) + \left(\frac{\partial r}{\partial c} x + \frac{\partial s}{\partial c}\right)$$

Rewriting,

$$-q(x) = (x^2 + bx + c) \cdot \frac{\partial q}{\partial c} + \left(\frac{\partial r}{\partial c} x + \frac{\partial s}{\partial c}\right)$$

and comparing with (7.13) we see that $\partial r/\partial c$ is the coefficient of the linear term in the remainder obtained in dividing $-q$ by $(x^2 + bx + c)$. Similarly, $\partial s/\partial c$ is the constant term of the same remainder. On the other hand, differentiating (7.13) with respect to b gives

$$-x \cdot q(x) = (x^2 + bx + c) \cdot \frac{\partial q}{\partial b} + \left(\frac{\partial r}{\partial b} x + \frac{\partial s}{\partial b}\right) \tag{7.15}$$

which shows $\partial r/\partial b$ and $\partial s/\partial b$ to be the linear and constant terms, respectively, after dividing $-xq$ by our trial factor. Thus we have a systematic procedure:

1. Divide $P(x)$ by our trial quadratic, $x^2 + bx + c$, to produce r_0, s_0 and $q(x)$.
2. Divide $-q(x)$ by our trial quadratic, $x^2 + bx + c$, to produce $\partial r/\partial c$ and $\partial s/\partial c$.
3. Divide $-x \cdot q(x)$ by our trial quadratic, $x^2 + bx + c$, to produce $\partial r/\partial b$ and $\partial s/\partial b$.
4. Solve the equations (7.14) for Δb, Δc.
5. $b_1 = b_0 + \Delta b$, $c_1 = c_0 + \Delta c$.
6. Go to step 1 with $x^2 + b_1 x + c_1$ as our new trial quadratic.

This process is second order in the neighborhood of a root, but—like most Newton processes—it can diverge in spectacular ways if employed in the vicinity of saddle points and extrema not involved with roots. It is therefore most useful for refining factors already approximately located.

Approximate location of quadratic factors

One way of locating factors very roughly is from a plot of $P(x)$. A minimum which does not quite touch the axis obviously could, by a minor change, produce two zeros. A tangent loop gives $(x - a_0)(x - a_0)$ for a factor; thus the picture leads to the form $(x - a_0 - bi)(x - a_0 + bi) = x^2 - 2a_0 x + (a_0^2 + b^2)$, where b^2 is small and is the height of the minimum above the axis. Other factors may distort and move the curve so that these geometric observations are no longer quantitatively correct—but in a world of continuous variations they may well provide an adequate start for a Newton iteration.

A more mechanical technique of locating quadratic factors uses both forward and backward division of our polynomial,

$$a_n x^n + a_{n-1} x^{n-1} + \cdots + a_1 x + a_0 \tag{7.16}$$

We begin with a trial quadratic divisor

$$x^2 + d_1 x + d_0 \tag{7.17}$$

and do a standard forward division to produce a quotient polynomial

$$q_{n-2} x^{n-2} + q_{n-3} x^{n-3} + \cdots + q_1 x + q_0 \tag{7.18}$$

of degree $n - 2$, as well as a linear remainder, which we shall not need. If we should now divide our original polynomial by our quotient (7.18), we would get our trial quadratic divisor (7.19) back again, so there is no point in such a maneuver. On the other hand, if we should do a *backwards division* of (7.16) by (7.18) we will get a quadratic quotient that generally will *not* be (7.17), and we have now completed one iteration of our algorithm. We begin again using the new quadratic as our trial divisor. Some care must be taken, since our division algorithms demand that the divisor be normalized so that the leading coefficient is unity. In the forward division this is the x^2 term, while in the backward division it is the constant term.

If there are no real roots this algorithm will converge to the complex root pair of *smallest modulus*, provided that this is a unique root pair. Thus if our polynomial has complex roots as shown in Figure 7.5, the algorithm will find the pair A, while in Figure 7.6 it will get very confused, both the root pairs A

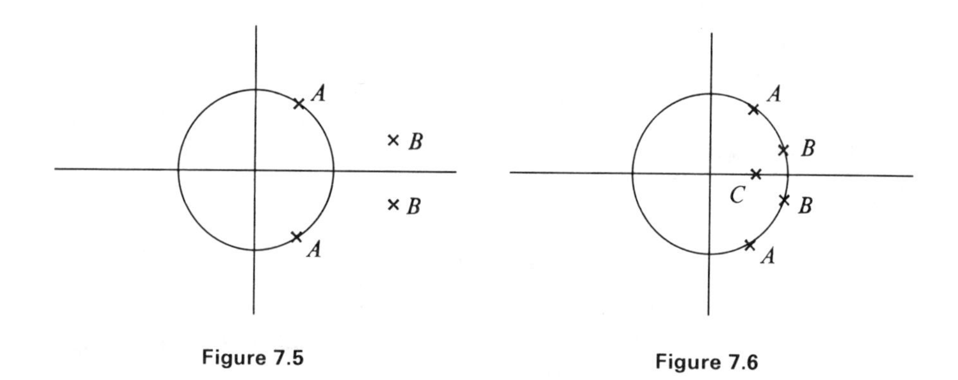

Figure 7.5

Figure 7.6

and B being equally distant from the origin. In this situation, however, it is obvious that moving the origin to C will promptly make the root pairs have unequal moduli, and a shift of all the roots of a polynomial by a known constant is easily accomplished by repeated synthetic division (p. 59). (Remember Horner's method from school?). Of course if there is a genuine *pair* of double complex roots, shifting the origin will do no good and our forward–backward algorithm will remain confused—but this is another example that supports our earlier remarks about there always being a polynomial sufficiently nasty to frustrate any system. We have to stop somewhere, and this seems to be a good place!

The Lehmer method

One of the displeasing aspects of all the previous polynomial root-finding strategies is their patchwork quality: they consist of unrelated algorithms that complement each other's deficiencies. Collectively they may ensure the discovery of roots with a high probability, but their fundamental disunity is anathema to the sensitive programmer or analyst. We would like *one* method that suffices for all polynomial roots, real or complex, single or multiple. Lehmer [1961] proposed such a method based on a subroutine that answers *yes* or *no* to the question "Is at least one root of our polynomial contained within a circle in the complex plane, center at z and of radius r?" Such a subroutine may be constructed and Lehmer shows how. For that part of the method we refer to the original article.

Assuming the availability of our question-answering subroutine, the strategy is to apply it to a circle in the complex plane, centered at the origin with unit radius. If a root is included, the radius is halved and the question repeated. Ultimately we will find an annular space, a ring, which includes at least one root but inside of which *no* roots are present. (If the unit circle contains no roots, the trial radius is doubled and the question repeated – with the same result.) The annulus containing a root has outer radius R and inner radius $R/2$. This is *Condition A*.

We now shift the center of our circle to some point near the center B of the annular band and choose a radius so that at least one sixth of the band is covered. If the root is not found, the center of our circle is rotated $\pi/3$ radians to C, so that the circle occupies the dashed position in Figure 7.7. In one of at most six

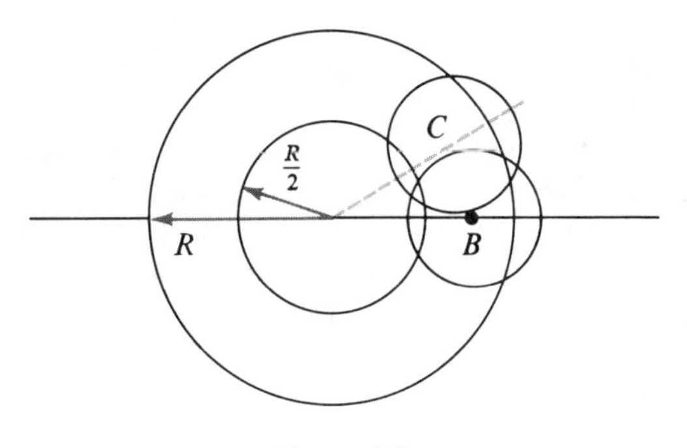

Figure 7.7

positionings our circle will contain a root. We then hold the center fixed and halve the radius. If no root is now enclosed we have defined a new annulus that holds a root; otherwise we halve again.

We are now clearly back at Condition A except our annulus is smaller and its center is in a different place. If we iterate the process we will gradually confine the root to a smaller and smaller annular geometry until finally we have located it with a precision that suffices. Since the basic question-answering routine is not bothered by multiple roots, it would seem that this iterative procedure might be the universal polynomial root finder that we seek. Alas no! Tests on a number of polynomials, both nasty and nice, show that Lehmer finds roots about as precisely as one of the unattractive standard packages but takes three times as long. (The package includes a stepping search leading to Newton in one dimension for real roots, Bairstow for quadratic factors, and a final Newton polishing.) Thus we are faced with one of those unpleasant triumphs of the efficiency of brute force. We can only take solace in that they seem to be rare.

Root-squaring methods

This chapter must close on a sour note of dissent. From the beginning of time an unbroken line of numerical analysts, starting with C. H. Graeffe [1837] and N. I. Lobaschevski, have advocated finding roots of polynomials by constructing other polynomials whose roots are the squares of the original. If there is a largest root, it will grow faster than its brethren and ultimately will stand alone in polished glory to be perceived at leisure by the admiring multitude. It is then only necessary to take as many square roots as we have taken squarings. Of course, the sign has been obscured, but a little care and observation en route will remedy *that* deficiency. If the "largest" root happens to be complex, or multiple, or multiple and complex, then the rules become different—and rather complicated, too—so that the whole process becomes a collection of different methods and loses the unity that it at first seems to possess. The rules for deciding which type of root we actually have are complicated. They are better suited, by far, to the human intelligence employed in hand computation than they are to the coded decision processes of the automatic stored-program computer. Unfortunately, the advocate of root squaring can cite an impressive list of authors who give these methods large amounts of space. We must, however, point out that most of them[1] are fascinated by the rather nice theorems the methods permit. We also comment on the paucity of evidence that the authors have ever compared root squaring with other methods on realistic polynomials.

Lin's method for quartics with complex roots

Thus far all our algorithms have been general, but we wish to give one technique that is quite specific: It splits a quartic with four complex roots into its two real quadratic factors. This problem occurs fairly often, especially after real roots have been removed from higher-degree polynomials. Lin's method divides quartics into two categories: those in which the roots are well separated and hence the factors are found easily by Newton's method in two dimensions, and those in which the roots are not well separated. It is precisely for this second, difficult, class that Lin works best.

[1] A conspicuous exception is the excellent objective article by Olver [1952] that supports root squaring, albeit entirely in a hand-computation context, where efficiency depends on prompt recognition of complicated patterns—clearly still a human prerogative.

The technique is iterative unless the quartic is exactly factorable into two identical quadratic factors or into two quadratic factors with identical constant terms. It is based on the assumption that the true factors are not far removed from one of these structures, and uses them as a first approximation.

If the normalized equation is written in the form

$$x^4 + A_3x^3 + A_2x^2 + A_1x + A_0 = 0 \tag{7.19}$$

we then compute

$$\begin{aligned} \alpha &= \frac{A_1}{2\sqrt{A_0}} \\ \beta &= A_2 - 2\sqrt{A_0} \\ \gamma &= \frac{A_3}{2} \end{aligned} \tag{7.20}$$

There are two main cases:

1. If $\beta > \alpha^2$ and/or $\beta > \gamma^2$, the roots are considered to be separated and the Newton method should converge to the quadratic factors.
2. If $\beta < \gamma^2, \alpha^2$ then we compute as follows:

$$\left.\begin{aligned} a_1 &= \gamma + \sqrt{\gamma^2 - \beta} \\ b_1 &= \gamma - \sqrt{\gamma^2 - \beta} \\ b_0 &= \frac{\sqrt{A_0}}{a_1}(\alpha + \sqrt{\alpha^2 - \beta}) \\ a_0 &= \frac{A_0}{b_0} \\ \beta' &= A_2 - (a_0 + b_0) \end{aligned}\right\} \text{ to give } (x^2 + a_1x + a_0)(x^2 + b_1x + b_0) \tag{7.21}$$

We then replace β by β' and repeat the cycle until successive values of the four parameters and β agree to the accuracy desired.

Special cases:

1. If $\alpha^2 = \gamma^2$, then $a_0 = b_0 = \sqrt{A_0}$.
2. If $\alpha^2 = \gamma^2 = \beta$, then $a_0 = b_0 = \sqrt{A_0}$ and a_1 and b_1 are exactly given by the first two equations of (7.21).

In either of these special cases no iteration is required.

Designing a system for public use

When we finally turn to the job of designing a subroutine for use by the public, we must stop arguing the pros and cons of the various techniques and actually choose between them. These choices are not simple, and a group of five experienced numerical analysts might well produce five different subroutines. Accordingly, the recommendations of your author must be taken as those of one person only, and if you have good reason for not following them, then by all means design your own system!

A system cannot do everything: There will always be some extremely nasty polynomials which will frustrate it and cause it to fail to find some of the roots. A system should handle most of the polynomials it is apt to encounter, but we cannot insist on perfection, lest the program become a monster that demands more than its share of library space and requires large amounts of running time looking for subtle difficulties that are not usually present. We must, however, insist that in the nasty situations the system should announce its difficulty, lest the trusting user be misled into thinking that all the roots have been found, and with the precision he desires.

More important, a system should be both flexible and easy to use. The user should not have to read detailed instructions (he won't) and he should be able to get the information about his polynomial that he wants without being forced to take too many things that he does not want. He should be able, for example, to ask for the largest real root without getting all the roots. Or he should be able to get all the real roots, along with a number that says how many there are, without having the routine find the complex ones as well. (He cannot, however, get the complex ones without the routine computing the reals – why?)

To cut down on the number of parameters that the user must supply to the subroutine, there should be a standard option that is implied if only the minimum information is given. Thus, if the user tells the routine only the degree of his polynomial and the name of the vector of coefficients, the routine should give him *all* the roots to some standard accuracy (or an error message if some difficulty occurs). If the user wants some subset of roots or a nonstandard accuracy, he must be prepared to ask for it by supplying additional information to the subroutine when he calls it. But the "standard" user should not be penalized by having to tell the accuracy and also the fact that he wants this or that type of root when all he really wants to know are "the roots of the polynomial."

As a general strategy, the routine should first seek the real roots – using a stepping search between ± 1.2 that also evaluates the derivative and finds the extrema, checking the value of the polynomial at those points. In this phase a list of approximate real roots is built up, as well as a special list of double or nearly double real roots and nearly real roots. In the second phase, these real roots are refined, by Newton's method, to the desired accuracy.

If complex roots are desired, the original polynomial is then reduced by dividing out the real roots, and a forward–backward iteration is carried out to locate, approximately, the complex pair of smallest modulus. Failure of this iteration to converge in n cycles should produce a shift of origin and a second attempt. (A second failure should produce an error message giving the news that there seem to be double complex root pairs and that the routine is not prepared to find these – but the roots thus far found are as follows, . . . , and that the reduced polynomial is . . . .) If forward–backward converges to two significant figures, a shift to Newton's method in two dimensions follows for the refinement of the quadratic factors and the reduction of the polynomial. This operation is repeated and ceases when only a quadratic factor remains. The quadratic factors are then given one cycle of Newton's two-dimensional method *with the original polynomial* and then solved for the complex root pairs. This pasteurization is necessary because inaccuracies in the successive roots that have been "removed" make the roots of the reduced polynomials poorer and poorer approximations to the roots of the original polynomial. We must therefore refer the individual roots to the original polynomial for a final polish before exhibiting them in public and certifying as to their accuracy.

Pathologic polynomials

No system will find all the roots of all polynomials correctly. We have already seen that multiple roots tend to confuse Newton's method in one or two dimensions, and the simple search may miss a double root even when it would certainly find an isolated one. But multiple roots are not the only difficulties. Wilkinson [1959] has given a particularly nasty example in the polynomial whose roots are $-1, -2, \ldots, -20$. This polynomial begins

$$x^{20} + 210x^{19} + \cdots + 20! = 0 \tag{7.22}$$

and it certainly has large coefficients. The roots, however, could scarcely be more isolated. He then applies a small change to the coefficient of x^{19} – he adds 2^{-23}, which is a change of *one* in the thirty-first binary digit or about one in the ninth significant decimal digit. Our intuition suggests that such a slight change could not seriously affect the roots of the polynomial (7.22), but our intuition is wrong! Accurate calculation of the roots shows them to be, to five decimals,

-1.00000	$-10.09527 \pm 0.64350i$
-2.00000	$-11.79363 \pm 1.65233i$
-3.00000	$-13.99236 \pm 2.51883i$
-4.00000	$-16.73074 \pm 2.81262i$
-5.00000	$-19.50244 \pm 1.94033i$
-6.00001	
-6.99970	
-8.00727	
-8.91725	
-20.84691	

We see that serious changes have appeared in all the roots from -9 to -20, with 10 of them becoming complex (Figure 7.8). Not even slightly complex, either; the imaginary parts are large. All these changes, it must be emphasized, are real. They are not mistakes in computing the roots; rather they are actual changes in the roots of our polynomial produced by changing one of its coefficients by one in the ninth significant figure. If we must find the roots of this polynomial, we must certainly perform all our arithmetic with an accuracy greater than nine significant figures, no matter what algorithm we use, and our algorithm must not let us stop before each root is determined to an accuracy sufficient not to perturb the further roots we have still to find.

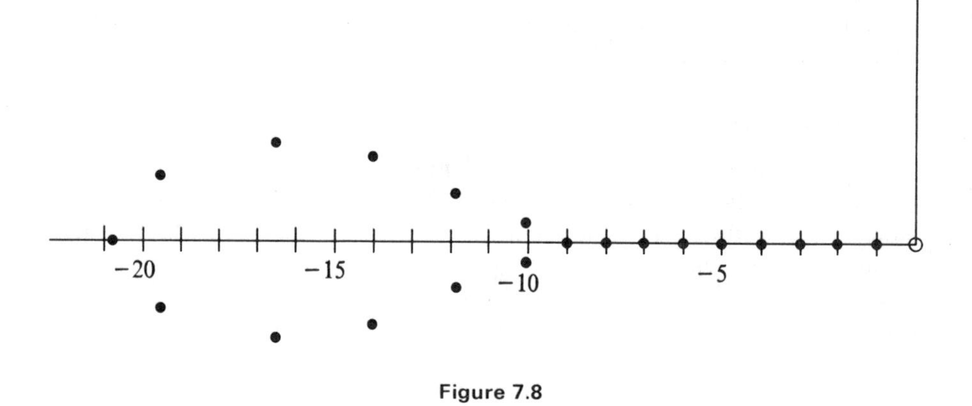

Figure 7.8

This example is horrifying indeed. For if we have actually seen one tiger, is not the jungle immediately filled with tigers, and who knows where the next one lurks? Wilkinson shows that the difficulty is caused by the regular spacing of the roots and observes that this is a more prevalent danger in practice than is a root of high multiplicity. We must, however, leave this topic here. Not very much is known about polynomial root instabilities, and the man who tries to compute many roots will certainly encounter occasional difficulties with even the best of systems. But the presence of nearly impossible problems should not deter us from attempting to solve the average problem well.

Problems

1. Make up several quartics having isolated and multiple real roots, then try to find them by using Newton's method. Note that slide-rule accuracy often suffices in the earlier stages of purely iterative methods.

2. If an automatic computer is available, write a program to carry out the steps of the previous problem. Be sure to print out all relevant quantities for each iteration.

3. Find the real quadratic factors of

$$x^4 + 0.048521x^3 + 5.0237x^2 - 0.24759x + 5.9824 = 0$$

4. Find a complex root of the preceding quartic by Newton's method in the complex plane.

5. By rewriting

$$x^4 - 25.7x^3 + 2.135x^2 - 1.5743x + 2.3579 = 0$$

as

$$x = 25.7 - \frac{2.135x^2 - 1.5743x + 2.3579}{x^3}$$

we have suggested an iterative algorithm that converges to one of the roots of this quartic. Carry out the algorithm, then investigate it analytically. When will it converge? Discuss its practicality compared with Newton's method.

6. Try Laguerre's method on the polynomials of Problems 5 and 3. Compare the total amount of computational labor via the two methods.

CHAPTER 8

EIGENVALUES

One of the major problems in the computation laboratory is the determination of the eigenvalues of a matrix. There are really several related problems, for the matrix may be symmetric and positive definite or it may have some far less obliging form. The customer, too, may want only the largest eigenvalue, or he may stubbornly insist on all of them—and with their corresponding eigenvectors yet! No single technique is proper for all these situations and, indeed, one need not look very far to find matrices for which *all* standard techniques fail. Our exposition must therefore be circumspect, being careful not to talk about all facets of these problems at once. In this chapter we shall discuss the relatively simple problem of finding the largest eigenvalue of a *symmetric* matrix by iteration techniques. In Part II we have a chapter that discusses the rotational techniques suitable for finding all the eigenvalues of reasonable matrices. For unreasonable matrices we refer the reader to the excellent 675-page book by Wilkinson [1965].

A physical problem

For those who like their eigenvalues set in a physical context, we include here a simple problem. Those who prefer their numerical processes untrammelled by physical reality may wish to skip on to the next section. We consider the in-line motions of three masses (Figure 8.1). Each is of the same mass M, but the

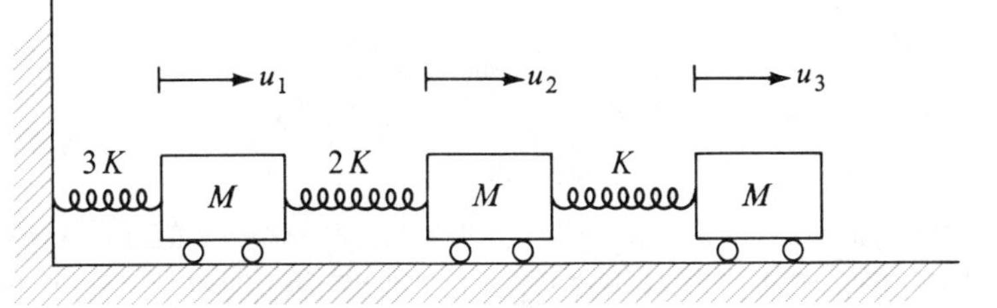

Figure 8.1. Three vibrating masses

springs that connect them differ in their spring constants in the ratios 3:2:1. There is no friction and gravity is not involved, the motions being horizontal. Coordinates are the locations of the three masses u_1, u_2, and u_3, each measured from its position when the entire system is unstressed and at rest. Newton's second law applied to each of the masses considered as a free body quickly gives the equations of motion as

$$M\frac{d^2u_1}{dt^2} = -3Ku_1 - 2K(u_1 - u_2)$$

$$M\frac{d^2u_2}{dt^2} = -2K(u_2 - u_1) - K(u_2 - u_3)$$

$$M\frac{d^2u_3}{dt^2} = -K(u_3 - u_2)$$

which simplify slightly to

$$-\frac{M}{K}\frac{d^2u_1}{dt^2} = 5u_1 - 2u_2$$

$$-\frac{M}{K}\frac{d^2u_2}{dt^2} = -2u_1 + 3u_2 - u_3$$

$$-\frac{M}{K}\frac{d^2u_3}{dt^2} = -u_2 + u_3$$

Since this is a system of three second-order differential equations, it needs six initial conditions. Physically, we see that we are free to specify the initial position and velocity of each mass. Further, since there is no damping, a motion once begun will never cease. This observation, coupled with the traditional guess-a-solution-and-try-it method for solving differential equations, leads us to look for *periodic* motions. In fact, we shall try a quite special motion: all three masses vibrating sinusoidally, in phase, and at the same unknown frequency ω. Such stringent conditions may seem too restrictive, but history assures us that our efforts will not be wasted. We try

$$u_i = x_i \cos \omega t \qquad i = 1, 2, 3$$

where the x_i are the three unknown amplitudes of vibration for the three masses. On substituting into the differential equations, the cosines all divide out—leaving us with a set of algebraic equations for the three amplitudes. The unknown frequency ω also appears, combined with M and K as

$$\lambda = \frac{\omega^2 M}{K}$$

to give

$$\begin{aligned} \lambda x_1 &= 5x_1 - 2x_2 \\ \lambda x_2 &= -2x_1 + 3x_2 - x_3 \\ \lambda x_3 &= -x_2 + x_3 \end{aligned}$$

which may also be written

$$\begin{bmatrix} 5-\lambda & -2 & 0 \\ -2 & 3-\lambda & -1 \\ 0 & -1 & 1-\lambda \end{bmatrix} \begin{bmatrix} x_1 \\ x_2 \\ x_3 \end{bmatrix} = \begin{bmatrix} 0 \\ 0 \\ 0 \end{bmatrix}$$

This set of linear algebraic equations is *homogeneous*; that is, its right side is zero. We note that it has the obvious solution

$$x_1 = x_2 = x_3 = 0$$

which merely states that our masses are sitting still, vibrating with zero amplitudes—a "solution" that satisfies the differential equations but almost nobody else. We want a nontrivial solution. We also know that homogeneous equations can have nontrivial solutions only if the determinant of the equations, D, is itself equal to zero. (If D is not zero, Cramer's rule gives only the zero solution and gives it as the ratio of two determinants, the upper one being zero. With the lower one also zero, at least we have an indeterminate form that could possibly have a nonzero value.)

Since we still have λ at our disposal we seek a value that will make

$$\begin{vmatrix} 5-\lambda & -2 & 0 \\ -1 & 3-\lambda & -1 \\ 0 & -1 & 1-\lambda \end{vmatrix} = 0$$

This is a cubic equation in λ with three real roots—and suddenly we have not just one peculiar solution available, we have three of them! We solve, obtaining λ_1, λ_2, and λ_3. If we now choose *one* of our special λ_i's we may substitute it back into the algebraic system, which is then solvable for the three amplitudes x_j. Since the system is homogeneous, any set of x's that solve it may be multiplied by a constant – that is, only the ratios of the three x's are set by the equations. *Physically* this says that if we had started the masses oscillating with bigger amplitudes, had kicked them twice as hard, then the same relative motions would have prevailed. *Algebraically* this says that we may choose any one of our x_i's to be a pleasant number, say 1, and then determine the others from the equations.

At this point we leave our masses happily vibrating in one of their three special peculiar modes. The λ's are basically the special frequencies and are, of course, the eigenvalues. With each λ, a scalar, there is a corresponding motion which is a collection of three amplitudes, an *eigenvector* $\mathbf{x}$ that is the solution of the algebraic equations when that λ has been substituted back into them. Thus we have three eigenvalues λ_i and three eigenvectors, each eigenvector having three components.

For those of our readers who are wondering about the general motion of our three masses, we shall only point out that we could have gone through this analysis using

$$u_i = x_i \sin \omega t$$

to get three more motions that are very much like the current ones, differing only in their phases. We are now possessed of six linearly independent solutions to a linear differential system with six degrees of freedom. Since our system is linear, the most general solution is merely a linear combination of these six independent solutions and we need to look no further. Formally the solution is a complicated mess, but fortunately in engineering vibration problems this general solution is seldom required. We are usually looking for the natural frequency of vibration and these, not too surprisingly, turn out to be the ω's – available directly from our λ's. So we may safely close our physical investigation here.

Basic concepts

The numerical techniques of this chapter sometimes have somewhat abstract geometric interpretations that can confer insight about their probable effectiveness. We therefore take this opportunity to review the essential facts about the eigenvalues of symmetric matrices. The symmetry is not essential to the techniques, but their geometric interpretation is simplified. Algebraically, the problem is to find a scalar λ and the corresponding vector $\mathbf{x}$ that solve the equation

$$Ax = \lambda x \tag{8.1}$$

when A is a given matrix—presumed symmetric in this chapter. These equations which are linear and homogeneous may also be written

$$(A - \lambda I)\mathbf{x} = 0 \tag{8.2}$$

in which form we can see that, apart from the trivial solution of $\mathbf{x}$ identically 0, we must require the determinant of (8.2) to be zero, that is,

$$\det(A - \lambda I) = 0 \tag{8.3}$$

before there can possibly be a solution vector $\mathbf{x}$. Even so, this $\mathbf{x}$ is determined only up to a proportionality factor, since the homogeneity of our equations permits any solution to be multiplied by an arbitrary constant and still remain a solution. Geometrically, our solution vector $\mathbf{x}$ has a unique direction but indeterminant length.

Before any eigenvector $\mathbf{x}$ may be found, however, an eigenvalue λ must be determined. With small matrices we may use (8.3) directly, as it defines a polynomial, $P_n(\lambda)$, of degree n (n is the number of rows in A) whose roots are thus the required λ's. Here our first complication arises: Generally we have not one but n solutions—n eigenvalues and with a different $\mathbf{x}$ for each λ. Our original equation might better be written

$$A\mathbf{x}_i = \lambda_i \mathbf{x}_i \qquad i = 1, 2, \ldots, n \tag{8.4}$$

but we shall frequently drop the subscript when no confusion should thereby arise. While the vectors $\mathbf{x}_i$ have no special magnitudes, the scalars λ_i most assuredly do, and we may talk about the largest λ_i or the smallest λ_i in the sense of either algebraic or absolute values. In this chapter we are mostly concerned with the *largest* λ and we shall frequently number them in decreasing order of algebraic magnitude, that is,

$$\lambda_1 \geq \lambda_2 \geq \lambda_3 \cdots \geq \lambda_n$$

so that our largest eigenvalue is λ_1 and our smallest is λ_n.

Once a correct λ_i is available from solving the characteristic polynomial (8.3), the vector $\mathbf{x}_i$ may be found by solving equation (8.2)—a process no more nor less difficult than that of solving any system of $n - 1$ linear simultaneous equations. Because of the homogeneity, one component of $\mathbf{x}_i$ may be picked arbitrarily to be unity and one of the equations may then be dropped from the system. The alleged solution may finally be checked by substituting it into the unused equation. If the student has never solved an eigenvalue problem in this fundamental way he should try the well-behaved matrix

$$\begin{bmatrix} 5 & -2 & 0 \\ -2 & 3 & -1 \\ 0 & -1 & 1 \end{bmatrix}$$

whose characteristic polynomial is

$$\lambda^3 - 9\lambda^2 + 18\lambda - 6 = 0$$

and whose solutions are shown in Table 8.1. We presume at least this amount of grubby experience in the sequel.

Table 8.1

λ_i	6.29	2.29	0.42
$\mathbf{x}_i$	1.00 −0.65 0.12	0.74 1.00 −0.77	0.25 0.58 1.00

This technique of turning a matrix into a polynomial, finding the roots of the polynomial, and then possibly solving n different sets of simultaneous equations of size $(n - 1)$ grows immediately from the algebraic definition of eigenvalues and eigenvectors—*but it is not the efficient way to compute them.* Other algorithms are shorter and more systematic. In this day of automatic digital computation, the second point is particularly important.

A geometric picture—the quadratic form

Every symmetric matrix has an associated *quadratic form* $Q(\mathbf{x})$ that is defined by

$$Q(\mathbf{x}) = \mathbf{x}^T A \mathbf{x} \tag{8.5}$$

where $\mathbf{x}$ is an arbitrary vector. This quadratic form is a scalar and if we demand that it takes on a constant value, c, it defines a conic surface in as many dimensions as $\mathbf{x}$ has components. If the main diagonal terms of A are large enough, and they are easily made so, we have an ellipsoid in n dimensions. By way of simple example, if

$$A = \begin{bmatrix} 2 & -1 \\ -1 & 1 \end{bmatrix} \quad \text{then} \quad Q(\mathbf{x}) = [x_1, x_2]\begin{bmatrix} 2 & -1 \\ -1 & 1 \end{bmatrix}\begin{bmatrix} x_1 \\ x_2 \end{bmatrix}$$

$$= 2x_1{}^2 - 2x_1x_2 + x_2{}^2$$

where, unlike our earlier usage, the subscripts denote components of the vector $\mathbf{x}$. Equating $Q(\mathbf{x})$ to unity defines a tilted ellipse centered at the origin of the coordinate system (x_1, x_2). Pictorially (Figure 8.2) we say that $\mathbf{x}$ is a *position vector*

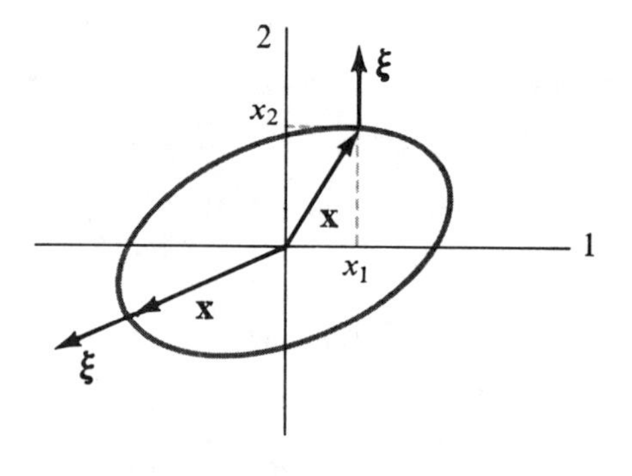

Figure 8.2

whose tip traces out the ellipse when constrained to move in accordance with

$$Q(\mathbf{x}) = 1$$

Another way of looking at our picture is to ask for the value of $Q(\mathbf{x})$ at every point. Once we have a value attached to every point in the (x_1, x_2) plane—a scalar field—our ellipse is simply the *level curve* that connects all the points where the value of the field is 1. Other ellipses, concentric with and similar to ours, connect the points where the value of the quadratic form is $\frac{1}{2}$, 5.3, or $\frac{2}{3}$. Moreover, at any point in the field we have not only our position vector $\mathbf{x}$ but a *gradient vector* $\boldsymbol{\xi}$ that points in the direction in which $Q(\mathbf{x})$ increases most rapidly—clearly perpendicular to the ellipse through the point. It is an exercise in differentiation of matrix equations to show that

$$\boldsymbol{\xi} = A\mathbf{x} \tag{8.6}$$

Finally we note from our picture that, for most points on the ellipse, the gradient vector points in a direction different from that of the position vector. We also note, however, that at four points the gradient vector *does* point along the position vector—and that these directions are the principal axes of the ellipse. To express the colinearity of the gradient vector with the position vector algebraically we multiply $\mathbf{x}$ by a proportionality constant λ to get

$$\boldsymbol{\xi} = \lambda\mathbf{x} = A\mathbf{x} \tag{8.7}$$

and hence we see that these principal axes are the *eigenvectors* of our matrix.

Thus the literature frequently refers to the eigenvalue problem as the principal-axis problem. Each eigenvalue, λ, is associated with the length of a principal axis (the precise function is unimportant to us here). Thus our first geometric insight: When both λ's are equal, the ellipse is a circle and there are no *unique* principal axes and hence no unique eigen*vectors*. We may expect computational difficulties if our algorithm is seeking to produce a unique vector. Further, if two λ's are nearly equal, then the principal axes are defined with difficulty, since the length of the position vector changes extremely slowly as it sweeps past a principal axis. Again we expect computational problems. A second geometric insight: The eigenvectors $\mathbf{x}$ of a symmetric matrix are mutually perpendicular, hence

$$\mathbf{x}_i^T \cdot \mathbf{x}_j = 0 \qquad i \neq j \tag{8.8}$$

Iterated multiplication

It we premultiply an arbitrary vector $\mathbf{y}$ by our matrix A, we produce a new vector $\mathbf{v}$ according to

$$\mathbf{v} = A\mathbf{y} \tag{8.9}$$

Generally it will differ from $\mathbf{y}$ in length and direction. In eigenvalue problems we care very little for lengths of vectors, but their directions are crucial. If it should happen that $\mathbf{y}$ were actually an eigenvector of A, then $\mathbf{v}$ would have the same direction, that is,

$$\mathbf{v} = \lambda\mathbf{y} = A\mathbf{y} \tag{8.10}$$

Further, it would be exactly λ times as big as $\mathbf{y}$ in every component—and, of course, would be λ times as long as $\mathbf{y}$. These remarks are correct but not overly illuminating. We must be more searching in our analysis. To simplify typography we describe a system of size *three*—matrix A is 3×3, all vectors have three components, there are three λ's and three corresponding eigenvectors $\mathbf{x}$. Thus our basic equation is

$$A\mathbf{x}_i = \lambda_i\mathbf{x}_i \qquad i = 1, 2, 3 \tag{8.11}$$

Any arbitrary vector may be expressed as a linear combination of the three eigenvectors $\mathbf{x}_i$; that is, if we knew the three eigenvectors of A we could then find three scalar coefficients b_1, b_2, and b_3 such that we could express our arbitrary vector, $\mathbf{y}$, as

$$\mathbf{y} = b_1\mathbf{x}_1 + b_2\mathbf{x}_2 + b_3\mathbf{x}_3 \tag{8.12}$$

This representation is possible because the eigenvectors of a symmetric matrix span the space, being mutually orthogonal. If we now premultiply both sides of (8.12) by A we have

$$A\mathbf{y} = b_1 A\mathbf{x}_1 + b_2 A\mathbf{x}_2 + b_3 A\mathbf{x}_3 \tag{8.13}$$

and we may then replace each $A\mathbf{x}_i$ by its corresponding $\lambda_i \mathbf{x}_i$. Factoring out λ_1 we have

$$A\mathbf{y} = \lambda_1 \left[b_1 \mathbf{x}_1 + \frac{\lambda_2}{\lambda_1} b_2 \mathbf{x}_2 + \frac{\lambda_3}{\lambda_1} b_3 \mathbf{x}_3 \right] \tag{8.14}$$

In a similar way if we again premultiply by A we obtain

$$A^2\mathbf{y} = \lambda_1^2 \left[b_1 \mathbf{x}_1 + \left(\frac{\lambda_2}{\lambda_1}\right)^2 b_2 \mathbf{x}_2 + \left(\frac{\lambda_3}{\lambda_1}\right)^2 b_3 \mathbf{x}_3 \right] \tag{8.15}$$

If we now consider repeated premultiplication of our arbitrary vector, we may think of each multiplication as changing our previous vector to a new one. In sympathy with this view we define an infinite sequence of vectors $\mathbf{y}_i$ by the

$$\mathbf{y}_{i+1} = A\mathbf{y}_i = A^{i+1}\mathbf{y}_0 \qquad i = 0, 1, 2, \ldots \tag{8.16}$$

Applying our resolution into components parallel to the eigenvectors of A to the nth iterate, we obtain

$$\mathbf{y}_n = \lambda_1^n \left[b_1 \mathbf{x}_1 + \left(\frac{\lambda_2}{\lambda_1}\right)^n b_2 \mathbf{x}_2 + \left(\frac{\lambda_3}{\lambda_1}\right)^n b_3 \mathbf{x}_3 \right] \tag{8.17}$$

In this analysis our numbering of the λ's and their corresponding $\mathbf{x}_i$'s has been arbitrary. We now choose to let λ_1 stand for the *largest* λ in *absolute value*; then all the fractions λ_k/λ_1 are less than unity in magnitude and the raising of them to successively higher powers makes these fractions become steadily smaller. Thus, as n increases, all the terms of (8.17) are suppressed except the first and we see that

$$\mathbf{y}_n = C_n \mathbf{x}_1 \tag{8.18}$$

Thus $\mathbf{y}_n$ approaches $\mathbf{x}_1$ in direction—and lengths are unimportant in the eigenvalue world.

The reader should try this iteration on the matrix

$$\begin{bmatrix} 5 & -2 & 0 \\ -2 & 3 & -1 \\ 0 & -1 & 1 \end{bmatrix}$$

starting with [1, 0, 0] and carrying out about five iterations, checking that the later iterates point in nearly the same direction by normalizing them so that their largest component is some convenient number, say 1. Then compare your vector with Table 8.1.

The iteration method for the largest eigenvalue

The analysis of the previous section has shown that repeated premultiplication of an arbitrary vector by A will generally cause that vector to become the eigenvector associated with the largest absolute eigenvalue of A. This sentence, which is a bit of a mouthful, is frequently and inaccurately rephrased to state that premultiplication causes convergence to the largest eigenvector. Since the eigenvalue problem is a homogeneous one, the length of vectors is unimportant and there certainly is no largest one, but it is to be hoped that this somewhat slipshod description will confuse nobody.

The essential fact about repeated premultiplication needs only a little dressing up to become a practical algorithm. The principal problem with unadorned premultiplication by A is the attendant multiplication of our surviving component by the factor $\lambda_1{}^n$, which usually causes $\mathbf{y}_n$ to grow, often very rapidly, while changing its direction rather slowly. This growth is easily stopped by inserting a *normalization* step that preserves the direction of $\mathbf{y}_n$ but reduces it to some standard length. The normalization may be applied whenever $\mathbf{y}_n$ has grown too big for comfort, but with automatic computation we usually normalize at each iteration. Thus we may compare two successive $\mathbf{y}$'s to see how small their difference is and hence to check whether or not we have probably achieved our desired eigenvector, $\mathbf{x}_1$. In hand computation the most convenient normalization is to make the largest component of our vector equal to unity. With automatic computation we usually prefer to treat all components symmetrically and hence choose to normalize the *length* of our vector to unity.

More important is the question: When does this iteration process fail? Looking at (8.17) we see that it may fail to converge to $\mathbf{x}_1$ either when some λ_k/λ_1 is equal to ± 1 or when b_1 is zero. The equality of λ_2 and λ_1 simply says that we have two "largest" eigenvalues, and our ellipse in this two-dimensional cross section is really a circle. Thus there is no unique eigenvector, but rather any two radii will serve as principal axes. The process will, indeed, converge to a radius vector. Specifically if λ_1 equals λ_2, it will converge to

$$\mathbf{y}_n = b_1\mathbf{x}_1 + b_2\mathbf{x}_2 \tag{8.19}$$

and it will converge rather quickly—giving the correct λ_1 in the bargain. If, however, we start our iteration with a different arbitrary $\mathbf{y}$, the iterations will converge to a different vector, a different radius of that same circular cross section of the ellipsoid. We shall once again get λ_1. If only λ_1 and λ_2 are equal, further repetitions with other starting vectors will give additional apparent eigenvectors, but they will all be linear combinations of the two we have already found.

The difficulties arising from randomly choosing an arbitrary vector in which b_1 is zero seem at first to be more serious, if less probable. Such an initial arbitrary vector is not arbitrary in the right way, in the sense that it happens to lack a component in the $\mathbf{x}_1$ direction. With no such component present our convergence should be to a different $\mathbf{x}_i$, but rounding errors usually guarantee that b_1 does not remain zero and hence this difficulty is largely illusory. It is true that infinitely precise calculation would produce convergence to $\mathbf{x}_2$ and give λ_2 – the second largest λ in magnitude. This may be seen by setting b_1 to zero and factoring out λ_2 to get

$$\mathbf{y}_n = \lambda_2{}^n\left[\left(\frac{\lambda_1}{\lambda_2}\right)^n b_1\mathbf{x}_1 + b_2\mathbf{x}_2 + \left(\frac{\lambda_3}{\lambda_2}\right)^n b_3\mathbf{x}_3\right] \tag{8.20}$$

or

$$\mathbf{y}_n = C_n\mathbf{x}_2 \tag{8.21}$$

where the last term in (8.20) has been eliminated by the usual attrition from the ratio of the two λ's becoming small and the first term has been eliminated by the definition of b_1. In practice, b_1 is merely very small, say ε, so that early iterations tend to give (8.20), where the factor λ_3/λ_2 has managed to reduce $\mathbf{x}_3$ to relative quiescence. The sequence of vectors $\mathbf{y}_n$ thus marches slowly toward apparent convergence – to $\mathbf{x}_2$. But then, instead of stopping, they seem to gather momentum and proceed with increasing speed to change direction, as $(\lambda_1/\lambda_2)\varepsilon$ is no longer small. Finally, to our dismay, the $\mathbf{y}_n$'s settle down comfortably at $\mathbf{x}_1$.

The real troubles with our iterative process occur when λ_1 and λ_2 are very close but not identically equal. We quickly get $\mathbf{y}_n$ into the space of $\mathbf{x}_1$ and $\mathbf{x}_2$ as all the other eigenvectors are eliminated by the λ factors, but elimination of $\mathbf{x}_2$ proceeds very slowly. At best, this means that we recognize that we have a slow algorithm. At worst, we may stop with a somewhat wrong λ_1 and a seriously wrong $\mathbf{x}_1$! Geometrically, our ellipse has a nearly circular section in the plane of (x_1, x_2) and the iteration is not very powerful in moving $\mathbf{y}$ into the $\mathbf{x}_1$ position. If our only interest is in getting λ_1 with some moderate accuracy, this geometry affords us no trouble – but we must remember that our supposed $\mathbf{x}_1$ may be quite bad. If we need $\mathbf{x}_1$, however, either as an answer in its own right or as a basis for finding further λ's and $\mathbf{x}$'s, we shall usually be in deep trouble.

A major problem of all slow iterations is the decision about when to stop. We have no sure mechanism for deciding whether we have already converged with a fast iterative process or are merely progressing slowly with a slow one. The present process is no exception.

Accelerating convergence by shifting

The eigenvalues of any matrix may all be shifted by the same arbitrary amount, s, by simply subtracting this amount from the main diagonal elements of A. We see that

$$A\mathbf{x} = \lambda\mathbf{x} = (\lambda - s)\mathbf{x} + s\mathbf{x}$$

hence

$$(A - sI)\mathbf{x} = (\lambda - s)\mathbf{x} \tag{8.22}$$

and

$$B\mathbf{x} = \mu\mathbf{x}$$

Thus we have our same eigenvalue problem back with the matrix and the eigenvalues changed in very simple ways, while our eigen*vectors* are not changed at all. We are therefore free to shift the eigenvalues of our matrix any time that we feel that the alteration might help us.

Suppose our matrix has eigenvalues lying at -3.9, 1, and 4, as shown in Figure 8.3. Clearly the largest λ is 4, but in magnitude the eigenvalue at -3.9 is

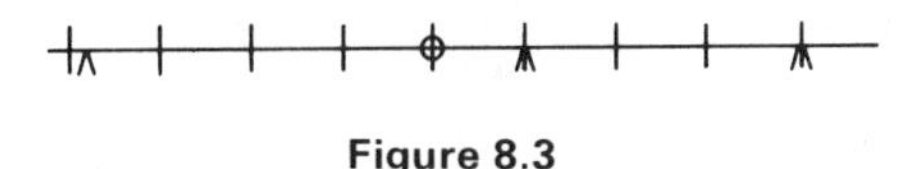

Figure 8.3

pretty close and the ratio $-3.9/4$ is so close to unity that the iterative process is not efficient. If, on the other hand, we add 3 to each λ, we now find the eigenvalues at -0.9, 4, and 7 and the crucial ratio is $\frac{4}{7}$—quite satisfactory. Thus the shift has improved the convergence property of our algorithm. Once we have located the shifted eigenvalue we merely subtract off the shifting constant to put the answer in terms of the original matrix.

The reader may protest that we seem to have used a good bit of knowledge about the location of the eigenvalues we seek in our picking of the shifting parameter, s. Is such a device useful, or is it just another instance of being able to produce an efficient method for solving a problem once you know the answer? The device of shifting is actually quite useful. Because it is simple and inexpensive, one may try it rather casually and abandon it whenever it does not seem to help. Thus if we have started an iteration that seems to be converging too slowly we may subtract some arbitrary s from the main diagonal of our matrix and continue the iterations. If the convergence speeds up, we may be tempted to shift somewhat further. If the convergence gets worse, we may try a shift in the opposite direction. Since $\mathbf{x}$ remains the same for all these shifts (provided we do not shift some other λ into prominence – see below), these various explorations are not wasted effort. Convergence of our vector toward $\mathbf{x}$ is merely speeded or slowed, not stopped. We are rather like the fisherman who moves tentatively to another part of the pond to see how the activity is there. If he likes it, he stays! If not, he moves again. This kind of flexibility is excellent in hand computation but can also be useful in automatic programs.

Accelerating convergence by Aitken's extrapolation

Whenever an iterated variable is approaching a limit in a regular way, we should try to use this regularity to bypass a number of the iterations, especially if the extrapolation is not very complicated. The problem in taking advantage of this obvious advice lies in the difficulty of recognizing regular behavior. Since there are many forms we need a pattern recognizer. Humans are fairly good at pattern recognition but digital computers are not. Thus with automatic computation we usually restrict our extrapolative attempts to one of the simplest of regular behaviors, *linear convergence*, which we hope we can teach our computer to recognize.

An iterative variable is said to converge linearly if the *error* is reduced by a constant factor at each iteration. Letting the limiting though unknown value of y be h, we may describe linear convergence to h by

$$(h - y_{n+1}) = c(h - y_n) \qquad |c| < 1 \tag{8.23}$$

By way of contrast, in *quadratic convergence* the new error depends on the *square* of the previous error,

$$(h - y_{n+1}) = c(h - y_n)^2$$

and the definition extends immediately to higher orders of convergence that are rarely encountered in practice.

Returning to a linearly converging process where our error is decreasing by some constant factor at each cycle, we have (8.23) and, after one more iteration,

$$(h - y_{n+2}) = c(h - y_{n+1}) \tag{8.24}$$

Eliminating c between (8.24) and (8.23), taking n as 0, and solving for $(h - y_2)$, we obtain

$$h - y_2 = -\frac{(\Delta_{1.5})^2}{\Delta_1{}^2} = -\frac{(y_2 - y_1)^2}{y_2 - 2y_1 + y_0} \tag{8.25}$$

so that we get h by adding this correction to y_2. To the extent that our convergence has actually been linear and our arithmetic accurate, we leap directly to the final answer. In practice, a few more cycles are almost always needed, and some test of the magnitude of the correction is desirable lest a spurious extrapolation actually worsen the iterate. Wild corrections can arise from unfortunate applications of (8.25) when the variable y is *changing* linearly (which is *not* convergence), as this makes the second difference, $\Delta_1{}^2$, equal to zero. Also, when y has very nearly converged to h, rounding errors are usually amplified by (8.25), especially by the $\Delta_1{}^2$ term, so as to make Aitken's extrapolation more harmful than helpful. Again, some discretionary test is needed.

The foregoing discussion is couched in terms of a single iterative variable y, but the entire argument succeeds equally well if the variable is a *vector*, $\mathbf{y}_i$. Thus in our iteration algorithm for the largest eigenvector we may and should employ this Aitken extrapolation scheme to speed convergence, although the shifting technique should be employed first. The warnings about testing for small and noisy second differences are at least as applicable to vectors as they are to scalar quantities, although we have an advantage that one component of a vector, if both small and noisy, may be replaced by a zero – assuming that the other components are stable enough to extrapolate firmly.

The smallest eigenvalue

Let us suppose that the largest λ is positive and that we have just found it. We may now shift all the eigenvalues negatively by approximately that amount. This shifts our known eigenvalue to the neighborhood of the origin and shoves the most negative λ out into prominence as the λ that now has the largest absolute value. Iteration with the shifted matrix will now converge to the λ that is algebraically the smallest and to its associated eigenvector. Thus the device of shifting makes the iteration method quite practical for finding both the largest and the smallest λ's of a matrix.

Unfortunately we must here add a slightly sour note. In many engineering eigenvalue problems the λ's are all positive and arranged in a somewhat logarithmic manner, being bunched toward the origin as in Figure 8.4. With this

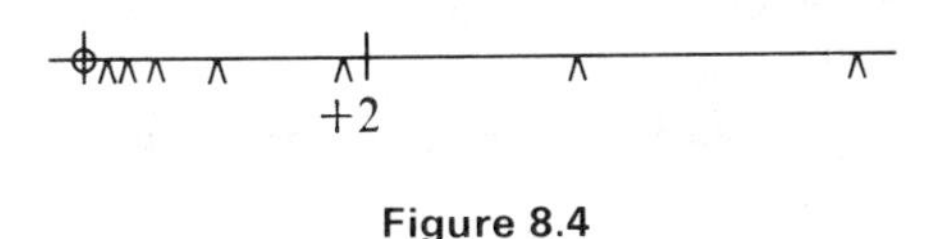

Figure 8.4

geometry, shifting in a negative direction will indeed bring the smallest λ to prominence, but it will do nothing to separate him from his closest neighbor – who is uncomfortably close. Thus the shift may merely lead to a highly inefficient iteration. For such configurations more drastic measures are needed. The obvious one is to invert the matrix A, although this may be expensive. We have

$$A\mathbf{x} = \lambda\mathbf{x}$$

hence

$$\mathbf{x} = \lambda A^{-1}\mathbf{x} \qquad \text{and} \qquad \frac{1}{\lambda}\mathbf{x} = A^{-1}\mathbf{x} \tag{8.26}$$

or

$$\mu\mathbf{x} = B\mathbf{x}$$

Inversion of the matrix gives the same eigenvectors again but inverts the eigenvalues. Thus λ's that were close to zero are now quite large and the narrow separations between the smallest λ's are thus magnified. A simple extension of this philosophy will allow any intermediate region to be brought into prominence for searching by the iteration technique. (The student should devise a method to find a λ known to lie near $+2$ in Figure 8.4 if the distribution of eigenvalues is believed to approximate the configuration shown there.) Except when rather precise and specialized knowledge is available, however, this approach to interior eigenvalues is less satisfactory than our rotational methods of Chapter 13.

Other eigenvalues by the power method

The repeated premultiplication of an arbitrary vector by our matrix A, called the iteration or power method, converges to the eigenvector associated with the largest eigenvalue, λ_1 – if it converges at all. Very often the convergence is rapid and the process is certainly the one to be tried first if this largest λ and its vector are the only information desired. We have also seen how the algebraically *smallest* λ_n and its eigenvector may be found from the same process merely by first shifting all the λ's sufficiently far in a negative direction by subtracting a suitable constant from the main diagonal of A.

We now ask if our process has any simple extensions that would permit us to find the other λ's – possibly even *all* λ's. The answer is yes, and no! The process can be extended to find the second largest λ and maybe the third largest, but at an increase in computational cost per λ and with a decrease in accuracy that combine to make the method rapidly unattractive relative to other algorithms. The breakeven point is a complicated function of the precision of our computer arithmetic and the size and condition of the matrix, but most experts will shy away from finding more than the three largest and the three smallest λ's by the power method.

We can make our iterations seek the second biggest eigenvalue, λ_2, most expediently by "removing" λ_1 and $\mathbf{x}_1$ from A. One technique, due to Hotelling, produces a new matrix with the same λ's except for λ_1, which is now zero. Thus

λ_2 has become the largest eigenvalue and our iterative process will seek it. This new matrix, B, is found by

$$B = A - \lambda_1 \mathbf{x}_1 \mathbf{x}_1^T \qquad \text{with} \qquad \mathbf{x}_1^T \mathbf{x}_1 = 1 \tag{8.27}$$

where $\mathbf{x}_1$ is the eigenvector of length unity corresponding to λ_1. Note that $\mathbf{x}\mathbf{x}^T$ is the backward or matrix product of two vectors. The student should verify that B satisfies the equation

$$B\mathbf{x}_i = \lambda_i \mathbf{x}_i \qquad i = 2, 3, \ldots, n$$

and also explore what happens when $\mathbf{x}_1$ is used. The orthogonality of the eigenvectors of symmetric matrices is essential here.

This *deflation* process, together with all deflation processes, suffers from any imprecision in the vector $\mathbf{x}_1$. The matrix B only has the same λ's and $\mathbf{x}_j$'s as A to the extent that $\mathbf{x}_1$ was accurately "removed." Since $\mathbf{x}_1$ is never accurately known, the remaining λ's and $\mathbf{x}$'s have been somewhat eroded by the deflation process. Thus λ_2 and $\mathbf{x}_2$ may be expected to be worse—the next deflation by them to produce a still more degraded matrix—and the whole process to get rapidly into trouble. Brute-force techniques such as double-precision computation for determining λ_1 and $\mathbf{x}_1$ will postpone the inevitable, but at an unattractive computational cost.

An alternative strategy to deflation is iteration with the original matrix, A, but on a vector that is perpendicular to $\mathbf{x}_1$, that is, a vector that is arbitrary except that its b_1 coefficient (8.20) is zero. Such a vector may easily be produced by *orthogonalizing* an arbitrary vector with respect to $\mathbf{x}_1$, and, provided our iterations are precise, will not lose its orthogonal property under multiplication by A. Since b_1 is zero, convergence to $\mathbf{x}_2$ will occur. This process will work but it suffers from the same precision troubles as deflation: If $\mathbf{x}_1$ is not precise, then b_1 will not be zero. Further, our iteration by A contains rounding errors that tend to erode the orthogonality of our vector even more, and frequent reorthogonalization with respect to $\mathbf{x}_1$ is necessary. The whole strategy is inferior to deflation because the crucial property is carried by the iterated *vector* rather than by the *matrix*. Thus our iterations on the vector further degrade the orthogonality when we work with a vector, whereas with deflation we merely have to live with the original imprecision of the matrix B.

Vector orthogonalization

If we insist on the inferior eigenvalue strategy, the vector orthogonalization may be performed in two ways. An arbitrary vector $\mathbf{y}$ may be orthogonalized with respect to a known vector in approximately $6n$ operations by the formula

$$\mathbf{z} = \mathbf{y} - \mathbf{x} \cdot \frac{\mathbf{x} \cdot \mathbf{y}}{\mathbf{x} \cdot \mathbf{x}}$$

where **x** is the known vector. Alternatively, we may orthogonalize an arbitrary vector with respect to the eigenvector **x** by premultiplying **y** by our original matrix suitably shifted by λ_1. Thus

$$\mathbf{z} = (A - \lambda_1 I)\mathbf{y}$$

will give us a suitably orthogonalized vector in $2n^2 + n$ operations. On the basis of numbers of operations the first method of orthogonalization seems distinctly more efficient, but the decision is not that simple. Our iterative algorithm tends to give us λ with many more correct figures than it gives for the components of **x**. Thus the orthogonalization process that uses lambda may well give a vector that is more nearly perpendicular than the cheaper process, which uses the less precise **x**. From the point of view of programming, the matrix multiplication process is also somewhat more attractive, since it uses the shifting and matrix multiplication routines that are already available.

Unsymmetric matrices

Much of the material in this chapter remains true for unsymmetric matrices. If the extreme eigenvalues are simple, isolated from their neighbors, then repeated premultiplication of an arbitrary vector will cause its convergence to the largest eigenvector. Shifting the other extreme eigenvalue into prominence is still a good tactic, permitting its expedient determination. Only when we turn to deflation and orthogonalization techniques for suppressing the vectors already found are we in trouble, for the eigenvectors of unsymmetric matrices are not usually orthogonal. Instead, they exhibit a biorthogonal relationship with the eigenvectors of the transposed matrix. This is already a sufficiently complicated arrangement so as to call the method into question on grounds of computational inefficiency for all but the extreme eigenvalues. Geometric interpretation of the quadratic form also becomes difficult, depriving us of a valuable intuitional tool. Finally, there is no necessity that the eigenvalues of unsymmetric matrices be real numbers. If the largest one happens to be complex, the iteration algorithm of this chapter will have its problems.

CHAPTER 9

FOURIER SERIES

The reasons for expanding functions in series of other functions are varied. The original function may be analytically intractable yet have an expansion in other functions that permits a desired operation – such as analytic integration. In another direction, a series expansion may grow naturally out of a partial differential equation "separated" in some standard coordinate system. Or a numerically specified function may need an analytic representation. Whatever the source, expansion in some kind of series is a frequent tool of the numerical analyst. The commonest series is, of course, the power series. Its utility stems from the fact that powers of x are easy to integrate, difference, and generally manipulate in the solution of larger problems. Moreover, the power series is usually the easiest to obtain. Unfortunately, its virtues stop there. The successive powers of x are not orthogonal to each other, their derivatives tend to get out of hand as n gets larger, and they do not approximate periodic functions very well.

The Fourier series complements the power series quite nicely on these points. Its terms *are* orthogonal, its derivatives neatly bounded, and it expresses periodic functions very well indeed! Both Fourier and power series, however, share one serious defect: They cannot handle functions with vertical or horizontal asymptotes. For such geometries we need a fraction bar somewhere – a denominator that can become large, or zero. If your function has such geometries, do not waste time wondering whether to expand in powers of x or in cosines, for neither will do the job well. Faced with a vertical asymptote you probably need to remove a singularity from something—for which we refer you to Chapter 15. Horizontal asymptotes succumb to a variety of treatments, but series expansion is rarely one of them.

In one sense Fourier series are merely the simplest of a class of orthogonal expansions. Analytic solutions to the simpler partial differential equations of mathematical physics almost always require expansion of some given boundary condition in a series of orthogonal eigenfunctions. Thus we have Fourier–Bessel series, Fourier–Jacobi series, series of Chebyshev polynomials, series of Legendre polynomials—all with strong family resemblances that are clearly discussed in many other places. In this chapter, however, we are primarily concerned with the numerical process of expanding more-or-less well-behaved functions in series of sines and cosines, which possess enough special properties to accord them specific treatment. Thus, having made our bow in the direction of the other orthogonal expansions, we shall here refer to them no more.

The role of symmetry

Since $\cos n\theta$ is an *even* function of its argument, it follows that a cosine Fourier series will be even and had better be used to expand even functions. Similarly, $\sin n\theta$ is odd and appropriate for odd expansions. The details of Fourier manipulations are slightly simpler if the even and odd expansions are treated separately—indeed, one usually splits a general function into the sum of an even and an odd function, which can always be done trivially,[1] and expands the two functions separately. Also in many practical problems the symmetry inherent in the physics limits us to the one expansion or the other.

Two kinds of Fourier series commonly occur in numerical work: the *infinite* Fourier series and the *finite* Fourier series. The infinite series is the one familiar to students of calculus and higher forms of analysis. Here the given function $f(\theta)$ is completely specified over one period and we desire to expand it into its series so as to make it analytically more tractable. There is no theoretical limit to the number of terms we may find and use. Our patience and the precision we desire will control. The finite series arises when the $f(\theta)$ to be expanded is specified at only a finite number N of points, usually evenly spaced around the circle. Here we have N pieces of information so we may compute only N coefficients. If the precision of this series is inadequate, we are stuck. Unless we can obtain $f(\theta)$ at some additional points, the N-term series will have to do.

[1] Given a general function $f(\theta)$, we define

$$g(\theta) = f(\theta) - f(-\theta)$$
$$h(\theta) = f(\theta) + f(-\theta)$$

and observe that $g(\theta)$ is odd while $h(\theta)$ is even.

The infinite even series is usually expressed as

$$f(\theta) = \frac{a_0}{2} + \sum_{n=1}^{\infty} a_n \cos n\theta \tag{9.1}$$

and the orthogonality of the cosines under integration over a period permits us to evaluate the a_n separately, a considerable saving of computational labor that is desirable though not essential. We have

$$a_k = \frac{1}{\pi}\int_{-\pi}^{\pi} f(\theta)\cdot\cos k\theta\cdot d\theta = \frac{2}{\pi}\int_0^{\pi} f(\theta)\cdot\cos k\theta\cdot d\theta \qquad k = 0, 1, \ldots \tag{9.2}$$

The factor of 2 under the a_0 coefficient of (9.1) is included so as to provide the universal definition (9.2) that holds for a_0 as well. [Some authors, preferring their universality in (9.1), omit the factor of 2 there—but then it pops up as a special formula in (9.2) for the computation of a_0.] The infinite odd series has no such peculiarities, being merely

$$f(\theta) = \sum_{n=1}^{\infty} b_n \sin n\theta \tag{9.3}$$

where the b_k coefficients are given by

$$b_k = \frac{1}{\pi}\int_{-\pi}^{\pi} f(\theta)\cdot\sin k\theta\cdot d\theta = \frac{2}{\pi}\int_0^{\pi} f(\theta)\cdot\sin k\theta\cdot d\theta \tag{9.4}$$

In both (9.2) and (9.4) the second form follows from noting that the total integrand is even.

If the integrals in (9.2) and (9.4) can be carried out analytically, the expansion is straightforward. If, however, we must perform these quadratures numerically, then we must be careful to take enough points—lest we try to squeeze out more information than we put in. Indeed, if it is the simple Fourier series expansion that we need, we should use the finite series, thereby going directly to summations and dispensing with irrelevant intermediate integrals of (9.2) and (9.4). If, however, we are attempting one of the more esoteric expansions, say a Fourier–Bessel expansion, which has no direct finite analogue, we have no choice but to approximate the integrals by some quadrature procedure over a finite set of points. But remember that N coefficients require at least N points!

The role of orthogonality

In matters computational, the property orthogonal is very closely wedded to the subject of efficiency. The expansion of a more or less arbitrary function into a systematic series of well-behaved functions is less work by far if the well-behaved functions happen to be mutually orthogonal. The principal saving arises from the fact, stressed repeatedly, that the coefficients of the expansion may then be calculated separately. Without orthogonality we must resort to solution of linear simultaneous algebraic equations, the size of the system being determined by the number of terms we seek for our expansion. The computational labor is of order k versus k^3. Further, having obtained an expansion of 17 terms, a decision to add 2 more terms for greater accuracy commits us to solve a 19×19 system, the work previously expended on the 17×17 system being largely irrelevant. With an orthogonal system, of course, we merely calculate the two additional coefficients.

In addition to these two manipulative efficiencies, orthogonal expansions also enjoy an optimal property that lends luster to their reputation and adds a modest increment to their already considerable computational superiority. Consider the formal expansion

$$f(\theta) = a_1 g_1(\theta) + \cdots + a_k g_k(\theta) + \cdots \tag{9.5}$$

where the functions $\{g_r(\theta)\}$ are mutually orthogonal and, though this is only a matter of convenient detail, normalized to unity according to the relations

$$\int_{-\pi}^{\pi} g_i(\theta) \cdot g_j(\theta) \cdot d\theta = \begin{Bmatrix} 0 & i \neq j \\ 1 & i = j \end{Bmatrix} \qquad i, j = 1, 2, \ldots \tag{9.6}$$

Suppose we decide to use the first k members of the set $\{g_r(\theta)\}$ and to expand $f(\theta)$ in them as well as we can by the criterion of the least-squared integrated residual. That is, we desire to choose the coefficients a_i in (9.5) so as to minimize the integral

$$S = \int_{-\pi}^{\pi} [f(\theta) - a_1 g_1(\theta) - \cdots - a_k g_k(\theta)]^2 \cdot d\theta$$

Focusing our attention temporarily on the coefficient a_j we differentiate S with respect to a_j, setting the derivative to zero. We have

$$0 = 2 \int_{-\pi}^{\pi} [f(\theta) - a_1 g_1(\theta) - \cdots - a_k g_k(\theta)] \cdot g_j(\theta) \cdot d\theta$$

which immediately becomes

$$\int_{-\pi}^{\pi} [a_1 g_1(\theta) + \cdots + a_k g_k(\theta)] g_j(\theta) \cdot d\theta = \int_{-\pi}^{\pi} f(\theta) \cdot g_j(\theta) \cdot d\theta$$

and, because of the orthogonality relation (9.6), subsides into merely

$$a_j \int_{-\pi}^{\pi} [g_j(\theta)]^2 \cdot d\theta = \int_{-\pi}^{\pi} f(\theta) g_j(\theta) \cdot d\theta$$

or

$$a_j = \int_{-\pi}^{\pi} f(\theta) \cdot g_j(\theta) \cdot d\theta$$

which is just the standard formula for evaluating the typical coefficient of a Fourier series. But here we were led to it not by a device of computational expediency but by a principle of error minimization. It tells us that our standard orthogonal expansion of $f(\theta)$ in the k functions $g_1(\theta), \ldots, g_k(\theta)$ gives us the *best* representation of $f(\theta)$ that we can get for any linear combination of those k functions. Thus we need not consider "economizing" our expansion, for it was already economized in the process of being born. The optimal property is due entirely to the fact that the set of g functions is orthogonal.

While this least-squares optimal property is desirable and comforting, we must be sure that we understand its limitations. It is true that we may not shorten a rather too long orthogonal expansion by some different method of computing the coefficients, but we might well find a different representation of $f(\theta)$ using the same, or fewer, $\{g_k(\theta)\}$ in a different arrangement – such as the *ratio* of two linear combinations, for example, or perhaps by using them multiplicatively in some imaginative way. The "best" expansion we have achieved is best within a very limited, though very useful class, the linear combination. And, of course, we might have done much better had we used a different set of functions, $\{h_k(\theta)\}$ in which to make our expansion.

How many terms are needed?

While there are impressive theorems that guarantee the existence of Fourier series for quite wild discontinuous functions, these theorems are uncomfortably silent about how many terms of the series will be needed to yield reasonable accuracy. Sines and cosines are remarkably continuous functions, so they may perhaps be pardoned if they rebel somewhat when asked to represent the functions of Figures 9.1 and 9.2. In one there are discontinuities in the function, in the other the first derivative is discontinuous at $\pm\pi$. Both functions are even and have cosine series representations (the theorems are, after all, correct) but

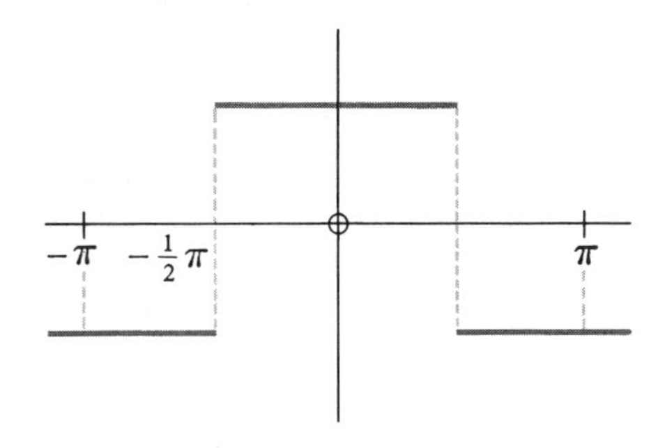

Figure 9.1

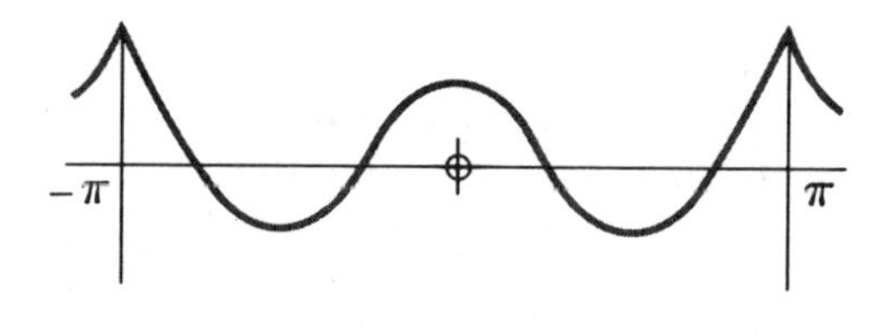

Figure 9.2

the series do not settle down to their functions very quickly—a form of cosinely protest.

Some idea of the number of terms needed for various configurations may be gained by integrating the formula (9.2) for a_k by parts a few times. If we assume $f(\theta)$ is continuous throughout the period (and hence, because of evenness, continuous everywhere), we have

$$\pi a_k = \int_{-\pi}^{\pi} f(\theta) \cdot \cos k\theta \cdot d\theta = \cancel{\left[f(\theta) \frac{\sin k\theta}{k} \right]_{-\pi}^{\pi}} - \frac{1}{k} \int_{-\pi}^{\pi} f'(\theta) \cdot \sin k\theta \cdot d\theta$$

$$= -\frac{1}{k}\left[f'(\theta)\left(\frac{\cos k\theta}{-k} \right) \right]_{-\pi}^{\pi} - \frac{1}{k^2} \int_{-\pi}^{\pi} f''(\theta) \cdot \cos k\theta \cdot d\theta$$

where the first term in brackets disappears at both limits because of the $\sin k\theta$. Generally $f'(\pi)$ will not be zero, so our first surviving term for a_k will be

$$(-1)^k \frac{2}{k^2} f'(\pi)$$

which says that the a_k decrease as k^{-2}. This is not too encouraging, for it says that as many as 1000 terms will be required for six-figure accuracy if k is the only agent causing convergence of the Fourier series. Faced with such facts we frequently settle for less accuracy. If, on the other hand, we are blessed with an even $f(\theta)$ that possesses *zero* first derivatives at $\pm\pi$, then the first surviving term in our repeated integration by parts will behave as k^{-4} (why not k^{-3}?)—a much pleasanter prospect that requires only 30 terms. Continuity of higher derivatives yields correspondingly higher rates of convergence, but such functions are correspondingly harder to achieve in practice. The convergence rate of the discontinuous even function may be explored by the pessimistic student. He will not be encouraged by his findings.

Fortunately, in many situations where Fourier series are appropriate, the formulator of the problem has some control over the function $f(\theta)$. He may, for example, still be able to represent it usefully not as a pure Fourier series but as the sum of (say) a parabola and a Fourier series. Thus, if $f(\theta)$ were given by Figure 9.2, a parabola may be subtracted off that is tangent to $f(\theta)$ at $\pm\pi$, thus leaving an even function with zero for its derivatives there to be expanded by Fourier series – yielding k^{-4} convergence instead of k^{-2}. Such devices should always be sought and used whenever possible.

Examination of the odd series coefficients yields similar results. We have

$$\pi b_k = \int_{-\pi}^{\pi} f(\theta) \cdot \sin k\theta \cdot d\theta = \left[f(\theta) \cdot \frac{\cos k\theta}{-k} \right]_{-\pi}^{\pi} + \frac{1}{k} \int_{-\pi}^{\pi} f'(\theta) \cos k\theta \cdot d\theta$$

and conclude that the odd series converges like $1/k$ unless $f(\pi)$ is zero – a result that can always be produced by subtracting off a linear function first (Figure 9.3). With $f(\pi)$ equal to zero, convergence of the odd series goes as

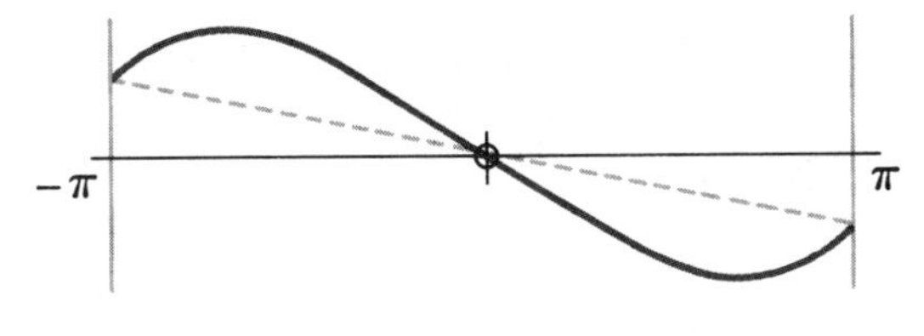

Figure 9.3

k^{-3}, and a cubic correction term before resorting to the Fourier series would yield k^{-5} – a good bargain if the intended use for the expansion can assimilate such terms.

The Gibbs phenomenon

When it is not possible to abolish a discontinuity in $f(\theta)$ then the Fourier series tends to overshoot the sudden rise and thereafter oscillate for quite

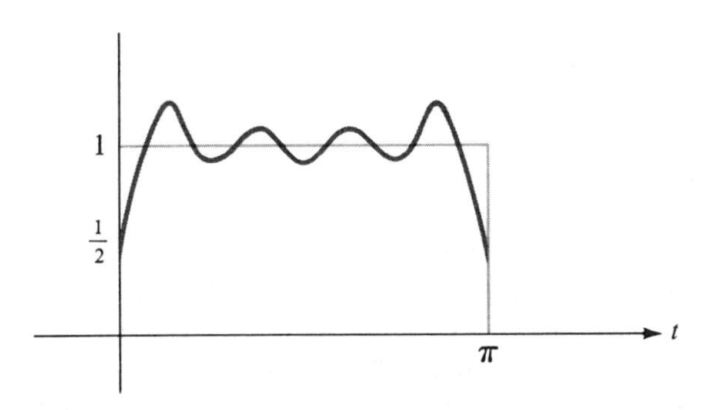

Figure 9.4. The Gibbs phenomenon

a while about the true function. For a square wave the result, qualitatively, looks like Figure 9.4. Taking more terms of the series does *not* get rid of this Gibbs phenomenon – indeed, in the limit the Fourier series converges to our square wave with "ears" about 9 per cent higher (Figure 9.5). This overshooting

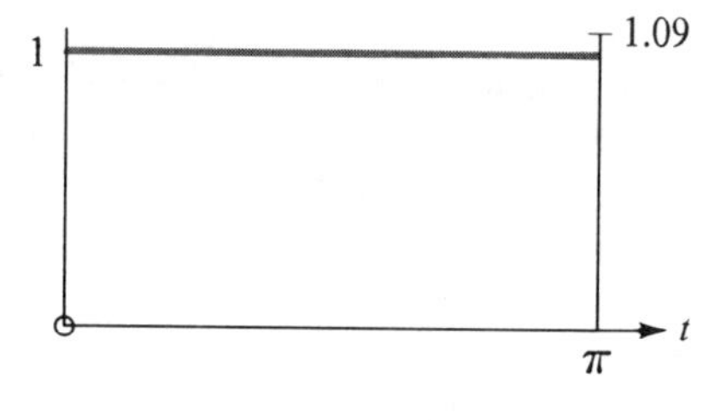

Figure 9.5

and subsequent damped oscillation are quite destructive of accuracy in the neighborhood of a discontinuity. While it is far preferable to abolish the discontinuity from $f(\theta)$ *before* expanding it and thereby to avoid the Gibbs phenomenon entirely, nevertheless it is possible to ameliorate the overshoot and oscillations by use of Lanczos's σ factors. We compute the a_k and b_k of the infinite Fourier series in the standard manner, but when we come to evaluate the term that we are using in the infinite series, we *evaluate*

$$f(\theta) = \frac{a_0}{2} + \sum_{n=1}^{m} \left[\frac{\sin(n\pi/2m)}{n\pi/2m}\right] \cdot (a_n \cos n\theta + b_n \sin n\theta)$$

where m is the last term in the infinite series that we are using. This σ factor greatly damps the Gibbs oscillations, reducing them by about a factor of 9. For an example and the derivation see Hamming [1962, p. 297] or Lanczos [1956].

Finite Fourier series

When the function $f(\theta)$ is given only numerically on a fixed set of points we should turn to finite Fourier series. If the points are evenly spaced around the circle, we have computational algorithms that are considerably more convenient and efficient than for the corresponding infinite Fourier series. The sine and cosine functions satisfy orthogonality relations with the respect to *summation* over these points that are quite analogous to the orthogonality relations with respect to integration. In addition, we may employ three-term recurrence relations to avoid the evaluations of sines and cosines of multiple angles. The procedure, very straightforward, seems sufficiently obscure to warrant some detail. We choose the even expansion.

Formally, the even finite Fourier series might simply be a truncation of its infinite brother (9.1) except for one peculiarity: If the number of *points* on which it is based is *even* (Figure 9.7), then both the coefficients a_N and a_0 are divided

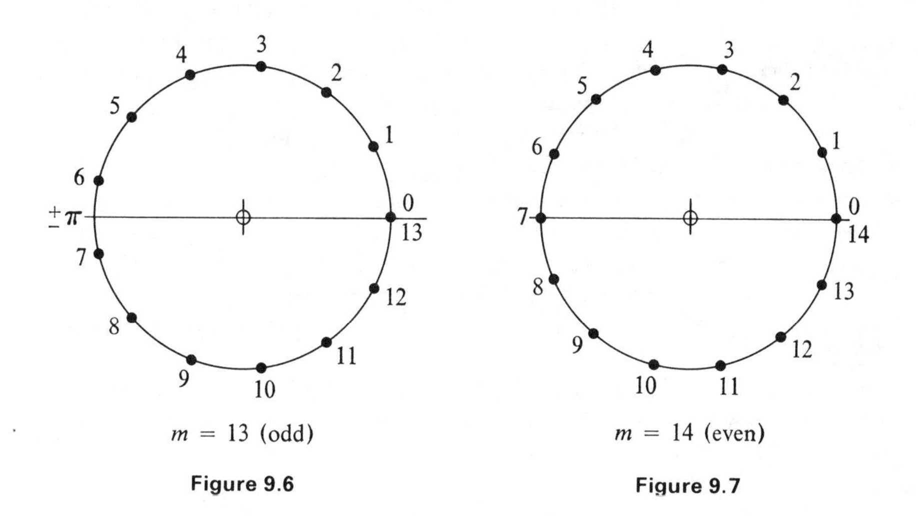

Figure 9.6

Figure 9.7

by 2 if we wish a common evaluation formula (9.9) to hold for all a_k. Thus we have

$$f(\theta) = \frac{a_0}{2} + \sum_{n=1}^{N-1} a_n \cos n\theta + \frac{a_N}{2} \cos N\theta \tag{9.7}$$

if m is even but we have

$$f(\theta) = \frac{a_0}{2} + \sum_{n=1}^{N} a_n \cos n\theta \tag{9.8}$$

if m is odd. [For odd *functions* $f(\theta)$ we have the sine series, but it has no divisions of any b_k by 2.] To evaluate the a_k we have

$$a_k = \frac{2}{m} \sum_{j=0}^{m-1} f(\theta_j) \cdot \cos k\theta_j \qquad k = 0, 1, \ldots, N \tag{9.9}$$

where

$$\theta_j = \frac{2\pi}{m} j \qquad j = 0, 1, \ldots, m - 1$$

and

$$N = \left[\frac{m}{2}\right] \tag{9.10}$$

Note that there are m distinct values of θ evenly distributed around the full circle (Figures 9.6 and 9.7) with the first point θ_0 always being taken at the origin in these formulas. The summation in (9.9) is over each point exactly once. The computations for odd and even numbers of points have minor differences that are best understood in terms of the diagrams.

Consider an odd number of points (13 in our example of Figure 9.6). We are making a cosine expansion, so $f(\theta)$ must be an *even* function. Because of this symmetry in f we see, on looking at Figure 9.6, that there are only seven independent values of f. The functional values f_1 through f_6 are duplicated by f_{12} through f_7, respectively. We can therefore expect to calculate only seven independent Fourier coefficients. Since our subscripts all begin with zero, this implies that N is 6. (The expression [13/2], being "the largest integer in 13/2," gives the correct result, but your author finds the argument easier to remember than the formula.) When we calculate a_k we can reduce our labor by noting that we need merely double the contribution of the points 1 through 6 and forget 7 through 12. Thus our formula becomes

$$a_k = \frac{2}{m}\left[f(0) + 2\sum_{j=1}^{N} f(\theta_j)\cdot\cos k\theta_j\right] = \frac{4}{m}\cdot\sum_{j=0}^{N} f(\theta_j)\cos k\theta_j - \frac{2}{m}f(0) \tag{9.11}$$

the second form being slightly neater later on.

When the number of points is *even* our argument leads to slightly different results. Figure 9.7 depicts 14 points around the circle. Even symmetry of $f(\theta)$, however, decrees that f_1 through f_6 are duplicated by f_{13} through f_8, respectively, so that we have only eight independent values of f. Hence we can compute only eight independent Fourier coefficients and N is 7. (Again $[m/2]$ gives the correct result.) This time in the calculation of a_k two points enter with weight 1 while the others enter with weight 2. We have

$$\begin{aligned} a_k &= \frac{2}{m}\left\{f(0) + 2\sum_{j=1}^{N-1} f(\theta_j)\cos k\theta_j + (-1)^k f(\pi)\right\} \\ &= \frac{4}{m}\sum_{j=0}^{N} f(\theta_j)\cdot\cos k\theta_j - \frac{2}{m}\left[f(0) + (-1)^k f(\pi)\right] \end{aligned} \tag{9.12}$$

To make sure he understands these points, the student should now go through the corresponding arguments for the expansion of *odd* $f(\theta)$ in sine series, verifying that the appropriate formulas are

$$f(\theta) = \sum_{n=1}^{N} b_n \sin n\theta \tag{9.13}$$

and

$$b_k = \frac{2}{m} \sum_{j=1}^{m-1} f(\theta_j) \cdot \sin k\theta_j \qquad k = 1, 2, \ldots, N \tag{9.14}$$

where

$$\theta_j = \frac{2\pi}{m} j \qquad j = 0, 1, \ldots, m - 1$$

and

$$N = \left[\frac{m-1}{2}\right] \tag{9.15}$$

Note that (9.15) is formally different from (9.10). In counting the number of independent values of $f(\theta_k)$ one should keep in mind that odd symmetry imposes the *fixed value of zero* on $f(0)$ and that $f(\pi)$, if it occurs, will always be multiplied by zero in the form of $\sin k\pi$ and hence cannot affect any coefficient in the series. Because of these facts we may be quite cavalier about our computational formulas obtaining

$$b_k = \frac{4}{m} \sum_{j=1}^{N} f(\theta_j) \sin k\theta_j \tag{9.16}$$

for both odd and even numbers of points, m.

Recursive calculation of the Fourier coefficients

In Chapter 1 we showed that the efficient way to evaluate finite Fourier cosine series

$$f(\theta) = \sum_{n=0}^{N} A_n \cos n\theta \tag{9.17}$$

when the coefficients A_n and the angle θ are known is to set up the recurrence

$$c_k = 2\cos\theta \cdot c_{k+1} - c_{k+2} + A_k \qquad k = N, N-1, \ldots, 0 \tag{9.18}$$

starting with

$$c_{N+2} = c_{N+1} = 0$$

and concluding with

$$f(\theta) = c_0 - c_1 \cdot \cos\theta \tag{9.19}$$

To achieve correspondence between (9.17) and (9.7) or (9.8), we set

$$A_n = a_n \qquad n = 1, 2, \ldots, N-1$$

but

$$A_0 = \frac{a_0}{2}$$

and

$$A_N = \begin{cases} \dfrac{a_N}{2} & \text{for } m \text{ even} \\ a_N & \text{for } m \text{ odd} \end{cases}$$

In this scheme we need evaluate only one cosine, and each step of the recurrence requires only one multiplication and two additions. The scheme follows directly from the three-term recurrence for the functions

$$\{\cos n\theta\} \qquad n = 0, 1, 2, \ldots$$

The formula (9.19) may be verified directly by eliminating the A_k's from (9.17) by the use of (9.18), collecting the coefficients of the several c_k, and then employing the identity

$$\cos(n-1)\theta - 2\cos\theta \cdot \cos n\theta + \cos(n+1)\theta = 0 \tag{9.20}$$

to show that almost everything disappears.

If we now look at our formula (9.9) for computing the Fourier coefficients, we see that it may easily be recast into a form that permits the same recurrence calculation to be employed. We merely rewrite

$$\cos k\theta_j = \cos\left(k\frac{2\pi}{m}\right)j = \cos j\varphi_k$$

and we see that φ_k is a known constant throughout the summation in (9.9). The $f(\theta_j)$ become the known coefficients A_j, so (9.9) may be written

$$a_k = \frac{4}{m}\left[\sum_{j=0}^{N} A_j \cos j\varphi_k\right] - \frac{2}{m}f(0)$$

and the quantity in brackets, being of the form (9.17), may be evaluated by the recurrence (9.18). This is, of course, the expedient method.

Since the functions $\{\sin n\theta\}$ also satisfy the same three-term recurrence (9.20) as the cosine—that is,

$$\sin(n-1)\theta - 2\cos\theta \cdot \sin n\theta + \sin(n+1)\theta = 0$$

we may easily construct an analogous method for evaluating odd finite Fourier series – and the scheme also applies directly to the calculation of their coefficients b_k through recasting (9.14) as

$$b_k = \frac{4}{m}\left[\sum_{j=1}^{N} B_j \sin j\varphi_k\right]$$

with

$$\varphi_k = \frac{2\pi}{m}k$$

and

$$B_j = f(\theta_j)$$

as before. The recurrence for the sine series

$$f(\theta) = \sum_{n=1}^{N} B_n \sin n\theta$$

is

$$c_k = 2\cos\theta \cdot c_{k+1} - c_{k+2} + B_k \qquad k = N, N-1, \ldots, 1$$

starting with

$$c_{N+2} = c_{N+1} = 0$$

and finishing with

$$f(\theta) = c_1 \cdot \sin\theta$$

An example of discontinuity removal

By way of simple example we take the even function $e^{-|x|}$ on ± 1. This function is even and continuous but its derivative has two discontinuities, the larger at the origin, the smaller at ± 1. Transferring our function to $\pm\pi$ we have

$$F(\theta) = e^{-|\theta/\pi|} \qquad F'(\theta) = -\frac{1}{\pi}e^{-\theta/\pi} \qquad 0 < \theta < \pi$$

and the slope at zero is $-\pi^{-1}$ while that at π is $-1/(\pi e)$. The trouble at the origin may be removed by adding a parabola

$$-\frac{1}{2}\left(1 - \frac{\theta}{\pi}\right)^2$$

This curve has the slope π^{-1} at the origin and a slope of zero at π, hence it fixes up one trouble spot without disturbing the other. A similar correction term,

$$\frac{e^{-1}}{2}\left(\frac{\theta}{\pi}\right)^2$$

has the slope zero at the origin and the slope of $(\pi e)^{-1}$ at π, thus repairing the other damage. Our final function for expansion is

$$F(\theta) = e^{-\theta/\pi} - \frac{1}{2}\left(1 - \frac{\theta}{\pi}\right)^2 + \frac{e^{-1}}{2}\left(\frac{\theta}{\pi}\right)^2 \qquad 0 < \theta \leq \pi$$

More generally, any finite discontinuity at θ_1 in the first derivative of an even function may be obliterated by adding a constant times the function

$$\sigma_1(\theta) = \begin{cases} \dfrac{\theta^2}{2\pi} + \dfrac{\pi}{2} - \theta_1 & 0 \leq \theta \leq \theta_1 \\[2ex] \dfrac{1}{2\pi}(\pi - \theta)^2 & \theta_1 \leq \theta \leq \pi \end{cases}$$

which is continuous but which has a simple jump of -1 in its derivative at θ_1 while having zero slope at both the origin and π. If our function has a first derivative jump of $+d_1$ at θ_1 we add $\sigma_1(\theta_1) \cdot d_1$ to compensate for it.

Returning to $e^{-|x|}$ we show in Table 9.1 the finite Fourier series *coefficients* for 20 points (a) without either singularity being removed, and (b) with both singularities removed. We also show the *errors* incurred in two Fourier series at $\theta = \pi/10$, which is one of the fitted points, and at two adjacent points that lie

Table 9.1. Finite Fourier series coefficients for $e^{-|x|}$ on ± 1—with and without corrections for two derivative discontinuities

n	$e^{-\lvert x\rvert}$	$e^{-\lvert x\rvert}$ with corrections
0	0.63264 72382	0.63212 09711
1	0.25397 96159	−0.02550 18057
2	0.03230 69410	−0.00079 13219
3	0.03284 03235	−0.00034 33369
4	0.00909 75034	−0.00005 06318
5	0.01363 33808	−0.00004 54132
6	0.00481 86023	−0.00001 03695
7	0.00860 22700	−0.00001 27451
8	0.00349 04475	−0.00000 38293
9	0.00700 46928	−0.00000 62749
10	0.00157 89858	−0.00000 13156

between the fitted points, $\pi/20$ and $3\pi/20$. These errors are given in Table 9.2 as functions of the number of terms of the finite Fourier series used. Naturally the error of the fitted point goes to zero when all terms of the series are used.

Table 9.2. Errors in the ten-term finite Fourier series representation of $e^{-|\theta/\pi|}$ according to the number of terms used[a]

Argument	No. of Terms Used	Error in $e^{-|\theta/\pi|}$	Error in $e^{-|\theta/\pi|}$ with Corrections
$\frac{\pi}{20}$	4	−0.00038 27768	0.00004 04654
	6	0.01208 97826	0.00000 22584
	8	0.01707 37390	−0.00000 47111
	10	0.01816 95144	−0.00000 56927
$\frac{2\pi}{20}$ (a fitted point)	4	0.03912 83780	−0.00007 99483
	6	0.02067 63949	−0.00002 41657
	8	0.00858 36775	−0.00000 75912
	10	0.00000 00011	−0.00000 00008
$\frac{3\pi}{20}$	4	0.01955 23754	−0.00005 50418
	6	0.00532 93563	−0.00001 30679
	8	−0.00599 08367	0.00000 26183
	10	−0.00917 09008	+0.00000 54670

[a] The function expanded in the last column had two discontinuities in its first derivatives removed before expansion. Note the considerable improvement over the untreated function. Also note, in Table 9.1, the much more rapid convergence of the Fourier series for the function after its first derivative discontinuities have been removed.

Exercises:

1. Find the function that removes a jump of -1 at θ_1 in the second derivative of an *even* function.
2. Find the function that removes a jump of -1 at θ_1 in the first derivative of an *even* function.

Finite orthogonal polynomials

Although the title of this chapter is Fourier Series, sines and cosines have gradually faded during our pursuit of expedient expansion algorithms for intractable functions. In their place have appeared orthogonality relations and three-term recurrence relations which turn out to be the useful properties that the trigometric functions possess. At this point we ask if perhaps those other useful functions, polynomials, might not be endowed with the virtues of orthogonality and recurrence relations, so as to make them more respectable than we implied at the beginning of the chapter. Our quest is limited to orthogonality with respect to summation over a finite discrete set of points $\{x_i, i = 1, \ldots, m\}$ for we already know Legendre, Chebyshev, Laguerre, and Hermite have lent their names to polynomials that are orthogonal with respect to *integration* over various standard intervals and with respect to various standard weight functions. All too often, however, we must conjure with functions defined by experiment over a discrete set of points. If the points are equally spaced, we can use the finite Fourier series expansion but we would like to be able to use polynomials as well. For some manipulations they are much more convenient than cosines.

Without delving into the proof, for which see Ralston [1965], we say immediately that polynomials $p_n(x_i)$ orthogonal with respect to summation over the points $\{x_i\}$ are easily generated from a three-term recurrence relation. More remarkable, the points need not be uniformly or even regularly spaced. *Any set of arbitrary points has its corresponding set of orthogonal polynomials.* The points may even be used with arbitrary weights, though here we shall ignore this embellishment, restricting ourselves to uniform weighting. Again, Ralston gives the more general treatment.

The fundamental recurrence relation is

$$p_{j+1}(x) = (x - \alpha_j) \cdot p_j(x) - \beta_j \cdot p_{j-1}(x) \tag{9.21}$$

where

$$p_0(x) = 1$$

and

$$p_1(x) = (x - \alpha_0) \qquad (\text{or } \beta_0 = 0)$$

These relations, being identities in x, hold for all x, not merely for the m points $\{x_i\}$.

The orthogonality relations have the form

$$\sum_{i=1}^{m} p_k(x_i) \cdot p_j(x_i) = \begin{cases} 0 & k \neq j \\ \gamma_k & k = j \end{cases}$$

and these, applied to (9.21), give

$$\gamma_j = \sum_{i=1}^{m} [p_j(x_i)]^2 \qquad\qquad j = 0, 1, \ldots$$

$$\alpha_j = \frac{\sum_{i=1}^{m} x_i [p_j(x_i)]^2}{\sum_{i=1}^{m} [p_j(x_i)]^2} \equiv \sum_{i=1}^{m} \frac{x_i p_j^2(x_i)}{\gamma_j}$$

$$\beta_j = \frac{\sum_{i=1}^{m} [p_j(x_i)]^2}{\sum_{i=1}^{m} [p_{j-1}(x_i)]^2} \equiv \frac{\gamma_j}{\gamma_{j-1}} \qquad j = 1, 2, \ldots$$

as the computational formulae. The calculations may be arranged in a convenient array

	x_1	x_2	x_3	$\cdots$	x_m	α	β	γ	x
$p_0(x_i)$	1	1	1	$\cdots$	1	α_0	0	m	1
$p_1(x_i)$	$(x_1 - \alpha_0)$	$\cdots$				α_1	β_1	γ_1	
$p_2(x_i)$						α_2	β_2	γ_2	
$\vdots$						$\vdots$	$\vdots$	$\vdots$	
$p_{m-1}(x_i)$									
	$f(x_i)$	$f(x_2)$	$f(x_3)$	$\cdots$	$f(x_m)$				

We may calculate $\gamma_0, \alpha_0, p_1(x_i), \gamma_1, \beta_1, \alpha_1, p_2(x_i)$, and so on, in order—using (9.21) for the several $p_j(x_i)$. Once the α_j and β_j are available, the $p_j(x)$ for any arbitrary given x may also be generated by the recurrence (9.21).

Returning to the expansion of a function $f(x_i)$ known at the given points, we write

$$f(x) = \sum_{k=0}^{r} c_k \cdot p_k(x) \qquad r \leq m$$

Multiplying by $p_k(x)$, summing over the x_i, and using the orthogonality relations we have

$$c_k = \sum_{i=1}^{m} f(x_i) \cdot \frac{p_k(x_i)}{\gamma_k} \qquad k = 0, 1, \ldots, r$$

We evaluate each c_k directly as the dot product of one row of the table with row vector $f(x_i)$ at its bottom.

Once the c_k are available we may evaluate $f(x)$ for any general x through use of a recurrence based on (9.21). We define

$$s_k = (x - \alpha_k)s_{k+1} - \beta_{k+1}s_{k+2} + c_k \qquad k = r, r-1, \ldots, 0$$

starting with

$$s_{r+1} = s_{r+2} = 0$$

and ending with

$$f(x) = s_0$$

As always with finite systems, we can evaluate no more than m coefficients, nor are there more than m orthogonal polynomials in our set.

Finite Fourier expansions on an irregular set of points

We have seen how to generate sets of polynomials orthogonal with respect to summation over arbitrary finite point sets $\{x_i\}$. The question then arises of obtaining finite Fourier expansions with equivalent freedom in the choice of the points. Fourier series certainly exist and can be found by pedestrian means. The equation

$$f(\theta) = \sum_{n=0}^{r} A_n \cos n\theta \tag{9.22}$$

may certainly be written out once for each angle in a set of $r + 1$ arbitrary angles

$$\{\theta_i\} \qquad i = 0, 1, \ldots, r$$

and then we have a set of $r + 1$ linear algebraic equations for the $r + 1$ unknowns $\{A_n\}$. We know that the cosines and sines are orthogonal with respect to integration over a circle and also with respect to summation over points *equally spaced* around the circle. It would seem too much to ask that they be orthogonal over a set of arbitrary points around the circle. So is there any hope for avoiding the solution of our $(r + 1, r + 1)$ system of linear equations? It seems that there is.

Our method depends on the observation that $\cos n\theta$ is a *polynomial* of degree n in $\cos\theta$—in fact, it is the Chebyshev polynomial of degree n. We define

$$\cos\theta_i = x_i \qquad i = 0, 1, \ldots, r \tag{9.23}$$

and then (9.22) may be rewritten

$$f(\cos^{-1} x_i) = \sum_{n=0}^{r} A_n T_n(x_i) \tag{9.24}$$

The points $\{x_i\}$ lie on $(-1, 1)$ but are otherwise arbitrary. By the methods of the last section we may define and evaluate a set of polynomials $\{p_k(x)\}$ that are orthogonal over this point set. Thus we may also write (9.24) as

$$f(\cos^{-1} x_i) = \sum_{n=0}^{r} A_n T_n(x_i) = \sum_{k=0}^{r} c_k p_k(x_i) \tag{9.25}$$

and we may evaluate the c_k easily because of the orthogonality of the $\{p_k(x_i)\}$ over $\{x_i\}$. Now any finite series of known polynomials may be rearranged into a series of equivalent degree in *another* series of polynomials, here the Chebyshev polynomials, to give the coefficients A_n. In fact, since we have the identity

$$\sum_{n=0}^{r} A_n \cos n\theta = \sum_{n=0}^{r} A_n T_n(x) = \sum_{k=0}^{r} c_k p_k(x) \tag{9.26}$$

we may use the orthogonal property of the Chebyshev polynomials to give a simple scheme for getting the A_n separately. We take a set of θ_j *equally spaced* around the circle, evaluate the right side of (9.26) for each of these new x's defined by

$$x_j = \cos \theta_j$$

and do the finite Fourier expansion for the A_n.

Finite Fourier series in exponential form

The reader has noticed that the formalisms of finite Fourier series have fragmented into four special cases depending on whether the function is even or odd and on whether the number of points at which it is sampled is even or odd. Easily forgotten factors of two keep cropping up and disappearing in ways that are theoretically explainable but manipulatively annoying. A considerably neater formalism attends the use of complex exponentials instead of sines and cosines. The factors of 2 and special cases disappear. Though the actual arithmetic labor is nearly the same, the regularities of the complex formalism permit one to write more compact computer programs and to take sophisticated advantages of internal symmetries should the number of sample points be factorable in one of several special ways. These are, fundamentally, the same tricks one finds employed in the hand-computational schemes for harmonic analysis that enjoyed considerable prominence around the turn of the century. The details do not belong here, but your author does wish to remind the reader that in almost all computational processes there are special tricks or gimmicks, if you will, that spell the difference between the efficient and the inefficient algorithm. For the longer finite Fourier series, say 16 to 4000 terms, the computational labor by pedestrian use of the recurrence schemes we have described above can be both expensive and progressively inaccurate. Recent work by Gentleman and Sande [1966], Tukey and Cooley [1965], and others has produced machine algorithms that compute long finite Fourier series in terms of factored shorter subseries. The regularity of the exponential formulation also permits efficiencies from such devices as storage of imaginary components with reversed subscripts. Taken together these special devices reduce the computational labor by *factors* of 4 to 40 and computational inaccuracies by *factors* of

30 to 5000. Clearly, anyone interested in finite Fourier series for many datum points is going to have to employ these devices whether he takes the trouble to understand them or not. Fortunately they are embodied in some computer programs available from standard sources.

In the same vein, we should note that all finite Fourier *transforms* are formally indistinguishable from finite Fourier *series*—this observation following directly from the impossibility of carrying out the operation of Riemann integration on a function defined only at a finite number of isolated points. Thus all the remarks about formulations and efficiencies in this section apply with equal force to Fourier transforms. Sande and Gentleman [1966] observe that digital filters, band-limited interpolation, Cauchy products for series multiplication, and the solution of Poisson's equation over simple two-dimensional geometries may all be cast into the formalism of finite Fourier series, thereby making their algorithms available, though not always efficiently, to these problems.

Exponential finite Fourier series

We begin with the familiar finite Fourier series for a general real function of the real variable θ. We have, if m is odd,

$$f(\theta) = \frac{a_0}{2} + \sum_{n=1}^{N} (a_n \cos n\theta + b_n \sin n\theta) \tag{9.27}$$

where the function is sampled at the m points

$$\theta_j = \frac{2\pi}{m} j \qquad j = 0, 1, \ldots, m-1 \tag{9.28}$$

equally spaced around the circle. There are $2N + 1$ constants to be determined, but the largest value N can have is

$$N_{\max} = \left[\frac{m-1}{2}\right]$$

Replacing the trigonometric functions in (9.27) by their exponential forms we have

$$f(\theta) = \frac{a_0}{2} + \sum_{n=1}^{N} \left[\frac{a_n}{2}(e^{in\theta} + e^{-in\theta}) - i\frac{b_n}{2}(e^{in\theta} - e^{-in\theta})\right]$$

$$= \frac{a_0}{2} + \frac{1}{2}\sum_{n=1}^{N} (a_n - ib_n)\, e^{in\theta} + \frac{1}{2}\sum_{n=1}^{N} (a_n + ib_n)\, e^{-in\theta}$$

We now replace θ by $-\theta$ in the first summation. In effect we now sum over the *other N* points of Figure 9.6. Since the a_n are the Fourier coefficients of an even function, they will not be affected by the change of θ to $-\theta$. The b_n, on the other hand, being coefficients of the odd expansion, will have their signs reversed. Thus we have

$$
\begin{aligned}
f(\theta) &= \frac{a_0}{2} + \frac{1}{2}\sum_{n=N+1}^{m-1} (a_n + ib_n)\, e^{-in\theta} + \frac{1}{2}\sum_{n=1}^{N} (a_n + ib_n)\, e^{-in\theta} \\
&= \frac{1}{2}\sum_{n=0}^{m-1} (a_n + ib_n)\, e^{-in\theta} = \sum_{n=0}^{m-1} c_n\, e^{-in\theta}
\end{aligned}
\tag{9.29}
$$

which says that the finite Fourier series for $f(\theta)$ may be written as a series with *complex* coefficients c_n on the complex exponential $e^{-in\theta}$, the summation being over all the points. We have absorbed the factor 2 in the definition of the c_n and used the fact that b_0 is irrelevant in the odd expansion, being multiplied by sin 0.

[The student should follow this argument through for m even, where the coefficient a_N in (9.27) also enjoys a factor of 2 beneath it in the manner of a_0, and for the same geometric reason. Be sure to redraw the figure!]

We may now apply the conventional orthogonality relations to (9.29) with respect to summation over the m points equally spaced around the circle. Thus

$$
\sum_{j=0}^{m-1} f(\theta_j)\cdot e^{ik\theta_j} = \sum_{n=0}^{m-1} c_n \sum_{j=0}^{m-1} e^{i(k-n)\theta_j} = \begin{cases} 0 & k \neq n \\ m\cdot c_n & k = n \end{cases}
$$

with

$$
0 \leq k < m \qquad 0 \leq n < m
$$

so that

$$
c_k = \frac{1}{m}\sum_{j=0}^{m-1} f(\theta_j)\cdot e^{ik\theta_j}
$$

which has none of the factors of 2 that hover around the previous treatment (p. 229) with cosines.

Some persons prefer to absorb the factor m into c_k, giving

$$
f(\theta_j) = \frac{1}{m}\sum_{n=0}^{m-1} C_n\cdot e^{-in\theta_j} \tag{9.30}
$$

$$
C_k = \sum_{j=0}^{m-1} f(\theta_j) e^{ik\theta_j} \tag{9.31}
$$

Using (9.28) and displaying 2π explicitly permits us to write (9.30) and (9.31) as

$$f(j) = \frac{1}{m} \sum_{n=0}^{m-1} C(n) \cdot e^{-2\pi inj/m} \tag{9.32}$$

$$C(n) = \sum_{j=0}^{m-1} f(j) \cdot e^{2\pi inj/m} \tag{9.33}$$

at which point we have achieved a pleasant symmetry. [But remember that

$$C(n) = m(a_n + ib_n)$$

when you unpack this compressed notation.]

Shifting notations only slightly further we let

$$t = j$$

$$\hat{t} = n$$

$$X(t) = f(j)$$

$$\hat{X}(t) = C(n)$$

$$N = m$$

Finally we rewrite (9.32) and (9.33) as

$$\begin{aligned} X(t) &= \frac{1}{N} \sum_{\hat{t}=0}^{N-1} \hat{X}(\hat{t}) \cdot e^{-2\pi it\hat{t}/m} \\ \hat{X}(t) &= \sum_{t=0}^{N-1} X(t) \cdot e^{2\pi it\hat{t}/m} \end{aligned} \tag{9.34}$$

which is the notation employed by Sande and Gentleman. It makes its point about symmetry most effectively, but your author finds it confusing to use. He loses carets too easily.[2]

Once we have expressed our finite Fourier series in exponential form we observe that, formally, it may be viewed as a *power series* in $e^{-i\theta}$. We have

$$f(\theta) = \sum_{n=0}^{m-1} c_n \cdot e^{-in\theta} = \sum_{n=0}^{m-1} c_n z^n$$

where

$$z = e^{-i\theta}$$

Such a series may be turned into a continued fraction and the corresponding rational function by straightforward algorithms. They are discussed in Chapter

[2] So do we. (*Signed*) The Typesetters.

11. If the Fourier series is slowly convergent because of essential nonpolynomic behavior, it often happens that the continued fraction will converge much more quickly and hence is the expedient computational form. In fact, continued fractions will often converge with arguments for which the corresponding power series diverges. This method of improving the convergence of Fourier series has been explored by Wynn [1966] with encouraging results.

Problems

1. Expand

$$y = 1 - t^4$$

as an even function over $(-1, 1)$ in a finite Fourier series. Compare its values at several points with the value of this expansion and also the value of the infinite Fourier series truncated to the same number of terms.

2. Remove the discontinuity in the derivative by subtracting off a suitable quadratic, then carry out the steps of Problem 1.

3. Make up an expansion problem involving an odd function, comparing the values of the expansion with those of the function at critical points. What differences need be remarked?

INTERLUDE: WHAT *NOT* TO COMPUTE

Do you ever want to *kick* the computer? Does it iterate endlessly on your newest algorithm that should have converged in three iterations? And does it finally come to a crashing halt with the insulting message that *you* divided by zero? These minor trauma are, in fact, the ways the computer manages to kick you and, unfortunately, you almost always deserve it! For it is a sad fact that most of us can more easily compute than think—which might have given rise to that famous definition, "Research is when you don't know what you're doing."

Although we have no desire to blunt the enthusiasm of the neophyte computor, many of whose excesses must be charged to the educational process, we would be remiss if we did not raise the spectre of uncritical wastage of computational resources by persons who are old enough to know better. We grudgingly concede that there are times when it is better to use the computer inefficiently than to saddle a professor with a laborious search for a better algorithm; nevertheless enough identifiable nonsense goes on in the computer room to justify a brief but hopefully cautionary exhibition. We begin with a personal experience.

It was 1949 in Southern California. Our computer was a very new CPC (model 1, number 1)—a 1-second-per-arithmetic-operation clunker that was holding the computational fort while an early electronic monster was being coaxed to life in an adjacent room. From a nearby aircraft company there arrived one day a 16 × 16 matrix of 10-digit numbers whose inverse was desired. We really had asked for it. Our matrix-inversion routine was new and barely tested, and the CPC represented a tremendous, if as yet unrealized, increase in computing capacity over a roomful of girls with desk calculators. We had offered to demonstrate our prowess—and in walked this matrix.

We labored for two days and, after the usual number of glitches that accompany any strange procedure involving repeated handling of intermediate decks of data cards, we were possessed of an inverse matrix. During the checking operations, which involved the multiplication of the inverse by the original, it was noted that, to eight significant figures, the inverse was the transpose of the original matrix! A hurried visit to the aircraft company to explore the source of the matrix revealed that each element had been laboriously hand computed from some rather simple combinations of sines and cosines of a common angle. It took about 10 minutes to prove that the matrix was, indeed, orthogonal! We were less than overjoyed, although—philosophically speaking—it was a good test of our inversion routine.

In this day of electronic computation one might argue for the uncritical inversion of such a matrix on the grounds that it only takes fractions of a second while the proof of orthogonality would require minutes and might not then succeed—indeed, the matrix might not even be orthogonal. Why not invert and be done with it? But in 1949 we who had striven mightily for two days had no such perspective. We were angry and with reason. Somebody had not done his proper homework and we had to suffer for it. Your author, in 1970, still opposes this kind of uncritical, shoot-from-the-hip computation. It is an outward and visible sign of an inward and intellectual deficiency. It epitomizes the "Why think? Let the computer do it" reaction that, unchecked, quickly undermines any critical review of either the direction or the value of an investigation. The computer is a precision tool. It should not be used as a bludgeon or a substitute for thought.

Formal mathematical training is a frequent if minor villain in the righteous struggle for efficient computer use. For example, every college sophomore has seen linear algebraic equation solving expressed in two lines as

$$Ax = b$$

so

$$x = A^{-1}b$$

It is not surprising that, faced with an equation set, he invokes a computer library program to invert his matrix A and then multiply it into b. The fact that this sequence of operations takes at least three times as much labor on the part of the computer as a direct solution of the original problem by the elimination of variable technique has never penetrated his consciousness. The glibness of the formal mathematical notation has obscured the realities of the arithmetic process.[1] The cure is simple enough—the student should be asked to solve a 4×4 system, by hand, both ways. But this sort of realism is currently unfashionable—arithmetic labor being deemed suitable only for machines.

At a more difficult but less pernicious level we have the inefficiencies engendered by exact analytic integrations where a sensible approximation would give a simpler and more effective algorithm. Thus

$$\int_0^{0.3} \sin^8\theta \cdot d\theta = [(-\tfrac{1}{8}\cos\theta)(\sin^4\theta + \tfrac{7}{6}\sin^2\theta + \tfrac{35}{24})(\sin^3\theta) + \tfrac{105}{384}(\theta - \sin 2\theta)]_0^{0.3}$$

$$= (-0.119417)(0.007627 + 0.101887 + 1.458333)(0.0258085) + 0.004341$$

$$= -0.0048320 + 0.0048341 = 0.0000021$$

manages to compute a very small result as the difference between two much larger numbers. The crudest approximation for $\sin\theta$ will give

$$\int_0^{0.3} \theta^8 \cdot d\theta = \tfrac{1}{9}[\theta^9]_0^{0.3} = 0.00000219$$

with considerably more potential accuracy and much less trouble. If several more figures are needed, a second term of the series may be kept.

---

[1] On the other hand, perhaps we should be glad he didn't resort to Cramer's rule (still taught as the practical method in some high schools) and solve his equations as the ratios of determinants – a process that requires labor proportional to $n!$ if done in the schoolboy manner. The contrast with $n^3/3$ can be startling!

In a similar vein, if not too many figures are required, the quadrature

$$\int_{0.45}^{0.55} \frac{dx}{1+x^2} = [\tan^{-1} x]_{0.45}^{0.55} = 0.502843 - 0.422854 = 0.079989 \simeq 0.0800$$

causes the computer to spend a lot of time evaluating two arctangents to get a result that would have been more expediently calculated as the product of the range of integration (0.1) by the value of the integrand at the midpoint (0.8). The expenditure of times for the two calculations is roughly 10 to 1. For more accurate quadrature, Simpson's rule would still be more efficient than the arctangent evaluations, nor would it lose a significant figure by subtraction. The student that worships at the altars of Classical Mathematics should really be warned that his rites frequently have quite oblique connections with the external world.

PERVERSE FORMULATIONS

Considerable computational grief arises from inappropriate or even perverse formulations of otherwise tractable problems. The man who insists on solving a boundary-value problem in ordinary differential equations via initial-value techniques may get away with it for a while, but sooner or later he will reap the trouble he has sown. At the very least he pays through inefficiency; at worst, through instability. An extreme example was recently posed by a mathematician who certainly knew better. He discusses the great difficulties of numerically integrating

$$\frac{d^4u}{dx^4} - 24\frac{d^3u}{dx^3} - 169\frac{d^2u}{dx^2} - 324\frac{du}{dx} - 180u = 0$$

subject to "initial" conditions

$$u(0) = u'(0) = 0$$
$$u''(0) = 2$$
$$u'''(0) = -12$$

which suffice to yield the *analytic* solution

$$u(x) = e^{-x} - 2e^{-2x} + e^{-3x} + 0 \cdot e^{30x}$$

—a solution that declines quietly away to zero as x becomes large. In any initial-value integration procedure, however, the hidden horrendously large exponential term will not have a coefficient of exactly zero. Rounding error will see to that. Thus it is meaningless to speak of *the initial-value solution that dies away*, just as it is a waste of time to devise methods for struggling to attain it. There is none, at least none that is stable, for there is no such problem. If a solution is wanted that goes to zero as x becomes large, then one is imposing a *boundary condition at infinity*. Boundary-value integration techniques will have no trouble in solving what is now a quite well-posed stable problem. The difficulties were created by a perverse formulation.

A lesser essay in perversity appears as the first example in a recent book on computing. It begins with the correct assertion that the solution to the differential equation:

$$\frac{dy}{dx} = \frac{x + y}{x - y} \qquad y(1) = 0$$

with the initial condition shown is

$$\ln(x^2 + y^2) = 2\tan^{-1}\frac{y}{x}$$

but then implies that plotting y versus x at small intervals of x between 1 and 2 is a formidable task—which it is not. If only *this formalism* is to be used, the labor is great, but with a moment's thought it becomes clear that x and y are not the proper coordinates and that a shift to the r and θ of polar coordinates will greatly simplify the solution, giving

$$r = e^{\theta} \qquad \text{or} \qquad \theta = \ln r$$

We may now easily take as many points as we wish and use the relations

$$x = r\cos\theta \qquad \text{and} \qquad y = r\sin\theta$$

to give our table in the original coordinates if, indeed, this is desirable. A final interpolation may be necessary to give exactly the x values we seek, but there is nothing that is nasty or complicated about the process. The difficulties were caused entirely by an unfortunate choice of variables, followed by insufficient analysis.

All too many problems fall into this pattern. People have a tendency to visualize their problems in the original variables even when other variables may simplify the equations and solutions. Usually this is a matter of habit rather than any virtue inherent in the variables; indeed, the variables that give the simpler formulation are often found more significant than the original ones when their physical interpretation is finally available.

POOR ENCODING OF STATISTICAL DATA

Although I have chosen to avoid for the most part the complexities of statistical calculations, their prevalence in the computer room provokes me to recall two of their classic if less graceful moments. The first was an educational survey of pupils in grades 1 through 8 that had been completed and recorded on punched cards before somebody decided to include kindergarten. *Zero* having been preempted for some other function, this new grade was encoded as 9—a fact soon conveniently forgotten. The subsequent regressions of age-sensitive factors against grade were, well, peculiar! And in the same vein, the person who encoded verbal responses of YES, NO, DON'T CARE as one, two, three, *respectively,* got the sorts of correlations she deserved.

At a somewhat trickier level, the person who wrote, but did not clearly document, a vector output subroutine so that it needed the *address* of the vector but the *number of items* to be put forth was quite mystified when irate customers who wanted 10 or 12 items were swamped with 24,371 of them—their number of items, *N*, being stored typically in location 24,371. FORTRAN, it must occasionally be remembered, always passes addresses to subroutines regardless of what its notation may imply to the casual observer. This last example is merely one of hundreds of troubles that are produced by programmers who are insensitive to the interests of the users of their programs. They fail to match properly their program impedances with probable human inputs. The subject is really part of computational language design that, in spite of its considerable importance, would take us too far afield if we pursued it further.

COLLOCATION, GALERKIN, AND RITZ

One of the sadder aspects of industrial development is the impoverishment of the handcraftsman by the mass-produced product. An old art dies out and, although a new one is born, the transition is painful and slow. Indeed, there is often a long period before it is clear that the new art will ever appear, threatening us with a loss of the old without a recompense in kind—only larger quantities of cheaper goods. In 1970 some numerical processes are in such a transition and it is with real regret that we must record the passing of techniques associated with several names, of which Galerkin will suffice as typical.

These techniques were used primarily to solve boundary-value problems in ordinary differential equations, usually *linear* ordinary differential equations. The basic strategy was to adopt a sequence of functions that satisfied the *boundary* conditions, string them together in a very finite series with arbitrary coefficients, then adjust these coefficients so as to satisfy the differential equation as well as possible. For example, with no handy table of Bessel functions and the problem

$$u'' + \frac{1}{x}u' + u = 0 \quad \begin{cases} u(\pi/2) = 1 \\ u(\pi) = 0 \end{cases}$$

we might try

$$\tilde{u}(x) = \sin x + a_1 \sin(2x - \pi) + a_2 \sin(4x - 2\pi)$$

Just how to adjust a_1 and a_2 is still an open question. We might pick two values of x and demand that at those values the function $\tilde{u}(x)$ satisfy the differential equation (collocation). But which values of x? The end points of our interval? Two intermediate points seem better, but even this is scarcely a precise prescription. Or should we demand a least-squared integral fit? (Galerkin.)
If we choose to fit the differential equation at the two intermediate points $\frac{2}{3}\pi$ and $\frac{5}{6}\pi$ we get

$$\tilde{u}(x) = \sin x - 0.11255 \sin(2x - \pi) - 0.0002 \sin(4x - 2\pi)$$

which gives, at $\frac{3}{4}\pi$, the value 0.595 versus the correct answer of 0.589. This is not too bad for fitting only two parameters, but if we had chosen the end points, $\frac{1}{2}\pi$ and π, we would have obtained

$$\tilde{u}(x) = \sin x - 0.25 \sin(2x - \pi) + 0.125 \sin(4x - 2\pi)$$

which gives 0.457 at this same midpoint—scarcely a respectable result.

By now the troubles inherent in this class of techniques are visible. Art, mostly in the form of intuition, still plays too large a part to permit consistent results when invoked by different persons. A good choice of approximating functions combined with a fortuitous selection of collocation points will often yield a quite good solution to the boundary-value problem. In the hands of a Feynman the technique works like a Latin charm; with ordinary mortals the result is a mixed bag. Most serious is the lack of any internal measure of success. Having chosen the series and fitted the parameters and evaluated the approximate solution, one is still left with more hope than knowledge. The results are certainly numbers, but do they have very much to do with the problem? This class of techniques produces unverified results and should be discouraged. For who can reliably detect a blunder, let alone a poor approximation? The development of internal error measures could restore the utility of these macro-methods for the boundary-value problem, but until they are available, don't tie up your computer with semiintuitional hand-calculational methods.

EXPONENTIAL FITTING

I am taller than Shakespeare;
I stand on his shoulders.
SHAW

Philosophers have opined that the inventions of printing and writing enable man to learn from the experience of his ancestors. As any computer center director knows, this theory is severely tested by that fraction of humanity that walks through his doors. Like tables of standard functions, most of the computational wisdom of previous generations seems to be regenerated rather than learned or inherited. (But hope springs eternal, so I shall finish this book.)

One of the perennial problems that plagues, among others, the analyzers of isotope decay is the fitting of data by a series of exponential functions. How much of A and how much of B, decaying at known rates a and b, are in the sample whose activity was sampled at several times in the historic past? This question is quite tractable. Computationally we are being asked to fit only the parameters A and B in the equation

$$y = Ae^{-at} + Be^{-bt}$$

when we have observed a sample at several times to produce a set of $\{t_i, y_i\}$ pairs. It is a simple least-squares fit that generally requires only a desk calculator.

Unfortunately there is a companion problem that looks only slightly more complicated—until you try it! We again have $\{t_i, y_i\}$ readings from a radioactive sample, *but the decaying materials are not known,* hence the decay rates a and b must also be fitted. The answer to *this* problem lies in the chemical rather than the computer laboratory, and the sooner the hopeful innocent can be sent there and away from the computer room, the better off everyone will be. For it is well known that an exponential equation of this type in which all four parameters are to be fitted is *extremely* ill conditioned. That is, there are many combinations of (a, b, A, B) that will fit most exact data quite well indeed (will you believe four significant figures?) and when experimental noise is thrown into the pot, the entire operation becomes hopeless. But those with Faith in Science do not always read the Book—and must be spanked or counselled. At the very least, keep them from obstructing Progress and the computer!

LARGE SETS OF LINEAR EQUATIONS

What fools these mortals be,
SENECA
(frequently requoted)

Whenever a person eagerly inquires if my computer can solve a set of 300 equations in 300 unknowns, I must quickly suppress the temptation to retort, "Yes, but why bother?" There *are*, indeed, legitimate sets of equations that large. They arise from replacing a partial differential equation on a set of grid points, and the person who knows enough to tackle this type of problem also usually knows what kind of computer he needs. The odds are all too high that our inquiring friend is suffering from a quite different problem: he probably has collected a set of experimental data and is now attempting to fit a 300-parameter model to it—by Least Squares! The sooner this guy can be eased out of your office, the sooner you will be able to get back to useful work —but these chaps are persistent. They have fitted three-parameter models on desk machines with no serious difficulties and now the electronic computer permits them more grandiose visions. They leap from the known to the unknown with a terrifying innocence and the perennial self-confidence that every parameter is totally justified. It does no good to point out that several parameters are nearly certain to be competing to "explain" the same variations in the data and hence the equation system will be nearly indeterminate. It does no good to point out that *all* large least-squares matrices are striving mightily to be proper subsets of the Hilbert matrix—which is virtually indeterminate and uninvertible—and so even if all 300 parameters were beautifully independent, the fitting equations would still be violently unstable. All of this, I repeat, does no good—and you end up by getting angry and throwing the guy out of your office.

Most of this instability is unnecessary, for there is usually a reasonable procedure. Unfortunately, it is undramatic, laborious, and requires thought—which most of these charlatans avoid like the plague. They should merely fit a five-parameter model, then a six-parameter one. If all goes well and there is a statistically valid reduction of the residual variability, then a somewhat more elaborate model may be tried. Somewhere along the line—and it will be much closer to 15 parameters than to 300—the significant improvement will cease and the fitting operation is over. There is no system of 300 equations, no 300 parameters, and no glamor. But a person has to know some statistics, he has to have a clear idea about the mechanisms by which the variability has entered his data, and he has to know the intended use for his fitted formula. It is infinitely easier to let a computer try to solve 300 equations and hope to put some sort of interpretation on the numbers, assuming one gets any, and be safe from criticism because the computer did it all. It is difficult to choose mathematical models to represent material phenomena and, while it is not easy to evaluate the parameters, the real difficulties are *statistical*, not computational. The computer center's director must prevent the looting of valuable computer time by these would-be fitters of many parameters. The task is not a pleasant one, but the legitimate computer users have rights, too. The alternative commits everybody to a miserable two weeks of sloshing around in great quantities of "Results" that are manifestly impossible, with no visible way of finding the trouble. The trouble, of course, arises from looking for a good answer to a poorly posed problem, but a computer director seldom knows enough about the subject matter to win any of those arguments with the problem's proposer, and the impasse finally has to be broken by violence—which therefore might as well be used in the very beginning.

MULTIPLE FOURIER SERIES

Another formulative danger flag is the multiple Fourier series, particularly in the hands of the physicist or the more theoretically trained engineer. For years the proper attack upon Laplace's equation and some of the associated simpler partial differential equations has been the separation-of-variables technique, which leads to rather complicated Fourier series expansions of the nonhomogeneous boundary conditions. The subject is taught under the guise of Potential Theory, Subsonic Flows, Elasticity, Electricity and Magnetism, Flow of Heat, and so on. It is a beautiful and still useful technique clearly expounded in many books, but it no longer is the only way to solve these problems nor is it apt to be the most efficient. More often than not, a replacement of the differential equation on a grid will give the desired answers faster. Thus when faced with detailed questions about multiple-precision programs for the evaluation of trigonometric functions of large arguments, the analyst would do well to get the petitioner to give him some idea of the original problem. Otherwise, both will become immersed in a forest of detail that will be both frustrating and inefficient. This stricture, however, is not clear cut: there are problems for which the Fourier series approach is expedient. Again, tact and judgement must rule. X-ray crystallographers, with their hours of computation on three-dimensional Fourier series, seem to be a legitimate exception, though even here we can't help feeling that there *ought* to be a better way.

BLACK-BOX DIAGNOSES

I will find where truth is hid, though it were
hid indeed within the centre.
POLONIUS

A favorite form of lunacy among aeronautical engineers produces countless attempts to decide what differential equation governs the motion of some physical object, such as a helicopter rotor. In the archtypical experiment, one measures both the input to the blade and its output—that is, motion. Arguing that the system is a "black box" to which they have the input and the output, they now wish to compute the transfer function, another name for the differential equation. The difficulty here is that many quite different differential equations can give rise to nearly the same outputs from identical inputs. Thus discriminating the "true" equation from its brethren is next to impossible. Computationally speaking, the problem is ill conditioned. I must quickly point out, however, that if one's purpose is to discover an empirical differential equation adequate for limited amounts of prediction, without overmuch worry about its ultimate physical reality, then this fitting approach, properly done with orthogonal functions, can be worthwhile—though still ill conditioned. But arguments about which differential equation represents the Truth, together with their fitting calculations, are wasted time.

RECURSIVE CALCULATIONS

Having hinted darkly at my computational fundamentalism, it is probably time to commit a public heresy by denouncing recursive calculations. I have *never* seen a numerical problem arising from the physical world that was best calculated by a recursive subroutine—that is, by a subroutine that called itself. I admit the idea is cute and, once mastered, it tends to impel its owner to apply it wherever possible—all questions of appropriateness aside. But these pressures should be resisted. Like Monte Carlo techniques, one can find suitable applications but they are not the ones that are usually in the computer realm. Liars may "figur," but enthusiasts compute recursedly —and they better be stopped at the door if you care about efficiency.

There are those who argue for recursive subroutines in situations [*Scientific American*, September 1967] that are complicated and not well understood. I feel that when a physical situation is complicated and not well understood, this is a good time for more thought, not recursive calculations. Indeed, I tend to conclude that recursive programming is to the logician what minimization techniques are to the engineer. (See Chapter 17.)

In combinatorial problems and in some compiler applications recursion seems to have a defensible utility, although it is not unchallenged even there. We opponents would claim that by recursion one is pushing one's responsibilities onto the computer instead of solving them directly. Of course if one is rich enough to afford the computer time . . . .

LARGE UNSYSTEMATIC COMPUTATIONS

The fault, dear Brutus, is not in our stars . . . .
SHAKESPEARE

The large complicated unsystematic computation that has no independent check poses a grim problem for the computer center director. The odds, generally speaking, are overwhelmingly against anything useful emerging, and one may all too confidently predict large losses of machine time, staff time, and tempers. Still, one does not enjoy saying "NO!" We can point to recently devised procedures whereby an exceedingly devout, punctilious, and energetic worker may verify the correctness of his program in the sense that it calculates what he asserts it calculates. The technique is laborious, is fraught with its own dangers—but it exists and can be used to reduce to acceptable levels the probability of a serious goof. The trouble, of course, is that 95 percent of the authors of large complicated unsystematic computations have never heard of these checking procedures and wouldn't use them if you pointed them out. The problem remains political.

PART II

DOUBLE TROUBLE

CHAPTER 10

THE EVALUATION OF INTEGRALS

In the solution of larger problems we often need to evaluate an integral that is a function of a parameter. We might, for example, require the value $F(b)$ of

$$F(b) = \int_0^\infty \frac{e^{-bt}}{1+t^2}\,dt$$

for some recently calculated value of b. This problem properly belongs under "Approximation of Functions," but it occurs sufficiently often to justify separate treatment. The variety of such problems is large. The devices used will depend in part on the precision with which $F(b)$ must be evaluated, but more importantly they depend on the ingenuity and experience of the evaluator. In this chapter we examine two of these integrals in detail, hoping to transmit some of the approaches that frequently are useful. Since many of the nonstandard functions encountered in engineering and scientific problems are defined by quadratures, we feel that the art needs exhibiting. We also encourage the reader to practice what we preach by doing the exercises, for there is a confusing variety that hides the essential similarities from those who merely read and run.

The problem seems well posed: $F(b)$ is the area under a curve (Figure 10.1) that clamps down onto the t axis very quickly, having started at unity. Simpson's rule could do the trick, even though we might feel a bit uneasy about using it on the tail, where the integrand has a horizontal asymptote. More serious, however, is the question of over-all computational efficiency. How much work is it to get one value of F given a b? Twenty to forty ordinates for Simpson are not unlikely, are even possibly optimistic, and the exponential function must be evaluated at each of them. The evaluation of e^x probably requires some 20 additions and multiplications, so we are talking about 600 arithmetic operations for a single value of F. If we need the function at all frequently, alternative methods seem desirable. A table with interpolation, once available, would give

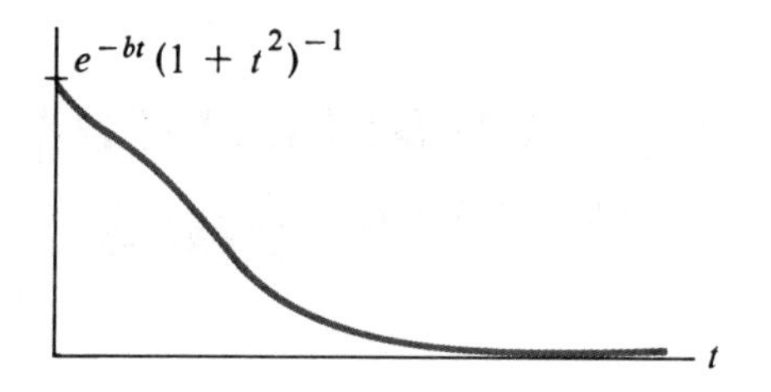

Figure 10.1

a minimum time approximation – about 10 multiplications. A representation of $F(b)$ as a power series or an asymptotic series, properly economized, would require about 20 multiplications – and considerably less memory space in the computer. Either the table or the series are far preferable to dogged quadrature for each evaluation.

The asymptotic series for large *b*

The commonest technique for generating asymptotic series, valid for *large* b, from integrals with exponential factors, is repeated integration by parts. For our current example, however, a more expedient method follows from the observation that most of the area under the integrand for large b arises when t^2 is very *small*, hence we can probably replace the $(1 + t^2)^{-1}$ factor with its series representation

$$\frac{1}{1+t^2} = 1 - t^2 + t^4 - t^6 + \cdots \tag{10.1}$$

even though it does not converge for t greater than unity. On the other hand, in the range where we really need it, this series is not an approximation, being exactly equal to $(1 + t^2)^{-1}$ there. We find

$$F(b) = \int_0^{\infty} e^{-bt}(1 - t^2 + t^4 - t^6 + \cdots)\,dt = \frac{1}{b}\left[1 - \frac{2!}{b^2} + \frac{4!}{b^4} - \frac{6!}{b^6} + \cdots\right] \tag{10.2}$$

which has all the appearances of an asymptotic series, and is.

For any given value of b, the terms of the series (10.2) decrease to a minimum, then grow without limit because of the factorial in the numerators. But since the series alternates in sign, the error is less than the first neglected term. For b equal to 5.0, the minimum is $4!/5^4$, or 0.038. If we can tolerate this error of 4 in the second significant figure, then the truncated series is adequate for arguments of five or greater. For b at 9, the series is considerably better, the minimum term being $8!/9^9$, or 0.00094, giving at least three significant figures. But these are not the accuracies that make us glad. If we are going to get the six or eight figures we have been taught to seek, further manipulation seems needed. Two courses are open:

1. To find a different asymptotic series, perhaps having a factor of e^{-b} outside it [a suggestion that arises from looking at the shape of the curve for $F(b)$ in Figure 10.2], or
2. Rephrase this series as a continued fraction or rational function, a form that more efficiently represents geometries like Figure 10.2 with its horizontal asymptote.

We must emphasize that these are merely suggestions to be explored. They may turn out, in fact, to have been bad ideas which, at the moment, seem reasonable.

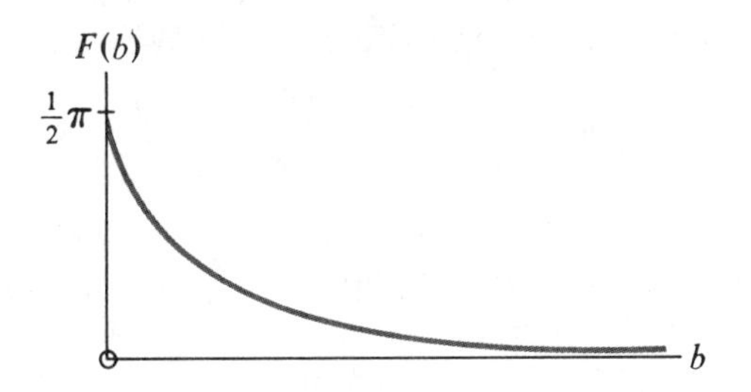

Figure 10.2

A direct approach to expressing the series in (10.2) as a rational function can be quickly tried. Letting y replace b^2 as the basic variable, we wish to write

$$1 - \frac{2!}{y} + \frac{4!}{y^2} - \cdots$$

as

$$\frac{y + r}{y + s}$$

or as

$$\frac{y^2 + ry + s}{y^2 + dy + u} \tag{10.3}$$

both of which approach unity as y becomes large. Exploring the cruder approximation first, we have

$$y + r = (y + s)\left(1 - \frac{2}{y} + \frac{24}{y^2} - \cdots\right) = y + s - 2 - \frac{2s}{y} + \frac{24}{y} + \cdots \tag{10.4}$$

and we wish to choose r and s so as to make the two sides of (10.4) an identity in y to as many terms as possible. Thus we find

$$\left.\begin{aligned} r &= s - 2 \\ 0 &= -2s + 24 \end{aligned}\right\} \quad \text{so} \quad \left\{\begin{aligned} r &= 10 \\ s &= 12 \end{aligned}\right. . \tag{10.5}$$

and the rational function approximation for $F(b)$ becomes

$$F_1(b) = \frac{1}{b} \cdot \frac{b^2 + 10}{b^2 + 12}$$

The approximation may be judged from the values shown in Table 10.1. We have three significant figures for b equal to 5 and two down at b equal to 3—a considerable improvement over the raw asymptotic series. Encouraged, we may try the more sophisticated approximant (10.3). We would obtain

$$F_2(b) = \frac{1}{b} \cdot \frac{3b^4 + 214b^2 + 1192}{3b^4 + 220b^2 + 1560} \tag{10.6}$$

The values are also shown in Table 10.1. We seem to have picked up about half a significant figure, a somewhat disappointing improvement. The next approximant requires the fitting of six parameters and is a lot of work. After solving the six simultaneous equations, we obtain

$$F_3(b) = \frac{1}{b} \cdot \frac{b^6 + 24{,}142b^4 + 946{,}776b^2 + 3{,}862{,}800}{b^6 + 24{,}360b^4 + 992{,}880b^2 + 5{,}342{,}400} \tag{10.7}$$

Again, as Table 10.1 shows, there is improvement, though not very much.

Table 10.1. Several approximations for $F(b) = \int_0^\infty e^{-bt}(1 + t^2)^{-1}\, dt$

b	F_1	F_2	F_3	F(correct)
3	0.3016	0.2961	0.2943	0.2920
5	0.1892	0.1884	0.1882	0.1881

Clearly the rational function is gradually being forced to yield respectable precisions, but not quickly enough. We must use some more systematic algorithm to produce a higher-order approximant which we then economize to shorten it as much as we can – having got clear indication here that the expressions are otherwise going to be inconveniently long. On the other hand, F_3 only requires seven multiplications and six additions for its evaluation, so we are encouraged when we compare it to the Simpson's rule approach with its many hundreds of operations. Our distress concerns the inexpedience in our method for getting the rational function rather than with its subsequent use.

But before getting involved with expedient techniques for generating rational functions from recalcitrant asymptotic series, perhaps we should see whether we can find approximations for $F(b)$ when b is *small*. The existence of an approximation for only part of the range might dampen our enthusiasm for the whole project. We shall first, however, look at some cruder methods when b is large.

Cruder methods for large *b*

We have already observed that most of the area under our integrand occurs for small values of t, especially when b is large. This fact suggests that we might seek approximations to $F(b)$ by replacing either the exponential by an algebraic quantity or the rational function by an exponential or algebraic function – the object being to permit analytic integration. Since there is no good systematic method for making these replacements, their success depends directly on the intuition, ingenuity, and patience of the approximator – and we should be very surprised if high-accuracy approximations are produced. Your author does not generally approve of this approach to approximation unless the functions being replaced are themselves empirical and of limited accuracy. But he includes an example here to illustrate the method and suggest the sorts of approximation to be expected.

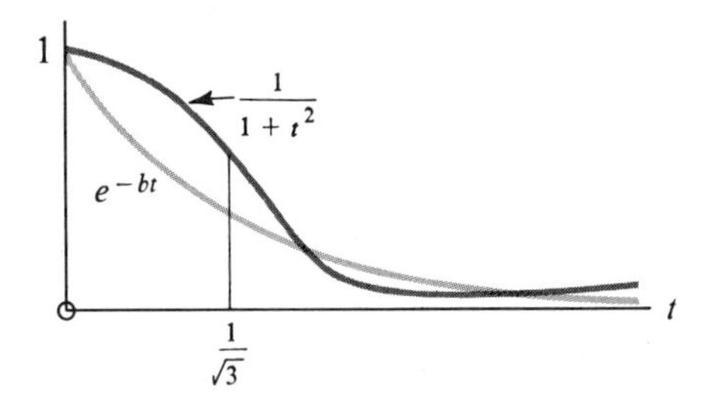

Figure 10.3

We shall get rid of the function $(1 + t^2)^{-1}$, replacing it by several functions that are compatible with e^{-bt} in the sense of permitting analytic integration. Examining the graphs in Figure 10.3 we see that most of the contribution occurs before the inflection point at $1/\sqrt{3}$. We might try replacing the left portion of $(1 + t^2)^{-1}$ by a parabola $(1 - 3t^2/4)$ stopping at the inflection point because no parabola ever had them. The tail from $1/\sqrt{3}$ to infinity can be more cavalierly treated, especially toward the right, so we suggest a declining exponential matched at the inflection point both for height and slope. Thus we approximate $F(b)$ by

$$I_1(b) = \int_0^{1/\sqrt{3}} e^{-bt}(1 - \tfrac{3}{4}t^2)\,dt + \int_{1/\sqrt{3}}^{\infty} e^{-bt} \cdot \tfrac{3}{4}\exp\left(\frac{1}{2} - \frac{\sqrt{3}}{2}t\right) dt$$

$$= \frac{1}{b}(1 - \tfrac{3}{4}e^{-b/\sqrt{3}}) + \frac{\sqrt{3}}{2b^2}e^{-b/\sqrt{3}} - \frac{3}{2b^3}(1 - e^{-b/\sqrt{3}}) + \frac{3}{2}\cdot\frac{e^{-b/\sqrt{3}}}{2b + \sqrt{3}}$$

The values of this approximation at 3 and 5 are shown in Table 10.2.

Table 10.2

b	I_1	I_2	Correct
3	0.295	0.2935	0.29196
5	0.1894	0.1890	0.18814

A different replacement uses the parabola only on $(0, \frac{1}{2})$, a straight line across the inflection point $(\frac{1}{2}, 1)$ and an exponential thereafter. It leads to

$$I_2(b) = \frac{1}{b}(1 - \tfrac{1}{2}e^{-b}) + \frac{e^{-b/2}}{5b^2}(1 + 3e^{-b/2}) - \frac{8}{5b^2}(1 - e^{-b/2})$$

which gives slightly better results, though still less than three significant figures at 3. It seems unlikely that variations on this theme will substantially improve upon it.

Seeking a series for small arguments

The perceptive reader may have noticed that our function

$$F(b) = \int_0^{\infty} \frac{e^{-bt}}{1 + t^2}\,dt = b \cdot \int_0^{\infty} \frac{e^{-s}}{b^2 + s^2}\,ds \tag{10.8}$$

is not entirely pleasant when b goes to zero. $F(0)$ exists and is equal to $\pi/2$, but

$F'(b)$ and all higher derivatives become infinite there, so that a Taylor's series is not available. Our function has a singular derivative at the origin and the strength of this singularity must be discovered if we are to deal competently with F for any small value of b.

While much information is available about F from its integral forms (10.8), the most useful representation is apt to be a differential equation. Accordingly, we seek one for F (as a function of b, of course, since t and s are only dummy variables of integration). We have

$$\begin{aligned} F'(b) &= -\int_0^{\infty} \frac{e^{-bt}t}{1+t^2}\,dt \\ F''(b) &= \int_0^{\infty} \frac{e^{-bt}t^2}{1+t^2}\,dt \end{aligned} \tag{10.9}$$

we thus see that

$$F'' + F = \frac{1}{b} = \int_0^{\infty} e^{-bt}\cdot dt \tag{10.10}$$

a most compact equation that needs some initial conditions to incorporate our knowledge about the nature of the singularity in the first derivative.

Having observed that $F(0)$ is $\frac{1}{2}\pi$ and that the first infinity occurs in $F'(b)$, we seek to ascertain just how fast that quantity grows as b approaches zero. Looking at

$$-F'(b) = \int_0^{\infty} \frac{e^{-bt}t}{1+t^2}\,dt$$

see that, when e^{-bt} is not present, trouble occurs at the upper limit. Choosing a t_1 large enough so that unity is negligible compared to ${t_1}^2$ we write

$$-F'(b) \approx \int_0^{t_1} \frac{e^{-bt}t}{1+t^2}\,dt + \int_{t_1}^{\infty} e^{-bt}\frac{dt}{t}$$

We let b approach zero in the first integral which becomes a finite constant, A_0, and we have

$$-F'(b) \to A_0 + \int_{bt_1}^{\infty} e^{-s}\frac{ds}{s} \approx A_0 + \int_{s_1}^{\infty} e^{-s}\frac{ds}{s} + \int_{bt_1}^{s_1} \frac{ds}{s}$$

The quantity s_1 is chosen small enough to permit e^{-s} to be replaced by *unity*. So finally

$$-F'(b) \to A_0 + A_1 + \ln s_1 - \ln b - \ln t_1 \to A_2 - \ln b$$

and we have that

$$\ln b - F'(b) \to \text{a constant, } A_2\text{, to be determined}$$

Now that we know how fast $F'(b)$ grows, we need an integral representation of $\ln b$ that can be combined with our integral for F'. Two standard integral forms are

$$\ln b = \int_1^b \frac{dt}{t} = \int_0^\infty (e^{-t} - e^{-bt})\frac{dt}{t} \tag{10.11}$$

The second one clearly being the tool we need, we write

$$F'(b) - \ln b = \int_0^\infty \left[e^{-bt}\left(\frac{1}{t} - \frac{t}{1+t^2}\right) - \frac{e^{-t}}{t}\right] dt \xrightarrow[b\to 0]{} \int_0^\infty \left[\frac{1}{t} - \frac{t}{1+t^2} - \frac{e^{-t}}{t}\right] dt \tag{10.12}$$

The first and third terms in our last integral get into equal but compensatory troubles at the lower limit, suggesting they should be combined into an expression that is well behaved there. At the upper limit, it is the first two terms that blow up, but together they cancel and are well behaved. This suggests dividing the interval of integration into (0, 1) and (1, ∞) so as to get

$$F'(b) - \ln b \to \int_0^1 \left[\frac{1 - e^{-t}}{t} - \frac{t}{1 + t^2}\right] dt + \int_1^\infty \left[\left(\frac{1}{t} - \frac{t}{1 + t^2}\right) - \frac{e^{-t}}{t}\right] dt$$

$$= \left[\int_0^1 \frac{1 - e^{-t}}{t} dt - \int_1^\infty \frac{e^{-t}}{t} dt\right] - \int_0^1 \frac{t \cdot dt}{1 + t^2}$$

$$+ \int_1^\infty \left[\frac{1}{t} - \frac{t}{1 + t^2}\right] dt$$

The last two integrals may be analytically integrated, both turning out to be equal to $\ln\sqrt{2}$, thereby canceling. (Alternatively, the dummy variable t may be replaced by $1/s$ in the last integral, whence it becomes identical to the other integral.) The bracketed expressions, clearly a constant that could be evaluated, may be found on page 1 of Jahnke-Emde [1951] to be γ, Euler's constant. Thus

$$F'(b) - \ln b \xrightarrow[b\to 0]{} \gamma = 0.5772\ldots$$

The series solution for small b

Returning to our differential equation

$$F'' + F = \frac{1}{b} \tag{10.13}$$

we observe that a term of the type $b \cdot \ln b$ is needed to produce $1/b$ after two differentiations without itself being infinite at zero. Let us now change our independent variable to x during the series solution operations. We try

$$G(x) = x \cdot \ln x \cdot W(x)$$

$$G'(x) = x \cdot \ln x \cdot W' + (1 + \ln x) \cdot W$$

$$G''(x) = x \cdot \ln x \cdot W'' + 2(1 + \ln x) \cdot W' + \frac{1}{x} \cdot W$$

and we observe that we get the $1/x$ term if W is a series whose leading term is unity. But at zero the value of $(G' - \ln x)$ approaches 1, which is not γ, nor does G approach $\frac{1}{2}\pi$. We need another term, so we try

$$G(x) = x \cdot \ln x \cdot W + R$$

$$G'(x) = x \cdot \ln x \cdot W' + (1 + \ln x) \cdot W + R'$$

$$G''(x) = x \cdot \ln x \cdot W'' + 2(1 + \ln x) \cdot W' + \frac{1}{x} \cdot W + R''$$

Now let

$$W = 1 + b_1 x + b_2 x^2 + \cdots$$

and

$$R = c_0 + c_1 x + c_2 x^2 + \cdots$$

On substituting into our differential equation (10.13) we have

$$\frac{1}{x} \equiv \ln x \cdot \begin{bmatrix} 2b_1 + 2 \cdot 2b_2 x + 2 \cdot 3b_3 x^2 + 2 \cdot 4b_4 x^3 + \cdots \\ +2b_2 x + 3 \cdot 2b_3 x^2 + 4 \cdot 3b_4 x^3 + \cdots \\ +1 \cdot x + b_1 x^2 + b_2 x^3 + \cdots \end{bmatrix} \begin{matrix} \text{from } 2W' \\ \text{from } xW'' \\ \text{from } xW \end{matrix}$$

$$+ \begin{bmatrix} \frac{1}{x} + b_1 + b_2 x + b_3 x^2 + \cdots \\ +2b_1 + 2 \cdot 2b_2 x + 2 \cdot 3b_3 x^2 + \cdots \\ c_0 + c_1 x + c_2 x^2 + \cdots \\ +2 \cdot 1c_2 + 3 \cdot 2c_3 x + 4 \cdot 3c_4 x^2 + \cdots \end{bmatrix} \begin{matrix} \text{from } W/x \\ \text{from } 2W' \\ \text{from } R \\ \text{from } R'' \end{matrix} \qquad \textbf{(10.14)}$$

and we have the conditions that

$$c_0 = \frac{\pi}{2}$$

$$c_1 = \gamma - 1$$

plus the further condition that all terms must vanish for all x. Clearly b_1 is zero, and we quickly see that b_3 must therefore also be zero, and $b_5, \cdots$. But b_2 is $-\frac{1}{6}$ and the other even b_k's are well determined by the necessity to get rid of the $\ln x$ term. With the b's all known, can we calculate the c's? Yes, for c_2 is fixed by the first column, c_3 by the second, etc. So the problem is solved, though the evaluation of the series coefficients may take some patience. We have

$$b_{2n} = (-1)^n/(2n+1)! \qquad c_{3n} = (-1)^n \frac{\frac{1}{2}\pi}{(2n)!}$$

$$c_{2n+1} = \frac{(-1)^n}{(2n+1)!}\left[c_1 - \frac{5}{2 \cdot 3} - \frac{9}{4 \cdot 5} - \frac{13}{6 \cdot 7} - \cdots - \frac{4n+1}{(2n)\cdot(2n+1)}\right]$$

so that the series is

$$F(x) = \frac{\pi}{2} \cdot \left[1 - \frac{x^2}{2!} + \frac{x^4}{4!} - \cdots\right] + x \cdot \ln x\left[1 - \frac{x^2}{3!} + \frac{x^4}{5!} - \cdots\right]$$

$$+ c_1\left[\frac{x}{1!} - \frac{x^3}{3!} + \frac{x^5}{5!} - \cdots\right] + \frac{x^3}{3!}\left(\frac{5}{2 \cdot 3}\right) - \frac{x^5}{5!}\left(\frac{5}{2 \cdot 3} + \frac{9}{4 \cdot 5}\right) \tag{10.15}$$

$$+ \frac{x^7}{7!}\left(\frac{5}{2 \cdot 3} + \frac{9}{4 \cdot 5} + \frac{13}{6 \cdot 7}\right) - \cdots$$

or, recognizing the series for $\sin x$ and $\cos x$,

$$F(x) = \frac{\pi}{2}\cos x + (\ln x + \gamma - 1) \cdot \sin x + \frac{x^3}{3!}\left(\frac{5}{6}\right) - \frac{x^5}{5!}\left(\frac{5}{6} + \frac{9}{20}\right)$$

$$+ \frac{x^7}{7!}\left(\frac{5}{6} + \frac{9}{20} + \frac{13}{42}\right) - \cdots \tag{10.16}$$

Closing the gap

We are now possessed of two representations for the function $F(b)$ that are relatively efficient to evaluate. The series that works so well for small arguments (10.16) begins to cause discomfort in the region of b equal to 3.0, where six terms give six significant figures. Larger arguments rapidly require more terms for the same accuracy, which is the minimal accuracy that should be accepted in preparing an approximation for general computer use. Further, the six figures we get arise through a subtraction of two nearly equal seven figure numbers that loses the first digit. (Without this loss we would be insisting on seven significant figures in our approximation!) At b equal to 5 we lose two figures through cancellations in the series. We probably do not want to use this representation for F beyond 3.

The asymptotic series (10.2) is miserable at 3.0 and, in fact, will give six significant figures only for arguments greater than 16. So we are faced with the gap

$$3 \leq x \leq 16$$

within which our series are not adequate. We reach first for continued fractions, since the shape of F, especially for larger arguments, is distinctly nonpolynomic. Using the quotient-difference algorithm (Chapter 11) we can turn both series into continued fractions, shown in Tables 10.3 and 10.4. For the series at the origin, the improvement if any is small: We may now get six-figure accuracy at 3.0 with five cycles of the P-Q algorithm or six "terms" of the continued fraction. The stability of the calculation is good.

With the asymptotic series the change is dramatic: We can now get five significant figures at the argument of 5.0 by using about 16 "terms" of the continued fraction, while at an argument of 8.0 we get five significant figures with only five terms or six figures with nine terms. By invoking the continued fraction representation we have markedly narrowed the gap to

$$3 \leq x \leq 8$$

Table 10.3. Coefficients of the continued-fraction representation of the asymptotic series for

$$\int_0^\infty e^{-bt}(1 + t^2)^{-1} \cdot dt,$$

together with the successive approximants at arguments of 5 and 3

	Coefficients	$b = 5$	$b = 3$
α_0	1.0		
α_1	2.0		
α_2	10.0		0.302
α_3	21.6	0.18764	0.285
α_4	39.733333	0.18840	0.295
α_5	60.961969	0.18798	0.2887
α_6	89.19094	0.188242	0.2943
α_7	120.07506	0.188074	0.2900
α_8	158.37848	0.188190	0.293460
α_9	198.93641	0.188107	0.290699
α_{10}	247.29366	0.1881687	0.293004
α_{11}	297.55326	0.1881225	0.291034
Average of last two values		0.1881456	0.29202
Correct values		0.1881428	0.291958

Table 10.4. A comparison of the series 10.15 and its continued fraction representation with $F(b)$ at 3

	Continued fraction		Series	
	Coefficients	$b = 3$	$b = 3$	No. of terms
α_0	1			
α_1	0.077			
α_2	−0.04744 7919			
α_3	0.00847 3271	0.28544		
α_4	−0.01515 2298	0.292206	0.283528	4
α_5	0.00344 1334	0.291974	0.292492	5
α_6	−0.00793 3426	0.291957	0.291934	6
α_7	0.00182 4727	0.291957	0.291959	7
	Correct value	0.291958		

but have not yet closed it. *Economization* of the rational function (continued fraction) representations of the two series would seem likely to close the gap by permitting greater accuracy with a smaller number of terms. A change in the independent variable might be even more helpful but, as we have indicated before, such changes depend largely on inspiration so are considerably harder to systematize. We have not tried that gambit here.

The useless formula

The differential equation

$$F'' + F = \frac{1}{b}$$

can be solved by variation of parameters to express F as

$$F(b) = \frac{\pi}{2}\cos b - \left[\cos b \cdot \int_0^b \frac{\sin t}{t}\,dt + \sin b \cdot \int_b^\infty \frac{\cos t}{t}\,dt\right] \tag{10.17}$$

This formulation of our function is another way to get the series representation. Since the two integrals are functions tabled in several standard references [AMS 55] (10.17) is also the practical formula for hand evaluation of F. When we consider automatic computation, however, this expression for $F(b)$ is virtually useless, since it requires the subcomputation of two nonstandard functions, one of them containing a singularity at the origin. We quote it here merely to stress how different in utility a particular expression may be for hand and automatic computations.

Another integral

Turning to another function, we consider

$$K(b) = \int_0^\infty \frac{e^{-s^2}}{b+s}\,ds = \int_0^\infty \frac{e^{-b^2t^2}}{1+t}\,dt \tag{10.18}$$

since it reinforces the lessons learned above while exhibiting some minor variations of its own. Our interest is in small values of b, where $K(b)$ blows up, as can easily be seen in its second form (10.18), where we have a $\ln(1+t)$ with an embarrassing upper limit. Seeking a differential equation we write

$$K'(b) = -2b \int_0^\infty \frac{e^{-b^2t^2}t^2}{1+t}\,dt = -2b \int_0^\infty e^{-b^2t^2}\left(t - 1 + \frac{1}{1+t}\right) dt$$

$$= -2b \int_0^\infty e^{-b^2t^2}(t-1)\,dt - 2b \cdot K(b)$$

Performing the two analytic integrations we have

$$K'(b) = -\frac{1}{b} + \sqrt{\pi} - 2b \cdot K$$

or

$$\boxed{K' + 2b \cdot K = \sqrt{\pi} - \frac{1}{b}} \tag{10.19}$$

Suspecting logarithmic behavior of K for small b, we note that this hypothesis satisfies the differential equation, since bK will clearly disappear to leave K' behaving like $-1/b$, the derivative of $-\ln b$. More carefully, we have

$$\int_b^1 K' \cdot db = \int_b^1 (\sqrt{\pi} - 1/b)\,db = \sqrt{\pi}(1-b) - \ln 1 + \ln b = K(1) - K(b)$$

Hence

$$K(b) + \ln b = \text{constant to be evaluated} = \Gamma$$

As before, we invoke the exponential integral representation for $\ln b$ to give

$$K(b) + \ln b = \int_0^\infty \left[\frac{e^{-b^2t^2}}{1+t} - \frac{e^{-bt}}{t} + \frac{e^{-t}}{t}\right] dt \tag{10.20}$$

At this point we must be careful. On letting b go to zero we get an evaluatable limiting integral that unfortunately does *not* equal the limit of the integral with b in it! Qualitatively, the trouble lies in the fact that, for any fixed b, the first term of the integrand in (10.20) disappears much faster with increasing t than does the second term. The troubles are all out at large t, which is where we get enough area for the integrals of these terms to become infinite. Proceeding cautiously we write

$$K(b) + \ln b = \int_0^1 \left[\frac{e^{-b^2t^2}}{1+t} - \frac{e^{-bt}}{t} + \frac{e^{-t}}{t}\right] dt + \int_1^\infty \left[\frac{e^{-b^2t^2}}{1+t} - \frac{e^{-bt}}{t}\right] dt + \int_1^\infty \frac{e^{-t}}{t} dt$$

There are no problems with the first integral as we let b go to zero, which we now do, but we would like $e^{-b^2t^2}$ in *both* terms inside the second integral before passing to the limit. Accordingly we subtract the appropriate integral, producing

$$K(b) + \ln b = \int_0^1 \left[\frac{e^{-b^2t^2}}{1+t} - \frac{e^{-bt}}{t} + \frac{e^{-t}}{t}\right] dt + \int_1^\infty e^{-b^2t^2}\left[\frac{1}{1+t} - \frac{1}{t}\right] dt + \int_1^\infty (e^{-b^2t^2} - e^{-bt})\frac{dt}{t} + \int_1^\infty \frac{e^{-t}}{t} dt$$

Changing the dummy variable of integration in the next-to-last integral from t to s/b we have

$$K(b) + \ln b = \int_0^1 \left[\frac{e^{-b^2t^2}}{1+t} - \frac{e^{-bt}}{t} + \frac{e^{-t}}{t}\right] dt + \int_1^\infty e^{-b^2t^2}\left[\frac{1}{1+t} - \frac{1}{t}\right] dt + \int_b^\infty (e^{-s^2} - e^{-s})\frac{ds}{s} + \int_1^\infty e^{-t}\frac{dt}{t}$$

We are finally ready to pass to the limit. We obtain

$$K(b) + \ln b \xrightarrow[b\to 0]{} \int_0^1 \frac{dt}{1+t} + \int_0^1 (e^{-t} - 1)\frac{dt}{t} + \int_1^\infty \left[\frac{1}{1+t} - \frac{1}{t}\right] dt + \int_0^\infty (e^{-s^2} - e^{-s})\frac{ds}{s} + \int_1^\infty e^{-t}\frac{dt}{t}$$

The first and third integrals may be formally integrated and found to cancel. The second and fifth integrals combine to form our old friend $-\gamma$, while the fourth integral turns out to equal $\gamma/2$ (Bierens de Haan). Thus we finally have

$$K(b) + \ln b \longrightarrow -\frac{\gamma}{2} = -0.2886\ldots \tag{10.21}$$

Clearly, intuition does not always suffice to resolve limiting process dilemmas!

Returning to the differential equation, we might try a direct series expansion, first having removed the negative $(\ln b + \gamma/2)$ to leave us with a variable without singularities at the origin. It is simpler, however, to define a new function $G(b)$, that does not have these singularities and then transform the differential equation into an equation for G. We try

$$K + \gamma/2 + \ln b = G$$

$$K' + \frac{1}{b} = G'$$

to get

$$G' - \frac{1}{b} + 2b\left(G - \frac{\gamma}{2} - \ln b\right) = \sqrt{\pi} - \frac{1}{b}$$

or

$$G' + 2bG = \sqrt{\pi} + \gamma b + 2b \ln b$$

—a dubious improvement. So we try again, this time including an e^{-b^2} factor that is apt to pop up in such problems. We define

$$K = \left(G - \frac{\gamma}{2} - \ln b\right)e^{-b^2} \tag{10.22}$$

whence

$$K' = \left(G' - \frac{1}{b}\right)e^{-b^2} - 2b\left(G - \frac{\gamma}{2} - \ln b\right)e^{-b^2}$$

so that

$$K' + 2bK = \left(G' - \frac{1}{b}\right)e^{-b^2} = \sqrt{\pi} - \frac{1}{b}$$

or

$$G' = \sqrt{\pi}\cdot e^{b^2} + \frac{1}{b}(1 - e^{b^2})$$

—a most gratifying simplification. In fact, we need only write out the series for

e^{b^2} and integrate from 0 to b to generate our series. We have

$$G(b) - G(0) = \sqrt{\pi}\int_0^b \left(1 + b^2 + \frac{b^4}{2!} + \frac{b^6}{3!} + \cdots\right) db$$

$$- \int_0^b \left(b + \frac{b^3}{2!} + \frac{b^5}{3!} + \cdots\right) db$$

$$= \sqrt{\pi}b\left[1 + \frac{b^2}{1!3} + \frac{b^4}{2!5} + \frac{b^6}{3!7} + \cdots\right]$$

$$- \left[\frac{b^2}{1!2} + \frac{b^4}{2!4} + \frac{b^6}{3!6} + \cdots\right]$$

Finally, we observe from (10.21) and (10.22) that

$$G(0) = K + \frac{\gamma}{2} + b \longrightarrow 0$$

so our final series for K is

$$K(b) = e^{-b^2}\left[-\frac{\gamma}{2} - \ln b + \sqrt{\pi}b \cdot \left(1 + \frac{b^2}{3 \cdot 1!} + \frac{b^4}{5 \cdot 2!} + \frac{b^6}{7 \cdot 3!} + \cdots\right)\right.$$
$$\left. - \frac{b^2}{2} \cdot \left(1 + \frac{b^2}{2 \cdot 2!} + \frac{b^4}{3 \cdot 3!} + \cdots\right)\right] \tag{10.23}$$

The form is not entirely reassuring, since it is basically an alternating series that we can expect to break down as b becomes only moderately large, the cancellation of significant digits being serious. But for small b, it works very nicely, all the troubles being concentrated in the function $\ln b$, for which computing centers presumably have standard algorithms.

For the record, the asymptotic series for $K(b)$ may easily be obtained from the first integral in (10.18) if we express $(b + s)^{-1}$ as a power series in s and then integrate term by term. We find

$$K(b) \sim \left\{\frac{\sqrt{\pi}}{2b}\left[1 + \frac{1}{2b^2} + \frac{1 \cdot 3}{(2b^2)^2} + \frac{1 \cdot 3 \cdot 5}{(2b^2)^3} + \cdots\right]\right.$$
$$\left. - \frac{1}{2b^2}\left[1 + \frac{1!}{b^2} + \frac{2!}{b^4} + \frac{3!}{b^6} + \cdots\right]\right\}$$

One last integral

Finally we consider

$$H(b) = \int_0^\infty \frac{e^{-bt}}{\sqrt{1+t}}\,dt$$

which also becomes infinite as b goes to zero. In the course of some preliminary explorations, we tried changing the dummy variable. Letting

$$1 + t = s^2$$

we have

$$H(b) = \int_1^\infty e^{-b(s^2-1)} \cdot \frac{2s}{s}\,ds = 2e^b \int_1^\infty e^{-bs^2} \cdot ds$$

With a further change of

$$s\sqrt{b} = t$$

we obtain

$$H(b) = \frac{2e^b}{\sqrt{b}} \int_{\sqrt{b}}^\infty e^{-t^2} \cdot dt = \frac{\sqrt{\pi}}{\sqrt{b}} \cdot e^b \cdot \operatorname{erfc}(\sqrt{b})$$

and we have expressed our function in terms of e^b and the complementary error integral, for which standard approximations are widely available. With a slight wave of the hand, our problem vanishes.

Other functions suitable for practice

If we consider the eight integrals of the form

$$\int_0^\infty \frac{\text{exponential factor}}{\text{denominator}}\,dt$$

where the numerator and denominators are chosen from Table 10.5, two of them transform directly into standard functions. The others pose various degrees of difficulty for those who seek the differential equation and the nature and strength of the singularity at the origin. As a partial aid, the differential equations are given in Table 10.5. Your author found the functions involving $\sqrt{1+t^2}$ to be rather difficult, and he had real troubles in the upper right corner (answer supplied upon request!).

Table 10.5

Denominator factors	Exponential factors	
	e^{-bt}	$e^{-b^2t^2}$
$\sqrt{1+t}$	$\sqrt{\frac{\pi}{b}}\,e^b \operatorname{erfc}(\sqrt{b})$	
$1+t$	$e^b \cdot Ei(b)$	$K' + 2bK = \sqrt{\pi} - 1/b$ $K + \ln b \longrightarrow \gamma/2$
$\sqrt{1+t^2}$	$H'' + \frac{1}{b}H' + H = \frac{1}{b}$ $H + \ln b \longrightarrow \ln 2 - \gamma$	$J'' + \frac{1}{b}J' - J = 0$ The Bessel function known as $K_0(b)$
$1+t^2$	$G'' + G = 1/b$ $G' - \ln b \longrightarrow \gamma$	$F' - 2bF = -\sqrt{\pi}$

Other functions that are fun to try are

(1) $$\int_0^{\infty} \frac{dx}{1+x^b}$$

which can be integrated in closed form—though not by every sophomore.

(2) $$\int_0^{\infty} e^{-b(t+1/t)}\frac{dt}{t}$$

which turns into one of the integrals of Table 10.5.

(3) $$\int_0^{\infty} e^{-b(t^2+1/t)}\frac{dt}{t}$$

this one being a real stinker!

CHAPTER

11

POWER SERIES, CONTINUED FRACTIONS, AND RATIONAL APPROXIMATIONS

This short chapter summarizes, in a convenient computational cookbook, some of the useful algorithms that permit us to pass from one standard computational form to another. Do you have a power series, but need a continued fraction? The quotient-difference algorithm is made for you. Or do you have a rational function but want a form that requires fewer multiplications to evaluate? Try the continued fraction in form 2.

A road map, Figure 11.1, shows several forms with their algorithmic connections. The arrows indicate that an algorithm is given for passing from one form to the other, usually in both directions but occasionally in one direction only. Empty arrows imply an easy general algorithm available for all orders of approximation; solid arrows indicate algorithms that are adequate up through P_3/Q_3—with generalizations either unavailable or too complicated to be very useful. The numbers alongside the arrows indicate the particular algorithm. Note that it is possible to go almost everywhere immediately from the rational function, except to the standard continued fraction (which is not often used for direct computation).

We shall take as standard forms

$$\frac{1 + a_1x + a_2x^2 + \cdots + a_nx^n}{1 + b_1x + b_2x^2 + \cdots + b_nx^n} = \frac{P_n(x)}{Q_n(x)} = \text{rational function} \tag{11.1}$$

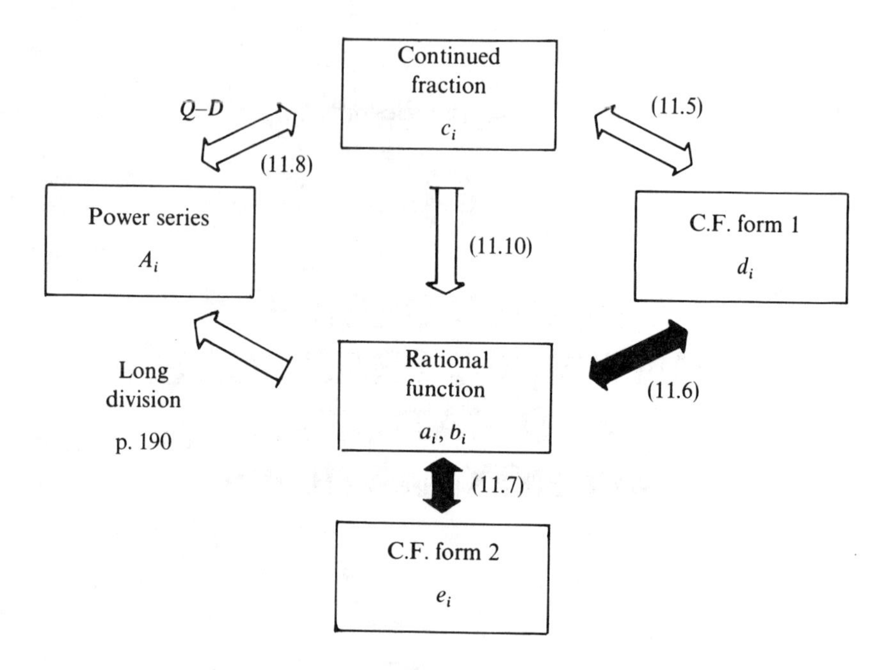

Figure 11.1

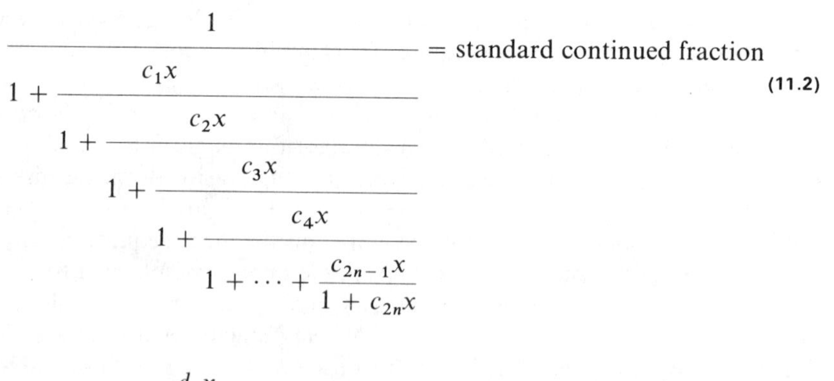

$$\cfrac{1}{1+\cfrac{c_1x}{1+\cfrac{c_2x}{1+\cfrac{c_3x}{1+\cfrac{c_4x}{1+\cdots+\cfrac{c_{2n-1}x}{1+c_{2n}x}}}}}} = \text{standard continued fraction} \tag{11.2}$$

$$1-\cfrac{d_1x}{1+d_2x-\cfrac{d_3x^2}{1+d_4x-\cdots-\cfrac{d_{2n-1}x^2}{1+d_{2n}x}}} = \text{continued fraction, computational form 1} \tag{11.3}$$

$$1 + \cfrac{e_1}{x + e_2 + \cfrac{e_3}{x + e_4 + \cdots + \cfrac{e_{2n-1}}{x + e_{2n}}}} = \text{continued fraction, computational form 2} \tag{11.4}$$

In all of these standard forms the number of essential parameters is the same, $2n$. It is obvious that any rational function can be brought into the given form, first by dividing the numerator and denominator by b_0 and then by factoring out the new a_0. All these forms, then, lack one multiplicative scale parameter, a_0, which must generally be inserted when fitting transcendental functions. When encoding the rational function for repeated evaluation on a computer, one multiplication may be saved by normalizing so that the coefficient of the highest power of x in either P or Q is unity, rather than the constant term.

Evaluative arithmetic labor

The several computational forms differ widely in the number of arithmetic operations required to evaluate them. They are summarized in Table 11.1.

Table 11.1

Form	Multiplications	Additions	Divisions	Total
RF	$2n$	$2n$	1	$4n + 1$
SCF	$2n$	$2n$	$2n$	$6n$
CF1	$2n + 1$	$2n$	n	$5n + 1$
CF2	—	$2n$	n	$3n$
PS	$2n$	$2n$	—	$4n$

The standard continued fraction is clearly the least desirable from the point of view of arithmetic labor. At the other extreme, the continued fraction computational form 2 is the most efficient. The rational function and the power series lie about halfway in between. These judgments are made under the assumption that arithmetic is carried out in floating-point mode, where all arithmetic operations require approximately the same amount of time. If the evaluation is programmed in fixed-point arithmetic, however, additions may be markedly faster than multiplications or division – considerably enhancing the superiority of computational form 2.

Although all the forms given in this chapter are *algebraically identical*, they are *not arithmetically equivalent* for they may differ markedly in their rounding properties. The finiteness of computer arithmetic often combines with essential canceling of significant digits through subtraction to produce much less precision in the final result from one computational form versus another. General rules are difficult to state, so the reader is advised to try his approximations quite thoroughly before releasing them for public use.

Power series from a rational function

From a rational function we may obtain a power series by simple long division of one polynomial by the other. The algorithm has been given in Chapter 7, page 190.

SCF to CF1

A simple algorithm permits us to pass in either direction between the standard continued fraction and computational form 1.

$$\begin{aligned} d_1 &= c_1 \\ d_2 &= c_1 + c_2 \\ d_3 &= c_2 c_3 \\ d_4 &= c_3 + c_4 \\ d_5 &= c_4 c_5 \\ d_6 &= c_5 + c_6 \\ &\vdots \end{aligned} \tag{11.5}$$

Note that an error in c_k will affect only two d coefficients, whereas an error in one of the d's will affect all the c's with equal or higher subscripts. In particular, integer c's guarantee integer d's, but the reverse is not true. The general implication is that passage from d to c will usually involve a loss in precision.

Rational function to computational form 1

The general algorithm to go directly from the rational function to computation form 1 is quite complicated. But in most practical applications we are interested in only four, six, or eight parameters. The algorithm in both directions is given below for these numbers of parameters.

$$\left.\begin{aligned} d_1 &= c_1 \\ d_2 &= c_1 + c_2 \\ d_3 &= c_2 c_3 \\ d_4 &= c_3 + c_4 \end{aligned}\right\} \text{so} \quad \left.\begin{aligned} b_1 &= d_2 + d_4 \\ b_1 - a_1 &= d_1 \\ b_2 - a_2 &= d_1 d_4 \\ b_2 &= d_2 d_4 - d_3 \end{aligned}\right\} \begin{gathered}\text{which may}\\ \text{also be}\\ \text{written}\end{gathered} \quad \begin{aligned} d_1 &= b_1 - a_1 \\ d_4 &= (b_2 - a_2)/d_1 \\ d_2 &= b_1 - d_4 \\ d_3 &= d_2 d_4 - b_2 \end{aligned} \tag{11.6}$$

For P_3/Q_3 the formula becomes

$$
\begin{aligned}
b_1 - a_1 &= d_1 \\
b_1 &= d_2 + (d_4 + d_6) \\
b_2 - a_2 &= d_1(d_4 + d_6) \\
b_2 &= d_2(d_4 + d_6) - d_3 + (d_4 d_6 - d_5) \\
b_3 - a_3 &= d_1(d_4 d_6 - d_5) \\
b_3 &= d_2(d_4 d_6 - d_5) - d_3 d_6
\end{aligned}
$$

which may be written

$$
\begin{aligned}
&d_1 = b_1 - a_1 &&; b'_1 = (b_2 - a_2)/d_1 \equiv d_4 + d_6 \\
&d_2 = b_1 - b'_1 &&; b'_2 = (b_3 - a_3)/d_1 \equiv d_4 d_6 - d_5 \\
&d_3 = d_2 b'_1 + (b'_2 - b_2) &&; b'_3 = d_2 b'_2 - b_3 \equiv d_3 d_6 \\
&d_6 = b'_3/d_3 \\
&d_4 = b'_1 - d_6 \\
&d_5 = d_4 d_6 - b'_2
\end{aligned}
$$

While passage from the rational function to the standard computational form 1 is always possible through the power series via long division, it is simpler to take the other route (Figure 11.1). In approximating an engineering function with a rational function we would seldom wish, as a practical matter, to go beyond P_3/Q_3. However, the algorithm for the P_4/Q_4 is available and I record it here for posterity. Again, both directions are quite practical.

$$
\begin{aligned}
b_1 - a_1 &= d_1 \\
b_1 &= d_2 + d_4 + d_6 + d_8 \\
b_2 - a_2 &= d_1[d_4 + d_6 + d_8] \\
b_2 &= d_2[d_4 + d_6 + d_8] + [(d_4 d_6 - d_5) + (d_6 d_8 - d_7) + d_4 d_8] - d_3 \\
b_3 - a_3 &= d_1[(d_4 d_6 - d_5) + (d_6 d_8 - d_7) + d_4 d_8] \\
b_3 &= d_2[(d_4 d_6 - d_5) + (d_6 d_8 - d_7) + d_4 d_8] + [d_4(d_6 d_8 - d_7) - d_5 d_8] \\
&\quad - d_3(d_6 + d_8) \\
b_4 - a_4 &= d_1[d_4(d_6 d_8 - d_7) - d_5 d_8] \\
b_4 &= d_2[d_4(d_6 d_8 - d_7) - d_5 d_8] - d_3(d_6 d_8 - d_7)
\end{aligned}
$$

$$d_1 = b_1 - a_1 \qquad b'_1 = (b_2 - a_2)/d_1 \equiv d_4 + d_6 + d_8$$

$$d_2 = b_1 - b'_1 \qquad b'_2 = (b_3 - a_3)/d_1 \equiv (d_4 d_6 - d_5) + (d_6 d_8 - d_7) + d_4 d_8$$

$$d_3 = d_2 b'_1 + (b'_2 - b_2) \qquad b'_3 = (b_4 - a_4)/d_1 \equiv d_4(d_6 d_8 - d_7) - d_5 d_8$$

$$d_4 = b'_1 - b'_4 \qquad b'_4 = (d_2 b'_2 + b'_3 - b_3)/d_3 \equiv d_6 + d_8$$

$$d_5 = d_4 b'_4 + b'_5 - b'_2 \qquad b'_5 = (d_2 b'_3 - b_4)/d_3 \equiv d_6 d_8 - d_7$$

$$d_8 = b'_6/d_5 \qquad b'_6 = d_4 b'_5 - b'_3 \equiv d_5 d_8$$

$$d_6 = b'_4 - d_8$$

$$d_7 = d_6 d_8 - b'_5$$

The general algorithm, taken from Maehly, is given by Spielberg [1961]. Its complexity will probably discourage all but specialists.

Computational form 2 from the rational function

To obtain computational form 2 from the rational function it is convenient to redefine our symbols by normalizing our rational function so that the highest powers have unit coefficients. Thus, *for this algorithm only*, the symbols a_i and b_i are defined for P_2/Q_2 by

$$\frac{x^2 + a_1 x + a_2}{x^2 + b_1 x + b_2}$$

thus effectively reversing the notation of our standard rational function where a_k is paired with x^k. For this fraction the computation form 2 coefficients may be computed from

$$\begin{aligned} e_1 &= (a_1 - b_1) \\ e_4 &= (a_2 - b_2)/e_1 \\ e_2 &= (b_1 - e_4) \\ e_3 &= b_2 - e_2 e_4 \end{aligned} \qquad (11.7)$$

For P_3/Q_3 the e coefficients are found from

$$e_1 = a_1 - b_1 \qquad b'_1 = (a_2 - b_2)/e_1$$

$$e_2 = b_1 - b'_1 \qquad b'_2 = (a_3 - b_3)/e_1$$

$$e_3 = (b_2 - b'_2) - e_2 b'_1 \qquad b'_3 = e_2 b'_2 - b_3$$

$$e_6 = -b_3'/e_3$$

$$e_4 = b_1' - e_6$$

$$e_5 = b_2' - e_4 e_6$$

(The correspondence with the formulas connecting d with a and b is obvious. These formulas are obtained directly by substituting e_i for b_i whenever i is even and by substituting $-e_i$ for b_i whenever i is odd.)

The quotient-difference algorithm

The quotient-difference algorithm of Rutishauser [1954] enjoys a number of uses which include passage between power series and continued fractions. If we construct a diagonal table[1]

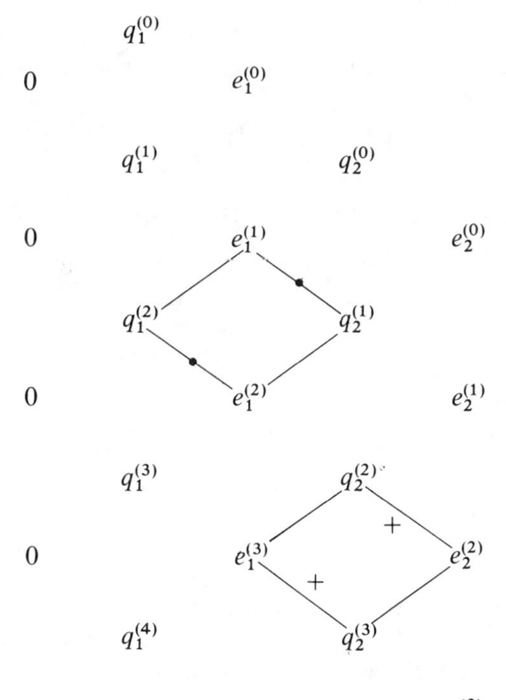

and impose the computational rules

$$\begin{aligned} q_{k+1}^{(n)} \cdot e_k^{(n)} &= q_k^{(n+1)} \cdot e_k^{(n+1)} \\ q_k^{(n)} + e_k^{(n)} &= q_k^{(n+1)} + e_{k-1}^{(n+1)} \end{aligned} \tag{11.8}$$

[1] *Natl. Bur. Std. (U.S.) Appl. Math. Ser. 49*, p. 36.

we may now build an entire table from either the first column or the top diagonal. If we examine the four elements at the corners of any elementary rhombus in our diagram we will find that they are connected by one of the two rules, either multiplicative or additive. The lines with dots and plus signs are intended as a mnemonic device rather more effective than the heavily subscripted formulas. Note that all the rhombi in any one vertical column obey the same rule. Thus if we are building out from the left, all the q elements are computed from the multiplicative relation, all the e's from the additive. If we are building downward from the top diagonal the reverse statement holds. Filling out the table from left to right is, however, frequently an unstable procedure.

If we now define our power series by

$$f(x) = \sum_{i=0}^{\infty} A_i x^i$$

and then compute the first column of our quotient-difference table from the relations

$$q_1^{(n)} = \frac{A_{n+1}}{A_n}$$

we can then generate the complete table. The corresponding continued fraction is given by

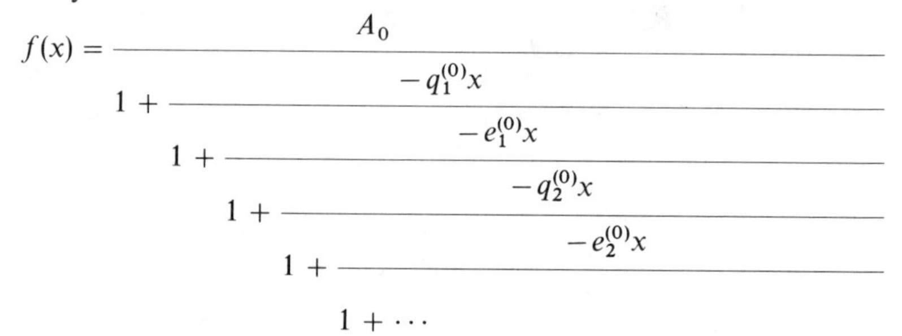

$$f(x) = \cfrac{A_0}{1 + \cfrac{-q_1^{(0)}x}{1 + \cfrac{-e_1^{(0)}x}{1 + \cfrac{-q_2^{(0)}x}{1 + \cfrac{-e_2^{(0)}x}{1 + \cdots}}}}}$$

Note the negative signs.

To obtain the power series from the standard continued fraction we reverse the procedure. The elements of the continued fraction are placed in the top diagonal, the quotient-difference table computed, and the left column is identified with the ratio of successive coefficients of the power series.

Rational function from standard continued fractions

If we take our standard continued fractions in the form

$$f(x) = \frac{\alpha_0|}{|\beta_0} + \frac{\alpha_1 x|}{|\beta_1} + \frac{\alpha_2 x|}{|\beta_2} + \cdots \tag{11.9}$$

which becomes our standard continued fraction if we set α_0 and all β_i to unity and α_i to c_i for all $i > 0$, we may then define

$$\left.\begin{aligned} P_m &= \beta_m P_{m-1} + \alpha_m \cdot x \cdot P_{m-2} \\ Q_m &= \beta_m Q_{m-1} + \alpha_m \cdot x \cdot Q_{m-2} \end{aligned}\right\} m \geq 2 \tag{11.10}$$

If we start our recursions with

$$\begin{aligned} P_0 &= \alpha_0 & Q_0 &= \beta_0 \\ P_1 &= \alpha_0 \beta_1 & Q_1 &= \beta_0 \beta_1 + \alpha_1 x \end{aligned} \tag{11.11}$$

then

$$f(x) = \frac{P_m(x)}{Q_m(x)}$$

This algorithm is used both arithmetically and algebraically. If we are given a value for x, we then generate a sequence of numerical values for P_m and Q_m – hence numerical values for $f(x)$. This is the effective way to evaluate our standard continued fraction since it permits the process to be continued until adequate convergence is visible in successive values of f. The arithmetic labor is $3n$ multiplications and $2n$ additions followed by a division. This is somewhat less than the other way of evaluating the standard continued fraction but somewhat more than the other way of evaluating an explicit rational function. A minor but annoying programming problem often arises during this arithmetic recursion out of the fact that the magnitudes of both $P_n(x)$ and $Q_n(x)$ usually increase very rapidly for the larger values of x which we are apt to be using. Thus, although the ratio P_n/Q_n will remain a quite reasonable number, the two parts separately will quickly overflow the capacity of the computer. The cure, easy but necessary, is to test for largeness and then divide all four current values $(P_n, Q_n, P_{n-1}, Q_{n-1})$ by the same arbitrary constant before carrying out the next step of the recurrence. Since the relations are linear in the P's and Q's, such normalizations have no effect on the desired ratio.

If our interest is *algebraic*, that is, we wish to generate the explicit coefficients of the rational function, the algorithm (11.10) above immediately leads us to the following recursion. Letting p_{nk} be the coefficient of x^k in P_n we compute

$$p_{nk} = p_{n-1,k} + c_n \cdot p_{n-2,k-1} \tag{11.12}$$

where the c_k are the parameters of our standard continued fraction. We start the recursion with

$$p_{n0} = 1 \qquad n = 0, 1, \cdots$$

and

$$p_{nk} = 0 \qquad \text{for } k > [n/2]$$

Note that the polynomial P_n is of degree $[n/2]$.

The corresponding algorithm for the coefficients of Q_n is given by

$$q_{nk} = q_{n-1,k} + c_n \cdot q_{n-2,k-1} \tag{11.13}$$

where the table is begun with

$$q_{n0} = 1 \qquad \text{for all } n, \quad q_{11} = c_1,$$

and

$$q_{nk} = 0 \qquad \text{for } k > \left[\frac{n+1}{2}\right]$$

Note that Q_n is a polynomial of degree $[(n+1)/2]$.

Problem

Take the standard continued fraction through c_4 for $\tan^{-1}x/x$ and derive from it the power series, the rational functions, and the computational form 1. Check your work by also obtaining computational form 1 from the rational function. Observe where significant figures disappear along your various routes and thereby deduce which paths are apt to be dangerous.

CHAPTER 12

ECONOMIZATION OF APPROXIMATIONS

The easily available approximations to a transcendental function are the truncated power series and continued fractions. Unfortunately, chopping the infinite tail off a power series does not endow the part that remains with any especially desirable properties. In particular, if we approximate $\cos x$ by the first six terms of its series, we are in effect deciding to use a very specific tenth-degree polynomial to represent this function without ever asking ourselves if we could possibly do better with a different tenth-degree polynomial. Why this one? Because it is available, of course! An argument in its favor, to be sure, but scarcely the sort of argument we have been offering in areas where computational efficiency and precision were our principal concerns.

This chapter discusses some of the techniques that seem useful once we admit that we want functional approximations which deliver as much information as possible per fitted parameter. We want the *best* tenth degree polynomial fit to $\cos x$, or the *best* continued fraction, or the *best* rational function with 10 parameters. But what do we mean by "best"? Do we want an approximation that is very, very good at one point near the middle of some range and then gets increasingly worse as we move away, and finally explodes into complete unusability? Seems silly, when stated that way, but any Taylor's series, including our cosine series, behaves this way—and we use them all the time. Your author would prefer to ascribe this phenomenon to human ignorance rather than to human laziness, though both are doubtless involved. For it must be admitted that the economization techniques of this chapter require a moderate amount of work and are not well enough known. But they can easily be automated, so we can some day expect to find computer laboratories equipped with subroutines to "economize" the user's favorite functional approximation – provided he knows what needs to be done. Besides, any approximation that will be used several millions of times deserves a little tender care and grooming.

The essential characteristics of a functional approximation appear most clearly in a graph of the *error* plotted as a function of the argument. We shall here use the simple error, the difference between the function and the approximation, rather than the percentage error, although a good case for the latter can often be made. Further, we shall assume that our approximation is going to be used over a finite rather than infinite range. This is the usual practice, although again examples of single approximations holding from zero to infinity do exist. For most functions they cannot be found, and so we settle for a more modest range, such as 0 to 3, leaving the rest of the arguments to be covered by a second and even possibly a third approximation. Since our approximations are for use with automatic computing machines, we would like accuracies commensurate with the single precision arithmetic of these machines – implying seven to nine decimals (or perhaps significant figures). Finally, having defined our requirements a little more carefully, we feel that the error a customer gets on using our approximation *anywhere* within its intended range should have some optimal property of smallness. Perhaps the *average squared error* over the range should be as small as possible (among all tenth-degree polynomials, that is). Or maybe we would like to assert that the *worst* error in the range is smaller than the worst error given by any other approximation of the same class. This is called the minimum maximum error – or *minimax* approximation. It is the widely favored criterion today, so we shall use it. For well-behaved functions this minimax property for the error curve implies that it oscillates as much as possible between *equal extreme values* (Figure 12.1). Such a curve is often called an *equal-ripple error curve.*

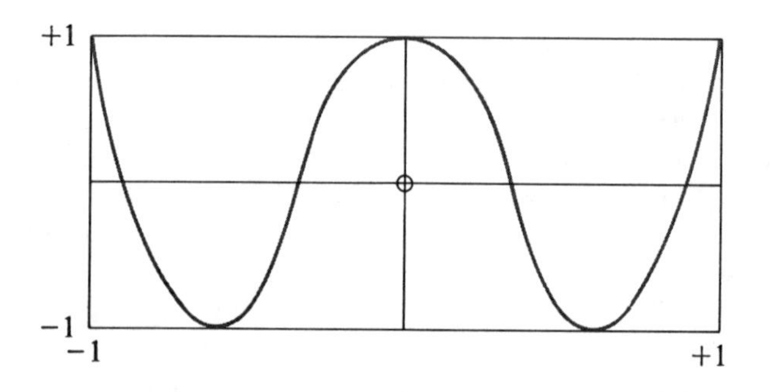

Figure 12.1. Error curve of standard form for an approximating polynomial of degree 3

The error term

While a precise expression of the error for any approximation is usually available only as a graph, good analytic approximations frequently converge sufficiently rapidly so that the first neglected term of some series is the principal part of the error. Since we usually are happy with any *approximately* equal-ripple approximation, we often seek to make this first neglected term of the series an equal-ripple polynomial—confident that the true minimax approximation to our function is not substantially better than the one we have thus obtained, and with less work. In this section we seek to produce series approximations whose first neglected term is an equal-ripple polynomial. We shall fall short in even this slightly less than optimal approach, but no matter. The results are often really quite good.

Economizing a power series

We illustrate the basic technique on the power series for $\cos x$ on the somewhat arbitrary range

$$0 \leq x \leq \pi$$

Since the function is even, we might say our arguments are less than π in absolute value. We shall also choose an accuracy of 10^{-6} as desirable—again an arbitrary choice for the purposes of demonstration.

The technique starts with a series that is completely adequate except for its length. Our efforts go into shortening this series without overmuch loss of precision. We begin with

$$\cos x = 1 - \frac{x^2}{2!} + \frac{x^4}{4!} - \cdots - \frac{x^{14}}{14!} + \frac{x^{16}}{16!} \tag{12.1}$$

which at π produces an error less than $\pi^{18}/18!$, or about 10^{-7}—clearly adequate. Dropping the x^{16} term is fatal since at π it equals about 10^{-5}. On the other hand, a plot of this term (Figure 12.2) shows that over most of the range of the argument it lies considerably below even our limit of 10^{-6} and it is only because of the spike out near π that we are forced to keep it in our truncated series approximation.

Except for a couple of scale factors Figure 12.2 is merely a plot of y^{16} on the range (0, 1), so to reduce a variety of similar problems to a common base we now make a change of variable—to be mandatory for all finite-range approximations, reducing the interesting arguments to (0, 1) or (−1, 1) if symmetries are present. Accordingly we define

$$y = \frac{x}{\pi}$$

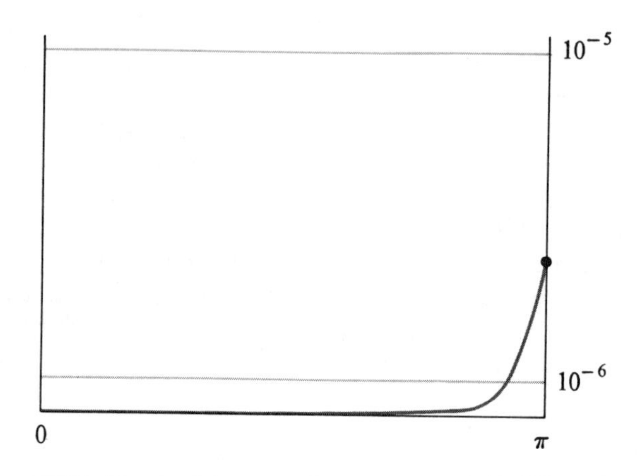

Figure 12.2. Sketch of the approximate error term $x^{16}/16!$ on $(0, \pi)$

and rewrite (12.1) as

$$\cos \pi y = 1 - \frac{\pi^2}{2!}y^2 + \frac{\pi^4}{4!}y^4 - \cdots - \frac{\pi^{14}}{14!}y^{14} + \frac{\pi^{16}}{16!}y^{16} \tag{12.2}$$

where

$$-1 \leq y \leq 1$$

Nothing has been changed. Each term of (12.2) contributes exactly the same numerical value towards the cosine that it did in (12.1). The advantage is that we now deal with a standardized variable y. And it is the y^{16} term that we would like to discard.

Geometrically, the crucial observation about y^{16} is that it *looks* very much like y^{14}, y^{12}, y^{10}, and many other lower powers of y—so much so that it seems reasonable to expect a linear combination of those lower powers to approximate the y^{16} curve very well. Thus perhaps we can replace the y^{16} term by a lower-degree polynomial in y that can be unobtrusively combined with the lower powers of y we already possess in (12.2). For this purpose we invoke the Chebyshev polynomials.

The Chebyshev polynomial $T_{16}(y)$ is

$$T_{16}(y) = 2^{15}y^{16} - 2^{17}y^{14} + 13 \cdot 2^{14}y^{12} - 11 \cdot 2^{14}y^{10} + 165 \cdot 2^9 y^8$$

$$-21 \cdot 2^{10}y^6 + 21 \cdot 2^7 y^4 - 2^7 y^2 + 1$$

so that

$$y^{16} = \frac{T_{16}(y)}{2^{15}} + 4y^{14} - \frac{13}{2}y^{12} + \frac{11}{2}y^{10} - \frac{165}{64}y^8 + \frac{21}{32}y^6 - \frac{21}{256}y^4 + \frac{1}{256}y^2 - \frac{1}{2^{15}} \tag{12.3}$$

Here we have expressed y^{16} as a linear combination of a Chebyshev polynomial of degree 16 and lower even powers of y. If we now discard the first term on the right side of (12.3) we make an equal-ripple error of 2^{-15}, because every Chebyshev polynomial oscillates exactly between ± 1 in amplitude. (They are actually cosine curves with a somewhat disturbed horizontal scale, but the vertical scale has not been touched.) If we now replace y^{16} in (12.2) by our equal-ripple approximation (12.3), we commit a maximum error in the $\cos \pi y$ of

$$\frac{\pi}{16!}\left(\frac{\pi}{2}\right)^{15} \qquad \text{or} \qquad 1.31 \times 10^{-10}$$

which is considerably less than the y^{18} term we discarded at the very beginning. Our approximation now consists of

$$\cos \pi y \approx \left[\begin{array}{l} 1 - \dfrac{\pi^2}{2!}y^2 + \dfrac{\pi^4}{4!}y^4 - \dfrac{\pi^6}{6!}y^6 + \cdots + \dfrac{\pi^{12}}{12!}y^{12} - \dfrac{\pi^{14}}{14!}y^{14} \\ -\dfrac{\pi^{16}}{16!2^{15}} + \dfrac{\pi^{16}}{16!2^8}y^2 - \cdots \quad \cdots - \dfrac{\pi^{16} \cdot 13}{16!2}y^{12} + \dfrac{\pi^{16}4}{16!}y^{14} \end{array}\right]$$

When we combine the terms, evaluating the numerical coefficients, we have altered the higher coefficients appreciably, the lower ones very little. The expression, in part, is

$$\cos \pi y = 0.99999\,99998\,7 - 0.99999\,99965\,938\frac{\pi^2}{2!}y^2 + \cdots + \frac{\pi^{12}}{12!}y^{12}(0.98550\,460) - \frac{\pi^{14}}{14!}y^{14}(0.83550\,659)$$

We have maintained separately the factor of $\pi^n/n!$ to permit easy return to our original variable, x, and to show the relation with the unmodified cosine series.

We would emphasize that in making this approximation, a polynomial of fourteenth degree, we commit an error of 10^{-10}, although if we create a different fourteenth degree approximation by merely dropping the $y^{16\text{th}}$ term of (12.2), our error is 10^{-5}. Not only are the approximations different, one is 100,000 times more accurate than the other. One pays a large price for the convenience of mere truncation!

Since the error we committed in abolishing the y^{16} term was well below our announced tolerance of 10^{-6}, we are now free to attempt replacement of the y^{14} term by the same technique. The maximum error incurred this time will be

$$\frac{\pi^{14}(0.84)}{14!2^{13}} = 1.21 \times 10^{-8}$$

still quite tolerable. We will thus reduce our approximation to an even polynomial of twelfth degree, but there the process will probably stop, for the next error is

$$\frac{\pi^{12}(0.828)}{12!2^{11}} = 0.79 \times 10^{-6}$$

and we may feel that this is a bit too much—though it clearly is a borderline decision. The final approximation is

$$\cos \pi y = 0.99999\,9989 - \frac{\pi^2 y^2}{2!}(0.99999\,9785) + \frac{\pi^4 y^4}{4!}(0.99999\,5790) - \cdots$$
$$- \frac{\pi^{10} y^{10}}{10!}(0.98277\,958) + \frac{\pi^{12} y^{12}}{12!}(0.82692\,539) \qquad \textbf{(12.4)}$$

over $-1 \leq y \leq 1$. Your author deliberately omits a few terms here (he has them at home!) because the cosine should not be approximated on the range $(-\pi, \pi)$. The efficient approximation strategy restricts the range to $(0, \pi/3)$, using the half-angle identities to get one's argument down there where the economized representation is very short indeed. Under this same strategy arguments near $\pi/2$ cause use of the economized series for the sine of the complementary argument. A full discussion of the considerable niceties that apply here is found in Kogbetliantz [1960]. The cosine was chosen for our exposition because it has a familiar series with which to display the economization technique. The economization was performed on a hand-cranked Curta calculator, which, with good tables, is adequate for the job. But such equipment is quite inadequate for the checking operation, which requires much greater precision. Since your author does not believe in doing well what should probably not be done at all, he has never expended the considerable labor required to check this approximation.

The error curve of (12.4) is approximately equal ripple, since it is mostly proportional to T_{14}, but with a small admixture of T_{16}. If only one stage of economization had been undertaken and if we ignore all terms originally truncated from the Taylor series to give (12.2), then our error would have been exactly one Chebyshev polynomial—exactly equal ripple. Further modification of the coefficients of (12.4) to give exactly an equal-ripple error curve is possible, but the reduction in the maxima would be slight, scarcely worth the bother.

These term-by-term economizations of power series can be automated, but care is needed lest machine rounding error creep in at delicate points. True error curves, which inevitably require subtraction of the approximant from the true value of the function, demand double-precision calculation of the true function if they are to be done internally on the computer.

A somewhat different but inferior approximation may be found if we replace y^2 by z in (12.2), then eliminate the last term by the Chebyshev polynomial $T_8(z)$, which throws the burden of representing this term back on the terms in z^6, z^4, z^2, and z^0. Similarly, the z^7 term may be removed. The approximation is

$$\cos \pi y \approx 0.99999\,99664 - \frac{\pi^2}{2!}y^2(1.00000\,23192) + \frac{\pi^4}{4!}y^4(1.00000\,02651)$$

$$-\frac{\pi^6}{6!}y^6(0.99993\,14304) + \frac{\pi^8}{8!}y^8(0.99997\,7144)$$

$$-\frac{\pi^{10}}{10!}y^{10}(1.00709\,566) + \frac{\pi^{12}}{12!}y^{12}(1.00446\,012)$$

but its maximum error is 1.6×10^{-6}, 100 times larger than that of (12.4), where *each* discarded term was absorbed more fully by the modification of *all* the available lower-order terms. Geometrically, y^{16} is more like y^{14} than y^{12}. When we use $T_{16}(y)$, a fair amount of the representation falls onto y^{14}. When we use $T_8(z)$, the principal load necessarily falls upon y^{12}. Looking at the problem slightly differently, $\cos \pi y$ is an even function, that is, symmetric about zero, and this symmetry is shared by all even Chebyshev polynomials. The replacement of y^2 by z gives an approximation for the cosine that holds only for positive z, so the use of the symmetric Chebyshev polynomials is a geometric mismatch over half the range. We could, of course, use the Chebyshev polynomials that have been orthogonalized over (0, 1), but this would give the same approximation (12.4) we originally achieved. The only advantage of such a procedure is the mechanical one of working with lower-degree polynomials. In hand computation, however, they undoubtedly reduce the human error rate and should therefore be used.

How much can be saved?

Before embarking on the sometimes tedious voyage through the reefs and islands of detail that comprise the land of economization, we might do well to answer questions about the advisability of it all. If economization requires more than a modicum of labor, are the savings going to justify the effort? Your author believes the answer is almost always "yes!", but he would prefer to help the reader to judge for himself.

In an attempt to summarize this complicated subject, we present three graphs prepared by Maehly [1956] that show the effects of economization on power series and continued fraction representations of three common functions, as well as the relative efficiencies of the uneconomized forms. In these figures the number of terms n is plotted horizontally, while the vertical axis shows the number of significant figures obtained by using n terms. Thus for e^x on the range shown in Figure 12.3 10 significant figures will be given by about 9 terms of the power series, 8 terms of the best-fit polynomial, 4.5 terms of the continued fraction, and 4 terms of the best-fit rational approximation. For this function the principal saving clearly is made by switching from

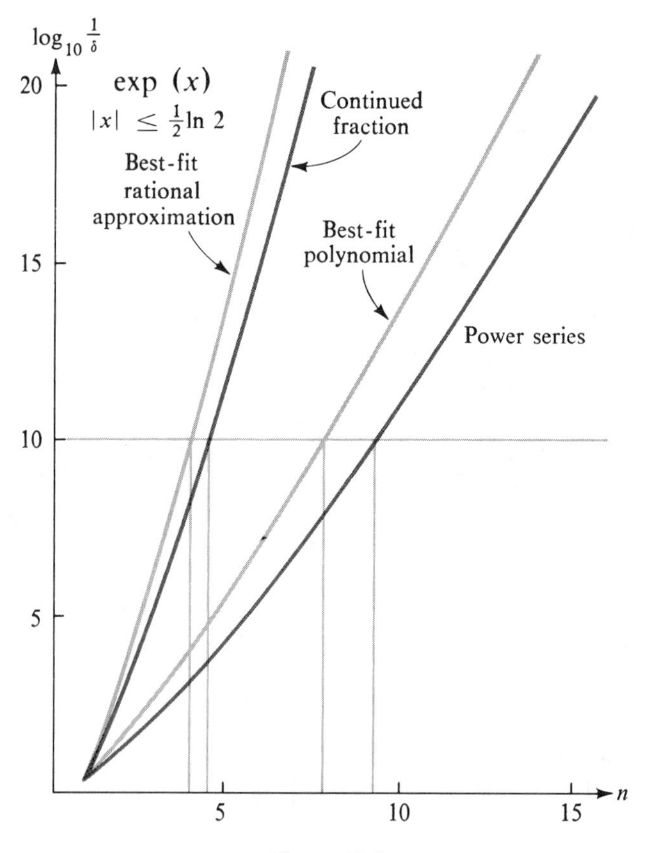

Figure 12.3

the power series to the continued fraction—economization adding comparatively little advantage to either form. (But even that little is doubtless worth saving for a function so frequent as e^x.)

Turning to the tangent function, Figure 12.4, we find that 10-figure accuracy requires 17 terms of the power series but only 8 terms of the best fit (that is, the economized) polynomial. The continued fraction may be truncated at 6 terms for the same accuracy, while the best-fit rational function will have 4 or 5 terms. Here economization produces startling savings in the series and smaller but eminently worthwhile advantages in the continued fraction.

Finally, in Figure 12.5 we show the arctangent, whose power series is absolutely abominable. (It takes 5000 terms to deliver five significant figures at x equal to unity and diverges for all larger arguments.) Surprisingly, the best-fit polynomial will give 10 significant figures with only 11 terms—better than the

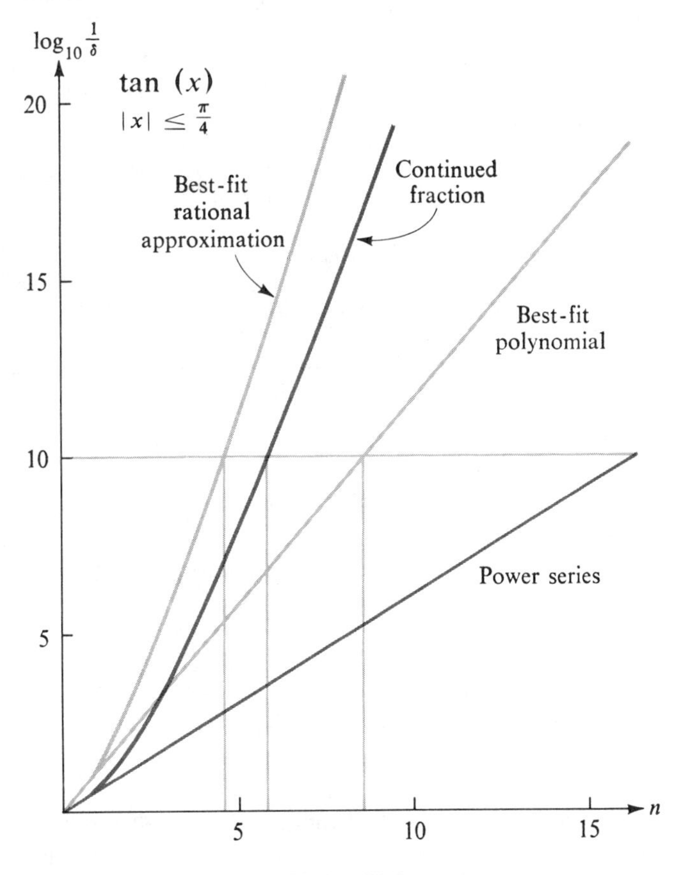

Figure 12.4

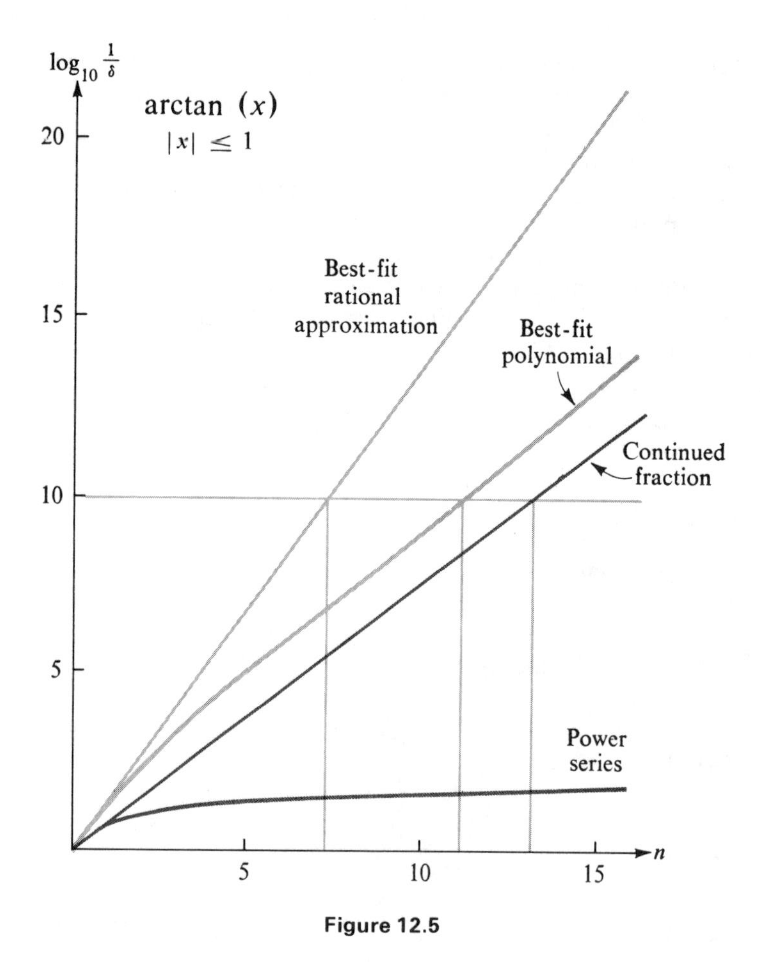

Figure 12.5

uneconomized rational function (continued fraction), which requires 13. With economization, however, we reduce the computational labor of a best-fit rational function to a mere seven terms. There is no argument here about the form to use!

By way of summary, these figures suggest that economization contributes most to those functions with rather slowly converging power series. Rational function approximations give their best improvement over the power series for functions that have horizontal or vertical asymptotes nearby. Thus for both the tangent and arctangent, power series are simply inappropriate. Economization helps restore the series to usefulness but that is rather like trimming enough of the corners off the proverbial square peg so it will fit into the round hole: It remains a patched-up job. Better to use the proper implement in the first place.

Curves for the cosine function are unfortunately not available, but one suspects that the rational function confers little advantage over the power series since the cosine wiggles rather like a polynomial in any case. Economization will yield some improvement, as shown above, but the series is already relatively efficient, so the gains are comparatively small. For a sufficiently restricted range, the exponential function is rather more polynomic than the tangent, if less so than the cosine. At least there are no asymptotes close by, but no maxima either. Both the power series and the continued fraction are fairly efficient, though the latter is twice as good as the former. For neither form does economization yield large dividends, although, as we must repeatedly emphasize, for the commonly used functions even small savings justify large efforts in preparing approximations for automatic computer use. In an engineering environment, your computer may be spending half its time evaluating the common functions—and at how much per hour?

Direct equal-ripple approximation techniques

Thus far we have economized truncated series to make them shorter. Although the resulting error curves of the economized series are not equal ripple, they come rather close and the simplicity of the economizing procedure recommends it. When a rapidly convergent power series is already at hand we probably shall be able to lop off only two or three terms, so the labor is not prohibitive. If a series converges more slowly or is not available, we may prefer to consider some method for generating an approximation directly. Again, equal-ripple errors are the ideal, even if they will only be approached rather than actually obtained.

One method, mathematically simple but usually difficult to effect, is the direct expansion of our function in a series of Chebyshev polynomials. If we express

$$f(x) = \tfrac{1}{2}c_0 + \sum_{k=1}^{\infty} c_k T_k(x) \tag{12.5}$$

we may invoke the orthogonality of the T_k functions to evaluate the c_k via the relation

$$\int_{-1}^{1} \frac{f(x) \cdot T_j(x)}{\sqrt{1 - x^2}}\, dx = c_j \qquad j = 0, 1, \cdots \tag{12.6}$$

Because of the singularities in the integrand at both limits, *numerical* quadrature requires a change in the dummy variable. More serious trouble arises because the higher $T_j(x)$ are highly oscillatory, causing considerable cancellation of areas with attendant loss of significant figures—the type of difficulty we learned to fear with alternating series. Thus (12.6) is useful for finding these c_j only indirectly, usually through analytic integrations. When it succeeds, however, the method is very efficient. We merely generate as many of the c_j as we need to get the desired degree of approximation. We may then convert the series of Chebyshev polynomials into a straight power series if we wish, although it is usually better to employ the recursive evaluation technique (Chapter 1) directly on the sequence of the $\{c_j\}$ themselves.

Criteria for direct approximation

Once we decide to fit a polynomial approximation directly to a function, our desire for some kind of minimal error leads us directly to the equal-ripple error-curve requirement. Since this curve is central to our strategy we would examine it briefly before proceeding with the practical fitting methods.

Throughout all discussions of economization one can easily become confused about n. Is it the order of the approximation, or the number of zeros on the error curve, or the number of humps on the error curve, or the number of parameters to be fitted? These quantities, being closely related, differ from each other by one or two and there is no accepted practice. Insofar as possible, in this chapter your author has tried to let n represent the *degree of the approximating polynomial*. Thus the number of parameters to be fitted is $n + 1$, as is the number of *zeros* on the error curve. This makes the number of true (interior) maxima and minima on our error curve (see Figure 12.1) equal to n; hence the total number of extrema is $n + 2$—the extra two occurring necessarily at the ends of the (finite) interval over which the approximation holds. Such a curve is sometimes called an "error curve of standard form." If all of its $n + 2$ extrema are equally high, our standard equal-ripple curve implies an optimal approximation. Occasionally special symmetries in the function and the approximation will cause the optimal error curve to have more than $n + 2$ extrema, but we shall ignore this possibility. For us, Figure 12.1 is the error curve we seek, $n + 2$ equal extrema alternately positive and negative, separated by $n + 1$ zeros.

A direct approach

Having a function, say

$$F(x) = \int_0^\infty \frac{e^{-xt^2}}{1 + t}\,dt$$

and having decided to use an approximating polynomial of degree n with $n + 1$ as yet undetermined coefficients, we might try taking $n + 1$ values of x, demanding that the approximation agree exactly with the function at these points. This method must produce an error curve that has $n + 1$ zeros at the chosen values, which we shall call $\{z_j\}$. For a sixth-degree approximation we might solve

$$a_0 + a_1 z_j + a_2 z_j^2 + \cdots + a_6 z_j^6 = F(z_j) \qquad j = 0, 1, \ldots, 6 \qquad \text{(12.7)}$$

If we choose these zeros $\{z_j\}$ to coincide with the $n + 1$ zeros of $T_{n+1}(z)$, the Chebyshev polynomial of degree $n + 1$, we might reasonably hope for our error curve to resemble $T_{n+1}(z)$ and hence to be approximately equal ripple. Since the error of any approximation is never exactly a polynomial of low degree, this method cannot give an exactly equal-ripple fit, but it can often come gratifyingly close. Accordingly, it is a good general strategy – especially for getting a first approximation that may then be improved, if necessary.

[*Two small details*: Since the function $F(x)$ shown above is only defined for positive x, we would be approximating it on $(0, b)$ rather than $(-b, b)$ and so the zeros of the Chebyshev polynomial $T_{n+1}(x)$ would have to be shifted to $(0, 1)$ by a suitable change of independent variable before being used. Further, since $F(x)$ is singular at 0, we would have to fit the well-behaved function $F(x) + e^{-x} \ln\sqrt{x}$. By ignoring such "small details" are most beginners' attempts at approximation quite thoroughly wrecked!]

Matching at extrema

The strategy of matching the approximation to the function at the $n + 1$ zeros of a Chebyshev polynomial seems somewhat indirect. We are not particularly interested in getting an error curve with those zeros. What we really want is a curve with equal extrema. So perhaps a better strategy would be to take our fitting equations at the location of the $n + 2$ extrema of the Chebyshev polynomial, insisting that the extrema all be of the same magnitude which, being unknown, is also a parameter to be fitted. This gives us the $n + 1$ undetermined coefficients of the approximation plus the one unknown magnitude of the error curve extrema, $n + 2$ parameters for $n + 2$ equations, all linear. Typically, for an approximation of degree 6, we would have

$$a_0 + a_1 x_j + a_2 x_j^2 + \cdots + a_6 x_j^6 - F(x_j) = (-1)^j \lambda \qquad j = 0, 1, \ldots, 7 \qquad \text{(12.8)}$$

where the eight $\{x_i\}$ are taken at the locations of the extrema of $T_7(x)$, suitably shifted. The $(-1)^j$ guarantees that the extrema alternate in sign, the unknown λ being their common magnitude.

Both of the above fitting strategies share several difficulties. At the conceptual level we have conceded that our error curve is not a polynomial at all and hence cannot be exactly a Chebyshev polynomial, even though we can hope to make it equal ripple if we work at it. Thus we might question the wisdom of fitting at x_j chosen from properties of some Chebyshev polynomial. Defense of these strategies lies, pragmatically, in their demonstrated usefulness and, more theoretically, in the observation that the Chebyshev polynomial in question has the desired extremal properties and that any other curve with these properties cannot be too different in its shape. More specifically, since extrema are flat, minor errors in our choice of the x_j should not be critical in determining the a_k of the approximation. The same observation is much less likely to be true of *zeros*, hence most approximators have a marked preference for fitting at the extrema. Maehly [1960], however, found matching at zeros to be quite feasible. He gives methods for altering the first set of $\{z_j\}$ based on the inequalities observed in the heights of the error-curve extrema associated with the approximation actually produced.

A second difficulty, tactical rather than strategic, arises while solving the fitting equations and involves precision problems. An essential characteristic of a good approximation is that the error be small. Thus λ in (12.8) is small. We further see that all quantities on the left of (12.8) will usually be human-sized numbers except possibly for the higher a_k. Thus the essential information by which the equations (12.8) are operating is obtained by subtracting large nearly equal quantities. If we hoped for λ to be near the limit of single-precision arithmetic, we will have to resort to double precision in solving (12.8), lest we find we are fitting rounding errors instead of information. At first glance this requirement for double-precision arithmetic seems avoidable if we use (12.7) to evaluate our approximation at the zeros – a clear advantage. Unfortunately, the difficulty is only postponed, for if equality of the extrema is to be the ultimate fitting criterion, the extrema ultimately must be computed – and computed in double precision. Of course, if you are willing to gamble that your first approximation will be good enough, you may solve (12.7) in single precision and use the result – untested. We do not advocate such gambles!

Remes's algorithm

Perhaps a solid fact (Remes's algorithm) would be reassuring in this miasma of qualitative opinion. Equal-ripple error approximations will be produced by an iterative process that, at the kth stage,

1. Fits by (12.8) at some set of points $\{x_j\}$ to obtain an intermediate approximant, A_k.
2. Finds the locations $\{x_j\}_{k+1}$ of the extrema of the error curve for A_k.
3. Goes to step (1), substituting $\{x_j\}_{k+1}$ for $\{x_j\}_k$.

The difficulties are the usual ones of precision and ill conditioning, but we have methods for coping with them. The ill conditioning of the parameter fitting equations may be alleviated by using Chebyshev polynomials for the basis of the equations rather than the powers of x as in (12.7) and (12.8). Thus we would better fit

$$\frac{c_0}{2}T_0(z_j) + c_1T_1(z_j) + \cdots + c_6T_6(z_j) = F(z_j) \qquad \textbf{(12.9)}$$

$\{z_j\}$ being the zeros of the previous error curve

or

$$c_0T_0(x_j) + c_1T_1(x_j) + \cdots + c_6T_6(x_j) - F(x_j) = (-1)^j\lambda \qquad \textbf{(12.10)}$$

$\{x_j\}$ being the extrema of the previous error curve

where the $T_k(x)$ are the standard Chebyshev polynomials if F is being fitted on $(-1, 1)$ and are the one-sided Chebyshev polynomials if $(0, 1)$ is the range.

Since Chebyshev polynomials are really cosines, we shall find that they satisfy a quite simple orthogonality condition with respect to *summation*. If we begin our iteration by taking the $\{z_j\}$ as the zeros of T_7, evaluating the c_k in (12.9) is quite easy. We have the relations

$$\sum_{j=0}^{n} T_k(z_j)\cdot T_l(z_j) = \left\{\begin{matrix} 0 & k \neq l \\ \dfrac{n+1}{2} & k = l \end{matrix}\right\} k, l < n+1 \qquad \textbf{(12.11)}$$

where

$$T_{n+1}(z_j) = 0 \qquad j = 0, 1, \ldots, n$$

so that

$$c_k = \frac{2}{n+1}\sum_{j=0}^{n} T_k(z_j)\cdot F(z_j) \qquad \textbf{(12.12)}$$

(For us, n is 6.) These relations are useful *only* for obtaining the first approximation, since subsequent cycles employ (z_j) that are not zeros of any Chebyshev polynomial. But the first approximation may be good enough. We still have that old semantic annoyance, familiar to fitters of cosine Fourier series, that c_0 has to be used in (12.9) with half-weight, as shown. Without this rule, the generality of (12.11) is fractured and we have to insert the factor of 2 there because, since T_0 is identically unity, it is obvious that

$$\sum_{j=0}^{n} T_0^2 = n + 1$$

Either we make the exception in (12.9) or in (12.11) or in (12.12)—somewhere that special factor of 2 must get in the zeroth coefficient.

Equations (12.10) cannot easily utilize the orthogonality relations, but expressing them in terms of T_k rather than the x^k improves their condition, giving more tractable arithmetic. It should therefore be done.

Locating the extrema

If we are trying to produce an equal-ripple curve by iterating the fitting process of the preceding section, we must find positions of the extrema for the error curve of the approximant. This operation is doubly difficult since it deals entirely with small quantities obtained by differencing the function and the approximant and is concerned with finding locations of the necessarily flat maxima and minima of this error curve. Double-precision arithmetic is mandatory merely to get the gross error curve. Locating the extrema is best done by evaluating the curve at three points that span the extremum, fitting a parabola, then using the location of the peak of the parabola as the position of the extremum. This tactic markedly reduces the number of points at which the $F(x)$ must be evaluated – usually an expensive operation, especially when great precision is required.

Once the locations of the new error curve extrema $\{x_j\}$ are in hand, (12.8) may be reentered directly for the next iteration. If, however, we are fitting our approximation via (12.7) we need a new set of points $\{z_j\}$ at which we will make our new error curve zero in the hope and expectation that it will thereby become more nearly equal ripple. Such a set of points is given [Maehly, 1963] via a set of corrections $\{\delta z_j\}$ computed from the heights $\varepsilon_j(x_j)$ of the error curve at the current extrema $\{x_j\}$. We solve the set of $n + 2$ linear equations

$$\ln|\lambda| + \sum_{k=0}^{n} \frac{\delta z_k}{x_i - z_k} = \ln|\varepsilon_i| \qquad i = 0, \ldots, n+1 \tag{12.13}$$

to give the $n + 1$ corrections $\{\delta z_k\}$ and the common error amplitude λ. The z_k in (12.13) are the old zeros to which the corrections must be applied.

Fitting the discrepancy

The direct techniques for finding equal-ripple approximations all are plagued by the need for extreme precision. While double- or triple-precision arithmetic is not prohibitive, it is expensive and attendant confusions usually dampen the enthusiasm of the approximator. We would prefer to avoid it.

When the function $F(x)$ has a convergent power series or continued fraction expansion, most multiple-precision arithmetic may be avoided, and the little that is still required comes at the far end where it does not interfere with the principal economization process. The reader has probably noticed that our economized representation of the cosine [p. 294] is just the standard series with slightly perturbed coefficients. If our fitting procedure could confine itself directly to the finding of the small perturbations, rather than the entire coefficients, single-precision arithmetic would be more than adequate. The essential idea here is to fit the *discrepancy* between the function and some easily obtainable approximant then add this fitted function as a correction to the approximant. Specifically, for cos x we might decide, as before, to use the terms through x^{12} in our equal-ripple approximation. Then we write

$$\cos \pi x - \left[1 - \frac{\pi^2}{2!}x^2 + \frac{\pi^4}{4!}x^4 - \cdots + \frac{\pi^{12}}{12!}x^{12}\right]$$
$$= \left[-\frac{\pi^{14}}{14!}x^{14} + \frac{\pi^{16}}{16!}x^{16} - \cdots\right] = [\text{discrepancy}] \tag{12.14}$$

If we now find an equal-ripple approximation to the discrepancy that is even and of twelfth degree, we may subtract it from both sides of (12.14), combining it with the standard truncated series on the left to turn that simple approximant into a good equal-ripple approximant. Since the discrepancy is small and since the approximation to it is also necessarily small, we shall be able to work with small quantities throughout. In effect, we remove the factor $\pi^{14}/14!$ from our calculations, dealing directly with the function

$$-x^{14}\left[1 - \frac{\pi^2}{15 \cdot 16}x^2 + \frac{\pi^4}{15 \cdot 16 \cdot 17 \cdot 18}x^4 - \cdots\right]$$

on $(-1, 1)$.

Because of the even symmetry, the proper tactic here is to confine our fitting to $(0, 1)$ by substituting y for x^2, then employ the one-sided Chebyshev polynomials up through the sixth degree.

The cosine again

To see how well the method worked, your author decided to seek the sixth degree, one-sided, approximation to the discrepancy by fitting at the seven zeros of the T_7 polynomial. For this first fitting, the orthogonality relations make the work light enough to employ a hand-cranked calculator. The labor required is:

1. Compute the seven zeros of T_7, which are merely

$$x_k = \cos\left[\frac{\pi}{14}(2k + 1)\right] \qquad z_k = \tfrac{1}{2}(x_k + 1) \qquad k = 0, 1, \ldots, 6$$

2. Evaluate the discrepancy for each z_k.
3. Evaluate the six unique values of $T_n(z_k)$ which are all cosines of multiple angles.
4. Sum the expressions obtained from the orthogonality conditions to get the coefficients of the Chebyshev series.

5. Check the adequacy of this approximation by evaluating it, and the discrepancy, subtracting them to get the error curve. (Only this step was performed on an electronic calculator.)

The results of this first fitting of the discrepancy were remarkably good. The error curve has the correct number of extrema and their values are

$$\left.\begin{array}{r} -1.056 \\ 1.054 \\ -1.047 \\ 1.039 \\ -1.029 \\ 1.016 \\ -1.022 \\ 1.014 \end{array}\right\} \times 10^{-4}$$

The peaks are not quite equal ripple, but it would take a very good eye to detect it from a graph of this error curve. The discrepancy itself goes from 0 to 1. All these figures must be multiplied by the constant factor

$$\frac{\pi^{14}}{14!} = 1.04638 \times 10^{-4}$$

which was omitted from the discrepancy during the fitting process. Thus the actual error in our approximation to the cosine will be nearly equal ripple with a maximum value of

$$(1.056 \times 10^{-4})(1.046 \times 10^{-4}) = 1.1 \times 10^{-8}$$

We felt no incentive to try a second iteration on this example, the results of the first having surpassed our most optimistic hopes. Perhaps it was just as well, since the locations for the zeros of the new error curve would no longer fall at the zeros of some Chebyshev polynomial, hence no orthogonality relations would be available, so we would have to solve equations (12.9) simply as a general set of seven linear equations in seven unknowns, $\{c_i\}$. Grim on a hand calculator!

The zeros of $T_7(x)$ are the roots of cos 7θ and hence occur at

$$x_j = \frac{\pi}{14}(2j + 1) \qquad j = 0, 1, \ldots, 6$$

The values of the lower degree $T_k(x_j)$ are simply cosines evaluated on a regular set of points, hence have values that are closely related. If we start to draw up a table (Figure 12.6) we soon see the connections. Designating the cosines of $\frac{1}{14}\pi$, $\frac{3}{14}\pi$, and $\frac{5}{14}\pi$ as a, b and c, respectively, and those with arguments $\frac{2}{14}\pi$, $\frac{6}{14}\pi$, and $\frac{4}{14}\pi$ by d, e, and g, we may then fill out the whole table. We only have to use the various symmetry relations. For example, the argument $\frac{15}{14}\pi$ is $\frac{1}{14}\pi$ beyond π, and hence its cosine is the negative of the cosine of $\frac{1}{14}\pi$. Thus the

j	$T_0(z_j)$	T_1	T_2	T_3	T_4	T_5	T_6
0	1	$\cos\frac{1}{14}\pi$	$\cos\frac{2}{14}\pi$	$\cos\frac{3}{14}\pi$	$\cos\frac{4}{14}\pi$	$\cos\frac{5}{14}\pi$	$\cos\frac{6}{14}\pi$
1	1	$\cos\frac{3}{14}\pi$	$\cos\frac{6}{14}\pi$	$\cos\frac{9}{14}\pi$	$\cos\frac{12}{44}\pi$	$\cos\frac{15}{14}\pi$	$\cos\frac{18}{14}\pi$
2	1	$\cos\frac{5}{14}\pi$	$\cos\frac{10}{14}\pi$	$\cos\frac{15}{14}\pi$	$\cos\frac{20}{14}\pi$	$\cos\frac{25}{14}\pi$	$\cos\frac{30}{14}\pi$
3	1	$\cos\frac{7}{14}\pi$	$\cos\frac{14}{14}\pi$	$\cos\frac{21}{14}\pi$	$\cos\frac{28}{14}\pi$	$\cos\frac{35}{14}\pi$	$\cos\frac{42}{14}\pi$
4	1						
5	1			(et cetera)			
6	1						

Figure 12.6

third entry under T_3 is $-a$. The third row consists of arguments of multiples of $\frac{1}{2}\pi$ and is thus trivially filled in. Finally, the bottom half of every column exhibits either odd or even symmetry according to whether the middle element is zero or ± 1. The arguments used for evaluating $f(x_j)$ have necessarily been translated to (0, 1) as described above, but the *values* of the $T_k(z_j)$ are not affected by the horizontal scales used. It is only necessary that they be evaluated at the *zeros* of T_7. Thus we may use the intuitively more satisfying range $(-\pi, \pi)$ for our arguments while evaluating the coefficients in Figure 12.7.

f_i	T_0	T_1	T_2	T_3	T_4	T_5	T_6
f_0	1	a	d	b	g	c	e
f_1	1	b	e	$-c$	$-d$	$-a$	$-g$
f_2	1	c	$-g$	$-a$	$-e$	b	d
f_3	1	0	-1	0	1	0	-1
f_4	1	$-c$	$-g$	a	$-e$	$-b$	d
f_5	1	$-b$	e	c	$-d$	a	$-g$
f_6	1	$-a$	d	$-b$	g	$-c$	e

Figure 12.7

Finally, the computational formulas may be written down by summing the products of each T_k column of Figure 12.7 with the first column. We get

$$c_1 = \tfrac{2}{7}[a(f_0 - f_6) + b(f_1 - f_5) + c(f_2 - f_4)]$$
$$c_2 = \tfrac{2}{7}[d(f_0 + f_6) + e(f_1 + f_5) - g(f_2 + f_4) - f_3]$$
$$c_3 = \tfrac{2}{7}[b(f_0 - f_6) - c(f_1 - f_5) - a(f_2 - f_4)]$$
$$c_4 = \tfrac{2}{7}[g(f_0 + f_6) - d(f_1 + f_5) - e(f_2 + f_4) + f_3]$$
$$c_5 = \tfrac{2}{7}[c(f_0 - f_6) - a(f_1 - f_5) + b(f_2 - f_4)]$$
$$c_6 = \tfrac{2}{7}[e(f_0 + f_6) - g(f_1 + f_5) + d(f_2 + f_4) - f_3]$$

where

$$a = \cos \tfrac{1}{14}\pi \qquad g = \cos \tfrac{1}{14}\pi$$
$$d = \cos \tfrac{2}{14}\pi \qquad c = \cos \tfrac{5}{14}\pi$$
$$b = \cos \tfrac{3}{14}\pi \qquad e = \cos \tfrac{6}{14}\pi$$

Finally,

$$c_0 = \tfrac{2}{7}\sum_0^6 f_i$$

Economized rational functions

Many functions are more efficiently approximated by rational functions than by power series. For these functions, the economization process must somehow be applied to the rational function or one of its alternative incarnations as a continued fraction. Computational practice favors some of the continued fraction forms because their evaluations require about half the number of multiplications and divisions needed by rational functions, but in this book we deal principally with the rational function because the concepts and difficulties emerge more clearly without being substantially different. An additional restriction will be to rational functions that have polynomials of the same degree in their numerator and denominator or to those that differ at most by unity. Thus

$$R_{m,k}(x) = \frac{\sum_{j=0}^{m} a_j x^j}{\sum_{j=0}^{k} b_j x^j} \qquad \text{with } |m - k| \le 1 \qquad b_0 = 1$$

While theoretically a rational approximation might have any combination of

(k, m), experience has shown that smallest error terms occur with the class that we are considering. We usually normalize a rational function by demanding that b_0 be unity. An additional simplification would remove a_0 as a factor from the numerator, thereby giving a new a_0 of unity also – but we do not generally bother.

As with the power series, economized rational functions can either be produced directly, from the function itself, or by tampering with the coefficients of an uneconomized rational function that one might just happen to have lying around. The details are somewhat messier, the equations to be solved are sometimes mildly nonlinear, but the central idea remains the same: We are trying to throw back onto the earlier part of our rational approximation most of the error that would have occurred had we merely truncated. The truncation error is largely of the form Cx^N, a notoriously inefficient shape for an error curve that we have castigated sufficiently elsewhere. Ideally, we would like to replace it with an equal-ripple curve. As before, we shall usually settle for something slightly less that we call an approximately equal-ripple error curve.

Direct methods for rational approximations

The direct methods for finding rational approximations to a given function $F(x)$ follow the same strategy as with polynomial approximations:

1. Given the approximate locations of the extrema, (or zeros) of the desired equal-ripple error curve, find the parameters of the rational approximation by solving a set of simultaneous equations.
2. Given the approximation to $F(x)$, find the location of the extrema of the (new) error curve – to be used in step 1 for the next iteration.

This iterative cycle (Remes's second algorithm) will converge provided one manages to begin it with a set $\{x_j\}$ of locations for the error-curve extrema that are close enough to the correct values. (By way of contrast, the polynomial version of this algorithm always converges, whatever the starting set $\{x_j\}$.) In practice, the locations of the extrema for the next higher degree Chebyshev polynomial are usually an adequate initial set to provide convergence.

A minor difficulty resides in the fact that the set of equations in step 1 are mildly nonlinear, requiring an iterative solution within the step. The major difficulty, as with direct polynomial approximations, is the need for multiple-precision arithmetic. And the cure, as before, is to avoid it by not finding the direct rational approximation at all, but to approximate the *discrepancy* between the early part of a truncated continued fraction and the function itself. To this end, Maehly [1963] has given a formula that expresses the "tail" of a continued fraction in a form that is convenient for direct approximation by a rational function. Whenever a continued fraction representation is readily available for $F(x)$, this strategy is clearly superior to all others, since it permits us to work with single-precision quantities during the laborious iterations of a fitting process and only requires double precision during the final combination with the original rational function of the fitted approximation to the discrepancy.

The equations for the rational approximations

If we write

$$R_{m,n}(x) = \frac{P_m}{Q_n} = \frac{\sum_{k=0}^{m} a_k x^k}{\sum_{k=0}^{n} b_k x^k} \qquad \text{with } b_0 \equiv 1 \tag{12.15}$$

we have $m + n + 1$ parameters to fit. Thus we may decide to have $R_{m,n}$ agree with our function $F(x)$ exactly at $m + n + 1$ points, creating zeros in the error curve at those points. Calling this set of zeros $\{z_j\}$, the fitting equations become

$$F(z_j) \cdot \sum_{k=0}^{n} b_k z_j^k = \sum_{k=0}^{m} a_k z_j^k \qquad j = 0, 1, \ldots, m + n \tag{12.16}$$

and we see that the $m + n + 1$ undetermined parameters a_k and b_k enter linearly.

If we follow the philosophy of attempting to fit at the *extrema* of our error curve instead of fitting it at the zeros, we are led immediately to the equations at the $m + n + 2$ prescribed points $\{x_j\}$

$$[F(x_j) - (-1)^j \lambda] \cdot \sum_{k=0}^{n} b_k x^k = \sum_{k=0}^{m} a_k x^k \qquad b_0 = 1 \qquad i = 0, 1, \ldots, m + n + 1 \tag{12.17}$$

where the $m + n + 2$ parameters consist of the a_k and b_k, as before, plus the common but unknown amplitude λ. These equations are not linear because λ and the b_k appear as products. On the other hand, λ always enters as a minor perturbation on the function F, so small errors in its value may be tolerable. This suggestion leads to a linearization of equations (12.17) by replacing λ with an assumed value, λ_0, wherever it is multiplied by an unknown b_k. Thus, splitting out the b_0 term, we have

$$[F(x_j) - (-1)^j \lambda] = \sum_{k=0}^{m} a_k x_j^k - [F(x_j) - (-1)^j \lambda_0] \cdot \sum_{k=1}^{n} b_k x_j^k \tag{12.18}$$

which is linear in the a's, b's, and λ. We now solve the set (12.18) with an assumed λ_0, getting a value of λ which is then substituted for λ_0 and the set is solved again. Experience indicates that this device is usually adequate, though it does nothing to ameliorate the need for multiple precision in the solution process. Both (12.16) and (12.18) suffer from the fact that they must produce the a's and b's directly to high precision. System (12.18) has the further liability that the essential driving information is the small discrepancy λ which presumably will not be much larger than the rounding error in single-precision arithmetic. These are the same difficulties that were encountered in direct polynomial approximation.

Using the discrepancy

If we have a convergent continued fraction for the function we wish to approximate, we may use its "tail" in a way analogous to using the tail of a power series. A formula given by Maehly [1963, but the article has a crucial misprint, see below] permits us to evaluate the *discrepancy* between our function and any rational approximation obtained by simple truncation of the continued fraction. Thus we may fit correction terms to this rational approximation that will include most of the discrepancy. Since the discrepancy and the correction terms are often small quantities, single-precision arithmetic will usually suffice.

We intend to seek a rational approximation P_n/Q_n to $f(x)$ that differs from f by an equal-ripple error, δ_N. Thus

$$\delta_N = f - \frac{P_n}{Q_n}$$

If we already have a rational approximation to f *of the same degree* (let it be called P/Q) that is obtained by truncating the continued fraction after the $2n$th term, we may write

$$\delta_N = \left(f - \frac{P}{Q}\right) - \left(\frac{P_n}{Q_n} - \frac{P}{Q}\right)$$

Thus

$$Q_n Q \delta_N = Q_n Q\left(f - \frac{P}{Q}\right) - (QP_n - PQ_n) \tag{12.19}$$

Denoting the discrepancy between f and the cruder approximation by δ and expressing the unknown P_n and Q_n as increments on P and Q, we have

$$\delta = f - \frac{P}{Q}$$

$$P_n = P + \Delta P$$

$$Q_n = Q + \Delta Q$$

so that (12.19) becomes

$$(Q + \Delta Q)Q\delta_N = (Q + \Delta Q)\cdot(Q\delta) - (Q\cdot\Delta P - P\cdot\Delta Q)$$

or

$$Q\cdot\Delta P - [P + Q\cdot\delta_N - (Q\delta)]\cdot\Delta Q + Q^2\delta_N = Q(Q\delta) \tag{12.20}$$

The quantity δ is given by

$$\delta = f - \frac{P_v}{Q_v} = \frac{x^v \prod_{i=0}^{v}(-\alpha_i)}{Q_v(x)\left[Q_{v-1}(x) + \dfrac{Q_v(x)}{f_{v+1}}\right]} \tag{12.21}$$

where

$$\frac{P_v}{Q_v} = \frac{\alpha_0 |}{|b_0 +} \frac{\alpha_1 x |}{|b_1 +} \frac{\alpha_2 x |}{|b_2 +} \cdots \frac{\alpha_v x |}{|b_v}$$

and the "tail"

$$f_{v+1} = \frac{\alpha_{v+1} x |}{|b_{v+1} +} \frac{\alpha_{v+2} x |}{|b_{v+2} +} \cdots \tag{12.22}$$

[Formula (12.21) in Maehly has $(-x)^{n+1}$ in the numerator, which is clearly wrong since α_0 has no x associated with it.] The quantities P_v, Q_v, and Q_{v-1} occur in the standard recursive evaluation of the continued fraction. The recursion may also be used to evaluate the tail, f_{v+1}, starting with α_{v+1} – but then we must be careful to remember the extra x from the first term of (12.22). In all the other formulas of this paragraph, Q stands for Q_v. Thus for any x we may evaluate δ or $(Q\cdot\delta)$ correctly.

We now introduce the equal-ripple requirement, deciding to fit (12.20) at the extrema of δ_N. Letting λ be the common magnitude of δ_N at the extrema, we have

$$Q\cdot\Delta P - [P + Q\cdot(-1)^j\lambda - (Q\delta)]\cdot\Delta Q + Q^2\cdot(-1)^j\cdot\lambda = Q(Q\delta) \qquad j = 0, 1, \ldots, N \tag{12.23}$$

which is a set of equations for the ΔP, ΔQ, and λ that is mildly nonlinear because of the occurrence of λ in the coefficient of ΔQ. Since it is a minor term here, however, we may call it λ_0, initially setting it to zero, before solving (12.23)– putting the value obtained for λ back in for λ_0, then solving again.

Example: We take the function

$$f = z \cdot e^{z^2} \cdot \int_z^{\infty} e^{-t^2}\,dt = \cfrac{1}{1 + \cfrac{\frac{1}{2}x}{1 + \cfrac{\frac{2}{2}x}{1 + \cfrac{\frac{3}{2}x}{1 + \cfrac{\frac{4}{2}x}{1 + \cdots}}}}} \qquad \textbf{(12.24)}$$

where $x = 1/z^2$ and seek an approximation on the range $(1, \infty)$ for z—which is $(0, 1)$ in x. Over this range our function is quite well behaved, descending monotonically from 1.0 to 0.758, so we might hope for a rather good approximation by merely quadratics for P and Q. Accordingly, we truncate (12.24) after $4x/2$, obtaining

$$\begin{aligned} P &= 1 + 4.5x + 2.0x^2 \\ Q &= 1 + 5.0x + 3.75x^2 \end{aligned} \qquad \textbf{(12.25)}$$

This approximation is perfect at the origin, degenerates rapidly to an error of 0.00145 at 1. We hope to produce a much better approximation P_2/Q_2 by computing

$$\begin{aligned} \Delta P &= r_0 + r_1 x + r_2 x^2 \\ \Delta Q &= \qquad s_1 x + s_2 x^2 \end{aligned}$$

from equations (12.23) evaluated at the extrema of the ultimate error curve δ_N. As a first guess for the locations of these extrema we take those of $T_5(x)$, the Chebyshev polynomial of degree 5, mapped onto (0, 1). They are, approximately:

$\{x_j\}$
0
0.095
0.345
0.655
0.905
1.0

Note that our normalization of P/Q required the constant term of Q to be unity, hence ΔQ contains only two parameters. Setting λ_0 to 0 and solving (12.23) we obtain line (1) in Table 12.1.

Table 12.1

	r_0	r_1	r_2	s_1	s_2	λ
(1)	-3.425×10^{-6}	-2.3888	-1.6812	-2.3909	-2.8333	-3.44840×10^{-6}
(2)	-5.148×10^{-6}	-2.3964	-1.6837	-2.3986	-2.8389	-5.1388×10^{-6}
(3)	-5.9296×10^{-6}	-2.4001	-1.6849	-2.4023	-2.8416	-5.9718×10^{-6}
(4)	-1.3861×10^{-5}	-2.30981	-1.65186	-2.31162	-2.76973	-1.3871×10^{-5}
(5)	$-1.5544445 \times 10^{-5}$	-2.3201397	-1.6547340	-2.3220775	-2.7766979	-1.5532×10^{-5}

Substituting λ for λ_0 in (12.23) and re-solving we get lines (2) and (3). At this point we evaluate our approximation to get its error curve δ_N – examining it to see just where its maxima really are. They seem to be at

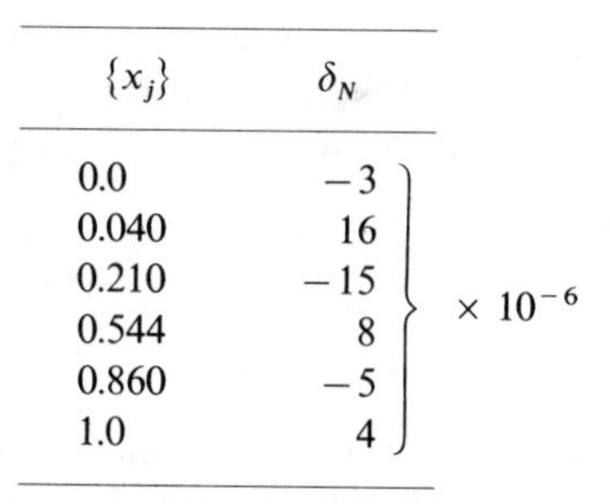

$\{x_j\}$	δ_N	
0.0	-3	$\times 10^{-6}$
0.040	16	
0.210	-15	
0.544	8	
0.860	-5	
1.0	4	

The associated extrema are scarcely equal, ranging from 3 to 16 in units of 10^{-6} – but many of us would settle for this sort of accuracy in an approximation.

If we wish to level off the errors we now use these new $\{x_j\}$, repeating the entire fitting process. After two cycles of substituting λ for λ_0 and resolving we get line (4). Again we evaluate the approximation to obtain the actual locations of the extrema of δ_N. A third fitting at these new extrema gives the parameters shown in line (5). This time the extrema are nearly equal being

$$\left.\begin{matrix} -8.8 \\ 8.8 \\ -9.1 \\ 9.1 \\ -9.6 \\ 9.3 \end{matrix}\right\} \times 10^{-6}$$

and we are content to stop. The final approximation is

$$z \cdot e^{z^2} \int_z^{\infty} e^{-t^2} \cdot dt = \frac{0.99998\,44556 + 2.1798603x + 0.3452660x^2}{1.0 + 2.6779225x + 0.9733021x^2}$$

with $x = 1/z^2$. The error is less than 10^{-5} over $(1, \infty)$. Since the function itself remains between 0.75 and 1.0 the approximation gives five significant figures.

Having written a computer program to evaluate the continued fraction, it is trivial to produce more precise approximations by reducing the range of x (say to $0, \frac{1}{4}$) or by increasing the degree of the rational function. The difficulties occur when the current δ_N must be produced in order to locate its extrema because this requires the evaluation of f at a rather large number of points—usually in double precision. This phase of the process consumes most of the computing time. A second difficulty arises with higher orders of approximation when the system of equations (12.23) becomes ill conditioned. Maehly gives suggestions for dealing with this problem.

Problems

1. Economize the series for the cosine on the sensible range $(0, \pi/3)$ using
   a. The throwing back of later terms of a truncated series onto the earlier ones.
   b. Direct evaluation of the corrections from the residual series. Attempt to provide at least six significant figures in your final approximation.

2. Economize the asymptotic series for $\sqrt{\pi} x e^{x^2} \operatorname{erfc}(x)$, page 18, on the range $1 \le x < \infty$. How many significant figures can you obtain with six terms?

3. If your interest is in $\operatorname{erfc}(x)$ instead of the function $x e^{x^2} \operatorname{erfc}(c)$, how many terms will be required to yield eight *decimal* accuracy? What changes will you make in your economization procedure from the one appropriate for Problem 2?

4. Economize the continued fraction representation for the appropriate function of $\operatorname{erfc}(x)$. Attempt to get the same number of significant figures as in Problem 2. Compare the length of the two economized approximations.

CHAPTER 13

EIGENVALUES II

When eigenvalues were first introduced back in Chapter 8 the interest was in the power method. That technique was particularly simple both to understand and to execute and it had the virtue of producing the largest eigenvalue and its associated eigenvector rather efficiently – provided there *was* a largest eigenvalue. The method could also be persuaded to find the smallest eigenvalue and, if you twisted its arm, it would yield the next-to-largest and next-to-smallest – though with decreasing efficiency. The farther one tried to push it, the more it resisted. Clearly the power method, effective as it might be for some eigenvalue problems, was not going to handle all of them.

In this chapter we shall deal with methods that can produce *all* the eigenvalues without overmuch bother about whether they are large or small or somewhere in between. Furthermore, we shall have large matrices in mind so that we are talking, typically, about 50 to 100 eigenvalues. Since finding all 100 eigenvalues of a 100×100 matrix is a lot of work, we are going to be continually concerned about the efficiency of our algorithm. Further, since large matrices require many arithmetic subtractions, we must worry about the *stability* of the algorithms lest we obtain numerical results that are, in fact, random numbers instead of eigenvalues.

The perspicacious reader may have noticed a curious avoidance of the word *eigenvector*, which is not accidental. The algorithms that expediently produce all 100 eigenvalues do not simultaneously find the corresponding eigenvectors, and so this calculation follows as a nearly separate and additional labor. Fortunately, the physical problems that lead to large eigenmatrices frequently do not require the vectors to be found, so that the postponing (perhaps indefinitely!) of the topic is not solely due to expedience. The worried reader may turn to page 355 if he needs reassurance that eigenvectors have not been entirely ignored.

The grand strategy

While algorithms for completely diagonalizing a general nondefective matrix exist, they require far too much computation. The practical strategies break the problem into two parts: reducing the original matrix to a specialized matrix having many zero elements but the same eigenvalues, followed by the finding of the eigenvalues for the specialized matrix. These specialized matrices are either tridiagonal or Hessenberg, which forms are illustrated in Figure 13.1. The

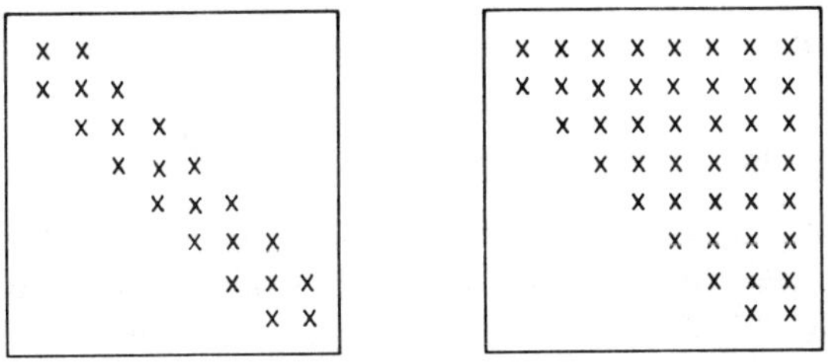

$$\begin{array}{cccccccc} x & x & & & & & & \\ x & x & x & & & & & \\ & x & x & x & & & & \\ & & x & x & x & & & \\ & & & x & x & x & & \\ & & & & x & x & x & \\ & & & & & x & x & x \\ & & & & & & x & x \end{array} \qquad \begin{array}{cccccccc} x & x & x & x & x & x & x & x \\ x & x & x & x & x & x & x & x \\ & x & x & x & x & x & x & x \\ & & x & x & x & x & x & x \\ & & & x & x & x & x & x \\ & & & & x & x & x & x \\ & & & & & x & x & x \\ & & & & & & x & x \end{array}$$

A tridiagonal matrix A Hessenberg matrix

Figure 13.1

desirability of this dual strategy becomes apparent if we consider the amount of work facing a man who would find all the eigenvalues of a large general matrix. The problem, by definition, is that of finding all the zeros of a high-degree polynomial—a process that is necessarily infinite, since there are no explicit formulas. Further, any serious manipulation with a general matrix, such as multiplication by another matrix, will usually require kn^3 arithmetic operations, where k depends on the particular manipulation. In contrast, manipulations with tridiagonal matrices tend to be proportional to n. Hesenberg matrices lie between, usually requiring kn^2 operations. But we can reduce a general matrix to tridiagonal or Hessenberg form in a *finite* number of arithmetic steps (the number of multiplications is close to n^3).

Thus the grand strategy is to postpone the infinite algorithm until we have reduced our matrix to one of the specialized forms—being happier about endlessly iterating an algorithm that costs kn multiplications per iteration than doing the same with one costing kn^3. When we consider that computing centers are regularly asked to find the eigenvalues of 200×200 matrices and larger, we realize that the difference between n^2 and n^3 is startling—not to mention n versus n^3! Only a socially irresponsible man would ignore such computational savings.

The two stages of our eigenvalue strategy are nearly independent. We shall discuss them separately. The finding of eigenvalues from tridiagonal or Hessenberg forms turns out, not very surprisingly, to be the familiar problem of finding the roots of a polynomial. The difficulties are already familiar, as are the remedies. Accordingly we shall spend most of our time in this chapter discussing the first stage: The reduction of a general matrix to a specialized form while carefully preserving the eigenvalues. It is here that the crucial questions about computational labor and computational stability arise. Often the methods that seem efficient are also unstable and hence ineffective. Since there is a variety of algorithms, each having its special utility, with no general superior tactic available, we shall be emphasizing our familiar theme of suiting the tool to the job.

The standard tools

All reductions of general matrices may be phrased in terms of *similarity* transformations. The matrix A is said to suffer a similarity transformation if it is pre- and postmultiplied by any other matrix and its inverse. Thus B and C are similarity transformations of A if

$$B = T^{-1}AT \qquad \text{or} \qquad C = TAT^{-1}$$

The only restrictions are the dimensionalities being compatible with the indicated multiplications, plus the existence of the inverse of T. Similarity transformations have the property of preserving eigenvalues. Starting with the eigenproblem equation

$$A\mathbf{x} = \lambda\mathbf{x}$$

where $\mathbf{x}$ is an eigenvector of A and λ is the corresponding eigenvalue we have

$$TA\mathbf{x} = \lambda T\mathbf{x}$$

and on defining

$$T\mathbf{x} = \mathbf{y} \qquad \text{or} \qquad \mathbf{x} = T^{-1}\mathbf{y}$$

we get

$$TAT^{-1}\mathbf{y} = \lambda TT^{-1}\mathbf{y} = \lambda\mathbf{y}$$

Thus we have the new eigenproblem

$$C\mathbf{y} = \lambda\mathbf{y}$$

with the new matrix C that is similar to A, new eigenvectors $\mathbf{y}$ that are simply related to $\mathbf{x}$ – but still the same old λ's.

The class of similarity transformations is very broad. Within it is included a much more restrictive type, the *orthogonal* transformation. Here the transforming matrix T is orthogonal, which implies that its transpose is also its inverse. We have

$$T^T = T^{-1}$$

and this property is enough to preserve symmetry through the similarity transformation. We note that the transpose of C is

$$C^T = (TAT^T)^T = TA^TT^T$$

which, if A is symmetric, becomes

$$C^T = TAT^T = C$$

Thus orthogonal transformations preserve both eigenvalues and symmetry, when the latter exists. They are also extremely stable. As one might suspect, any transformation that has such admirable properties must also suffer some drawbacks. Orthogonal transformations tend to require more arithmetic than their competitors. Thus we will tend to use them for those matrices that are poorly behaved or when other methods happen to fail.

The Jacobi plane rotation

One of the simplest orthogonal transformations is the plane rotation, usually called the Jacobi transformation. We discuss it first in spite of the fact that no efficient eigenvalue technique employs it. Its simplicity recommends its expositional place, together with the observation that an elaboration of it gives one of the practical techniques.

The transformation matrix is simply a unit matrix with four elements altered, as shown in Figure 13.2. Any two of the diagonal elements have been changed to some value c less than unity in magnitude, while the elements $-s$ and s have been added to form the corners of a square with the c elements.

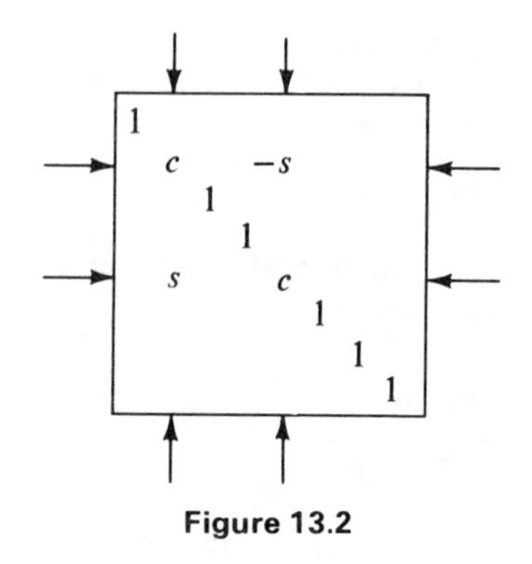

Figure 13.2

A normalization relation

$$c^2 + s^2 = 1$$

holds between c and s, leaving only one value free to be chosen—although the *positions* of the c elements are also at our disposal. The inverse of T is its transpose—which interchanges s with $-s$. The full plane rotation consists of matrix multiplications that can be indicated schematically by

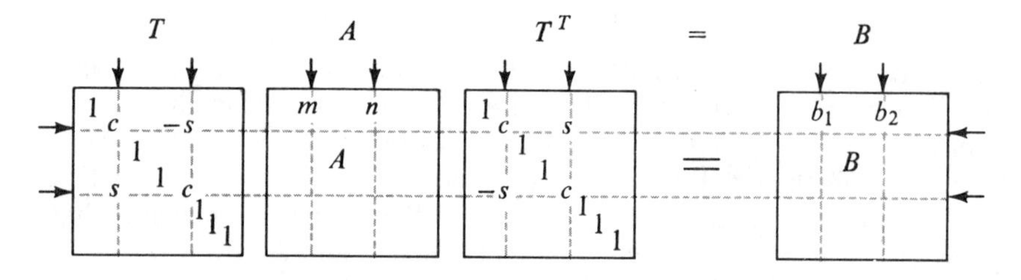

in which c has been chosen to annihilate the element of A sitting in the position corresponding to that of s or $-s$ in T. If we write general elements into A and B and carry out the indicated matrix multiplications, we get the explicit algorithm. Perhaps surprisingly, we find that only those elements of A are altered which lie in the two rows and the two columns that contain the c's. All other elements of A pass into their corresponding positions in B unchanged. Further, except for the four elements at the corners of the square, the rule of formation for the b's from the a's in the special columns is particularly simple, being merely a linear combination of the two a's sharing a common row (or column) that cuts the special columns (or rows). Thus b_1 and b_2 in Figure 13.3 are each linear combinations of m and n, the weights being c and s. The rules for the four elements from the square are more complicated. The student should verify

	↓			↓		
a_{11}	$ca_{12} - sa_{15}$	a_{13}	a_{14}	$sa_{12} + ca_{15}$	a_{16}	a_{17}
→ $ca_{21} - sa_{51}$	b_{22}	$ca_{23} - sa_{53}$	$ca_{24} - sa_{54}$	b_{25}	$ca_{26} - sa_{56}$	$ca_{27} - sa_{57}$
a_{31}	$ca_{32} - sa_{35}$	a_{33}	a_{34}	$sa_{32} + ca_{35}$	a_{36}	a_{37}
a_{41}	$ca_{42} - sa_{45}$	a_{43}	a_{44}	$sa_{42} + ca_{45}$	a_{46}	a_{47}
→ $sa_{21} + ca_{51}$	b_{52}	$sa_{23} + ca_{53}$	$sa_{24} + ca_{54}$	b_{55}	$sa_{26} + ca_{56}$	$sa_{27} + ca_{57}$
a_{61}	$ca_{62} - sa_{65}$	a_{63}	a_{64}	$sa_{62} + ca_{65}$	a_{66}	a_{67}
$\vdots$	$\vdots$	$\vdots$	$\vdots$	$\vdots$	$\vdots$	$\vdots$

Figure 13.3

the formulas given in Figure 13.3 for the 7×7 system with the rotation performed on the (2, 5) columns. Note that if A is symmetric, then B is also—as required by the orthogonality of the transformation.

Inadequacy of Jacobi for matrix reduction

Although we may produce zeros quite effectively with the plane rotation of Jacobi the transformation is not practical for reducing a general symmetric matrix to tridiagonal or any other specialized form, because it permits no strategy that does not destroy earlier zeros as it produces the later ones. Consider for a moment how we might seek to use the Jacobi transformation to produce tridiagonal form. We are free to choose any two columns and their corresponding rows; the zero will then appear at the two intersections that do not lie on the principal diagonal. Thus if we choose the last two rows and columns (Figure 13.4) we will produce the zeros shown—and we will generally alter *all* the elements in those two rows and columns. If we now seek to zero the other elements lying, for example, in the last column, our next Jacobi transformation would have to involve that column and the zero currently sitting in the (3, 4) position

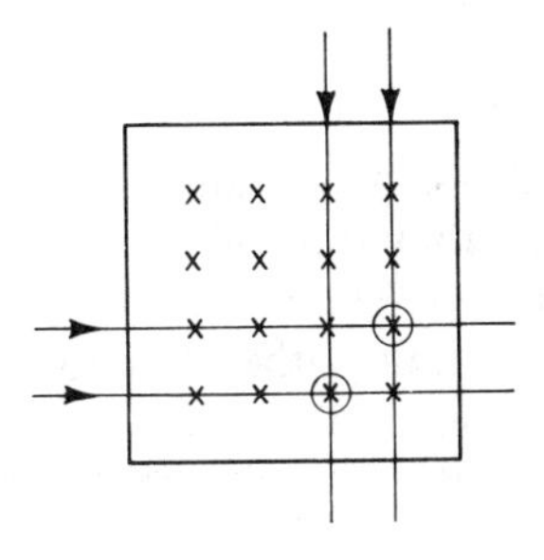

Figure 13.4

would be destroyed. Thus we have no hope of reaching any specialized form simply by annihilating the offending elements with Jacobi transformations via a one-element, one-transformation strategy.

It *is* true that these Jacobi transformations always reduce the sum of squares of the off-diagonal elements, piling the substance of the matrix upon the principal diagonal, so that in this weak sense they nudge the matrix persistently toward diagonal form. But this process is usually far too slow to interest us and is, in theory, infinite. In practice it takes about $6n^3$ multiplications before we are willing to consider the off-diagonal elements to be zero, and with large matrices this is too much work. The other strategies outlined in this chapter will do much better.

The Givens reduction—a finite strategy

Although the plane rotation of Jacobi is not immediately useful, a slight modification of it allows a practical finite strategy. Instead of seeking to produce zeros at the corners of the square in the plane-rotation pattern, we choose the parameter c so as to zero an element that is *not* one of the four corners. Thus, if we seek to make the $(1, 4)$ element of Figure 13.5 vanish [and $(4, 1)$ as well, for our matrices are all symmetric in this chapter and plane rotations preserve symmetry], we may use rows 1 and 2 (Figure 13.5), rows 1 and 3, rows 2 and 4, or rows 3 and 4 (Figure 13.6)—together with the corresponding columns. The

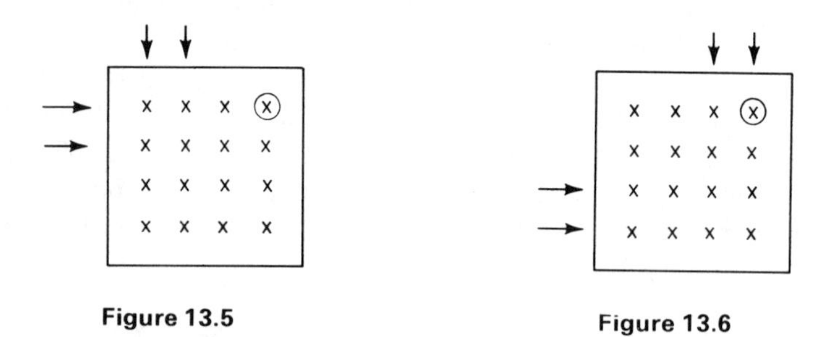

Figure 13.5

Figure 13.6

new elements that lie on the same column and on the two chosen rows are formed simply as linear combinations of the two old elements in the same positions. Thus the new value of a and b in Figure 13.7 are simply linear combinations of the old values sitting in the same places. Only the four elements lying in the corners of the square have more complicated formulas—and they are of minor concern to us here. This simple linear combination rule confers the important property that *if a and b were both already zero, they will remain zero* under any choice of c and s. Thus if we can find a strategy that produces, for example, a row of adjacent zeros, subsequent transformations aimed at elements in other rows have a good chance of leaving the old zeros undisturbed. Let us be more specific by taking an example.

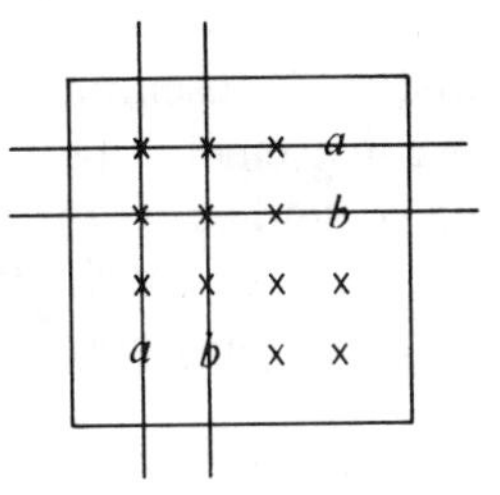

Figure 13.7

In Figure 13.8(a) we used a plane rotation with columns 3 and 4 to produce zeros in the (1, 4) and (4, 1) positions. Next, Figure 13.8(b), we use columns 2 and 3 to produce a zero at (1, 3) – a transformation that does not disturb (1, 4) since that position is not on either of the crucial rows or columns. If, however, we now invoke a transformation on columns 3 and 4 again, but choosing c so as to produce a zero in the (2, 4) position, Figure 13.8(c), we see that the two "new"

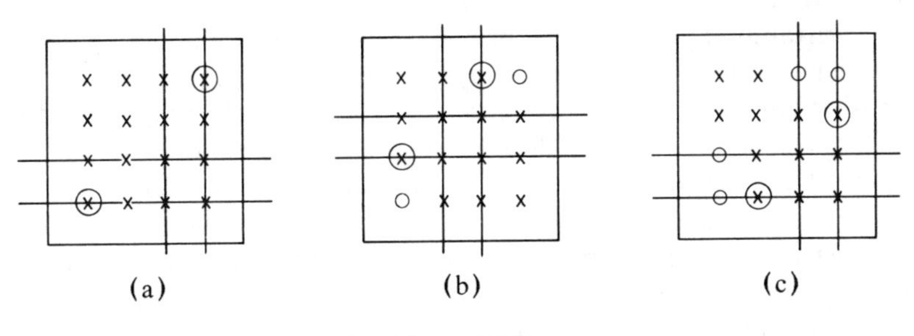

Figure 13.8

elements in the top row will each simply be linear combinations of the two old zeros – and hence will again be zeros. In this 4×4 example we have now arrived at tridiagonal form, and a few trials should easily convince one that any further plane rotations will destroy at least one pair of zeros. [Try columns 2 and 4 to give a zero at (3, 4) noting the trouble at (2, 4) and (1, 4).]

This production of tridiagonal form clearly takes only $(n-1)(n-2)/2$ plane rotations, is extremely stable, and requires approximately $4n^3/3$ multiplications. Its only disadvantage lies in the fact that there is a similar strategy that accomplishes the same tridiagonal form with half the labor! We discuss it below as Householder's method.

The Householder reduction

Like the Givens technique above, Householder produces zeros via orthogonal transformations that are extremely stable. Unlike Givens, however, Householder produces them a row (and column) at a time. Naturally the transformation is rather more complicated than a simple plane rotation. We derive it later (p. 327). In this section we are only concerned with the larger computational plan.

The basic tool is a symmetric orthogonal matrix, P, designed to annihilate *all elements* of a given vector $\mathbf{x}$ *except the first* when $\mathbf{x}$ is premultiplied by P. Thus we want

$$P\mathbf{x} = k\mathbf{e}_1 \tag{13.1}$$

where $\mathbf{e}_1$ is the unit vector having 1 as its first element and zeros thereafter. The scalar k is necessarily the length of $\mathbf{x}$ (why?). Let us be quite clear: A vector $\mathbf{x}$ is explicitly given to us and from it we construct a symmetric orthogonal matrix P that satisfies (13.1). This matrix P may then be used in a similarity transformation with our matrix A and if $\mathbf{x}$ was, for example, a column of A we may thus have produced a lot of zeros in that column. Or will we? We must proceed cautiously here.

It is certainly true that if we consider the entire first column of A to be $\mathbf{x}$ and construct P accordingly, then PA gives us a new matrix having k in the (1, 1) position and zeros in the rest of the first column. Unfortunately, PA is *not* a similarity transformation, and if we would preserve eigenvalues we must postmultiply PA by P. (Since P is orthogonal, its transpose is also its inverse. Since it is symmetric, it is also its own transpose—hence its own inverse. A most peculiar matrix!) The multiplication $(PA)P$ may be transposed to read $P(PA)^T$ and in this form we see (Figure 13.9) that there is no reason why the first column should be annihilated nor the zeros in the first row be preserved—and they won't! We will end with a symmetric matrix but will have lost our zeros.

$$(PA)^T = \begin{bmatrix} k & 0 & 0 & 0 & 0 \\ y & y & y & y & y \\ y & y & y & y & y \\ y & y & y & y & y \\ y & y & y & y & y \end{bmatrix}$$

Figure 13.9

The correct strategy is less ambitious: We attempt to zero only the bottom $(n - 2)$ elements of the first column. This maneuver may be phrased in terms of our P tool by constructing a partitioned matrix T consisting of a single unit element in the (1, 1) position, zeros as shown in Figure 13.10, and P in the rest

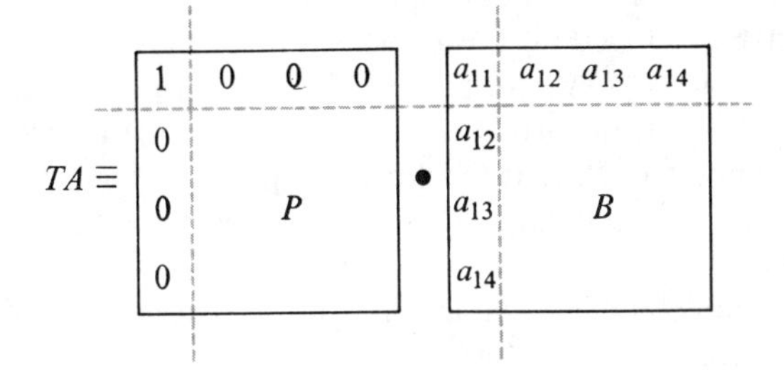

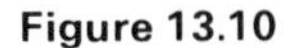
Figure 13.10

of the matrix. We choose $\mathbf{x}$ to be the bottom $(n - 1)$ elements of the first column of A. Thus PA produces the correct zeros (Figure 13.11) *and the postmultiplication by T does not destroy them* (Figure 13.12).

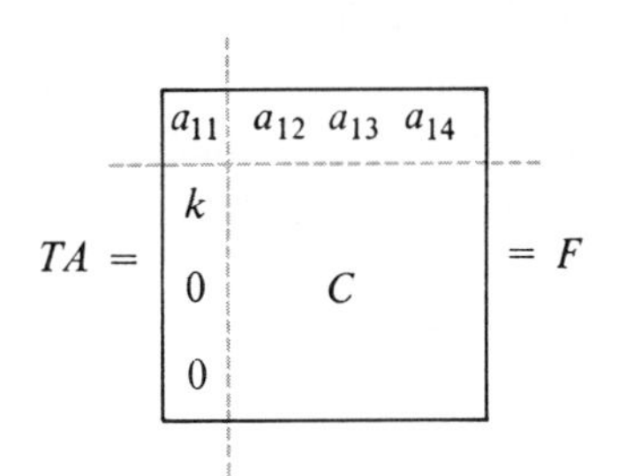

Figure 13.11

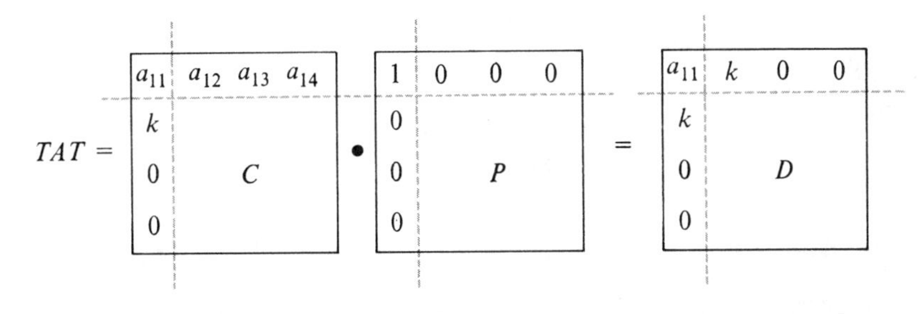

Figure 13.12

Any postmultiplication by T will leave the first column unchanged, just as any premultiplication will leave the first row unchanged. Calling TA the matrix F we may see that the postmultiplication FT produces the necessary zeros, either by arguing that symmetry is preserved by an orthogonal transformation or by noting that $(FT)^T$ is TF^T and hence it is equivalent to a premultiplication acting on the same first column that appeared in PA.

We now have produced all the zeros in the first column and row that we need for tridiagonal form, so we turn to attack the second column. The technique is the same: Produce a transformation matrix that is partitioned (Figure 13.13) and contains a submatrix P_2 that we design to produce the zeros we need. The structure of the other elements in T preserves everything in the first row and column. Thus P_2 acting on the vector that consists of the bottom two elements of the second column changes them into k_2 and 0, respectively. [More generally, P_2 acts on the bottom $(n - 2)$ elements in the second column.]

Since our example is a 4×4 matrix, we have now produced tridiagonal form, but for a larger matrix the process would have to be continued. At each stage P_k is one order smaller, acts on the bottom $n - k$ elements of the kth column, and zeros all but the top one of those elements. The blocks of zeros and the unit matrix that fill out the T transformation conspire to leave the earlier columns and rows undisturbed. Thus the whole process takes only $n - 2$ of these transformations to reach tridiagonal form. The total labor contains $2n^3/3$ multiplications, which is half that of the Givens' strategy. For symmetric matrices the Householder technique is the unchallenged efficient stable strategy and can be recommended without qualification—a peculiarly gratifying situation that is rare in the complex compromised world of numerical computation.

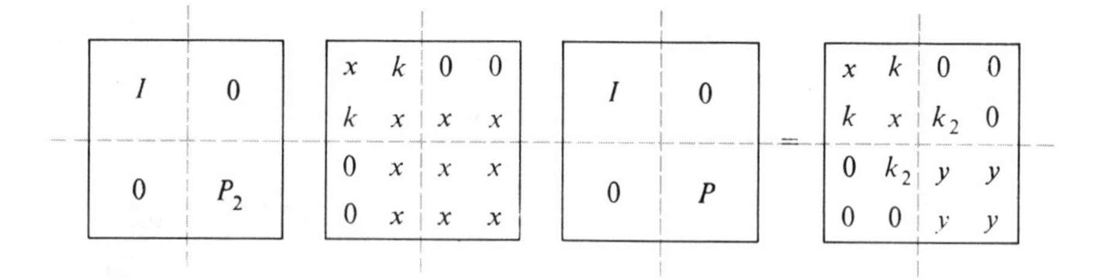

Figure 13.13

The construction of P for Householder's orthogonal transformation

If we are given a vector $\mathbf{x}$ our object is to construct a symmetric orthogonal matrix P such that $P\mathbf{x}$ is a vector with all its elements except the first equal to zero. We may write this statement as

$$P\mathbf{x} = k\mathbf{e}_1 \tag{13.2}$$

where $\mathbf{e}_1$ is a unit vector whose first element is 1 and all the other elements 0. Householder showed that a satisfactory matrix P could be constructed in the form

$$P = (I - 2\mathbf{w}\mathbf{w}^T) \qquad \mathbf{w}^T\mathbf{w} = 1 \tag{13.3}$$

where $\mathbf{w}$ is a unit vector yet to be determined. Note that $\mathbf{w}\mathbf{w}^T$ is the backward, or matrix, product of two vectors, rather than the dot product. From (13.3) we can deduce the symmetry of P and, less obviously, its orthogonality. For

$$PP^T = PP = (I - 2ww^T)(I - 2ww^T) = I - 4ww^T + 4w(w^Tw)w^T = I$$

and thus P, which is its own transpose, is also its own inverse. Hence P is orthogonal. From (13.2) we see that k is the length of $\mathbf{x}$, via

$$(P\mathbf{x})^T(P\mathbf{x}) = k^2$$

$$\mathbf{x}^T(P^TP)\mathbf{x} = k^2$$

$$\mathbf{x}^T\mathbf{x} = k^2$$

Following Wilkinson [1960], we now define

$$\boxed{S^2 = x_1{}^2 + x_2{}^2 + \cdots + x_n{}^2 = k^2} \tag{13.4}$$

so

$$\boxed{\pm S = k}$$

If we write out

$$(I - 2\mathbf{w}\mathbf{w}^T)\mathbf{x} = k\mathbf{e}_1$$

component by component, we have

$$\begin{aligned} x_1 - 2w_1(\mathbf{w}^T\mathbf{x}) &= k = \pm S \\ x_2 - 2w_2(\mathbf{w}^T\mathbf{x}) &= 0 \\ &\vdots \\ x_n - 2w_n(\mathbf{w}^T\mathbf{x}) &= 0 \end{aligned} \tag{13.5}$$

Denoting the dot product

$$\mathbf{w}^T\mathbf{x} = K$$

we rewrite (13.5) as

$$\begin{aligned} 2Kw_1 &= x_1 \mp S \\ 2Kw_2 &= x_2 \\ &\vdots \\ 2Kw_n &= x_n \end{aligned} \tag{13.6}$$

(the form of this step is important). Squaring and adding the equations (13.6) we have

$$S^2 \mp 2x_1S + S^2 = 4K^2(w_1{}^2 + w_2{}^2 + \cdots + w_n{}^2) = 4K^2$$

We may thus compute $2K^2$ from

$$\boxed{2K^2 = S^2 \mp x_1S}$$

Define $\mathbf{u}$ by

$$\boxed{\mathbf{u}^T = [x_1 \mp S, x_2, x_3, \ldots, x_n]}$$

taking the sign that makes $x_1 \mp S$ *large* in absolute value. Then observe that

$$\mathbf{w} = \frac{\mathbf{u}}{2K}$$

and hence

$$\boxed{P = I - \frac{\mathbf{u}\mathbf{u}^T}{2K^2}}$$

The boxes surround the computational formulas.

Application of P in factored form is more efficient than constructing the explicit P and then postmultiplying it by $\mathbf{x}$. As the following example shows, with factoring we need only deal with vector inner products instead of having to generate, store, and use the matrix. Hence our labor is of order $2n^2$ instead of n^3.

Example: If

$$\mathbf{x}^T = [5 \quad 1 \quad -3 \quad -4 \quad 2 \quad 3]$$

then

$$k^2 = S^2 = 64$$

so that

$$S = \pm 8$$

and

$$\mathbf{u}^T = [5+8 \quad 1 \quad -3 \quad -4 \quad 2 \quad 3]$$

with

$$2K^2 = 64 + 40 = 104$$

Applying P in factored form to $\mathbf{x}$ we have

$$P\mathbf{x} = \mathbf{x} - \frac{1}{2K^2}\mathbf{u}\mathbf{u}^T\mathbf{x} = \begin{bmatrix} 5 \\ 1 \\ -3 \\ -4 \\ 2 \\ 3 \end{bmatrix} - \frac{1}{104}\begin{bmatrix} 13 \\ 1 \\ -3 \\ -4 \\ 2 \\ 3 \end{bmatrix}[13 \quad 1 \quad -3 \quad -4 \quad 2 \quad 3] \cdot \begin{bmatrix} 5 \\ 1 \\ -3 \\ -4 \\ 2 \\ 3 \end{bmatrix}$$

Combining the final vectors first:

$$P\mathbf{x} = \begin{bmatrix} 5 \\ 1 \\ -3 \\ -4 \\ 2 \\ 3 \end{bmatrix} - \frac{1}{104}\begin{bmatrix} 13 \\ 1 \\ -3 \\ -4 \\ 2 \\ 3 \end{bmatrix} \cdot 104 = \begin{bmatrix} -8 \\ 0 \\ 0 \\ 0 \\ 0 \\ 0 \end{bmatrix} = -8\mathbf{e}_1$$

Clearly this order of arithmetic events is much more efficient than forming the 6×6 matrix $\mathbf{u}\mathbf{u}^T$ or $\mathbf{u}\mathbf{u}^T/2K^2$. Indeed, to store P one need store only the vector $\mathbf{u}$ and the scalar $2K^2$. We have here one of many examples where matrix notation, though very useful for derivations and proofs, can lead the unwary computor astray.

Reduction of banded symmetric matrices

Partial differential equations with the Laplacian or biharmonic operators can lead to matrices that are symmetric and *banded* in the sense of Figure 13.14.

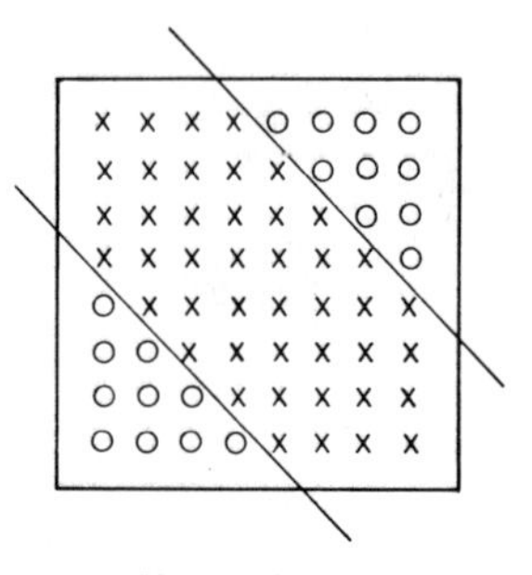

Figure 13.14

They are not tridiagonal, yet their form is close enough to tridiagonal to suggest that general reduction methods may be wasteful. The following strategy, based on a succession of Givens' transformations is efficient in reducing these banded symmetric matrices.

The grand strategy is to annihilate the outermost bands first, starting with the uppermost element. Each annihilation reduces one symmetric pair of elements to zero by a Givens transformation that involves *two adjacent columns* (and the corresponding rows)—the right-hand column containing the element to be annihilated. For our sample matrix the first transformation would invoke columns 3 and 4 (Figure 13.15). The circled x's become 0 because that is the

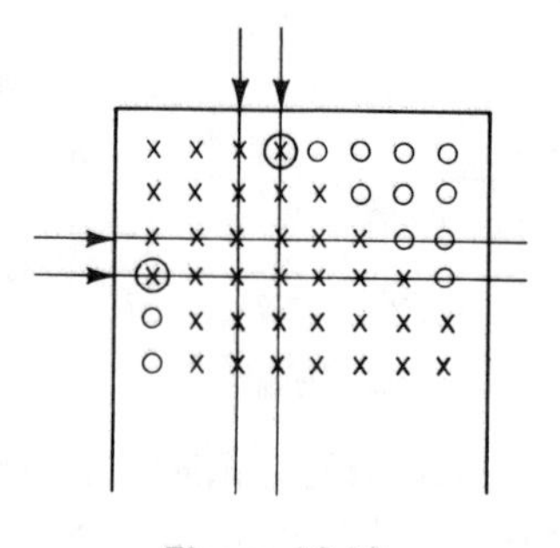

Figure 13.15

way we construct the transformation. But generally all elements on the two rows and columns are changed. The exceptions are those pairs of elements, such as those of the last two columns, that are *both* zero. Since the Givens transformation replaces all pairs of elements from the same column by linear combinations of the two old elements from the same positions, pairs of zeros reproduce themselves. In the pictured matrix, however, one old zero is destroyed—the (3, 7) element – since it is paired with a nonzero element. [Of course, its symmetric (7, 3) zero elements is also destroyed, but for expository convenience in this section we speak only about the elements in the upper triangle.]

The destruction of an already existing zero necessitates a substrategy: Whenever a zero element outside the current band is destroyed by a Givens transformation, the *next* transformation should be aimed at annihilating this new nonzero element. The tactics are unchanged: We use adjacent columns with the right-hand one containing the critical elements. In Figure 13.15 we lost the (3, 7) element, so our next transformation uses columns 6 and 7

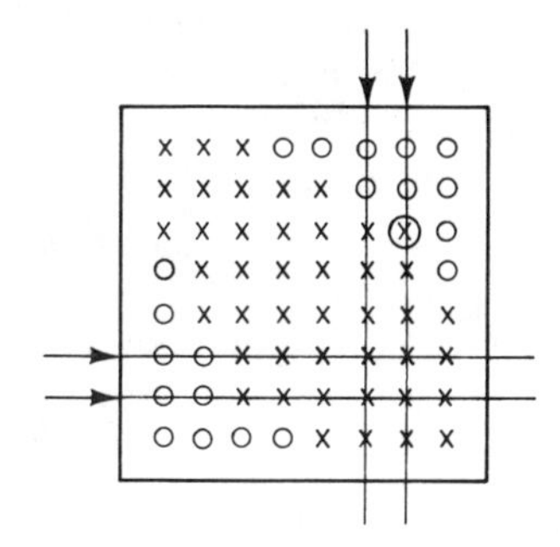

Figure 13.16

(Figure 13.16). This time we successfully get rid of an element without producing another, so we may now return to our main strategy, which points to the (2, 5) element as next on the band to be eliminated. Again we will destroy an old zero (4, 8) and be diverted to a cycle of the substrategy, but no other elements of our current band will get us into this trouble. With big matrices the zeroing of one element of a band may start a fairly long chain of destruction and restoration of previously zero elements, but the chain finally terminates because the new destruction always takes place well to the right of the columns of the transformation and hence the sequence runs into the right-hand boundary of the matrix and is broken.

After eliminating an entire band from the matrix, we start on the next diagonal band, continuing until only a tridiagonal form remains. A total of $4n^2\left(m - \dfrac{1}{m}\right)$ multiplications is required, where $(2m + 1)$ is the number of bands in the original matrix.

Evaluation of the characteristic polynomial for tridiagonal form

Although the eigenvalues of an $n \times n$ matrix are *defined* to be the roots of the nth-degree characteristic polynomial formed from the matrix in a certain way, nobody seriously considers finding this polynomial explicitly and then solving for its roots. Not that there aren't algorithms enough for computing the coefficients of the polynomial! There are, but they are too much work. And then one must still face the difficulties of finding the roots of a polynomial of high degree. Since this path is very unattractive, we shall say nothing further about it.

In contrast with producing the polynomial explicitly, a more reasonable strategy is to *evaluate* this same polynomial directly from the matrix for some trial eigenvalue. If our guess is good, the value of the polynomial will be small; if our guess is exact, the value of the polynomial will be exactly zero, except for rounding errors in the evaluating arithmetic. For full matrices, evaluating the determinant requires about $n^3/3$ multiplications—an inauspiciously large number that we would prefer to avoid. But with the tridiagonal form the work drops to a most cooperative $7n$. Even for Hessenberg form the labor is not unreasonable, being proportioned to n^2. Hence our interest in algorithms for evaluating the characteristic polynomial directly from these special forms.

We write our characteristic polynomial as a determinant. The notation is defined by

$$p_n(\lambda) = \begin{vmatrix} \alpha_1 - \lambda & \beta_2 & & & & \\ \gamma_2 & \alpha_2 - \lambda & \beta_3 & & & \\ & \gamma_3 & \alpha_3 - \lambda & \cdot & & \\ & & \gamma_4 & \cdot & & \\ & & & \cdot & \alpha_{n-1} - \lambda & \beta_n \\ & & & & \gamma_n & \alpha_n - \lambda \end{vmatrix}$$

Figure 13.17

Expanding by minors on the last column gives a term $(\alpha_n - \lambda) \cdot p_{n-1}(\lambda)$ and one that is the product of β_n and another determinant of order $n - 1$. (See Figure 13.18.) Expanding this last determinant by its bottom row, which contains only γ_n, we have

$$p_n(\lambda) = (\alpha_n - \lambda) \cdot p_{n-1}(\lambda) - \beta_n \cdot \gamma_n \cdot p_{n-2}(\lambda) \tag{13.7}$$

which is a three-term recurrence for $\{p_i(\lambda)\}$. As may be seen from the definition (Figure 13.17),

$$p_1(\lambda) = (\alpha_1 - \lambda) \qquad \text{and} \qquad p_2(\lambda) = (\alpha_2 - \lambda)p_1(\lambda) - \beta_2\gamma_2$$

$$p_n(\lambda) = (\alpha_n - \lambda) \cdot \begin{vmatrix} \alpha_1 - \lambda & \beta_2 & & & & \\ \gamma_2 & \alpha_2 - \lambda & \beta_3 & & & \\ & \gamma_3 & \alpha_3 - \lambda & \cdot & & \\ & & \gamma_4 & \cdot & \cdot & \beta_{n-1} \\ & & & & \cdot & \alpha_{n-1} - \lambda \end{vmatrix}$$

$$- \beta_n \cdot \begin{vmatrix} \alpha_1 - \lambda & \beta_2 & & & & \\ \gamma_2 & \alpha_2 - \lambda & \beta_3 & & & \\ & \gamma_3 & \alpha_3 - \lambda & \cdot & & \\ & & \gamma_4 & \cdot & \cdot\ \beta_{n-2} & \\ & & & & \alpha_{n-2} - \lambda & \beta_{n-1} \\ & & & & & \gamma_n \end{vmatrix}$$

Figure 13.18

so if we now define

$$p_0(\lambda) = 1$$

then (13.7) is a recurrence formula that may either

1. Be used *algebraically* to generate a sequence of explicit polynomials $\{p_i(\lambda)\}$ in λ, the last of which is the characteristic polynomial of our tridiagonal matrix, or
2. Be used *arithmetically* with a specific value of λ to generate a sequence of *numbers* that are the values of the $\{p_i\}$ for the specific λ.

It is the *arithmetic* use that shall concern us here, for it permits economic evaluation of the characteristic polynomial for any trial value of λ.

Armed with a reasonable algorithm for evaluating $p_n(\lambda)$, we may now invoke any of the methods for finding roots of polynomials that depend only on knowing functional values. Thus we might step along in λ, evaluating p_n until it changes sign and then close in on the root by cutting the interval in half—the method of bisection or binary chop. Or, having bracketed a root, we might prefer False Position or the secant method. None of these strategies appeals strongly, for they converge somewhat slowly. To improve upon them we must acquire more information about our hidden polynomial. One direction we may go is to compute the first derivative of p_n, and possibly the second, in order to employ quadratically convergent root seekers such as Newton's method. These derivatives are easily computed from algorithms given by formal differentiation of (13.7). We have

$$p_n'(\lambda) = (\alpha_n - \lambda) \cdot p_{n-1}'(\lambda) - \beta_n \cdot \gamma_n \cdot p_{n-2}'(\lambda) - p_{n-1}(\lambda) \tag{13.8}$$

with

$$p_1' = -1 \qquad \text{and} \qquad p_0' = 0$$

Also

$$p_n''(\lambda) = (\alpha_n - \lambda)p_{n-1}''(\lambda) - \beta_n \cdot \gamma_n \cdot p_{n-2}''(\lambda) - 2p_{n-1}'(\lambda) \tag{13.9}$$

with

$$p_2'' = 2 \qquad \text{and} \qquad p_1'' = 0$$

so that the labor in getting p_n' or p_n'' is approximately equal to that for evaluating p_n itself.

A quite different piece of information is already available to us in the sequence of numbers $\{p_i(\lambda)\}$ that we have evaluated at any particular λ in getting through to p_n. For our $\{p_i\}$ are a Sturmanian sequence of polynomials, and hence the *number of changes in sign* as we progress from p_0 to p_n gives us the *number of roots of* p_n *that are smaller than* λ. This information is very valuable in any stepping search, guaranteeing that we will not overstep a pair of eigenvalues without noticing them because p_n did not change sign. We may actually use this number of sign changes as our primary information, employing it with a binary chop to locate our eigenvalues (or groups of eigenvalues) to any required precision. The method has the objectionable slowness of the binary chop's linear convergence, but it has admirable stability in the presence of multiple and nearly multiple roots. There are times when we are willing to pay for stability.

Zeros in the tridiagonal form

If the tridiagonal form has a zero element on a sub- or superdiagonal, the recurrence (13.7) breaks down. The final characteristic polynomial is expressible as the product of two polynomials of lower degree, both of which may be found directly from the tridiagonal form. We need only partition our determinant as in Figure 13.19 and evaluate the two subproblems. [In Figure 13.19, $p_2 = (\alpha_1 - \lambda)(\alpha_2 - \lambda) - \beta_2\lambda_2$ and $p_3 = (\alpha_3 - \lambda) \cdot p_2$. β_3 is irrelevant, since it occurs in our polynomial only as the product $\beta_3\gamma_3$ which here is zero. The zeros of p_3 are clearly those of the polynomial p_2 and $(\alpha_3 - \lambda)$.] This splitting of the big problem at an unpredictable place into two smaller ones can materially reduce the total computational labor, but it can also be an administrative nuisance within an automatic computer program. We may therefore react in either of two ways: Welcome the labor saving by seeking and encouraging zeros and then performing the splitting they imply, or avoid the programming troubles by seeking zeros on the matrix diagonals in order to replace them with small, hopefully harmless, quantities that permit a unified treatment of our matrix.

$$p_3(\lambda) = \left|\begin{array}{cc|c} \alpha_1 - \lambda & \beta_2 & \\ \gamma_2 & \alpha_2 - \lambda & \beta_3 \\ \hline & 0 & \alpha_3 - \lambda \end{array}\right| = (\alpha_3 - \lambda) \cdot p_2(\lambda)$$

$$p_2(\lambda) = \begin{vmatrix} \alpha_1 - \lambda & \beta_2 \\ \gamma_2 & \alpha_2 - \lambda \end{vmatrix}$$

Figure 13.19

While zeros do not necessarily imply multiple roots, multiple roots *do* imply superdiagonal zeros. This observation is of dubious value, for Wilkinson [1965] exhibits matrices with very close eigenvalues that do not produce the zeros that exact equality would demand. Thus, considering the finiteness of computational arithmetic, inferences about whether or not to expect superdiagonal zeros are at least undependable. But a zero, once observed, is a fine pragmatic reason for splitting the problem.

Laguerre's method

One of the more attractive isolated methods for roots of polynomials is from Laguerre. It has cubic convergence to simple real roots, always converges to some root (possibly complex) from *any* starting approximation, will converge to a complex root from a real first approximation if complex arithmetic is permitted, and is easily modified to suppress roots already found. The price one pays for these advantages is that degree of computational complexity and numerical indeterminancy associated with all algorithms requiring second derivative information. In the eigenvalue problem, where we can usually pick up some structural information about the locations and nature of the roots from the tridiagonal form, Laguerre's method can be employed to good advantage. In a more general, unstructured polynomial root finder, its virtues are less apparent. We have given a somewhat restricted derivation of the method in Chapter 7 (p. 187). For a more sophisticated treatment that yields more of the underlying structure the student should see Wilkinson [1965].

Finding all the eigenvalues—suppression versus deflation

What use shall be made of eigenvalues already found when we seek the remaining ones? With explicit polynomials, the known roots are always removed so as to reduce the problem to one of progressively lower degree. This strategy reduces the total amount of labor and, perhaps more importantly, makes the later roots easier to locate by making them more isolated. Neighbors, even of the best sort, always interfere a little. Our matrix eigenvalue problem, however, has carefully avoided producing an explicit polynomial. The relentless pressures for stability and computational efficiency have caused us to transform the original matrix into a tridiagonal form that has the same characteristic polynomial (still hidden) and now we seek its roots directly from this form. But for all its unusual appearance, the problem remains that of finding roots of a high-degree polynomial—a task that is far from trivial even when assisted by free information from the Sturmanian sequence. We would therefore seek to employ the roots thus far found to lighten the burden of finding the rest.

Known roots may be used in either of two ways to simplify our further task: We may *remove* the known root, thus creating a polynomial of lower degree that contains the remaining roots, or we may *suppress* the effect of a known root without actually changing the polynomial. Removing an eigenvalue is called *deflating the matrix*. It reduces the amount of further computation required by reducing the size of the system, but since the removed root is never accurate, the reduced system is also never accurate. Thus the later roots are perturbed, sometimes seriously. Unless the deflation strategy is pursued with considerable care, our computational savings can vanish—consumed by some sort of root-polishing process that will be required to restore our later roots to respectable accuracy. Wilkinson [1965] gives deflation processes, but with a lack of enthusiasm that causes your author to say no more about them.

Root *suppression* reduces neither the size of our matrix nor the labor of evaluating the characteristic polynomial. But it frequently does reduce the total *number* of iterations required to evaluate the later roots to a specified accuracy. It is therefore eminently worthwhile. The techniques are the familiar ones already discussed in the chapters on polynomials (7) and on transcendental equations (2), so we only mention them here. The function

$$u(\lambda) = \frac{p_n(\lambda)}{\lambda - \lambda_j} \tag{13.10}$$

permits us to use Newton's method rather well by expediently removing the effect of a nearby root. We find that we may compute the reciprocal of the Newton correction for the u root from the reciprocal of the usual Newton correction for the p root. We have

$$\frac{u'}{u} = \frac{p'}{p} - \frac{1}{\lambda - \lambda_k} \tag{13.11}$$

and we see that the effect of the known root λ_k falls off quickly with its distance from our trial value of λ. This technique for root suppression can be used without modifying our evaluation of the tridiagonal form. If we are using a technique requiring second derivative information, the useful relation between u and p is

$$\frac{u'^2 - u''u}{u^2} = \frac{p'^2 - p''p}{p^2} = \frac{1}{(\lambda - \lambda_k)^2} \tag{13.12}$$

Since it is this combination of derivatives and function that usually appears in higher-order methods, such as Laguerre's, the modification to seek roots of u is again nearly trivial.

A problem of precision loss arises in both (13.11) and (13.12) if the root being sought is nearly or exactly double, since the term p'/p is nearly or exactly numerically indeterminate at this point. This observation does not constitute a criticism of the method, however, since the numerical indeterminacy merely reflects the lowered precision by which the close roots of the problem may be determined whatever the algorithm. The essential fact about double or nearly double roots is the flatness of the polynomial curve in the region—a flatness that may severely limit the precision with which x may be found when $p_n(x)$ may only be represented by (say) seven decimal digits within a computer word. If this precision will not suffice, double-precision arithmetic for evaluating $p_n(x)$ is the only recourse. Changing to a different algorithm is mere wishful thinking.

Quadratic factors from the tridiagonal form

If complex eigenvalues are expected from a real but unsymmetric tridiagonal form, the appropriate strategy seeks the real quadratic factors of the characteristic polynomial. The algorithms use as inputs the linear remainders after each of two divisions by our trial quadratic factor and they yield two corrections to this factor. When a polynomial is explicitly available, generalized synthetic division gives the necessary inputs (Chapter 7, p. 190), but when the polynomial is presented implicitly as a tridiagonal, or even as a Hessenberg matrix, we may still obtain the necessary numbers by an easily derived technique. The successive principal minors of our tridiagonal determinant, the polynomials $p_i(\lambda)$, satisfy the recurrence

$$p_n(\lambda) = (\alpha_n - \lambda)p_{n-1}(\lambda) - \beta_n\gamma_n p_{n-2}(\lambda) \tag{13.13}$$

where

$$p_0 = 1$$

If we now consider dividing *each* of these polynomials by our trial divisor $(\lambda^2 - E\lambda - F)$ to produce a linear remainder $(A_n\lambda + B_n)$, we have the identity in λ,

$$p_n(\lambda) = (\lambda^2 - E\lambda - F)q_n(\lambda) + A_n\lambda + B_n \tag{13.14}$$

Substituting into (13.13) we have

$$\begin{aligned} p_n &= (\lambda^2 - E\lambda - F)q_n + A_n\lambda + B_n \\ &= (\alpha_n - \lambda)[(\lambda^2 - E\lambda - F)q_{n-1} + A_{n-1}\lambda + B_{n-1}] \\ &\quad -\beta_n\gamma_n[(\lambda^2 - E\lambda - F)q_{n-2} + A_{n-2}\lambda + B_{n-2}] \end{aligned} \tag{13.15}$$

whose right side may be rewritten

$$\begin{aligned} &(\lambda^2 - E\lambda - F)[(\alpha_n - \lambda)q_{n-1} - \beta_n\gamma_n q_{n-2} - A_{n-1}] \\ &\quad +[(\alpha_n - E)A_{n-1} - B_{n-1} - \beta_n\gamma_n A_{n-2}]\lambda \\ &\quad +[\alpha_n B_{n-1} - \beta_n\gamma_n B_{n-2} - FA_{n-1}] \end{aligned}$$

Equating the corresponding factors of our symbolic division .

$$\begin{aligned} q_n &= (\alpha_n - \lambda)q_{n-1} - \beta\gamma q_{n-2} - A_{n-1} \\ A_n &= (\alpha_n - E)A_{n-1} - \beta\gamma A_{n-2} - B_{n-1} \\ B_n &= \alpha B_{n-1} - \beta\gamma B_{n-2} - FA_{n-1} \end{aligned} \tag{13.16}$$

Observing that

$$p_1 = \alpha_1 - \lambda$$

we have the initial values

$$\begin{aligned} &q_0 = q_1 = 0 \\ &A_0 = 0 \qquad A_1 = -1 \\ &B_0 = 1 \qquad B_1 = \alpha_1 \end{aligned}$$

We may now generate a sequence of q's, A's, and B's to yield the required A_n and B_n. A second division, this time of q_n, by the trial quadratic will give two more recursions for the principal parameters of the Bairstow process. Defining

$$q_n = (\alpha_n - \lambda)t_n + C_n\lambda + D_n \tag{13.17}$$

then

$$\begin{aligned} t_n &= (\alpha_n - \lambda)t_{n-1} - \beta\lambda t_{n-2} - C_{n-1} \\ C_n &= (\alpha_n - E)C_{n-1} - \beta\lambda C_{n-2} - D_{n-1} \\ D_n &= \alpha_n D_{n-1} - \beta\lambda D_{n-2} - FC_{n-1} - A_{n-1} \end{aligned} \tag{13.18}$$

with the initial conditions

$$C_0 = C_1 = D_0 = D_1 = 0$$

A similar analysis yields algorithmic equations for evaluating the remainder terms when the polynomial is encoded as a Hessenberg matrix. Wilkinson [1965] gives the details.

Multiple eigenvalues

Since the eigenvalue problem is equivalent to finding some or all of the roots of a high-degree polynomial, we should anticipate the troubles that we by now associate with that discipline. Multiple or near-multiple roots, in particular, usually require special treatment. If we are seeking our roots from a tridiagonal form, however, we have some advantages not vouchsafed unto the analyst who has only an explicit polynomial. Most importantly, we have the root count from the Sturmanian sequence that enables us to tell whether or not a root is reasonably isolated and, if it is not, how many roots lie bunched together. This information permits very precise application of tools appropriate to the configuration. We know, for example, how to modify Newton's method for double or triple roots, but with explicit polynomials we never know the multiplicity of the root in sufficient time to make the modifications. More important, through Sturm we can recognize an isolated root, thereby permitting a carefree application of Newton that would be a daring and dangerous adventure for an explicit polynomial where such information was lacking.

If we are seeking a root that is known to lie near one already found, then we should seek the root of

$$u(\lambda) = \frac{p_n(\lambda)}{\lambda - \lambda_0} \tag{13.19}$$

This stratagem speeds convergence, is simple to apply. Since

$$\frac{u'}{u} = \frac{p'}{p} - \frac{1}{\lambda - \lambda_0} \tag{13.20}$$

it offers a particularly easy modification to Newton's method. It even works if the root λ_0 is double, although the computation of (13.20) gets rather wild here, so tests to prevent overflow should be incorporated in any automatic program.

If eigenvalues are *known* to be double, then any methods that depend only on evaluations of p_n and p_n' necessarily give trouble through indeterminate forms. Since the indeterminacy actually occurs only *at* the multiple root, we may be able to live with it – approaching closely enough to say that we have found the root. A better strategy, however, is to switch to a root seeker that emphasizes second-derivative information as the first derivative sinks quietly toward oblivion. Three terms of a Taylor's series will quickly provide such a method.

Unsymmetric matrices

The eigenworld of unsymmetric matrices is not a pleasant place. While we may have real eigenvalues there is no guarantee of them. The presence of some complex-conjugate root pairs in the characteristic polynomial suggests that its graph will have bumps that could confuse root seekers like Newton which employ derivative information. Then, too, since unsymmetric matrices can degenerate in several more ways, tendencies toward computational instability are more likely. In general we can expect more worries and more work.

On the other hand, the essential shape of our problem remains the familiar one we have explored so fully with symmetric matrices. Computational logistics demand that we reduce the general matrix to a specialized form, Hessenberg this time, rather than tridiagonal, before launching on some infinite algorithmic journey to find the eigenvalues. Again we have a highly systemmatic algorithm for evaluating both the characteristic polynomial and its derivatives from the Hessenberg form, permitting us to use all the standard tricks for polynomial roots. The only change is an increased degree of tactical caution that is required by the greater geometric variety that we should be prepared to handle. In particular, we may have to seek quadratic factors and, having found them, suppress them while seeking more. These ideas are not new, though they imply more care and labor. The absence of Sturmanian information about the number of roots is a distinct handicap.

Orthogonal reductions

The Householder transformations that reduce symmetric matrices to tridiagonal form will reduce unsymmetric ones to Hessenberg form. The same zeros appear below the subdiagonal and for the same reasons. Only the superdiagonal zeros fail to appear, and some computational savings that depended on symmetry are no longer available. But the efficient tactics still include the application and storage of the transformations in vector, rather than matrix, form. The method remains dependably stable—and requires a sizable number of operations. The only new element introduced by the lack of symmetry is the possibility that more efficient but less stable direct reduction methods may now be reasonable. We say more about them in the next section.

The Givens technique can also be used, but since its computational cost is double of that of Householder, there seems to be little need. The Jacobi plane rotation still suffers from its congenital cannibalization of previously produced zeros, hence is still not a candidate for producing Hessenberg form.

Direct reduction to Hessenberg form

Hessenberg form may be produced most directly by a slight modification of the elimination philosophy that is used in solving linear equations. Instead of completely triangularizing our matrix by eliminating *all* the coefficients below the diagonal we only zero those below the *sub*diagonal. Then, since eliminations do not of themselves preserve eigenvalues, we must perform additional calculations to turn our operation into a similarity transformation. We may easily do this by formulating the elimination of our elements from one column as a matrix multiplication (Figure 13.20), after which we need only postmultiply by

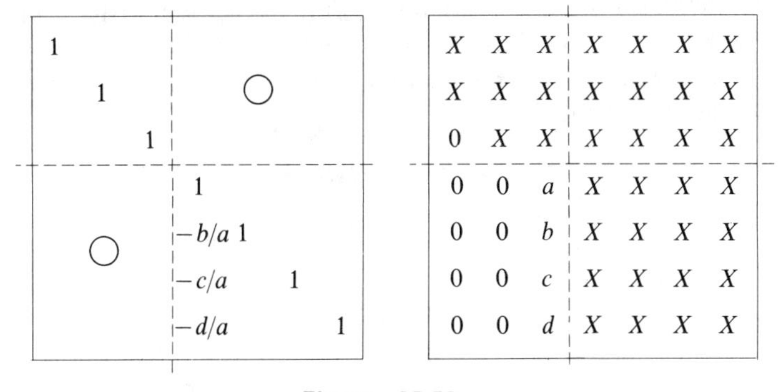

Figure 13.20

the inverse of the transformation. (If we are too greedy and try for complete triangularization, we find that the postmultiplication wrecks the zeros produced by the premultiplication—hence our interest in Hessenberg form which does not suffer such destructions.) Consider the elimination of elements b, c, d from the third column in Figure 13.20. Clearly the first matrix shown embodies the standard elimination arithmetic. The inverse matrix is shown in Figure 13.21.

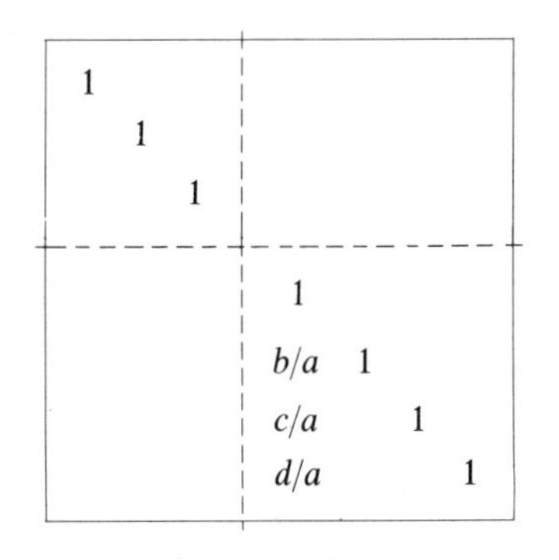

Figure 13.21

Since the transformation is *not* orthogonal we must beware of instability. Just as in Gaussian elimination, this algorithm is usable only if we take care to avoid small pivot elements. Thus we must incorporate row interchanges at each stage to bring the largest element from below the diagonal of the active column into the subdiagonal position before creating zeros in that column. The technique is described by Wilkinson [1965, p. 353] for automatic computers.

While the above description is couched in terms of already familiar operations and points up the essential properties of the algorithm, it also focuses the attention on a sequence of intermediate matrices that hold no interest for us and cause some unnecessary computational steps. We may construct the minimal-work algorithm by a different formulation. We wish to produce a Hessenberg matrix H from a given unsymmetric matrix A while producing the transformation matrix N as shown in Figure 13.22. This transformation is obviously a similarity one,

$$AN = NH \tag{13.21}$$

where N has the form

$$\begin{bmatrix} 1 & & & & \\ 0 & 1 & & & \\ 0 & X & 1 & & \\ 0 & X & X & 1 & \\ 0 & X & X & X & 1 \end{bmatrix}$$

Figure 13.22

By writing out the equations implied by the matrix multiplications of (13.21) we obtain the elimination algorithm *without interchanges*. As always with elimination algorithms, however, interchanges are essential for stability. The technique for incorporating interchanges into this algorithm is also given by Wilkinson [1965, p. 359].

Hessenberg to tridiagonal form

We have just given algorithms for reducing a general matrix to *upper* Hessenberg form. If our "general" matrix just happened to be in lower Hessenberg form already and if we used an algorithm that did not disturb the zeros in the upper triangle while producing new ones in the lower triangle, we will obtain tridiagonal form. The direct method *without interchanges* has this property of leaving the upper zeros undisturbed and hence could be used. Wilkinson reports that it is almost always unstable (because small pivots can no longer be avoided) but that if it is performed in double precision it can be a useful technique.

Thus our grand strategy for unsymmetric matrices now has another choice: to reduce all the way to tridiagonal form in the face of possibly serious instability and the necessity for double-precision arithmetic, or stopping at Hessenberg form. The gains in efficiency for root seekers by operating on tridiagonal form instead of merely Hessenberg form are, of course, considerable. A price contrast of $2n$ versus $n^2/2$ is indeed tempting. Wilkinson suggests gambling even if, once in a while, we will lose and presumably have to go back to Hessenberg form again.

Inner-product accumulation

In the constant battle against arithmetic erosion of information by the multitudinous operations required with large matrices, one feature of automatic computers can be put to good use: the double-precision accumulator. Most electronic computers, like most desk calculators, develop the product of two d-digit numbers in a $2d$-digit accumulator or product register. Thus it is relatively easy to accumulate a long string of such products, rounding to single precision only when the final sum has to be stored. This computational tactic should be used wherever possible in operations with large matrices.

Most operations with large matrices, such as matrix-vector multiplication or Gaussian triangularization, consist primarily of products of vectors – the sums of products of number pairs. Loss of precision occurs when serious cancellation of digits takes place in these sums (the vector components can, of course, be negative). And any loss that has occurred becomes more serious when a small imprecise result is used as a divisor in a later stage of the algorithm. We guard against the divisions by so-called *pivoting* techniques, which try to shove small intermediate results out of the main stream of the computation, or to postpone their use to the latest possible moment so as to affect as few other numbers as possible.

If the dot product of two vectors is small even though their components are not, then cancellation of significant digits is inevitable. We may be able to mitigate the effects of the precision loss by pivoting, but precision loss there must be. Thus when two vectors are nearly perpendicular, their dot product will lose significant figures from the left. What we wish to stress here, however, is the double-length accumulation gambit to cut down on *unnecessary* rounding noise – trouble that enters our numbers at the least-significant or right-hand end, rather than significant-figure trouble that enters from the left. A dot product of two vectors each of length 200 will have either one rounding or 200, depending on whether or not the double-length accumulation technique is used. Two hundred roundings combine statistically, often canceling one another, but sometimes they reinforce and are then able to make several of the least significant digits in the sum essentially random. This loss is unnecessary and hence intolerable in situations where we are already struggling to preserve accuracy in the face of essential cancellations at the other end of the number among the most significant figures. All algorithms for dealing with large matrices should be encoded to take advantage of the double-precision equipment provided by the designers. Usually this entails the recoding, by hand, of only a few instructions at several key points—but it does require the use of machine language and some knowledge of the computer arithmetic.

Evaluating the characteristic polynomial of a Hessenberg form

We may evaluate the determinant of a Hessenberg matrix for a trial value of λ by a method rather similar to that used with tridiagonal form (p. 332). We wish to find the numerical value of

$$P_n(\lambda) = \begin{vmatrix} a_{11}-\lambda & a_{12} & a_{13} & a_{14} & \cdots & a_{1,n-1} & a_{1n} \\ b_1 & a_{22}-\lambda & a_{23} & a_{24} & & a_{2,n-1} & a_{2n} \\ & b_2 & a_{33}-\lambda & a_{34} & & a_{3,n-1} & a_{3n} \\ & & b_3 & a_{44}-\lambda & & a_{4,n-1} & a_{4n} \\ & & & b_4 & & \vdots & \vdots \\ & & & & & a_{n-1,n-1}-\lambda & a_{n-1,n} \\ & & & & & b_{n-1} & a_{nn}-\lambda \end{vmatrix} \tag{13.22}$$

$$\mathbf{C}_1 \quad \mathbf{C}_2 \quad \mathbf{C}_3 \quad \mathbf{C}_4 \quad \cdots \quad \mathbf{C}_{n-1} \quad \mathbf{C}_n$$

where all the quantities in the determinant are known numbers. Our strategy is to find a linear combination of the first $n - 1$ columns, which, when added to the last column, reduces that column to zeros in all but its top component. Algebraically, we would like to find a set of β_i's such that

$$\mathbf{C}_n + \sum_1^{n-1} \beta_i \mathbf{C}_i = k(\lambda) \cdot \mathbf{e}_1 \tag{13.23}$$

where the $\mathbf{C}_i$'s are the columns of our determinant. If we replace the last column of (13.22) by its new value we do not change the numerical value of the determinant, and it now looks like

$$P_n(\lambda) = \begin{vmatrix} a_{11} - \lambda & a_{12} & a_{13} & \cdots & a_{1,n-1} & k \\ b_1 & a_{22} - \lambda & a_{23} & & a_{2,n-1} & 0 \\ & b_2 & a_{33} - \lambda & & a_{3,n-1} & 0 \\ & & b_3 & & a_{4,n-1} & 0 \\ & & & & \vdots & \vdots \\ & & & \ddots & & \\ & & & & a_{n-1,n-1} - \lambda & 0 \\ & & & & b_{n-1} & 0 \end{vmatrix}$$

On expanding this form by minors, using the last column, we obtain

$$P_n(\lambda) = k(\lambda) \cdot \begin{vmatrix} b_1 & a_{22} - \lambda & a_{23} & \cdots & a_{2,n-1} \\ & b_2 & a_{33} - \lambda & & a_{3,n-1} \\ & & b_3 & & a_{4,n-1} \\ & & & & \vdots \\ & & & \ddots & a_{n-1,n-1} - \lambda \\ & & & & b_{n-1} \end{vmatrix}$$

but this determinant is triangular and hence is the product of its diagonal elements; thus

$$P_n(\lambda) = \pm k(\lambda) b_1 b_2 \cdots b_{n-1} \tag{13.24}$$

We can find the β_i successively using the equations (13.23):

$$\begin{aligned} \beta_{n-1} b_{n-1} + (a_{nn} - \lambda) &= 0 \\ \beta_{n-2} b_{n-2} + \beta_{n-1}(a_{n-1,n-1} - \lambda) + a_{n-1,n} &= 0 \\ \vdots \qquad\qquad & \vdots \\ \beta_1 b_1 + \beta_2(a_{22} - \lambda) + \beta_3 a_{23} + \cdots + \beta_{n-1} a_{2,n-1} + a_{2n} &= 0 \end{aligned}$$

so that finally we get k from

$$\beta_1(a_{11} - \lambda) + \beta_2 a_{12} + \cdots \beta_{n-1} a_{1,n-1} + a_{1n} = k$$

and may now evaluate $P_n(\lambda)$ from (13.24). A small amount of arithmetic labor is saved if we generate our Hessenberg form with all the subdiagonal elements, b_i, identically equal to 1.

LR algorithms

Barring unfortunate accidents of degeneracy, any matrix may be decomposed into the product of a lower unit triangular matrix and an upper triangular matrix. Symbolically we have

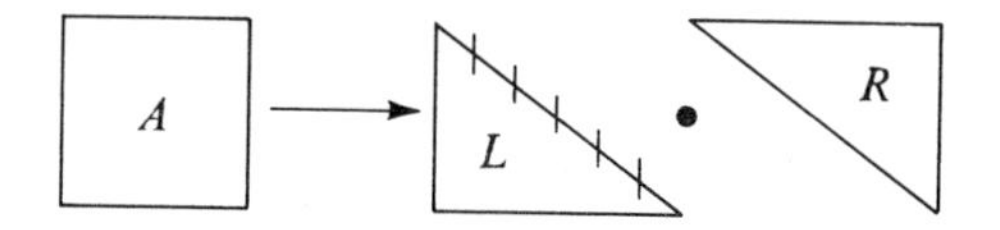

where the symbols are meant to convey the presence of 1's on the principal diagonal of L and no such specialization of R. The algorithm effecting such separation is given immediately by filling our schematic with components and writing out the implied matrix multiplication. It is precisely the Gaussian triangularization by elimination of variables, familiar to every schoolboy. For stability it is usually necessary that row interchanges be performed to bring large pivot elements to the diagonal, but here we shall assume that (fortuitously) our matrix happens to be arranged in the correct order – since we are not at the moment concerned with such details that are, admittedly, essential in practice.

If we now *reverse the order* of our factor matrices and *remultiply* them, we create another square matrix, completing one cycle of an iteration. We have

$$A_s \to L_s R_s \qquad \text{then} \qquad R_s L_s \to A_{s+1} \tag{13.25}$$

We see that the iteration preserves eigenvalues, for it is a similarity transformation:

$$A_{s+1} = R_s L_s = L_s^{-1} A_s L_s \tag{13.26}$$

It will be proved below that the sequence of $\{A_s\}$ converges to upper triangular form, hence reveals the eigenvalues on the principal diagonal. A necessary and sufficient corollary is the convergence of L_s to a unit matrix, and it is this fact that will be proved. But first we would examine the *utility* of the algorithm.

For a general square matrix, symmetric or unsymmetric, the separation requires $n^3/3$ multiplications, the recombination another $n^3/3$. Since the algorithm produces eigenvalues it is necessarily infinite, and we have steadily warned against engaging in infinite iterations with $0(n^3)$ multiplications per iteration. We still do. The *LR* algorithm is not practical for general matrices. But it has modifications that *are* practical with specialized matrices.

If we write out the triangular decomposition algorithm (13.25) we find that triangles of zeros in the lower left corner of A are transmitted identically to the same elements of L. Likewise, triangles of zeros in the upper right corner of A turn up unchanged in R. Further, when L and R are remultiplied in the reverse order, our zeros reappear so that an original *banded* matrix A_s has its form reproduced in A_{s+1}. In particular, tridiagonal and Hessenberg matrices keep their tridiagonal and Hessenberg forms under the LR algorithm. Since factorization of tridiagonal matrices is $0(n)$ and that of Hessenberg $0(n^2)$, we just might hope that the LR algorithm would compete effectively for finding eigenvalues of these specialized forms. It will turn out that it does but only after elaborating the algorithm with a number of devices for suppressing roots as they are found, accelerating convergence of the iterations, and splitting the problem into two or more subproblems whenever possible – in short, by using all the tricks that we need for finding roots of a high degree polynomial.

We are now in the position of hoping to use LR on specialized matrices, since the volume of work is of the right order of magnitude. Still the nagging problem of interchanges remains. Without interchanges, we know that the Gaussian triangular decomposition is unstable. *With* interchanges, we seem to risk loss of our specialized forms. Trial of a few examples, however, will soon convince the sceptical that zeros below the diagonal are preserved by LR with interchanges, while zeros above the diagonal are gradually lost. Thus Hessenberg form is preserved, though tridiagonal is not. Accordingly, we might expect LR_wI to be useful for unsymmetric matrices that have already been reduced to Hessenberg form. It is.

The *QR* algorithm

Returning to the basic algorithm (13.25), we note that *any* factorization into two matrices that are then multiplied in the reverse order constitutes a similarity transformation. For a practical algorithm we need (1) convergence, (2) stability, and (3) efficiency. The Gaussian triangular (LR) decomposition is not the world's most stable one, though it rates high on efficiency, and the upper triangular form of the R matrix turns out to be the important contributor to the convergence. Another decomposition, an exceedingly stable one, is given by *orthogonal* transformations. Here we have

$$A_s \to Q_sR_s$$

where R_s is still upper triangular but Q_s is square and orthogonal. Now we have

$$A_{s+1} = R_sQ_s = Q_s{}^TA_sQ_s$$

with no worries about stability. Furthermore, symmetry is preserved and, since no interchanging is needed, there is no disruption of useful patterns. A trial will reveal that all *symmetric* banded forms of A_s are transmitted safely to A_{s+1} (though unsymmetric ones are again eroded above the diagonal). Thus we might be tempted to try a QR algorithm on a symmetric tridiagonal form. It will do quite well, though there is still another member of this family, the Cholesky decomposition, that is more efficient and just as stable for this, the best of all specialized forms.

The Cholesky algorithm

Positive definite symmetric matrices may be factored into triangular matrices that are transposes of each other. We have

$$A_s = L_s L_s^T \tag{13.27}$$

and the decomposition is often called the square-root factorization. It is extremely stable, never requires interchanging to avoid small pivots, and requires the least calculational labor of all decomposition, largely because of the symmetry. Positive definiteness, however, is essential lest complex elements appear in the factors. This restriction is not serious, for all symmetric matrices have real eigenvalues, and one may add a constant to all the eigenvalues simply by adding that same constant to the principal diagonal of the matrix. (Positive definiteness only requires all the eigenvalues to be positive.) Thus the Cholesky version of LR is the favorite algorithm of the family for symmetric matrices—adjusted if necessary to ensure positive eigenvalues.

Convergence of *LR*

To demonstrate the convergence of the LR algorithm we must show that

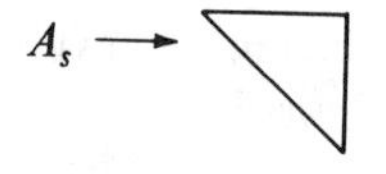

but since $A_s = L_s R_s$, this is equivalent to showing that $L_s \to I$. It is this latter condition that we shall prove, but we need some mechanical preliminaries. We have

$$A_{s+1} = R_s L_s = L_s^{-1} A_s L_s = L_s^{-1} L_{s-1}^{-1} \cdots L_2^{-1} L_1^{-1} A_1 L_1 L_2 \cdots L_{s-1} L_s \tag{13.28}$$

and we observe that

$$L_1 L_2 \cdots L_{s-1} L_s A_{s+1} = A_1 L_1 L_2 \cdots L_s \tag{13.29}$$

Defining

$$\left.\begin{aligned} T_s &= L_1 L_2 \cdots L_s = \\ U_s &= R_s R_{s-1} \cdots R_1 = \end{aligned}\right\} \text{by construction} \tag{13.30}$$

we have

$$T_sU_s = L_1L_2\cdots(L_sR_s)\cdots R_2R_1 = \underline{|L_1L_2\cdots L_{s-1}A_s|}R_{s-1}\cdots R_2R_1$$

$$= \underline{|A_1L_1L_2\cdots L_{s-1}|}R_{s-1}\cdots R_2R_1$$

where the last equality follows by replacing the underlined expression via (13.29). Repeating this logic we finally obtain

$$(A_1)^s = T_sU_s \tag{13.31}$$

and T_s is seen to be the lower unit triangular decomposition of the sth power of our original matrix. Since

$$T_{s-1}^{-1}T_s = L_s \tag{13.32}$$

we need only prove that T_s approaches *any* limit in order to imply that L_s becomes a unit matrix and hence that A_s becomes upper triangular.

From

$$Ax_i = \lambda_i x_i \qquad \text{and} \qquad A^T y_i = \lambda_i y_i$$

we have

$$AX = XD \qquad \text{and} \qquad A^T Y = YD \qquad \text{where } D = \begin{vmatrix} \lambda_1 & & & & \\ & \lambda_2 & & & \\ & & \cdot & & \\ & & & \cdot & \\ & & & & \cdot \\ & & & & & \lambda_n \end{vmatrix}$$

Also we know that

$$X^T Y = I \qquad \text{hence } A = XDX^{-1} = XDY^T$$

Further, since $A^s x_i = \lambda_i^s x_i$ it follows that

$$\boxed{A^s = XD^sY^T} \tag{13.33}$$

If we now factor X and Y^T into the same type of triangular factors that we have been using throughout, we have

$$A^s = L_xR_xD^sL_yR_y = L_xR_x(D^sL_yD^{-s})D^sR_y \tag{13.34}$$

It is the central expression in (13.34) that is crucial:

$$D^s L_y D^{-s} = \begin{bmatrix} \lambda_1{}^s & & \\ & \lambda_2{}^s & \\ & & \lambda_3{}^s \end{bmatrix} \begin{bmatrix} 1 & & \\ l_{21} & 1 & \\ l_{31} & l_{32} & 1 \end{bmatrix} \begin{bmatrix} \lambda_1{}^{-s} & & \\ & \lambda_2{}^{-s} & \\ & & \lambda_3{}^{-s} \end{bmatrix}$$

$$= \begin{bmatrix} 1 & & \\ \left(\dfrac{\lambda_2}{\lambda_1}\right)^s l_{21} & 1 & \\ \left(\dfrac{\lambda_3}{\lambda_1}\right)^s l_{31} & \left(\dfrac{\lambda_3}{\lambda_2}\right)^s l_{32} & 1 \end{bmatrix}$$

If $|\lambda_1| > |\lambda_2| > \cdots |\lambda_n|$, then the subdiagonal elements are gradually suppressed and we may write

$$D^s L_y D^{-s} = I + E_s \qquad \text{where } E_s \longrightarrow 0$$

then

$$A^s = L_x R_x (I + E_s) R_x^{-1} R_x D^s R_y = L_x (I + F_s)(R_x D^s R_y)$$

where $F_s \to 0$ and the last parenthesis is *upper triangular*.

Finally we see that $(A)^s$ is tending to a lower unit triangular matrix L_x that does not depend on s, that is, $T_s \to L_x$. Thus $L_s \to I$ and $A_s \to \nabla$, as required.

LR tactics for tridiagonal form

Tridiagonal form enjoys one peculiar property: Its superdiagonal elements are arbitrary and can always be made unity by a trivial similarity transformation. Since it thus contains only $2n - 1$ pieces of useful information, the algorithms that work with it have only kn multiplications or divisions. Because of these advantages it is often good strategy to reduce the wider banded matrices to tridiagonal form before invoking the *LR* or *QR* type of algorithm. Methods for such reductions have been given by Rutishauser [1959] and are described on page 329.

Here we wish to give two explicit algorithms of the *LR* (Cholesky) and *QR* types for diagonalizing tridiagonal forms, but we first shall document some useful facts. A general tridiagonal matrix A may have its superdiagonal row set to any arbitrary collection of nonzero numbers we wish by a diagonal transformation of the form

$$A_2 = BAB^{-1}$$

We see that

$$A_2 = \begin{vmatrix} c & & \\ & d & \\ & & e \end{vmatrix} \cdot \begin{vmatrix} d_1 & \beta_2 & - \\ \gamma_2 & d_2 & \beta_3 \\ - & \gamma_3 & d_3 \end{vmatrix} \begin{vmatrix} c^{-1} & & \\ & d^{-1} & \\ & & e^{-1} \end{vmatrix}$$

$$= \begin{vmatrix} cd_1c^{-1} & c\beta_2 d^{-1} & - \\ d\gamma_2 c^{-1} & dd_2 d^{-1} & d\beta_3 e^{-1} \\ - & e\gamma_3 d^{-1} & ed_3 e^{-1} \end{vmatrix} \tag{13.35}$$

If we now choose c arbitrarily, the desired value for the (1, 2) position determines d^{-1} and then our desired value for the (2, 3) position fixes e^{-1}. Equation (13.35) also shows that A_2 may be made symmetric provided the products $\beta_i\gamma_i$ are not negative. Next we write out the factorization

$$\begin{bmatrix} \alpha_1 & \beta_2 & \\ \gamma_2 & \alpha_2 & \beta_3 \\ & \gamma_3 & \alpha_3 \end{bmatrix} = \begin{bmatrix} 1 & & \\ l_2 & 1 & \\ & l_3 & 1 \end{bmatrix} \cdot \begin{bmatrix} u_1 & \beta_2 & \\ & u_2 & \beta_3 \\ & & u_3 \end{bmatrix}$$

$$\text{or} \quad \begin{matrix} \textcircled{u_1} = \alpha_1 & \textcircled{l_2}\, u_1 = \gamma_2 \\ l_2\beta_3 + \textcircled{u_2} = \alpha_2 & \textcircled{l_3}\, u_2 = \gamma_3 \\ l_3\beta_3 + \textcircled{u_3} = \alpha_3 & \end{matrix}$$

and quickly find that the superdiagonal passes unchanged into the R matrix. If we now compute the LR algorithmic iteration we have

$$\begin{bmatrix} u_1 & \beta_2 & \\ & u_2 & \beta_3 \\ & & u_3 \end{bmatrix} \cdot \begin{bmatrix} 1 & & \\ l_2 & 1 & \\ & l_3 & 1 \end{bmatrix} = \begin{bmatrix} a_1 & \beta_2 & \\ c_2 & a_2 & \beta_3 \\ & c_3 & a_3 \end{bmatrix}$$

$$\text{or} \quad \begin{matrix} a_1 = u_1 + \beta_2 l_2 & \\ a_2 = u_2 + \beta_3 l_3 & c_2 = u_2 l_2 \\ a_3 = u_3 & c_3 = u_3 l_3 \end{matrix}$$

We note that again the superdiagonal survives unscathed. More important, we may eliminate the l_i from our formulas to get

$$\begin{matrix} u_1 = \alpha_1 & a_1 = u_1 + \beta_2\gamma_2/u_1 & \\ u_2 = \alpha_2 - \beta_2\gamma_2/u_1 & a_2 = u_2 + \beta_3\gamma_3/u_2 & c_2 = u_2\gamma_2/u_1 \\ u_3 = \alpha_3 - \beta_3\gamma_3/u_2 & a_3 = u_3 & c_3 = u_3\gamma_3/u_2 \end{matrix}$$

and we see that the β_i enter only in the products $\beta_i\gamma_i$ — as was indeed also true in the evaluation of the characteristic polynomial from the tridiagonal form (p. 322).

Our algorithm becomes simpler if we first set all the β_i to unity via (13.35). Then we have only $n - 1$ divisions and $n - 1$ multiplications per iteration — the irreducible minimum. The condition of positive definiteness requires that we be careful about shifting eigenvalues to speed convergence, but that tactic is easily incorporated into our algorithm [Wilkinson, 1965, p. 565] and should be used for really efficient convergence.

A similar minimal algorithm, but for *symmetric* tridiagonal matrices, lies in the QR family. Wilkinson [1965, p. 567] gives its derivation. The cost is $3n$ divisions and $2n$ multiplications. Our original tridiagonal matrix has diagonal elements α_i and off-diagonal elements β_i while the next iterate has a_i and b_i, respectively. The intermediate quantities γ_i, p_i, c_i, s_i are never negative.

$$
\begin{aligned}
\gamma_i &= \alpha_i - u_{i-1} \\
p_i &= \gamma_i^2/c_{i-1} \qquad (\text{if } c_{i-1} \neq 0) \\
&= c_{i-2}/\beta_i^2 \qquad (\text{if } c_{i-1} = 0) \\
b_i^2 &= s_{i-1}(p_i + \beta_{i+1}^2) \qquad (i \neq 1) \\
s_i &= \beta_{i+1}^2/(p_i + \beta_{i+1}^2) \\
c_i &= p_i/(p_i + \beta_{i+1}^2) \\
u_i &= s_i(\gamma_i + \alpha_{i+1}) \\
a_i &= \gamma_i + u_i
\end{aligned}
$$

These equations are executed *seriatim*, for $i = 1, 2, \ldots, n$, starting with $c_0 = 1$ and $u_0 = 0$. Also the value of β_{n+1} and α_{n+1} are both taken as zero.

Because of the newness of these algorithms (1955, 1961) the verdict of massive experience is unavailable. Thus we cannot say whether the LR and QR complete reductions are preferable to seeking the eigenvalues of tridiagonal matrices via evaluations of the characteristic polynomial. Since no clear advantage has yet emerged, however, one is tempted to suspect that the differences are second-order — depending on the distribution of the roots and the needs of the user.

We should point out that under the LR and QR iterations the various eigenvalues do not appear at the same rate. The operative mechanism, as in the old power method, suppresses subdiagonal elements by multiplying them with $(\lambda_i/\lambda_{i+1})^s$. Thus a large eigenvalue separation gives a good suppression factor, which may be enhanced by judicious temporary shifting of all the λ's. After a λ has fully developed, the problem can be deflated so that the subsequent iterations require less work. While Wilkinson gives some automatic prescriptions that seem reasonable, one is left with the impression that efficient eigenvalue liberation via LR or QR still contains more than a modicum of art. It is rather like casting in troubled waters for elusive fish. Some people do it better than others. This opinion (and it is certainly nothing more than opinion) does nothing to help or damn the LR technique, for the alternatives also still depend on art for their efficiency. A man may become quite good at finding roots of polynomials encoded as tridiagonal matrices with the help of Laguerre's method, suppression of previously found roots, etc., and still be quite poor if required to use some less familiar technique. The formulation of such art into precise computer programs is a worthy challenge for contemporary numerical analysts.

Summary of work loads

In an attempt to distinguish the general strategies for eigenvalues, as well as the peculiar competences of our several algorithms, we here tabulate their principal properties, placing special emphasis on the amount of work that each requires. Generally our measure is the number of multiplications, although we comment additionally if the square-root labor is noticeable.

The finite part of our strategy, reduction to a specialized form, is covered in Table 13.1. By way of contrast, the Jacobi plane rotation technique seems usually to take $12n^3$ multiplications to reduce a general nonsymmetric matrix to triangular form, or $6n^3$ to bring a symmetric matrix to diagonal form. Since both of these configurations represent a complete solution to the eigenvalue problem, they should be compared to one of our finite strategies followed by one of the infinite strategies that actually finds the eigenvalues.

Table 13.1. Number of multiplications in the reduction to specialized form

	Symmetric		Nonsymmetric
	General to tridiagonal	Banded to tridiagonal	General to Hessenberg
Givens	$\frac{4}{3}n^3$ $+n^2/2$ square roots	$4n^2(m - 1/m)$	$\frac{10}{3}n^3$
Householder	$\frac{2}{3}n^3$ $+n$ square roots	–	$\frac{5}{3}n^3$
Elimination with interchanges	–	–	$\frac{5}{6}n^3$

Turning to the infinite strategy that takes a trial value and then evaluates the characteristic polynomial to see if it is, indeed, zero there, we find that each evaluation takes the number of multiplications given in Table 13.2. The advantage of tridiagonal form is immediate and overwhelming, but Hessenberg form is not to be disparaged if tridiagonal is unavailable.

Table 13.2. Number of multiplications to evaluate, for one trial λ, the characteristic polynomial given as a determinant

General determinant	$n^3/6$
Hessenberg determinant	$n^2/2$
Tridiagonal determinant	$2n$

If we contemplate using Newton's method to select our next trial λ, we need not only the value of the characteristic polynomial but also that of its derivative. Since each of these takes about $2n$ multiplications from tridiagonal form and since we might expect perhaps five cycles from Newton to liberate one root, we estimate $20n$ multiplications per root. Assuming we want all n roots, this leads to a total labor for a general symmetric matrix of

$$\tfrac{2}{3}n^3 + 20n^2$$

which is considerably better than the $6n^3$ that the Jacobi routine in the average computer laboratory is apt to require.

Turning to the alternative infinite strategy of the *LR* family, we find that the efficiency of the several algorithms is governed largely by their ability to preserve specialized patterns of zeros. Thus *LR* with interchanges requires fewer

Table 13.3. Number of multiplications per iteration in the *LR* family

	Symmetric		Nonsymmetric	
	Banded	Tridiagonal	Hessenberg	General
QR	BS $3nm^2$	TD $5n$	H $4n^2$	$2n^3$
LR with interchanges	—	—	H n^2	n^3
Cholesky	BS nm^2	TD $2(n-1)$		

multiplications than QR, but messes up banded symmetric matrices and is thus not suitable for them. Table 13.3 gives the details, showing work load and the properties that are preserved. The last column gives the statistics on general matrices—emphasizing the unsuitability of this algorithmic family for such use.

Simple comparisons of arithmetic labor per iteration are useful for avoiding gross misapplications of our algorithms, but they should not be pushed too far. Laboratory experience shows that eigenvalues of Hessenberg matrices are usually produced to some standard accuracy with fewer QR iterations than with LR. Thus the QR labor is not four times as great, in spite of Table 13.3.

Eigenvectors II

The problem of finding eigenvectors arises far less often than that of finding eigenvalues—and this is perhaps fortunate. For the eigenvalue methods that succeed best with large matrices do not directly produce the vectors, and thus we must expend considerable additional computational energy if we would have our vectors too. As usual, the formal statement of the problem is simple enough. We have found one (or all) of the eigenvalues, λ_i, of the system

$$A\mathbf{x}_i = \lambda_i \mathbf{x}_i \tag{13.36}$$

and we must now solve the linear homogeneous system of algebraic equations

$$(A - \lambda_i I)\mathbf{x}_i = 0 \tag{13.37}$$

for the corresponding vector $\mathbf{x}_i$. Because of the homogeneity we are free to pick the value of one *component* of $\mathbf{x}_i$ and to reserve one equation from the system, whereafter we have a well-posed problem in linear equation solving. Well posed, but not necessarily well conditioned; and therein lies the trouble. But we get ahead of our story.

The strategies for finding eigenvalues of large systems divide the operation into two parts: reduction of the matrix to tridiagonal or Hessenberg form by transformations that preserve the eigenvalues, followed by the extraction of the eigenvalues from the specialized form. The strategy for eigenvectors mirrors these two operations: We find eigenvectors of the specialized form and then parlay them back up to be the eigenvectors of the original matrix. The alternative strategy of solving equation (13.37) directly with the original matrix involves too many operations that duplicate the labor already expended in getting the specialized form.

The conceptually simpler algorithms are the ones that turn our eigenvectors of S, the specialized matrix, back into those of A. We note that S was produced from A by a sequence of similarity transformations $\{T_i\}$ which may be collectively called T. We thus have

$$T_k^{-1} \cdots T_2^{-1} T_1^{-1} A T_1 T_2 \cdots T_k = T^{-1} A T = S$$

Defining

$$T\mathbf{y} = \mathbf{x} \tag{13.38}$$

we have

$$A\mathbf{x} = \lambda\mathbf{x} \qquad \text{or} \qquad T^{-1}A\mathbf{x} = \lambda T^{-1}\mathbf{x}$$

$$T^{-1}AT\mathbf{y} = \lambda T^{-1}T\mathbf{y} = \lambda\mathbf{y}$$

$$S\mathbf{y} = \lambda\mathbf{y}$$

and so we see that $\mathbf{y}$ is an eigenvector of S. Equation (13.38) turns it into $\mathbf{x}$, the eigenvector of A that we require. Computationally, we need only perform the successive combination of

$$T_1 \cdot T_2 \cdots T_k \to T$$

as the successive T_i are produced to have available the final T for use with $\mathbf{y}$. We carefully avoid saying the successive *matrix multiplications* $T_1 T_2 \cdots$ for the transformations T_i usually arise in factored form as products of vectors. We save much arithmetic labor by using these forms to produce the current combination H_c. We form

$$H_c = T_c \cdot H_{c-1}$$

but we let T_c operate on H_{c-1} in its factored form. H_c is thus the current transformation matrix T that will get us back to the eigenvector $\mathbf{x}$ from the eigenvector $\mathbf{y}$ of the current partially reduced matrix. The amount of computation is usually small compared to the labor of finding the T_i themselves and the storage requirements are those of an additional matrix. No serious stability or other computational difficulties arise here. Our troubles lie in finding the $\mathbf{y}_i$ from S.

Eigenvectors from specialized matrices

Formally, we may obtain the eigenvector $\mathbf{u}_k$ of the tridiagonal matrix S corresponding to the eigenvalue λ_k from the homogeneous equation system

$$(S - \lambda_k I)\mathbf{u}_k = 0 \tag{13.39}$$

This strategy, however, forces us to throw out one equation while setting the corresponding variable to unity—an aesthetically distressing choice that seems needlessly arbitrary and makes us wonder if choosing a different equation might not yield significantly different components for $\mathbf{u}_k$. Quite often it does. Wilkinson [1965] gives some sobering examples of equation systems that yield the correct eigenvector very well if the proper equation is discarded (which fixes the order of computation for the components) but which give wildly wrong vectors otherwise. The chance that we would make the proper choice is 1 in n – scarcely adequate even in the most adventurous computer laboratory. Accordingly, both he and we prefer a more nearly symmetric iterative strategy that brings to bear the information carried by all n equations: the *inverse iterated eigenvector*.

The underlying principle of the inverse iterative algorithm arises from the observation that the solution, $\mathbf{x}$, of the linear equation system

$$(S - \lambda I)\mathbf{x} = \mathbf{b} \tag{13.40}$$

for a general matrix S, random vector $\mathbf{b}$, and any value of λ close to λ_k tends to resemble the corresponding eigenvector $\mathbf{u}_k$. Further, if this approximate $\mathbf{u}_k$ is substituted for $\mathbf{b}$ on the right side of (13.40) and the system solved again, an even closer approximation to $\mathbf{u}_k$ is produced. The process succeeds best if λ is equal to λ_k, but only requires that it be closer to λ_k than to some other eigenvalue. Neither is the initial choice for $\mathbf{b}$ overly critical. Like all first approximations, a good one reduces the number of iterations required, but a poor one almost never defeats the process. The iterations are computationally inexpensive since $(S - \lambda I)$ may be inverted once and then used as often as needed. In practice, two or three iterations suffice.

Let us prove our assertions by considering the solution to (13.40). We expand both $\mathbf{b}$ and $\mathbf{x}$ as linear combinations of the eigenvectors $\mathbf{u}_i$ to get

$$(S - \lambda I)\left(\sum_i \gamma_i \mathbf{u}_i\right) = \mathbf{b} = \sum_i \beta_i \mathbf{u}_i$$

or

$$\sum_i \gamma_i(\lambda_i - \lambda)\mathbf{u}_i = \sum_i \beta_i \mathbf{u}_i$$

so that

$$\gamma_i = \frac{\beta_i}{\lambda_i - \lambda} \quad \text{and hence} \quad \mathbf{x} = \sum_i \frac{\beta_i \mathbf{u}_i}{\lambda_i - \lambda} \tag{13.41}$$

From (13.41) we see that if λ is near to λ_k and not near to any other eigenvalue and if β_i is not extraordinarily small, then $\mathbf{x}$ is nearly $\mathbf{u}_k$ – as asserted. Further, if $\mathbf{b}$ is replaced by $\mathbf{x}$ the effect is to square the factor $(\lambda_k - \lambda)$, and each iteration increases this factor by one more power. The rate of convergence is clearly large if λ_k is an isolated eigenvalue.

While the above discussion holds for general matrices S, our concern for efficiency has decreed that S will be tridiagonal, Hessenberg, or possibly banded. Thus we should solve (13.40) with algorithms that are stable and efficient for the form of our matrix. With tridiagonal and Hessenberg forms, we may use Gaussian elimination (triangularization) with row interchanges to avoid small pivots. At the kth stage there will only be two rows involved, the kth and $(k + 1)$st rows, all lower elements in the kth column already being zero. For Hessenberg matrices each iteration will require $n^2/2$ multiplications while tridiagonal need only $6(n - 1)$.

Example of interchanging

An example seems desirable, the interchanging technique being required in several algorithms. We take the equation system provided by our oscillating mass problem (p. 205), having subtracted an approximate λ of 6.0 (the correct value is 6.29). The right side is arbitrary.

$$\begin{bmatrix} -1 & -2 & \\ -2 & -3 & -1 \\ & -1 & -5 \end{bmatrix} \begin{bmatrix} 1 \\ 0 \\ 0 \end{bmatrix}$$

Immediately, the first column demands an interchange, recorded below the column.

$$\begin{bmatrix} -2 & -3 & -1 \\ -1 & -2 & \\ & -1 & -5 \end{bmatrix} \begin{bmatrix} 0 \\ 1 \\ 0 \end{bmatrix}$$
$$(1, 2)$$

Standard elimination technique follows, the multiplication constant m_2 being recorded to the left of the row.

$$\begin{array}{r} \\ \hline m_2 = 0.5 \\ \\ \end{array} \begin{bmatrix} -2 & -3 & -1 \\ \hline 0 & -0.5 & 0.5 \\ & -1 & -5 \end{bmatrix} \begin{bmatrix} 0 \\ \hline 1 \\ 0 \end{bmatrix}$$
$$(1, 2)$$

Again an interchange is needed because of the second column. We leave m_2 in place, obtaining

$$\begin{array}{r} \\ \hline m_2 = 0.5 \\ \\ \end{array} \begin{bmatrix} -2 & -3 & -1 \\ \hline & -1 & -5 \\ 0 & -0.5 & 0.5 \end{bmatrix} \begin{bmatrix} 0 \\ \hline 0 \\ 1 \end{bmatrix}$$
$$(1, 2) \quad (2, 3)$$

The next multiplication constant, m_3, also happens to be 0.5 and is duly recorded while the elimination is performed.

$$\begin{array}{l} \\ m_2 = 0.5 \\ m_3 = 0.5 \end{array} \left[\begin{array}{ccc} -2 & -3 & -1 \\ & -1 & -5 \\ & 0 & 3.0 \end{array}\right] \left[\begin{array}{c} 0 \\ 0 \\ 1 \end{array}\right] \longrightarrow \left[\begin{array}{c} \frac{7}{3} \\ -\frac{5}{3} \\ \frac{1}{3} \end{array}\right] = \left[\begin{array}{c} 1 \\ -0.714 \\ 0.143 \end{array}\right] \quad \text{cf.} \left[\begin{array}{c} 1 \\ -0.645 \\ 0.122 \end{array}\right]$$

$$(1, 2)\ (2, 3)$$

The back solution is now performed exactly as if no interchanging had taken place. (We have not altered the definitions of our unknowns, merely altered the order in which our equations were employed. In effect we rearranged them *before* we solved them.) The answer vector finally is normalized for comparison with the correct eigenvector.

We now carry out the same process except that we use our current vector as the right-hand side. The elimination operations on the matrix need not be repeated, but the interchanges and m's must be applied to the new right side.

$$\begin{bmatrix} 1 \\ -0.714 \\ 0.143 \end{bmatrix} \xrightarrow{(1,2)} \begin{bmatrix} -0.714 \\ 1 \\ 0.143 \end{bmatrix} \xrightarrow{m_2} \begin{bmatrix} -0.714 \\ 1.357 \\ 0.143 \end{bmatrix} \xrightarrow{(2,3)} \begin{bmatrix} -0.714 \\ 0.143 \\ 1.357 \end{bmatrix}$$

$$\xrightarrow{m_3} \begin{bmatrix} -0.714 \\ 0.143 \\ 1.286 \end{bmatrix} \xrightarrow[\text{solution}]{\text{back}} \begin{bmatrix} 3.579 \\ -2.288 \\ 0.429 \end{bmatrix} \xrightarrow[\text{ized}]{\text{normal-}} \begin{bmatrix} 1 \\ -0.639 \\ 0.120 \end{bmatrix}$$

Carrying out the operation once more,

$$\begin{bmatrix} 1 \\ -0.639 \\ 0.120 \end{bmatrix} \xrightarrow{(1,2)} \begin{bmatrix} -0.639 \\ 1 \\ 0.120 \end{bmatrix} \xrightarrow{m_2} \begin{bmatrix} -0.639 \\ 1.320 \\ 0.120 \end{bmatrix} \xrightarrow{(2,3)} \begin{bmatrix} -0.639 \\ 0.120 \\ 1.320 \end{bmatrix}$$

$$\xrightarrow{m_3} \begin{bmatrix} -0.639 \\ 0.120 \\ 1.260 \end{bmatrix} \xrightarrow[\text{solution}]{\text{back}} \begin{bmatrix} 3.440 \\ 2.220 \\ 0.420 \end{bmatrix} \longrightarrow \begin{bmatrix} 1 \\ -0.645 \\ 0.122 \end{bmatrix}$$

We obtain the correct eigenvector, even though the λ we subtracted from our matrix was not 6.29 but merely 6.0.

Eigenvector for close eigenvalues

A brief reflection upon the geometry of a quadratic form will reveal that eigenvectors corresponding to multiple eigenvalues are not uniquely determined. The ellipsoid has circular cross sections in that subspace and hence *any* direction in that cross section is an axis. By the same token, pathologically close eigenvalues imply cross sections that, while ellipsoidal, are very nearly circular. Thus their eigenvectors are very poorly determined. This is an essential geometric fact and no amount of algorithmic thrashing will change it. Thus we should not be surprised to learn that even inverse iteration, efficient as it is for other eigenvectors, will not precisely determine eigenvectors for pathologically close eigenvalues. The computational problem remains unsolved. We currently settle for rather crude component values of such vectors.

CHAPTER 14

ROOTS OF EQUATIONS

Apologia pro opera nostro

If there exists any one reliable algorithm for finding the roots of transcendental equations it is yet to be found. We have a variety of medicines that work with varying degrees of potency (including zero!), but the state of the art still precludes the confident writing of computational prescriptions without having looked over the patient rather closely. In Chapter 2 we discussed the one-dimensional problem. The principal advice given there might tersely be summarized as:

1. Search the 1-space until you detect the presence of a root; then,
2. Switch to an appropriate algorithm, probably Newton's, to find the root accurately.
3. If possible, remove the root, then go to step 1.

There were other details. The search can usually be greatly simplified if one first examines the geometry of the function; correlatively, stepping searches without a priori information are possible but not particularly attractive; and multiple roots require special algorithms if efficiency (though not convergence) is to be preserved.

When we turn to *systems* of equations, seeking their roots in two or more dimensions, the advice is much the same though harder to apply. A search for a lost dog on a foggy country lane is much easier than a search for the same dog in a foggy field, and if the lost pet is a squirrel in the forest, his three-dimensional capability complicates the task by still another order of magnitude. As for finding ghosts, reputed to have at least four-dimensional existences, we defer to our British colleagues, pleading inexperience.

A system of three functions of three independent variables occasionally has a transparent geometry that locates the regions of its roots, and we should certainly seek such help – but all too often it remains a rather obscure system of three functions of three independent variables. If we had enough computer time (that is, money) we might search the 3-space by small steps with some hope of finding at least most of the roots, but the prospect does not please.

If your author were writing another typical book on Numerical Analysis, he would content himself with depicting Newton's method in n dimensions, examine its error terms, make a recommendation or two, then pass quietly on to more tractable topics. The student would thereby gain a false sense of security that could only evolve into an unpleasant conviction that the author had never tried his own medicine on real problems. At the risk of frightening off the timid, he prefers to make quite clear his conviction that *roots of systems of equations almost always require special methods* specifically tailored to the system at hand. To aid the bold he attempts to classify the types of problems and to point out their salient features that might profitably be used in constructing such special methods – whether or not such methods currently exist.

Lest the reader be too depressed by admittedly gloomy pronouncements concerning the difficulties of locating roots of systems of equations, we would emphasize that most such systems arriving at a computation laboratory are not isolated problems. Usually there is some experimental process, each experiment generating a set of equations that we are asked to solve. Hence we have a whole *family* of equation systems, each member rather similar to the last. The experience gained in solving the first system naturally accrues to the solver of the second – be he man or machine. Indeed, this is the single most important source of information about the probable location of the roots: They are close to the roots of a similar system solved yesterday or perhaps only a few seconds ago. Thus one important piece of advice is to *order the equation systems* on their parameter values so that each system follows one that is near to it, in some sense, in the parameter space, then work hard on *one* typical system by hand and by machine, using intuition, analysis, or even a dart board, until this typical system is solved. The experience gained in messing around with the one system is invaluable for understanding the structure of the others. Further, it provides some reasonable starting values for the searches that will probably be the first steps of the next problem – reducing, if not entirely eliminating, costly n-dimensional wanderings through foggy Numberland.

Types of problems

Since the algorithms by which we locate roots usually have clear geometric interpretations, it seems useful to classify root problems by their geometries. Taking our cue from two-dimensional systems we find it useful to distinguish between problems that have a large number of clustered roots versus those with isolated roots. These terms are, of course, relative. We can always expand the scale of a region so that roots that were once clustered are now isolated, provided that they were not actually multiple roots. But every problem has a natural scale, a region outside of which we have no immediate interest, and we use the term *clustered* versus *isolated* in the sense of a bird's-eye view that embraces most of this interesting region. Thus, one of our first examples in Chapter 2, the intersections of

$$y = \tan x$$

$$y = \frac{\alpha}{x}$$

gives us the archtype of the isolated root problem. There is here, indeed, an infinity of roots, but it is unlikely that we will have an interest in more than the first five or ten. On this range not only are the roots isolated, but we have very good information about their approximate location—slightly northeast of the points on the x axis at $2n\pi$. More precisely, they lie near

$$x = 2n\pi + \frac{\alpha}{2n\pi}$$

$$y = \frac{\alpha}{2n\pi}$$

With this kind of starting information we have no other difficulties; any sensible root finding algorithm will succeed admirably.

Clustered roots

As an example of clustered roots consider Figure 14.1, where the roots are shown as large dots. These are the solutions of the system

$$f(x, y) = (y - \tan^{-1}x)\left(y + \frac{b}{x}\right) = 0$$

$$g(x, y) = [y - (x^2 + 1)^{-1}]\left[\sin\left(\frac{4}{x^2 + y^2 + a}\right)\right] = 0$$

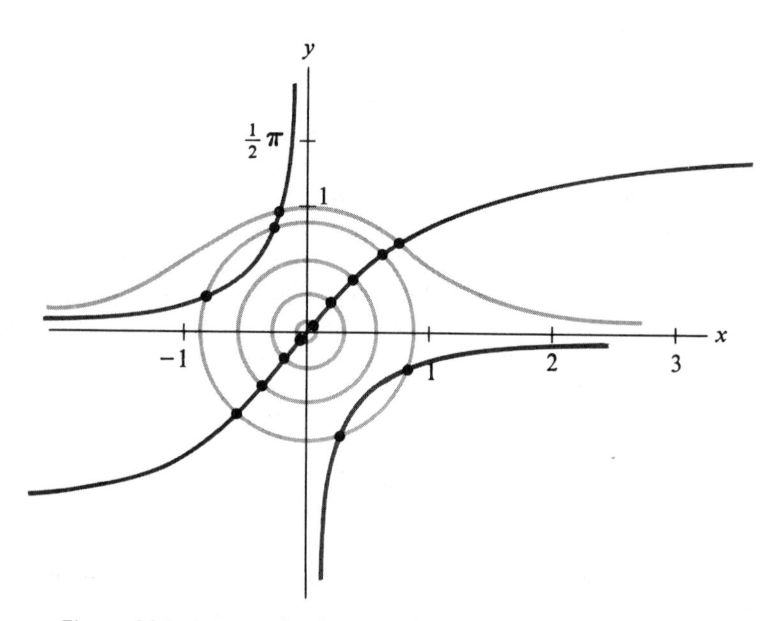

Figure 14.1. An example of an equation system with clustered roots

This example is, of course, artificial. Each function was constructed by multiplying together two simpler functions to acquire the union of their zero curves. In the unlikely event that one should encounter the problem in real life, the reasonable strategy is to break both f and g into their separate factor functions, then solve the four subproblems that arise by taking one subfunction from each set. In this way the clustering of the roots is sharply reduced, though not eliminated.

Still, the example can be instructive. The geometric difficulties inherent in clustered root configurations are ones of mutual interference. If we think of Figure 14.1 as a contour map on which only the sea-level lines are shown, we feel that the mountains and valleys of our map cannot be very high or steep. Since it is the slopes of these mountains that most frequently drive our root-seeking algorithms, we are deprived of our principal mechanism except possibly in the immediate vicinity of each root. A slope $\frac{1}{2}$ inch away from the root that we seek will probably have very little to do with *that* root and only a little to do with some of the nearer roots. The surface just wiggles too rapidly to contain useful macro information. This problem, however, does contain one geometric feature that can be employed to find the roots—assuming that the factored structure is not discovered. One curve, $\tan^{-1}x$, proceeds steadily through the figure encountering many of the roots. We can therefore employ a curve follower to march along the arctangent, testing the sign of the other function as it goes. We thus reduce the finding of this set of roots to a *one*-dimensional problem, a highly desirable simplification. The remaining roots, it happens, all lie on another curve, the hyperbola, whose two branches may be followed in the same spirit. Sensible ways of following two-dimensional curves have been proposed by Kuiken [1968], which involve modifications of the Runge-Kutta scheme to ensure stability—that is, a slight step off the curve produces pressures that force one back onto the curve again. It is a sensible answer to an unlikely problem.

Unvisualizable roots

A third type of geometry might be called the *unvisualizable*. Consider the least-squares fitting problem in which 10 pairs of points (x_i, y_i) are to be fitted by the three-parameter (a, b, C) equation

$$y = e^{-ax} + Ce^{-bx} \tag{14.1}$$

in such a way that the sum of the squared deviations is minimized. We have

$$S(a, b, C) = \sum_{i=1}^{10} (y_i - e^{-ax_i} - Ce^{-bx_i})^2$$

from which we immediately get the three "normal" derivative equations

$$\begin{aligned} \sum_{i=1}^{10} F_i \cdot x_i e^{-ax_i} &= 0 \\ \sum_{i=1}^{10} F_i \cdot x_i e^{-bx_i} &= 0 \\ \sum_{i=1}^{10} F_i \cdot e^{-bx_i} &= 0 \end{aligned} \tag{14.2}$$

where

$$F_i = y_i - e^{-ax_i} - Ce^{-bx_i}$$

In this problem the root is isolated, but the equations (14.2) that determine it are quite ill conditioned. The student might try his hand at solving the system (14.2) by any method he fancies—first making up a set of (say) 10 (x, y) pairs from (14.1) using values of x on the range $(0, 1)$ and the values of $(1, 3, 4)$ for his actual values of (a, b, C). He will be surprised to find how flat his $S(a, b, C)$ surface really is and how nearly zero it can be for trial values of (a, b, C) that are nowhere near the values he used to construct his (x, y) data.

Quite aside from the fact that this is a poorly posed problem, your author must agree with the man who complains that he cannot visualize a priori the geometry of the three surfaces defined by (14.2). After some computation a picture begins to emerge but it is not very helpful. The problem is fairly described as unvisualizable.

The utility of Newton's method

For isolated n-dimensional root geometries combined with starting values that are noticeably nearer one root than any of the others, almost any reasonable method will converge to the root. If explicit first derivatives are calculable, Newton's method is our first choice. It is straightforward and converges quadratically. This rapidity of convergence will usually more than offset the additional calculational labor required by its n^2 first derivatives over methods that require merely the n functions at $n + 1$ or more points. Indeed, the labor of evaluating the derivative of a function at the same time and place as the function itself usually costs only a small additional fraction of computer time. The principal labor in computing a function goes into evaluations of subfunctions such as e^{-x}, $\cos ax$, and so on, that are also the subfunctions of the derivative and hence need not be reevaluated.

When derivative information cannot be computed directly we must, of course, fall back on methods that do not require it. False Position (*regular falsi*) in several dimensions is one reasonable possibility with specialized iterative techniques lurking just behind. *Generalized* functional iterations, by which we refer to methods of form

$$x = F(x)$$

that take no cognizance of the specialized geometry of the particular problem, are not as satisfactory in the sense that they converge more slowly without usually delivering an appreciably wider region of convergence than the other schemes.

When roots are clustered, there seems no *general* method for getting into the immediate neighborhood of the root short of a massive detailed search on the n-dimensional cube, although many problems have special geometric structures that permit more efficient searches in fewer dimensions—surface or line crawling along one or more of the functions, for example (p. 379). Once the root has been approached quite closely, of course, Newton's method reasserts its usefulness in extracting the root to high accuracy, but this is not our principal concern. Getting close to a particular root in a cluster of roots is the real problem. For this job neither Newton nor anybody else seems particularly suited.

Roots by minimization—a warning

Probably the most attractive *formal* solution to finding roots of systems of equations is to convert the problem to that of minimizing a positive function.

Anyone can take the system

$$\begin{aligned} f_1(x, y, z) &= 0 \\ f_2(x, y, z) &= 0 \\ f_3(x, y, z) &= 0 \end{aligned} \tag{14.3}$$

and observe that the function

$$S(x, y, z) = f_1{}^2(x, y, z) + f_2{}^2(x, y, z) + f_3{}^2(x, y, z) \tag{14.4}$$

has its zeros in one-to-one correspondence with the solutions of (14.3). Furthermore, S is otherwise positive—so all we have to do is start somewhere and go downhill to the nearest minimum. If the minimum is zero we have a solution to (14.3). Then we begin somewhere else and repeat the process until we have all the solutions. It sounds easy, too easy!

The troubles are manifold:

1. Unsophisticated minimizing procedures will tend to find some minima repeatedly, ignoring others that often are close by. Thus much of our computational labor is apt to be redundant.
2. Since we have no assurance at any time from the minimizing process that we have found all the solutions, we tend to keep looking forever. A priori information that there are exactly 13 roots will relieve but not remove this type of trouble because:
3. False zeros tend to be accepted in the belief that a small minimum is not zero only because of computational imprecision.
4. Minimizing procedures are inherently inefficient because (a) they get confused by saddle points, and (b) the valley around the true zero can be very flat for a long distance, thereby providing poor gradients to drive the procedure quickly or even in the proper direction. See Chapter 17.

In spite of these difficulties, your author must admit that, as a court of last resort, minimizing methods will have to be condoned simply because some problems yield to nothing else. But the man who turns to use them should first make every effort to apply root-finding processes specific to his problem. Otherwise, he may receive an unpleasantly large computational bill at the end of the month.

Newton's method in *n*-dimensions

Formally, Newton's method in n dimensions is identical with its one-dimensional version except that there are n functions in n independent variables.

We wish to solve the system

$$\begin{aligned} f_1(x_1, x_2, \ldots, x_n) &= 0 \\ f_2(x_1, x_2, \ldots, x_n) &= 0 \\ \vdots \qquad\quad & \vdots \\ f_n(x_1, x_2, \ldots, x_n) &= 0 \end{aligned} \tag{14.5}$$

or, more succinctly in vector notation,

$$\mathbf{f}(\mathbf{x}) = 0 \tag{14.6}$$

Having a current point $\{a_i\}$ firmly in hand we merely expand each function about $\mathbf{a}$ in n-space via Taylor's series, retaining only linear terms. Then we set this expansion equal to zero, giving

$$\begin{aligned} 0 &= f_1(\mathbf{a}) + \frac{\partial f_1}{\partial x_1}\delta_1 + \frac{\partial f_1}{\partial x_2}\delta_2 + \cdots + \frac{\partial f_1}{\partial x_n}\delta_n \\ 0 &= f_2(\mathbf{a}) + \frac{\partial f_2}{\partial x_1}\delta_1 + \frac{\partial f_2}{\partial x_2}\delta_2 + \cdots + \frac{\partial f_2}{\partial x_n}\delta_n \\ &\vdots \\ 0 &= f_n(\mathbf{a}) + \frac{\partial f_n}{\partial x_1}\delta_1 + \frac{\partial f_n}{\partial x_2}\delta_2 + \cdots + \frac{\partial f_n}{\partial x_n}\delta_n \end{aligned} \tag{14.7}$$

where the δ_i represent deviations from $\mathbf{a}$. After solving this set of linear equations for the $\{\delta_i\}$ we compute our new approximation to the root from

$$x_i = a_i + \delta_i \qquad i = 1, \ldots, n$$

For some purposes expository convenience is well served by more compact notations and the system (14.7) is often written

$$\begin{aligned} 0 &= f_1 + (\mathbf{grad\ f}_1)^T \cdot \boldsymbol{\Delta} \\ 0 &= f_2 + (\mathbf{grad\ f}_2)^T \cdot \boldsymbol{\Delta} \\ &\vdots \\ 0 &= f_n + (\mathbf{grad\ f}_n)^T \cdot \boldsymbol{\Delta} \end{aligned} \tag{14.8}$$

where

$$\boldsymbol{\Delta} = \{\delta_i\}$$

and

$$\mathbf{grad\ f}_j = \left\{\frac{\partial f_i}{\partial x_i}\right\} \qquad i = 1, \ldots, n$$

or even more compactly, though less usefully, by

$$0 = \mathbf{f}(\mathbf{a}) + G \cdot \mathbf{\Delta}$$

where G is the square matrix whose rows are the vectors $(\mathbf{grad\ f}_i)^T$. In two dimensions, however, your author prefers to use small literal subscripts to denote partial differentiation, in which Newton's method may be written

$$0 = f_0 + f_x \cdot \delta + f_y \cdot \varepsilon$$

$$0 = g_0 + g_x \cdot \delta + g_y \cdot \varepsilon$$

– a less generalizable but more appealing notation.

Geometrically, Newton's method fits tangent (hyper)planes to each surface in $(n + 1)$-space. It then solves for *their* intersections instead of the intersection of the original surfaces. If the planes represent the surfaces well in the region between where we are and the root we seek, the method will succeed admirably. Unfortunately, a blind planar replacement at a random point in $(n + 1)$-space has little chance to be a good fit, so Newton's method finds its chief use *after* we have got ourselves close to a root. The difficulties of locating roots in n-space without prior knowledge have been commented upon earlier.

One dimensional Newton's method in *n*-space

If we are given a particular direction in n-space, a vector $\mathbf{v}$, then we may apply Newton's method to *one* function. The problem is one-dimensional, so in spite of superficial n-dimensional appearances we need not fear it. The seeking of a zero of $f_1(x_1, \ldots, x_n)$ along the direction $\mathbf{v}$ is a quite ordinary problem. Formally we have

$$f_1(\mathbf{x}) = f_1(\mathbf{a}) + \mathbf{grad\ f}_1 \cdot \mathbf{\Delta} + \tfrac{1}{2}\mathbf{\Delta}^T Q \mathbf{\Delta} + \cdots \qquad \mathbf{\Delta} = \lambda \mathbf{v} \tag{14.9}$$

but now both $\mathbf{grad\ f}_1$ and $\mathbf{v}$ are known. Thus Newton's method here consists of solving for the scaler λ in

$$0 = f_1(\mathbf{a}) + (\mathbf{grad\ f}_1 \cdot \mathbf{v})\lambda \tag{14.10}$$

Indeed, we may even retain the quadratic terms in (14.9). The square matrix Q is the array of all second partial derivatives of f:

$$Q = \begin{bmatrix} \dfrac{\partial^2 f}{\partial {x_1}^2} & \dfrac{\partial^2 f}{\partial x_1\, \partial x_2} & \cdots & \dfrac{\partial^2 f}{\partial x_1\, \partial x_n} \\ \vdots & & & \vdots \\ \dfrac{\partial^2 f}{\partial x_n\, \partial x_1} & \dfrac{\partial^2 f}{\partial x_n\, \partial x_2} & \cdots & \dfrac{\partial^2 f}{\partial {x_n}^2} \end{bmatrix} \tag{14.11}$$

Thus the quadratic form in (14.9) becomes merely

$$\tfrac{1}{2}\mathbf{v}^T Q\mathbf{v} \cdot \lambda^2 = c\lambda^2$$

which is a simple number, c, multiplying λ^2. If the labor of evaluating the n^2 second partial derivatives is not too severe, we may find this quadratic term useful—especially near minima in f_1 along $\mathbf{v}$. Remember, however, that one may always *estimate* c by evaluating f_1 at the three equally spaced points along $\mathbf{v}$, then, taking the second difference

$$\frac{\Delta^2 f}{2\delta^2} \approx c$$

where δ is the spacing. This tactic proves to be much more economical in large numbers of dimensions.

"Removing" a known root

In one-dimensional problems with several roots we shall often do well to remove a root as soon as it has been found, especially if other roots are expected to lie near by. This removal prevents the presence of the known root from distracting the algorithm while it seeks the next root. The device is straightforward: We define a new function

$$f_1(x) = \frac{f(x)}{x - x_0}$$

where x_0 is the known root. The Newton correction for seeking a simple root of f_1 is closely related to the same term for the function f. We have

$$\frac{f_1'}{f_1} = \frac{f'}{f} - \frac{1}{x - x_0}$$

which effectively suppresses simple roots, although alternative computations must be provided if x_0 is expected to be a multiple root. (Why?)

In two and higher dimensions, we have no strictly analogous device for removing known roots. The difficulties arise because of the diversity of geometries exhibited by roots in several dimensions. Consider Figure 14.2 with a root at the intersections of

$$f(x, y) = g(x, y) = 0$$

Since the geometry of the $f(x, y)$ and $g(x, y)$ surfaces is not shown, we are tempted to assume them as depicted in Figure 14.3—surfaces that pass simply through the (x, y) plane to produce the curves of Figure 14.2. But Figure 14.2

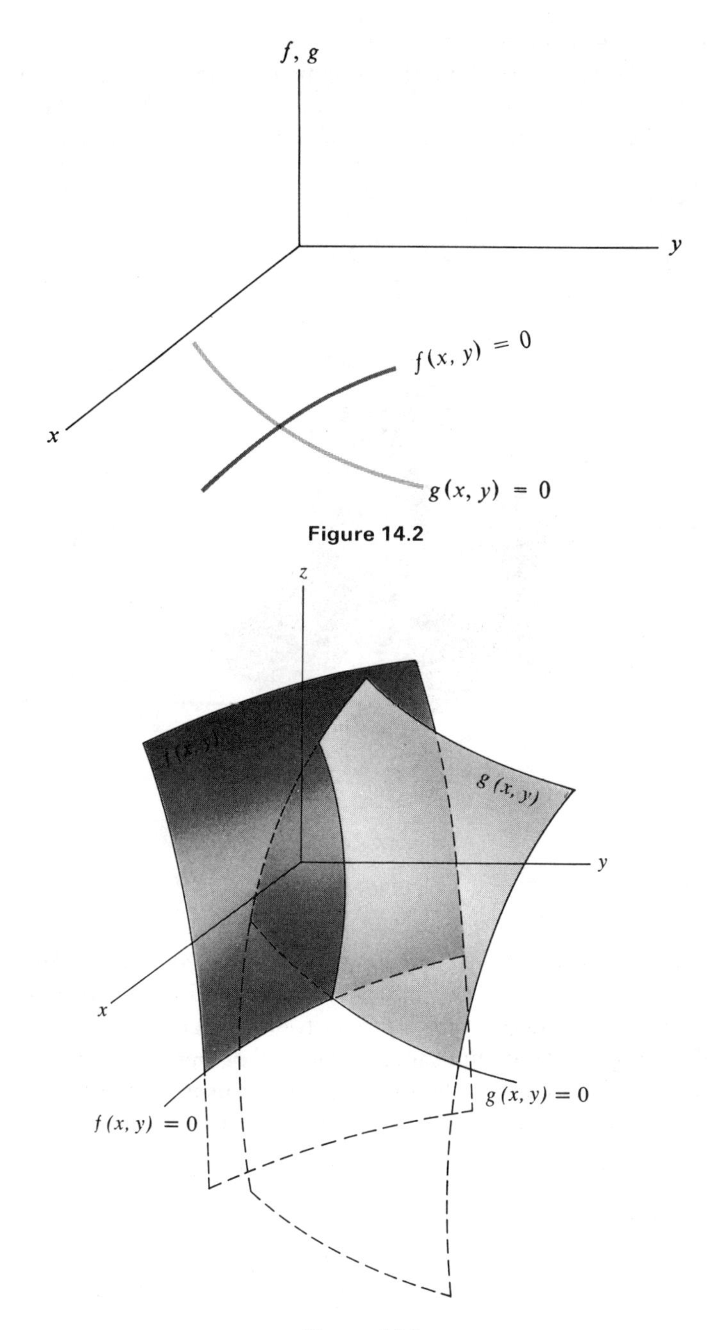

Figure 14.2

Figure 14.3

could just as well have been produced by the geometry of Figure 14.4 – in which $g(x, y)$ is never negative, touching the (x, y) plane tangentially in all directions along the curve $g = 0$. For this geometry our root is clearly double; a slight lowering of the $g(x, y)$ surface splits the $g = 0$ curve into *two* nearly parallel

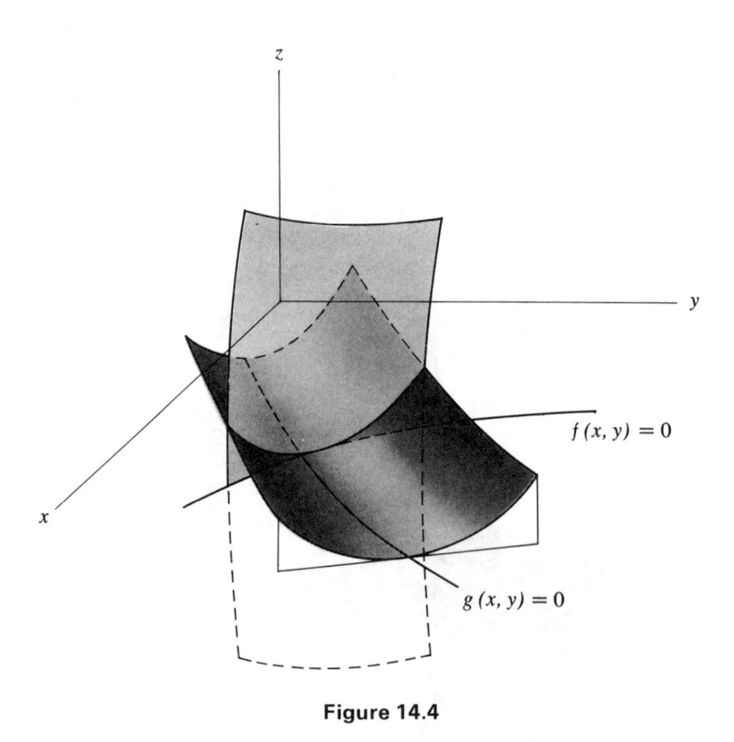

Figure 14.4

curves – each intersecting $f = 0$ in a simple root. Since one of our purposes in removing a root is to permit the finding of its twin if the root is multiple, we must be careful. Dividing $g(x, y)$ by a proper root factor is all right, but dividing $f(x, y)$ will remove the entire multiple root instead of merely one of the siblings.

More immediately arises the question of the proper factor by which to divide. If we look at the Taylor expansion around the root,

$$f(\delta, \varepsilon) = f_x \cdot \delta + f_y \cdot \varepsilon + (\text{quadratic terms } \delta, \varepsilon)$$

we might decide that

$$f_1(x_0, y_0) = \frac{f(x, y)}{(x - x_0)f_x + (y - y_0)f_y}$$

was suitable. If the root is simple, it certainly makes $f_1(x_0, y_0)$ equal to unity there. Unfortunately, it creates a *line* in the (x, y) plane (Figure 14.5) along which $f_1(x, y)$ is infinite. This line is tangent to $f(x, y) = 0$ at the old root and hence, if the curve is fairly straight, introduces very unpleasant singular geometries quite close to locations where the other roots are most apt to occur.

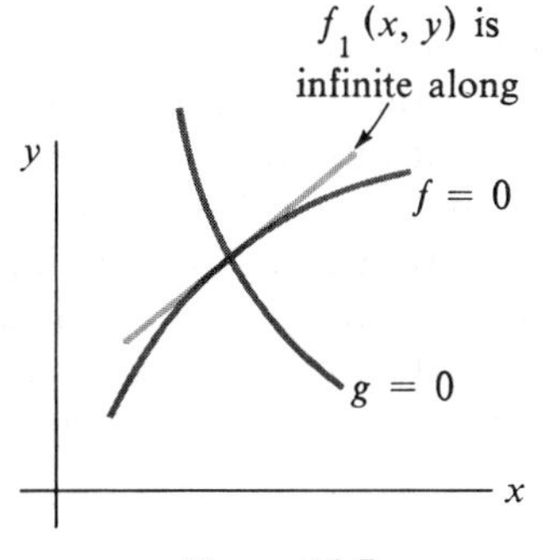

Figure 14.5

We can swing our line around to be normal to f by using

$$(x - x_0)f_y - (y - y_0)f_x$$

as the divisor, but we still have "removed" a *line* from the (x, y) plane, where we really want to remove only a *point*. Similar remarks hold, of course, for $g_1(x, y)$ should we opt for removal of our root through that function. Or should we, perhaps, remove the root from both functions?

A more reasonable strategy seems to be simple removal of the point (x_0, y_0), or rather a small disc centered at (x_0, y_0) whose radius R_0 is at one's disposal (Figure 14.6). We can then define our new function to have its old

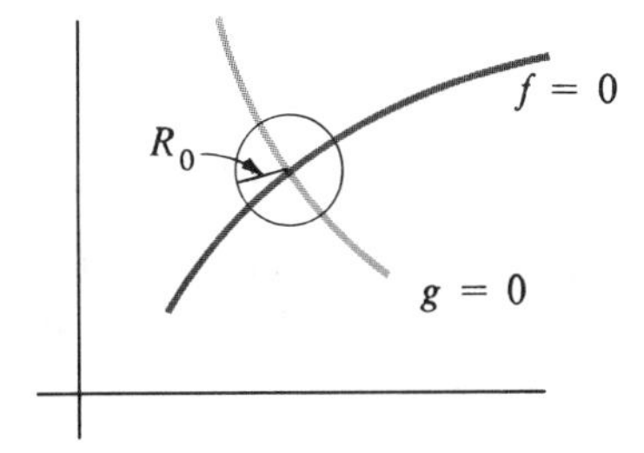

Figure 14.6

The corresponding version of false position replaces the surface

$$z = f(x, y)$$

by a plane that coincides with it at three points (Figure 14.9). The intersection of this plane with the (x, y) plane temporarily replaces $f(x, y) = 0$ in our original problem. Similarly, another plane replaces the g surface and the

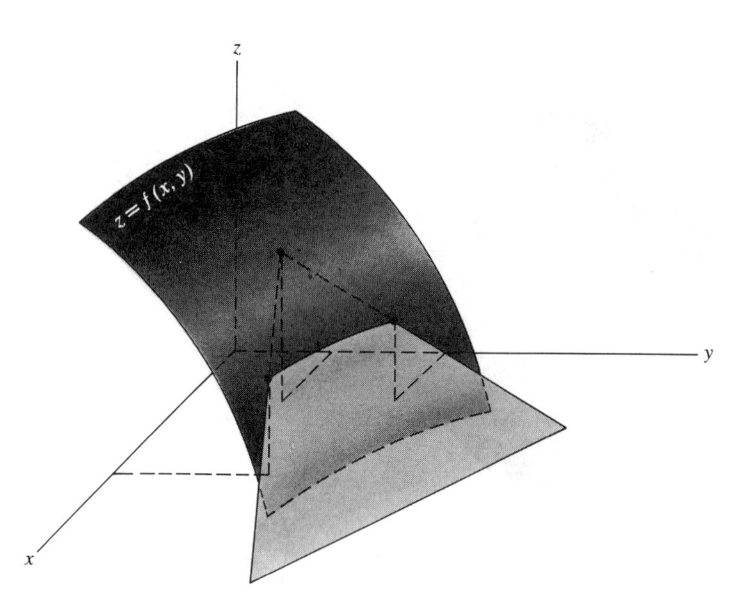

Figure 14.9. Fitting a plane to a surface at 3 points

problem then solved is the common intersection of these two planes with (x, y) plane. As with Newton's method, there is no logical necessity that the three points used for fitting be the same for both surfaces, but this choice considerably simplifies the algorithm. We adopt this version in the sequel. Our algorithm is:

1. Using three given points in the (x, y) plane, fit planes to each of the two surfaces, f and g, and determine the intersection of these two planes with the (x, y) plane—a point (x_4, y_4).
2. Throw away one of the first three points, and substitute the new point for it. Repeat steps 1 and 2 to (hopefully) convergence.

The process just described still contains several degrees of freedom in the way that we choose the point to discard. As is geometrically obvious, the process fails through indeterminacy if all three points in the (x, y) plane ever become collinear, and presumably a nearly collinear condition produces near failure via instability. So we might wish to discard that old point which maximizes some measure of noncollinearity. Alternatively, in keeping with the false-position strategy of always bracketing the root between a plus and a minus functional value, we might insist that our three points never all lie on one side of either of the two curves. Intuitively, these two criteria will tend to reinforce one another, so we initially adopt the latter as easier to implement.

The two curves divide the (x, y) plane into four "quadrants" with the root at their "origin." While we may not know the exact location of the boundaries, any point may be easily placed in its proper quadrant merely by examining the

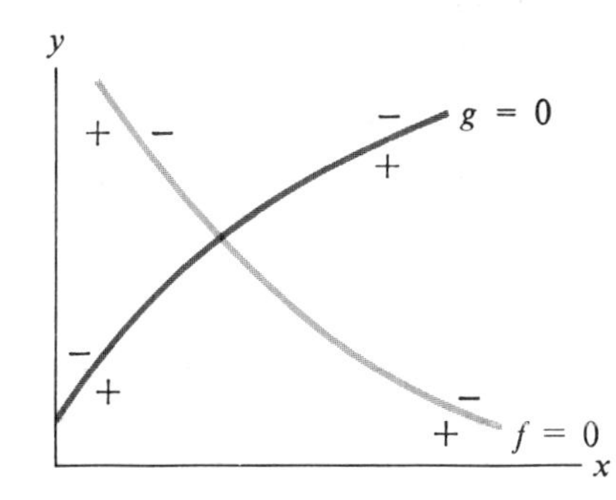

Figure 14.10. The division of the (x, y) plane into quadrants by the curves $f(x, y) = 0$ and $g(x, y) = 0$

signs of f and g there (Figure 14.10). Any group of three points that we admit to the first step of our algorithm will be either:

1. Distributed in three of the four quadrants, or
2. Distributed in two *opposite* quadrants.

Any other distribution would place all three on one side of at least one of our curves, violating the fundamental requirement of false position that the root be bracketed. If three quadrants are occupied, either the fourth point will fall in the fourth quadrant, in which case any of the old points may be discarded, or it will fall in a quadrant that already contains one point, whence that old point may be discarded. If two opposite quadrants are occupied, the new point may fall in an already occupied quadrant, in which case a point in this same quadrant may be discarded, or it will fall in an unoccupied quadrant, in which case any one of the old points may be discarded. These rules still leave some choice that one might desire to resolve by invoking the measure of noncollinearity. Such a energy in staying close to the function where it holds no particular interest for us. The technique is slow, sure, and expensive. In one-dimensional root seeking we can devise more efficient algorithms that do not significantly sacrifice stability for all but pathologic functions. These have been discussed in Chapter 2 and they usually terminate with an application of Newton's method. As the number of dimensions increases, however, the geometries for which Newton is effective are progressively rarer, so we become more and more reconciled to the cost of the slower but safer algorithms. Just when to abandon bold Linear Leaps for timid Curve Crawlers is an imprecise economic decision. For a set of similar problems the answer must usually be found by judicious experimentation with one of the set.

The geometric strategy of curve crawling is easy to describe. Once we possess a point that lies in the surface $f_1 = 0$ we wish to move to the second surface $f_2 = 0$ without leaving the first. Our current point A lies in some surface f_2 = constant (shown as $f_2 = 1.73$ in Figure 14.11). An obvious gambit

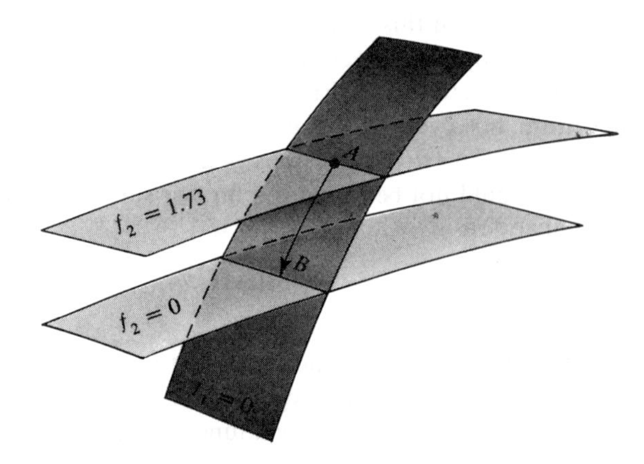

Figure 14.11

is to take the negative gradient of f_2 (which would usually lead us to $f_2 = 0$ quickly if we had no other constraints)—but first removing from it any component that is parallel to the gradient of f_1 at A. Without this component the remaining vector is tangent to the f_1 surface, and hence small motions along it will keep us on $f_1 = 0$ but will also decrease f_2. After a small step we re-evaluate the two gradients and repeat. Formally, we orthogonalize a negative gradient $(-\mathbf{g}_2)$ with respect to the gradient $\mathbf{g}_1$ to get a marching direction, $\mathbf{m}$. We may write

$$\mathbf{m} = -\mathbf{g}_2 + \frac{\mathbf{g}_2 \cdot \mathbf{g}_1}{\mathbf{g}_1 \cdot \mathbf{g}_1}\mathbf{g}_1 \tag{14.15}$$

and, if the components of **m** are m_1, m_2, and m_3, we wish to integrate

$$\left.\begin{aligned} \frac{dx}{d\lambda} &= m_1 \\ \frac{dy}{d\lambda} &= m_2 \\ \frac{dz}{d\lambda} &= m_3 \end{aligned}\right\} \quad \text{or} \quad \frac{d\mathbf{x}}{d\lambda} = \mathbf{m} \tag{14.16}$$

where the second form has been written in terms of the position vector **x** with components (x, y, z). Since

$$\mathbf{g}_1 = \left\{ \frac{\partial f_1}{\partial x} \quad \frac{\partial f_1}{\partial y} \quad \frac{\partial f_1}{\partial z} \right\}$$

and

$$\mathbf{g}_2 = \left\{ \frac{\partial f_2}{\partial x} \quad \frac{\partial f_2}{\partial y} \quad \frac{\partial f_2}{\partial z} \right\}$$

we simply use (14.15) and (14.16) in whatever differential equation integrator we prefer and let it go until f_2 becomes zero (at B in Figure 14.11, but see Figure 14.13).

Since our example is 3-dimensional, the next phase is to curve crawl up the intersection of $f_1 = 0$, $f_2 = 0$. Here we take any vector that points more or less toward $f_3 = 0$, $(-\mathbf{g}_3)$ is apt to be our choice, and remove from it those components that are perpendicular to the two surfaces in which we wish to remain. Here we must orthogonalize with respect to $\mathbf{g}_1$ and $\mathbf{g}_2$ *simultaneously* for $\mathbf{g}_1$ and $\mathbf{g}_2$ are not generally themselves mutually perpendicular. We thus wish to find scalers α and β such that an **m** defined by

$$\mathbf{m} = \mathbf{g}_3 + \alpha\mathbf{g}_1 + \beta\mathbf{g}_2$$

is perpendicular to both $\mathbf{g}_1$ and $\mathbf{g}_2$. Taking the dot product of **m** with $\mathbf{g}_1$ and $\mathbf{g}_2$ separately we have

$$a_{11}\alpha + a_{12}\beta = a_{13}$$

$$a_{12}\alpha + a_{22}\beta = a_{23}$$

where

$$a_{ij} = \mathbf{g}_i \cdot \mathbf{g}_j$$

Thus we can solve for (α, β), then compute **m** whose three components get used in (14.16) as before. Since our example is only three dimensional, **m** could have

We must emphasize that these philosophies need considerable elaboration before they become usable strategies. Even stable curve crawling cannot find a root that is not there. Consider Figure 14.13 where the sphere intersects with both of the other surfaces but at no place do all three surfaces meet.

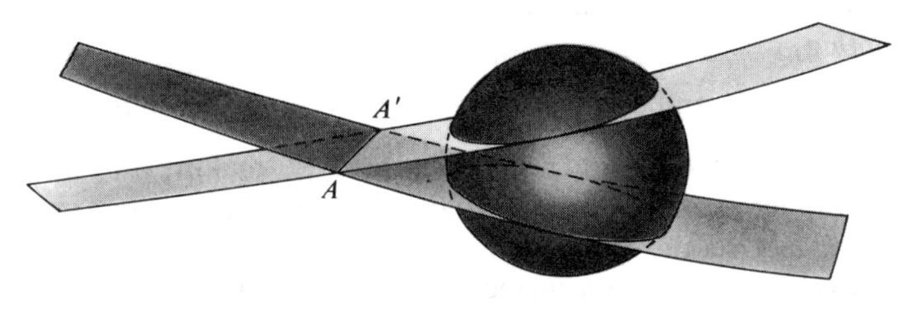

Figure 14.13

An automatic curve crawler could keep busy happily circumnavigating the globe for quite a while unless there were some tests for repetition. Likewise, if the process happened to descend to the intersection AA' it could crawl a long distance unless some allowable region were specified and our present location tested for being inside it.

Descent to a nearby subspace

In the course of seeking a new surface

$$f_{k+1}(\mathbf{x}) = 0$$

in n-space while attempting to maintain contact with the k surfaces

$$f_l(\mathbf{x}) = 0 \qquad l = 1, \ldots, k \tag{14.17}$$

already found, we shall frequently find that our algorithm has wandered somewhat astray. In its eagerness to reach f_{k+1} it probably has moved along a reasonable direction, a straight line that points toward the immediate goal. This line may have been carefully chosen to be tangent to the k surfaces already attained, but to go off on a tangent is often a wasteful procedure—though perhaps not as bad as the common usage of the phrase might suggest. It all depends on how far off we go. If our k surfaces are slowly curving and we pursue our tangent direction with moderation, we shall leave but not lose sight of the surfaces we desire to retain. For such geometries a reasonable strategy is to pursue the tangent line with some boldness, then take corrective action to get back on those k surfaces (14.17) that now lie nearby rather than at our feet. In so doing we shall often make faster progress than with the more conservative strategy that tries never to leave any of the k surfaces, for such timidity virtually assures that only small steps will be possible. Of course, if we fly off too far on our tangent line we shall lose touch with everything, wasting our previous progress toward whatever roots may have lain locally ahead. We have here the Scylla and Charybdis that lurk in all iterative processes: the inefficiency of the Slow Crawl versus the inefficiency of the Destructive Leap. We need a golden mean, a Cautious Canter, that unfortunately we do not know how to prescribe in precise terms.

To return to the nearby k-space

$$f_j(\mathbf{x}) = 0 \qquad j = 1, \dots, k \tag{14.18}$$

from which we have somewhat drifted, our fundamental device is Newton's method. The difficulty is that we wish to move to a k-subspace of our n-space. Putting it differently, we must determine n independent variables and we have only k conditions. The most straightforward tactic is to construct $n - k$ additional conditions in the form of planes that lie perpendicular to the k-space and to each other. Adding such a set of planes in effect creates a temporary solution or root to our problem that is nearby and hence is easy to find – and it lies on the surfaces (14.18) already previously attained. Thus one execution of Newton's method should correct the drift from these surfaces that was occasioned by enthusiastic following of a tangent direction slightly earlier.

Thc construction of our planes is straightforward. We observe that

$$b_1(x_1 - a_1) + b_2(x_2 - a_2) + \cdots + b_n(x_n - a_n) = 0$$

is a plane through the point $\{a_i\}$ with the normal vector $\{b_i\}$ where i runs from 1 through n. Since the point $\mathbf{a}$ is our current location, we only have to find a suitable vector $\mathbf{b}$ in order to write down the equation of such a plane. We illustrate for five dimensions in which we have strayed slightly from three surfaces and hence need two planes. We possess the gradients of our three surfaces. We call them $\mathbf{g}_1$, $\mathbf{g}_2$, and $\mathbf{g}_3$. We now choose a random vector $\mathbf{m}$ and make it orthogonal to these three, calling its result $\mathbf{r}$. We have

$$\mathbf{r} = \mathbf{m} - \alpha_1\mathbf{g}_1 - \alpha_2\mathbf{g}_2 - \alpha_3\mathbf{g}_3 \tag{14.19}$$

where the α_i are scalars to be found. Dotting (14.19) with respect to $\mathbf{g}_1$ and demanding that $\mathbf{r} \cdot \mathbf{g}_1$ be zero, we have

$$\mathbf{m} \cdot \mathbf{g}_1 = (\mathbf{g}_1 \cdot \mathbf{g}_1)\alpha_1 + (\mathbf{g}_1 \cdot \mathbf{g}_2)\alpha_2 + (\mathbf{g}_1 \cdot \mathbf{g}_3)\alpha_3$$

Similarly we obtain

$$\mathbf{m} \cdot \mathbf{g}_2 = (\mathbf{g}_1 \cdot \mathbf{g}_2)\alpha_1 + (\mathbf{g}_2 \cdot \mathbf{g}_2)\alpha_2 + (\mathbf{g}_2 \cdot \mathbf{g}_3)\alpha_3$$

$$\mathbf{m} \cdot \mathbf{g}_3 = (\mathbf{g}_1 \cdot \mathbf{g}_3)\alpha_1 + (\mathbf{g}_2 \cdot \mathbf{g}_3)\alpha_2 + (\mathbf{g}_3 \cdot \mathbf{g}_3)\alpha_3$$

which may be solved for the α_i's and then **r** computed from (14.19). Our plane is

$$r_1(x_1 - a_1) + r_2(x_2 - a_2) + \cdots + r_5(x_5 - a_5) = 0$$

or, more concisely,

$$\mathbf{r} \cdot \mathbf{\Delta} = 0$$

The normal vector for the next plane comes from a random vector **n** and orthogonalization first with respect to **r**. The resultant vector is then orthogonalized with respect to the $\{\mathbf{g}_i\}$ as above by defining

$$\mathbf{s} = \mathbf{n} - \beta_1 \mathbf{g}_1 - \beta_2 \mathbf{g}_2 - \beta_3 \mathbf{g}_3$$

and then dotting through to give three equations for the three β's. The rounding errors in solving for the betas will erode the orthogonality of **s** with respect to **r**, but this is not a crucial erosion. Approximate orthogonality is quite sufficient to make life easy for the Newton step that is to come. (If we needed strict orthogonality with respect to **r** we could reorthogonalize at this point, or do the orthogonalization with respect to the **g**'s first—since it is always the latest orthogonalization that will be the best.)

The final corrective Newton's method step is then taken via the equations

$$-f_i(\mathbf{a}) = \mathbf{grad\ f}_i \cdot \mathbf{\Delta} \qquad i = 1, 2, 3$$

$$0 = \mathbf{r} \cdot \mathbf{\Delta}$$

$$0 = \mathbf{s} \cdot \mathbf{\Delta}$$

with

$$\mathbf{\Delta} = \{x_j - a_j\} \qquad j = 1, \ldots 5$$

This is a linear system of five equations for the five hopefully small discrepancies that will take us back onto the three surfaces

$$f_i(\mathbf{x}) = 0 \qquad i = 1, 2, 3$$

Specialized iterative techniques

Because general techniques like Newton's method lack robustness in higher dimensions, we must compensate by ferreting out all the information we can about the approximate locations and density of the roots we seek. In this process we may well acquire enough insight into our equations to permit the use of simple *iterations* to find the roots. Such iterations are often easier to program and, if high accuracy is not our aim, are about as efficient as the more general techniques. Indeed, we may have to employ the iterations to get close enough for the general methods to be stable. We here give some examples of iterative root finding, our emphasis being on gaining as much geometric information about our equations as possible before launching the computational attack. The topic, while fun, is not comfortably systematic so our discussion is largely by example.

Roots of $ze^z - b$ for positive real b

We shall seek the *complex* roots of

$$ze^z = b \tag{14.20}$$

for b real and positive. Our strategy may have to depend on the size of b but for the moment we shall be optimistic, mentally considering b to be somewhere between 1 and 10 – say 2. Tackling the easy things first, we look for real roots. As soon as we recast the equation into the form

$$\frac{x}{b} = e^{-x}$$

and sketch it (Figure 14.14) we immediately see that for fixed b only one real root is possible. Furthermore, we need not worry about how to find it – almost any of the classical methods being directly applicable. Newton is probably our official recommendation.

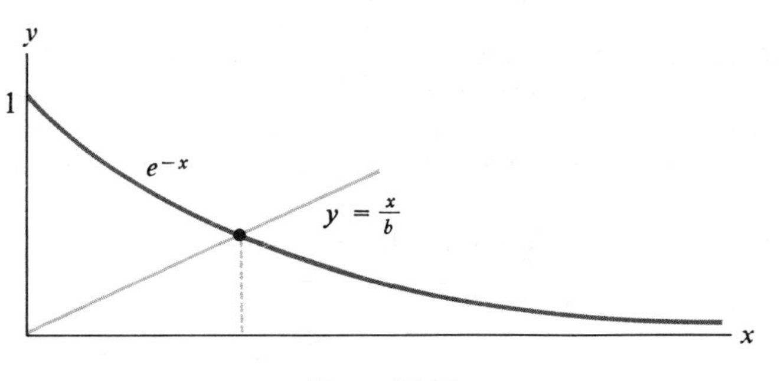

Figure 14.14

Turning to the complex roots we note that we have a well-behaved function of the single complex variable z. Should we, perhaps, use Newton's method here, too? Formally we may write

$$z_{i+1} = z_i - \frac{z_i e^{z_i} - b}{z_i e^{z_i} + e^{z_i}} = z - \frac{z - be^{-z}}{z + 1} = \frac{z^2 + be^{-z}}{z + 1}$$

with your author favoring the middle of the three forms. The only problem is where to begin? Newton's method, like all quadratically convergent methods, is apt not to converge at all unless started somewhere near the desired root—and at the moment we haven't a clue. (For hand computation, complex arithmetic seems more annoying than the tabular lookups our next algorithm will employ, but in an era of automatic computers this point is minor.) We need to explore our function after rewriting it in more familiar terms.

We have the choice of writing z in rectangular or polar form. Half a minute's experimentation will reveal that polar notation is the more useful here. We obtain

$$re^{i\theta} \cdot e^{r\cos\theta} \cdot e^{ir\sin\theta} = b$$

which gives

$$re^{r\cos\theta} = b \tag{14.21}$$

$$\theta + r\sin\theta = 0 \tag{14.22}$$

The second equation (14.22) is interesting, for it does not involve b. Rewriting it as

$$-r = \frac{\theta}{\sin\theta} \tag{14.23}$$

and examining the graph of Figure 14.15 we see that, since we take r positive, no root can occur before π. Further, if there is a root in the $(\pi, 2\pi)$ region, r must be larger than approximately $\frac{3}{2}\pi$—and it might well be close to this value since the minimum value of this branch of the curve occurs nearby, so the curve is rather flat.

Rewriting the other equation (14.21) we have

$$\frac{r}{b} = e^{-r\cos\theta} \tag{14.24}$$

and we see that r/b is positive and generally, though not necessarily, *smaller* than unity. For our first root (taking b as 2) we roughly estimate 4.8/2, or 2.4. The only way the exponential in (14.24) can be bigger than unity is for $\cos\theta$ to be negative, that is, $\pi < \theta < \frac{3}{2}\pi$. The next possible region for roots is out near $\frac{7}{2}\pi$ and, generally, near $(4n - 1)\pi/2$.

As θ goes from $\frac{3}{2}\pi$ toward π, $\cos\theta$ goes from 0 toward -1, while r, after a slight decrease, increases monotonically from 4.6, so we can obviously generate any positive exponent in (14.24) from 0 on up. Since the exponential e^{kr} starts at 1 and increases a lot faster than r, a unique crossing will take place—showing that only one complex root occurs on this branch of Figure 14.15. Similar

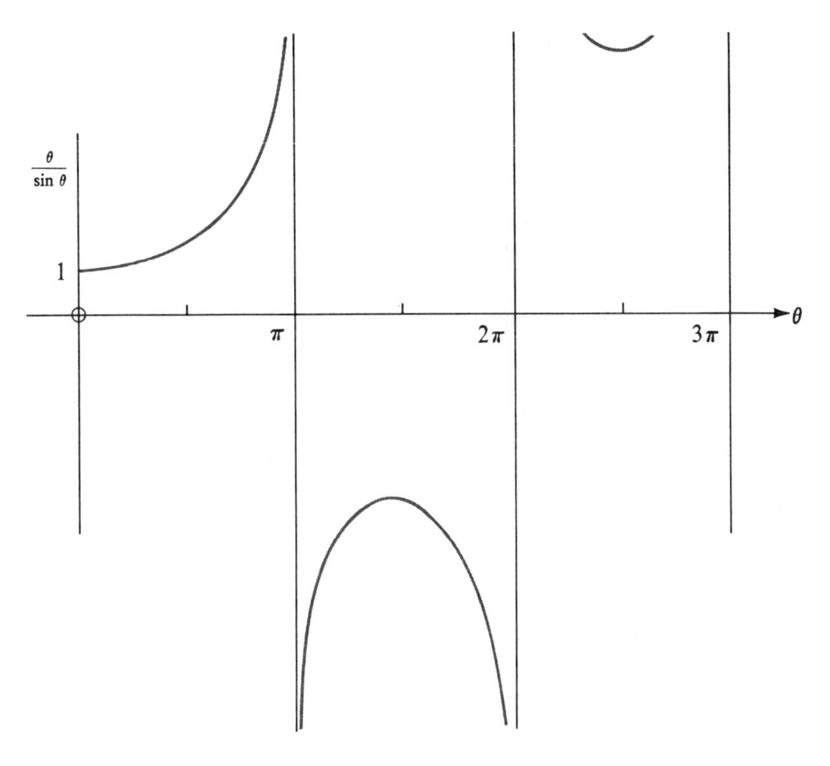

Figure 14.15

analysis shows that we shall find isolated roots, one on each branch centered at $(4n - 1)\pi/2$.

Since θ is always *smaller* than the odd multiple of $\frac{1}{2}\pi$, it is expedient to change our angular variable. We set

$$\theta = \tfrac{3}{2}\pi - \alpha = 4.71240 - \alpha$$

so our equations become

$$r = \frac{4.71240 - \alpha}{\cos \alpha} \tag{14.25}$$

and

$$\sin \alpha = \frac{\log(r/b)}{r} \tag{14.26}$$

for the slightly complex w that we actually have. Thus we try

$$\cosh x = \frac{u}{\cos y} = \frac{1.5}{\cos y}$$
$$\sin y = \frac{v}{\sinh x} = \frac{0.1}{\sinh x} \tag{14.31}$$

We obtain

(y)	$\cos y$	$\cosh x$	$\sinh x$	$\sin y$	(x)
0	1.000	1.500	1.12	0.089	
	0.996	1.506	1.125	0.0887	
	0.99605	converged			

The process is rapid, as expected. (The x and y are irrelevant during the calculation since we may calculate $\cos y$ from $\sin y$ and $\sinh x$ from $\cosh x$ by Pythagorean rules.)

A slightly different value of w,

$$w = 0.75 + 0.1i$$

leads to an entirely different iteration. This w clearly is close to 0.75, which *cannot* be solved by a real value of z because the minimum of $\cosh x$ is unity. Hence the $\cos y$ factor must be operative – and one suspects that it is controlling. Here we try

$$\cos y = \frac{0.75}{\cosh x}$$
$$\sinh x = \frac{0.1}{\sin y} \tag{14.32}$$

We obtain

(x)	$\cosh x$	$\cos y$	$\sin y$	$\sinh x$
0	1.00	0.75	0.662	0.151
	1.0114	0.7415	0.670	0.149
	1.0111	0.7417	converged	

and again the iteration is rapidly convergent – though it is a different iteration from the previous example.

We now turn to

$$w = 1.5 + 1.0i$$

If we decide to try (14.31) on it, we obtain

$\cos y$	$\cosh x$	$\sinh x$	$\sin y$
1.000	1.50	1.12	0.89
0.47	3.20	3.02	0.33
0.743	1.59	1.24	0.81

(14.31a)

and it is clear that the process is quite slowly convergent. We can speed it up by judicious averaging—but this is an artifice. We would prefer an iteration that moves more quickly. The trouble arises from the fact that, in the ranges we are using, all factors move about equally. Thus a change in $\cos y$ changes $\sin y$ about as much in the opposite direction. Only if y is small or near $\frac{1}{2}\pi$ do we get the disproportionate changes we prefer. With $\sin x$ and $\cos x$ we have a similar phenomenon—only near *zero* or $\frac{1}{2}\pi$ are they different—and herein lies our clue. For our values of x it is $\tanh x$ that changes slowly. This suggests the iteration

$$\tan y = \frac{\tan \varphi}{\tanh x} \qquad \sinh^2 x = R^2 - \cos^2 y \qquad \text{where} \begin{cases} \tan \varphi = \dfrac{v}{u} = 0.667 \\ R^2 = u^2 + v^2 = 3.250 \\ \tanh x_0 = 1 \end{cases} \tag{14.33}$$

Noting that

$$\cos^2 y = (1 + \tan^2 y)^{-1}$$

We obtain the iteration

$\tanh x$	$\tan y$	$\cos^2 y$	$\sinh^2 x$
1.000	0.667	0.690	2.560
0.850	0.785	0.620	2.630
0.850	converged		

which is very good indeed. The important equation is the first one of our pair. The second could have been either

$$\sinh x = \frac{v}{\sin y}$$

or

$$\cosh x = \frac{u}{\cos y}$$

with reasonable results, although a somewhat more rapid iteration is achieved with the equation we actually used because of the smallness of $\cos^2 y$ relative to R^2.

The obvious fourth iteration

$$\tanh x = \frac{\tan \varphi}{\tan y} = \frac{0.1}{\tan y}$$
$$\cos y = R^2 - \sinh^2 x \tag{14.34}$$

succeeds nicely for

$$w = 0.70 + 0.07i$$

where we obtain

$\sinh^2 x$	$\cos^2 y$	$\tan y$	$\tanh x$	
0	0.4949	1.01	0.099	
0.0099	0.4850	1.03	0.097	
0.0095	0.4835	1.033	0.097	converged

Here R^2 is approximately equal to 0.494 and $\tan \varphi$ is small (about 0.1). These figures suggest a small x as well as a small y. But an earlier iteration (14.32) would have done as well.

In all our manipulations to set up convergent iterations we strive to place insensitive terms on the right, then solve for the variable on the left. Thus a poor guess for the variables on the right give us better estimates for the variables on the left. Our present example is typical in that it requires different iterations for different ranges of the parameter. It is not so typical in that it permits a great variety of rearrangements that make it fairly easy to find convergent iterations.

Still another approach to the same problem would be to reduce it to a single equation in a single unknown and apply Newton's method—having first acquired some idea about probable value for the desired root through the methods used above.

Convergence of an iteration—analytics

We have been using iterations of the form

$$x_{i+1} = f(x_i, y_i)$$
$$y_{i+1} = g(x_i, y_i) \tag{14.35}$$

and we now ask about conditions under which these iterations converge. Intuitively, we know that a change in x_i should produce a *smaller* change in x_{i+1} – that is, $|\partial f/\partial x| < 1$ – and similarly for Δy acting thru g. But also we must worry about the cross effects, the Δy_{i+1} produced by a change in x_{i+1}, and so on. The criteria are not simple.

Some light is shed if we expand f and g about the solution point (a, b). We have

$$x_{i+1} = f(a,b) + \Delta x \cdot f_x + \Delta y \cdot f_y + \cdots$$

$$y_{i+1} = g(a,b) + \Delta x \cdot g_x + \Delta y \cdot g_y + \cdots$$

where we have used f_x to represent $\partial f/\partial x$ and the i subscripts on the Δ quantities have been dropped as unimportant. Noting that

$$f(a,b) = a$$

$$g(a,b) = b$$

we may write

$$\begin{aligned} x_{i+1} - a &= \Delta x \cdot f_x + \Delta y \cdot f_y + \cdots \\ y_{i+1} - b &= \Delta x \cdot g_x + \Delta y \cdot g_y + \cdots \end{aligned} \tag{14.36}$$

Assumption A: We now assume that the neglected second-order and higher terms are unimportant, which, in practice, is to assume that we are already rather close to the solution point – certainly nearer to (a, b) than to any other significant place such as another root, or a pole, or a saddlepoint. For most practical problems this crucial assumption vitiates this analysis for establishing any convergence from a distance. The left side of (14.36) is the error at the end of the first iteration, whereas the vector $(\Delta x, \Delta y)$ on the right is that same error before the iteration. We rewrite (14.36) as

$$\begin{bmatrix} \Delta x_{i+1} \\ \hline \Delta y_{i+1} \end{bmatrix} = \mathbf{e}_{i+1} = \begin{bmatrix} f_x & f_y \\ g_x & g_y \end{bmatrix} \begin{bmatrix} \Delta x_i \\ \hline \Delta y_i \end{bmatrix} = \frac{\partial(f, g)}{\partial(x, y)} \cdot \mathbf{e}_i$$

where the matrix is the Jacobian of our iterative transformation, evaluated at the solution point. It is thus a strictly constant numerical matrix and our iterative process is seen to be the repeated premultiplication of an arbitrary (though hopefully small) vector $\mathbf{e}$ by this matrix. Generally

$$\mathbf{e}_n = J^n \mathbf{e}_0 \tag{14.37}$$

We know that this operation leads rather quickly from an arbitrary vector to a most specific vector: the eigenvector of J corresponding to the eigenvalue of J that is largest in magnitude. Equation (14.37) is, in fact, the classical iteration procedure (Chapter 8) for finding the largest eigenvalue except for one crucial point: The normalization step has been omitted! Thus we know that

$$\mathbf{e}_n \to \lambda^n \cdot \mathbf{j}$$

where λ is the largest eigenvalue of J in absolute value and $\mathbf{j}$ is its corresponding eigenvector. If $|\lambda| > 1$, then $|e_n|$ does not shrink toward zero as a good error vector should. If, on the other hand, $|\lambda| < 1$ there is a real hope that our iterative process will converge, for e_n certainly goes to zero and our only doubts concern the influence of those neglected higher-order terms in (14.35). We shall pursue our analysis with a deep sense of obligation to Assumption A and the corresponding realization that our results have all too limited relevance to most real problems except in the immediate neighborhood of a solution.

We now explore the requirement that the largest eigenvalue of J be less than unity in magnitude. The eigenvalues λ are determined by

$$\begin{vmatrix} f_x - \lambda & f_y \\ g_x & g_y - \lambda \end{vmatrix} = 0 = (f_x g_y - f_y g_x) - (f_x + g_y)\lambda + \lambda^2$$

which may be rewritten

$$\lambda^2 - T\lambda + D = 0$$

by letting T stand for the trace of the Jacobian and D for its determinant. If we demand that λ equal 1 we see that

$$D = T - 1$$

which line appears in Figure 14.16. Likewise, if λ equals -1 we have

$$D = -T - 1$$

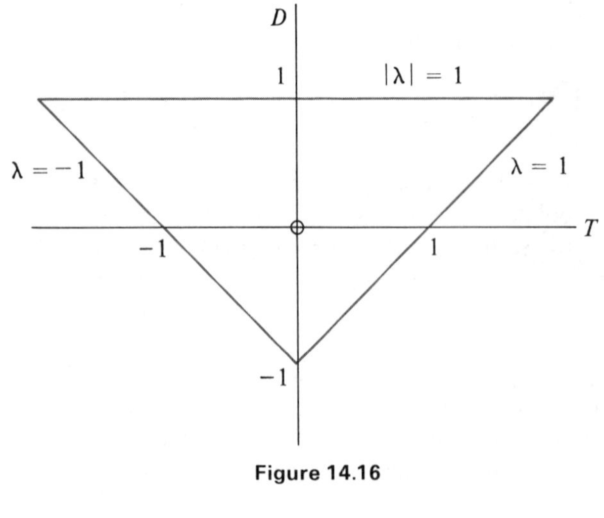

Figure 14.16

and finally if λ is complex but $|\lambda|$ equals 1 we have that D also is 1, so the region *within the triangle* of Figure 14.16 is the place where all the λ's have absolute value of less than unity.

Since

$$T = f_x + g_y$$

and

$$D = f_x g_y - f_y g_x$$

we can test any proposed iteration to see if the point (T, D) lies inside our triangle. Both T and D are functions of x and y, but they might be slowly varying functions, so the fact that we are not evaluating them at or near the unknown root may not be too harmful. One point is clear: We had better not start an iteration at a place where (T, D) lies clearly outside our triangle. Instability will almost surely follow, and swiftly.

Some values of (T, D) for our several iteration schemes are

Equation	T	D	
(14.31)	0	0.015	These iterations all converge very quickly,
(14.32)	0	0.024	and their (T, D) points are very close to the
(14.33)	0	0.024	origin
(14.31a)	0	0.040	At the root
–	0	0.160	At the starting point

The first three iterations, which converged very quickly, have their (T, D) points very close to the origin, which would correspond to a zero eigenvalue. The slowly converging iteration (14.31a), however, also has its (T, D) point well inside the triangle, although its D value at the starting point is a decimal order of magnitude larger than our other processes. Thus we see that an apparently safe value of (T, D) does not guarantee rapid convergence.

Another analytic approach

Let us return to our algorithm (14.35), writing

$$x_2 = f(x_1, y_1).$$

By subtracting this from the actual solution obtain

$$a - x_2 = f(a, b) - f(x_1, y_1)$$

which may be rewritten

$$a - x_2 = \frac{f(a, b) - f(x_1, b)}{a - x_1} \cdot (a - x_1) + \frac{f(x_1, b) - f(x_1, y_1)}{b - y_1} \cdot (b - y_1)$$

The fractions are standard finite-difference approximations to the first partial derivatives $\partial f/\partial x$ and $\partial f/\partial y$, which we prefer to denote by f_x and f_y, giving

$$a - x_2 = f_x(a - x_1) + f_y(b - y_1) \tag{14.38}$$

provided we understand our derivatives to be evaluated at some unknown point in the intervals (a, x_1) and (b, y_1), respectively. Exactly analogous operations give

$$b - y_2 = g_x(a - x_1) + g_y(b - y_1) \tag{14.39}$$

At this point everything is still exact although the equalities are a bit fuzzy because we only know approximately where the four partial derivatives have been evaluated to make them equal to their finite-difference representations. But no matter, for we are about to do possible violence by taking absolute values throughout and then add! We get

$$|a - x_2| + |b - y_2| \leq (|f_x| + |g_x|) \cdot |a - x_1| + (|f_y| + |g_y|) \cdot |b - y_1|$$

where the equality can hold only if all the products in (14.38) and (14.39) were composed of terms with like signs. If we now let the largest absolute partial derivative be denoted by $M/2$ and worsen our inequality by replacing all four of them by this bound, we may write

$$|a - x_2| + |b - y_2| \leq M \cdot (|a - x_1| + |b - y_1|)$$

which expresses the error in (x_2, y_2) in terms of the error in (x_1, y_1) and a bound on the four partial derivatives. Clearly if this bound is less than $\frac{1}{2}$ in absolute value in the region bounded by (a, b, x_1, y_1) we will get a convergent iteration. The condition is not necessary but it is sufficient. Unfortunately, with most practical problems the regions in which the four partials are small are such small areas around the sought roots as to make them rather unlikely to be hit by one's first guess at the roots. But one can frequently *construct* functions f and g to satisfy our sufficient condition—as we show in the next section.

Constructing a convergent iteration

Any computer programmer must behold the great variety of transcendental equations with misgiving. Surely one ought not to be forced to write a separate program for every problem. Yet, for practical purposes, this is what we currently must do. The few algorithms that at first seem general turn out to be among our weaker weapons, in a battle where all the strength that we can muster is sorely needed. But, on the theory that a weak weapon is better than none, we shall describe one fairly general technique for constructing convergent iterations for a pair of simultaneous transcendental equations.

We observed above that convergence of both Newton's method and the iterative algorithm

$$\begin{aligned} x_{i+1} &= f(x_i, y_i) \\ y_{i+1} &= g(x_i, y_i) \end{aligned} \tag{14.40}$$

was guaranteed provided the several first partial derivatives were small enough – the smaller, the better, and certainly smaller than $\frac{1}{2}$ in absolute value in the vicinity of the root. One general device is to construct a pair of functions with this property. We write

$$\begin{aligned} x_{i+1} &= x + a[f(x, y) - x] + b[g(x, y) - y] = F(x, y) \\ y_{i+1} &= y + c[f(x, y) - x] + d[g(x, y) - y] = G(x, y) \end{aligned} \tag{14.41}$$

Here we see that any solution of (14.40) is certainly a solution of (14.41) for all values of the undetermined parameters of a, b, c, and d. We may now determine these four parameters by the four conditions that the first four partial derivatives of F and G be zero at some point (x, y) that hopefully is near a root. Since the parameters enter linearly, our calculation of them poses no problems and the program is quite general, since it does not depend in any critical way on the particular functions f and g. We presumably need only provide a subroutine that evaluates f, g, and their four first partial derivatives.

The difficulty with our method arises, as with most methods in two (or more) dimensions, from a frequently rapid variation in our functions and their derivatives as we move around the (x, y) plane. Thus, although we may easily produce an F and G that are very well behaved at (x_0, y_0), they may behave quite badly a small distance away. If our current strategy is to be successful we must not only give F and G small partial derivatives in some region, but the region for which those derivatives are small must include the root we seek. We are casting a small disc upon the (x, y) plane, hoping to cover a root with it. The disc represents the region within which our algorithm will converge. Some algorithms tend to have larger discs than others but nobody can give precise estimates a priori about their sizes. Your author's quite limited experience with this particular algorithm has not been especially reassuring. (On the other hand, his experience with other algorithms has tended to be worse!) The example that follows was chosen to illustrate the troubles. The functions are very steep, so *any* method that depends in a critical way on sizes of derivatives will have difficulty.

Consider

$$x^{10} + y^{10} = 2^{10}$$
$$e^x - e^y = 1 \tag{14.42}$$

We write

$$x = x + a(x^{10} + y^{10} - 2^{10}) + b(e^x - e^y - 1) = F$$
$$y = y + c(x^{10} + y^{10} - 2^{10}) + d(e^x - e^y - 1) = G \tag{14.43}$$

and adjust our parameters to produce zero partial derivates of F and G at (1, 1). We get

$$F_x = (10x^9)a + (e^x)b + 1 = 0$$
$$F_y = (10y^9)a - (e^y)b \qquad = 0$$

which at (1, 1) give

$$a = -0.950$$
$$b = -0.184$$

Similarly, we find

$$c = -0.050$$
$$d = +0.184$$

and thus our iteration becomes

$$x_{i+1} = x_i - 0.050(x^{10} + y^{10} - 2^{10}) - 0.184(e^x - e^y - 1)$$
$$y_{i+1} = y_i - 0.050(x^{10} + y^{10} - 2^{10}) + 0.184(e^x - e^y - 1) \tag{14.44}$$

which, for (1, 1) gives

$$x_{i+1} = 1 - (0.050)(-512.0) + 0.184 = 26.8$$
$$y_{i+1} = 1 - (0.050)(-512.0) - 0.184 = 26.4$$

which is absurd! The function x^{10} is just too steep to be controlled very far and the root is not close to (1, 1) in terms of the geometry around the root.

If we evaluate our four parameters at (2, 2) a different story emerges. We have

$$a = c = -0.0001$$
$$b = -d = -0.075$$

and the iteration at (2, 2) gives

$$x_{i+1} = 2 - (0.0001)(2^{10}) + 0.075 = 2 - 0.102 + 0.075 = 1.973$$
$$y_{i+1} = 2 - (0.0001)(2^{10}) - 0.075 = 2 - 0.102 - 0.075 = 1.823 \tag{14.45}$$

We now have the choice of iterating with the same coefficients or of recomputing them. Using the same coefficients we obtain

$$x_{i+1} = 1.973 - (0.0001)(275.2) - 0.075(0.00182) = 1.9454$$

$$y_{i+1} = 1.823 - (0.0001)(275.2) + 0.075(0.00182) = 1.7956$$

and we are clearly converging quite nicely. The correct root is (1.93, 1.773).

In contrast to the foregoing, we might consider the simple iteration of the original equation (14.42) in the form

$$x = (2^{10} - y^{10})^{1/10}$$

and

$$u = y - x = \ln(1 - e^{-x}) \tag{14.46}$$

where

$$y = x + u$$

which gives

x	y
1.953	1.800
1.916	1.757
1.937	1.782
1.925	1.768
1.932	1.776
1.929	1.772

If we plot these points we see that we are oscillating along a diagonal line with slow convergence. Averaging will speed the process since the average value of two successive points is close to the root. Although this iteration with averaging is substantially worse in computational labor than the construction and use of the more highly convergent iteration, it is easier to program. We suspect that personal preferences will divide sharply over this example.

Double-root strategy

In higher dimensions, as in one dimension, double and nearly double roots will be found more efficiently if special tactics are used. The variety of double-root geometries, however, makes their confident detection and identification less likely than in the one-dimensional problem, so we may question the wisdom of trying to incorporate these special techniques in a general root-solving package.

If, however, we must repeatedly solve a limited class of problems in which nearly double roots frequently occur, then we probably should study the local geometry, devising special methods that put its features to favorable use rather than fighting stubbornly on with an unsympathetic general method. We illustrate here with the commonest geometry in two variables.

If we examine a plot of the curves

$$f(x, y) = 0$$
$$g(x, y) = 0$$

we recognize three configurations in which double roots might lurk nearby. In Figure 14.17 we have a true double root at A, two distinct but close roots at B, C,

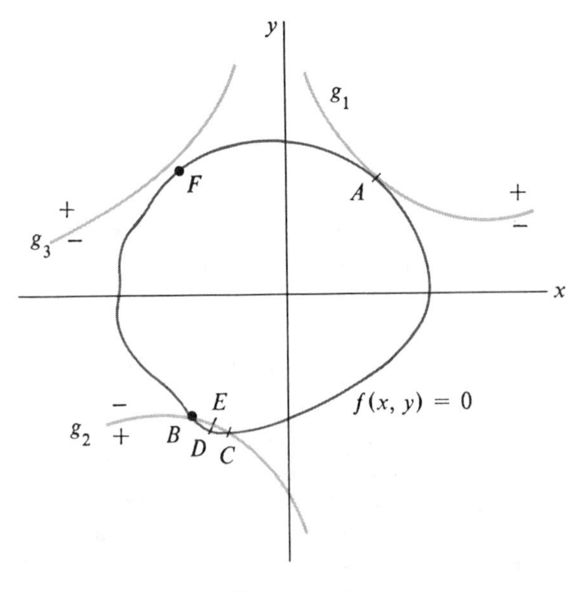

Figure 14.17

and a near miss at F. If our sketch is based on incomplete information – the evaluation of $f(x, y)$ and $g(x, y)$ over a grid, for example – we realize that none of these geometries is accurately represented and hence we would accord all three the same suspicious treatment. In each of them there is a special point on the f curve at which we would like to evaluate the $g(x, y)$ function (or vice versa) to see if it is positive, negative, or zero – thereby showing the local geometry to be that of D, F, or A, respectively. Thus we shift our tactics from seeking the root(s) to that of seeking this special point. This tactical shift is the two-dimensional analogue of our temporary shift in one dimension to finding the *minimum* of $f(x)$ when we suspect the nearly double root geometry of Figure 14.18. The purpose behind the shift is threefold: The special point is easier to find directly

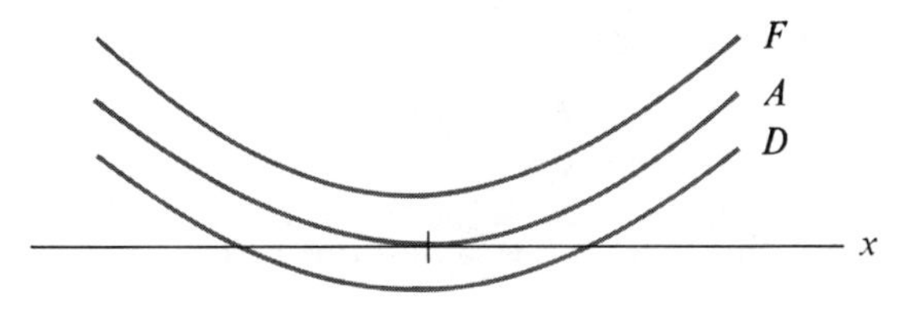

Figure 14.18

than are the roots; once the point is found, the three root geometries are easily distinguished by a functional evaluation there; the roots may then be found comparatively easily as *deviations* from this special point.

In two dimensions the point we seek satisfies two conditions: It lies on $f(x, y) = 0$ and the level line of the g function through it is *tangent* to the f curve there. If we look at a plot (Figure 14.19) of the curves

$$g(x, y) = c$$

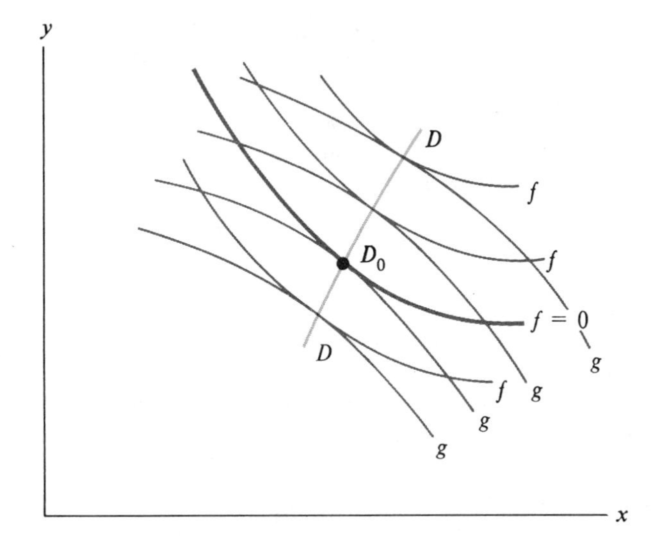

Figure 14.19

for several small values of c we realize that one of these has a double root with our f curve (at D_0). This is the point we need to find.

Along any curve on which $g(x, y)$ remains constant we have

$$\Delta g = \frac{\partial g}{\partial x}\Delta x + \frac{\partial g}{\partial y}\Delta y = 0$$

so the slope of the level curve is

$$\frac{\Delta y}{\Delta x} = -\frac{g_x}{g_y}$$

Demanding that both curves have the same slope gives

$$\frac{f_x}{f_y} = \frac{g_x}{g_y}$$

or

$$f_x g_y - f_y g_x = |J| = 0 \tag{14.47}$$

This is precisely the condition that *prevents* solution of our original root problem through linear replacement of the curves, so we are encouraged. When the regular Newton's method breaks down, this method is at its best. Geometrically, (14.47) defines a curve that is generated by the double roots of

$$f(x, y) = b \quad \text{with} \quad g(x, y) = c$$

as the parameters b and c are continuously varied (Figure 14.19). As such, it points locally in the direction of the common gradient of the two functions, intersecting each of the curves at right angles. Thus equation (14.47) and one level line give a very well-defined intersection that is quickly found by standard processes such as Newton's. It is only the subsequent steps to get the nearby roots that are slightly unpleasant. We illustrate below.

Double roots—number bashing versus finesse

Throughout this book the reader has been exhorted to look carefully at the shape of his problem and then choose the appropriate tool. In this day of cheaper automatic computation it is just barely possible that this advice can be overdone. Most persons will not need to be told that they might better let a converging iterative process go on for 15 milliseconds more rather than spend several hours writing and debugging an auxiliary process that could save 12 of those 15 milliseconds. Dogged repetition of a slowly convergent process can, at times, be the economic strategy—even though your author makes the point reluctantly. Of course, if the inefficient process lies at the heart of five nested loops where it constitutes an appreciable component of the labor, saving those 12 milliseconds can be the saving of thousands of dollars. It depends, as they say in the infantry, on the situation and the terrain.

Our double-root strategy of the previous section provides an illustration. We consider the equations (which can also be solved as a one-variable problem)

$$f(x, y) = x^2 + 4y^2 - 4 = 0 \tag{14.48}$$

$$g(x, y) = x^2 + y^2 - 8x - 2y + R = 0 \qquad R = 12.1969\,3244 \tag{14.49}$$

The parameter R has been chosen to produce nearly a double root in the first quadrant. There are no other roots, as a rough sketch of the ellipse and circle will quickly reveal. Since the sketch (the student should draw it at this point) also suggests that the roots, if any, will be close, the quantity

$$D = f_x g_y - f_y g_x = -12xy - 4x + 64y = 0 \tag{14.50}$$

is solved jointly with (14.48) by Newton's method. Since the D and f curves intersect essentially at right angles, convergence to 12 significant figures occurs in four iterations, starting from $(2, -0.01)$. We have the solution

$$\begin{aligned} x &= 1.9620\,0715\,180 \\ y &= 0.1939\,8964\,939 \end{aligned} \tag{14.51}$$

for the point D_0 and the knowledge that roots, if any, are separated by this point roughly as a bisector. At this point g_0 is $-0.0000\,0002\,540$, so there *are* two roots (consider your sketch again). Expanding f and g about the point (14.51) and equating the expansions to zero we have the local equations for our two curves (Figure 14.20).

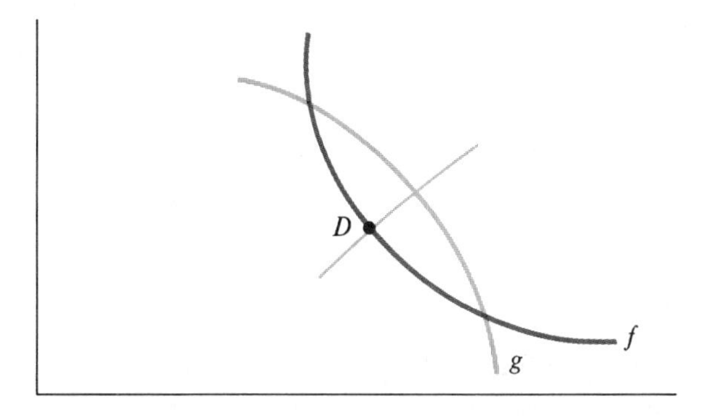

Figure 14.20

$$\begin{aligned} 0 &= f_x \cdot \delta + f_y \cdot \varepsilon + \tfrac{1}{2} f_{xx} \cdot \delta^2 + \tfrac{1}{2} f_{xx} \cdot \varepsilon^2 \\ -g_0 &= g_x \cdot \delta + g_y \cdot \varepsilon + \tfrac{1}{2} g_{xx} \cdot \delta^2 + \tfrac{1}{2} g_{yy} \cdot \varepsilon^2 \end{aligned} \tag{14.52}$$

(A more general problem would also have f_{xy} and g_{xy} terms.) Multiplying the first equation by g_y, the second by f_y, subtracting and using the fact that D is zero at our point of expansion, we eliminate both linear terms, obtaining

$$2g_0 f_y = (g_y f_{xx} - f_y g_{xx})\delta^2 + (g_y f_{yy} - f_y g_{yy})\varepsilon^2 \tag{14.53}$$

Thus

$$\varepsilon = \pm\sqrt{\frac{-2g_0}{[g_{yy} - f_{yy}(g_y/f_y)] + [g_{xx} - f_{xx}(g_y/f_y)] \cdot (\delta/\varepsilon)^2}} \tag{14.54}$$

while the first of equations (14.52) gives us

$$\frac{\delta}{\varepsilon} = -\frac{f_y}{f_x} - \frac{\varepsilon}{2f_x}\left[f_{yy} + f_{xx}\left(\frac{\delta}{\varepsilon}\right)^2\right] \tag{14.55}$$

These equations must be solved iteratively, starting with (14.55) by ignoring its small second term. We have

$$\frac{\delta}{\varepsilon} \approx -0.3954\,92240$$

[If δ/ε had been greater than unity we would have solved (14.53) for δ and (14.52) for ε/δ.]

Substituting into (14.54) we have

$$\begin{aligned} \varepsilon &= \pm\frac{(0.0001\,59374)\sqrt{2}}{\sqrt{[2 - 8(-1.0387286)] + [2 - 2(-1.0387286)](0.39549)^2}} \\ &\qquad\qquad\qquad\qquad\qquad\qquad\qquad\qquad (0.156414) \\ &= \pm 0.0000\,681198 \end{aligned}$$

which tentatively gives

$$\delta = \mp 0.0000\,269408$$

On using our two values of ε in (14.55), together with the approximate value of $(\delta/\varepsilon)^2$, we obtain

$$\frac{\delta}{\varepsilon} = \begin{cases} -0.39556\,4394 \\ -0.39542\,0086 \end{cases}$$

the upper figure corresponding to the positive value of ε.

Since the value of δ/ε has changed appreciably—indeed, it has bifurcated into two values—we substitute them back into (14.54) to get better value of ε.

The change is slight, our values being

$$\varepsilon = +0.0000\,681190 \qquad \text{and} \qquad -0.0000\,681205$$

and it does not appear useful to recompute δ/ε for a third time. Thus we have

$$\delta = -0.0000\,2694545 \qquad \text{and} \qquad +0.0000\,2693621$$

We thus obtain the roots

$$\begin{aligned} x &= 1.9619\,802063 & & 1.9620\,340880 \\ & (1.9619\,802068) & \text{and} \quad & (1.9620\,340875) \\ y &= 0.1940\,577684 & & 0.1939\,215289 \\ & (0.1940\,577672) & & (0.1939\,215301) \end{aligned}$$

The correct roots (in parentheses) show our values to be very accurate—with at least nine correct significant digits—and, unfortunately, they may also be obtained merely by using Newton's method directly on (14.48) and (14.49)! True, it takes 13 iterations per root and one must take care to start on opposite sides of the root region lest the same root be found a second time, but 22 extra iterations is a small price to pay to save programming these (δ, ε) equations. We shall have to be carrying out a great many nearly double root calculations to justify using this essentially hand technique on an automatic computer. The mere thought of 22 additional iterations by hand is, of course, horrendous. One's mental reference frame is crucial here.

In closing this warning about being penny-wise, pound-foolish in automatic computation your author observes that most persons do not need it. They are quite content to let inefficient iterations run on—for hours—rather than look for alternative and more appropriate techniques.

Problems

1. For what values of c will the curves

$$u^2 + 3v^2 = c^2$$
$$u^2 - uv + v^2 = \tfrac{1}{4}$$

be tangent?

2. Find a value of b such that

$$x^4 + 2y^4 = 1$$
$$x^2 + 2y^2 = 1 + b$$

have double roots. Locate the roots.

3. Do the equations

$$y + b = \cos kx$$
$$k^2x^2 + y^2 = 1$$

have four clustered roots or only 2 near $x = 0$ for b equal to 0.0001?

4. Sketch the essential geometry of

$$y = \cos x$$
$$\left(\frac{y + h}{1 + h}\right)^2 + \left(\frac{x}{a}\right)^2 = 1$$

for $1 < a < \pi/2$ and $0 < 1 + h < a$. For $a = 1.25$ and $h = 0.10$ find the roots by iteration; by Newton's method.

5. How many solutions will the equation pair

$$y = \frac{1}{x^2 + 1} - \varepsilon$$
$$2y = e^{-(x-1)}$$

have for various values of ε? Find the critical values, together with their corresponding roots.

6. Find the intersections of

$$y = 1.001 - 0.5x$$

with

$$y = (x^2 + 1)^{-1}$$

7. Consider the equation

$$\tanh bx = e^{-1/x^2}$$

a. If b is 0.26, find the real roots, if any.
b. For what value of b is there at least one double root?

8. For what value of b is

$$y = be^{-x}$$

tangent to

$$x^4 + 2y^4 = 1$$

9. Find a such that the witch

$$y = \frac{1}{x^2 + a^2}$$

be tangent to the superellipse

$$x^4 + 2y^4 = 1$$

Can it be made osculatory?

10. Solve

$$x^4 + 4y^4 = 6$$
$$x^2 y = 1.6787$$

first as a two-dimensional problem. Then eliminate y and solve as a one-dimensional problem. Compare your relative difficulties.

CHAPTER

15

THE CARE AND TREATMENT OF SINGULARITIES

The two most persistent difficulties in numerical computation are the disappearance of significant figures by subtraction of nearly equal quantities and the production of wrong answers by polynomic formulas applied where they should not be used. Furthermore, these troubles often occur together, compounding their confusions. In this chapter we wish to emphasize the trouble caused by functions that become infinite or acquire an infinite derivative, *the computational singularity*, but we cannot entirely ignore its companion in disaster, *the disappearing digits*—so both topics appear. Neither topic is new to this book and if the student wonders why he should read still more about infinite integrals, let the author assure him that the medicine seems necessary. The standard reaction of most persons to a computational singularity is to *avoid* it. They hope that if they don't look at it, it will go away. A more profitable approach is to seek it out as a challenging and interesting puzzle to be solved, or, at worst, a diabolical obstruction to be overcome. As with most puzzles, there are some standard tricks that help considerably. This chapter, by example, is devoted to them.

We already have observed that

$$\int_0^a f(x) \cdot \ln(x) \cdot dx \qquad (15.1)$$

is unpleasant at its lower limit even if $f(x)$ is behaving essentially like a constant there. The integrand is infinite, though the integral may well exist and be of quite modest size. We should not apply a polynomial quadrature formula, such as Simpson's rule, directly; the integrand has a most unpolynomic vertical asymptote at zero. The condition of our integral may be improved through integration by parts. Assuming that $f(x)$ can be differentiated we have

$$[(x \ln x - x) \cdot f(x)]_0^a - \int_0^a f'(x) \cdot (x \ln x - x) \cdot dx \tag{15.2}$$

where our $\ln x$ now appears carefully wedded to x—which dominates it as x approaches zero. If $f'(x)$ has not suddenly become unpleasant, our integral is now well enough behaved so that standard quadrature formulas may represent it quite adequately. We only must be sure to evaluate the potentially indeterminate form $x \cdot \ln x$ as a unit whenever x is very small, and not as two pieces to be joined later. Geometrically, the trouble with (15.1) is the very nonpolynomic

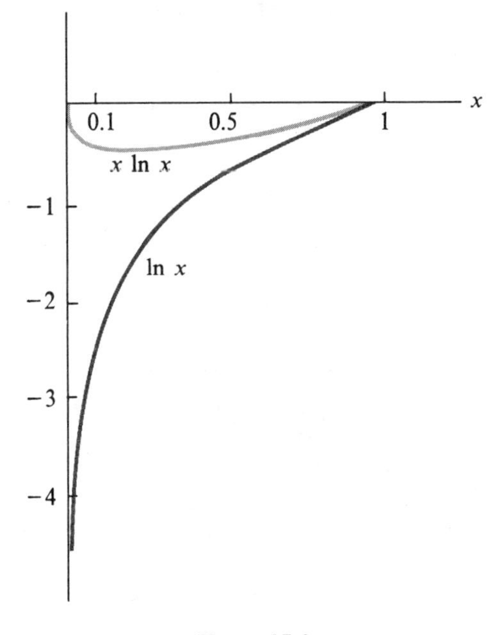

Figure 15.1

shape of $\ln x$ near the origin (Figure 15.1). The function $x \cdot \ln x$ (also Figure 15.1) is much more easily represented by a polynomial in this region, even though its *slope* is still vertical at the origin.

The integral

$$I(a) = \int_0^a \frac{dx}{\sqrt{\sin x}} \qquad a < \pi \tag{15.3}$$

likewise is finite, although its integrand becomes infinite like $x^{-1/2}$ at the origin. We have several options that will work. Conceptually simplest is the replacement of $\sin x$ by merely x in a subdivision of the range of integration designed to keep this replacement an adequate representation of the trigonometric function. Thus we have

$$I(a) = \int_0^\varepsilon \frac{dx}{\sqrt{x}} + \int_\varepsilon^a \frac{dx}{\sqrt{\sin x}} \tag{15.4}$$

and we evaluate the first integral analytically. The second integral is better behaved than it was in (15.3) but its integrand is still a rather nonpolynomic function. We would prefer a more effective method.

Rewriting (15.3) as

$$I(a) = \int_0^a \sqrt{\frac{x}{\sin x}} \cdot \frac{dx}{\sqrt{x}} \tag{15.5}$$

and then integrating by parts gives

$$I(a) = \left[2\sqrt{x} \cdot \sqrt{\frac{x}{\sin x}}\right]_0^a - \int_0^a \sqrt{x}\left(\frac{\sin x}{x}\right)^{1/2} \cdot \frac{\sin x - x\cos x}{\sin^2 x}\,dx \tag{15.6}$$

The factor $\sin x/x$ is well behaved at the origin, as is the more complicated factor at the right of the new integrand. True, both factors are indeterminate at the lower limit, but a quick insertion of the first three terms of the sine and cosine series will reveal their values there and give a formula adequate for evaluating the factors nearby, should our polynomial quadrature formula require it. This strategem (15.5) is sometimes called "dividing out the singularity."

Still another device is "subtracting out the singularity." We write (15.3) as

$$I(a) = \int_0^a \frac{dx}{\sqrt{x}} + \int_0^a \left(\frac{1}{\sqrt{\sin x}} - \frac{1}{\sqrt{x}}\right) dx \tag{15.7}$$

and integrate the left term analytically. The second integral may be treated by Simpson's rule after we rewrite the integrand to avoid the loss of precision inherent in the direct subtraction. We have

$$I(a) = 2\sqrt{a} + \int_0^a \frac{\sqrt{x} - \sqrt{\sin x}}{\sqrt{x \cdot \sin x}}\,dx$$

$$= 2\sqrt{a} + \int_0^a \frac{(x - \sin x)\,dx}{\sqrt{x \sin x}(\sqrt{x} + \sqrt{\sin x})} \tag{15.8}$$

$$= 2\sqrt{a} + \int_0^a \frac{\dfrac{1}{\sqrt{x}}\left(1 - \dfrac{\sin x}{x}\right)}{\sqrt{\dfrac{\sin x}{x}}\left(1 + \sqrt{\dfrac{\sin x}{x}}\right)}\,dx$$

Here the $\sin x/x$ terms in the denominator are well behaved, while a series replacement for them in the numerator will show the second numerator factor to be a well-behaved $\frac{1}{6}x^2$ for small values of x. Thus it can more than suppress the $x^{-1/2}$ there.

One final method is to substitute $\sin^2\theta$ for $\sin x$ in (15.3) to give

$$I(a) = 2\int_0^b \frac{d\theta}{\sqrt{1 + \sin^2\theta}} \qquad b = \sin^{-1}(\sqrt{\sin a}) \tag{15.9}$$

with a suitably altered upper limit. This integrand is very well behaved.

Offstage singularities

Sometimes the point at which our function becomes infinite is not quite in the range of the problem. Thus we might have

$$I(a, \varepsilon) = \int_\varepsilon^a \frac{dx}{x \sin x} \tag{15.10}$$

where ε is small. At *zero* the integrand and the integral are both infinite, but fortunately we are not being asked to go that far with our quadrature. Here, the wishful avoider has great scope for hope. He can plausibly argue (wrongly—but no matter!) that since the singularity in the integrand will never be reached he may ignore it. This is, of course, nonsense. A vertical asymptote at zero imparts a quite nonpolynomic geometry nearby, and if ε is at all small the integrand will be poorly fitted by standard quadrature formulas precisely where the contribution to the area is greatest. A proper treatment is, again, subtracting off the singularity. We write

$$I(a, \varepsilon) = \int_\varepsilon^a \frac{dx}{x^2} + \int_\varepsilon^a \left(\frac{1}{x \sin x} - \frac{1}{x^2}\right) dx = \left(\frac{1}{\varepsilon} - \frac{1}{a}\right) + \int_\varepsilon^a \frac{x - \sin x}{x^2 \sin x}\,dx \tag{15.11}$$

where the effect of the singularity has now been successfully encapsuled in the first term, leaving the new integrand with respectable polynomic behavior near the lower limit. It is, of course, nearly indeterminate in this functional form so the usual alternative form for small arguments must be generated by series substitution. Otherwise, serious loss of significant digits through subtraction, subsequently amplified through division by a small number of the order of x^3 will wreck a strategy that thus far has been eminently sound. We finally have

$$I(a, \varepsilon) = \frac{1}{\varepsilon} - \frac{1}{a} + \frac{1}{6}\int_{\varepsilon}^{\delta} \frac{x}{\sin x}\left(1 - \frac{x^2}{20} + \cdots\right) dx + \int_{\delta}^{a} \frac{x - \sin x}{x^2 \sin x} dx \tag{15.12}$$

with δ chosen to give adequate accuracy for the truncated series.

Singularities of the form $(b^2 - x^2)^{-1/2}$

In most of the previous examples the strategy has been to isolate the usually quite real singularity in a simple form that is capable of being treated analytically. We now turn to a class of singularities that need not exist—and hence are "removable" by suitable transformations of the independent variable. In Chapter 2 we encountered one: The integral

$$\int_0^a \frac{e^{-x}}{\sqrt{x}} dx = 2\int_0^{\sqrt{a}} e^{-y^2} dy$$

was made tractable by replacing x with y^2. Every sophomore is familiar with another: the trigonometric substitution. Thus we may remove the singularity present in

$$\int_0^b \frac{e^{-x} \cdot dx}{\sqrt{b^2 - x^2}} = \int_0^{\pi/2} e^{-b \sin \theta} \cdot d\theta$$

by letting x become $b \sin \theta$—and our integrand is no longer infinite at the upper limit. In calculus texts the trigonometric transformation is usually advanced as a manipulative device to permit analytic integration, but its principal service is really this singularity suppression—invaluable alike to analytic and numeric manipulation.

But suppose we have *two* such annoying factors in our integrand! Consider

$$\int_0^1 \frac{dt}{\sqrt{(1 - t^2)(b^2 - t^2)}} \tag{15.13}$$

with b^2 not very much bigger than unity. With a cosine substitution we may remove either, *but not both*, of the denominator factors. Of course, if b is considerably larger than unity (say 3) the second factor may be tolerated, but a nearby spike is troublesome. And then consider the closely related integral

$$\int_1^b \frac{dt}{\sqrt{(b^2 - t^2)(t^2 - 1)}} \tag{15.14}$$

in which both singularities are on the range of integration. Again, if b is far from unity we may split the problem in the middle and deal separately with the local singularities in each part. But a b near 1 causes trouble over the whole range. Here we seem fated to choose between two evils.

Fortunately there is an analytic tool available specifically for the simultaneous removal of these two singularities – the Jacobian elliptic functions. They seldom appear in undergraduate curricula and hence are shunned as esoteric or specialized or possibly just difficult. None of these terms is accurate or even fair. They satisfy identities quite similar to those of the standard trigonometric functions, which they resemble in their periodicity and graphical geometry. The elliptic functions are easily available in AMS 55 [Abramowitz and Stegun, 1964], which gives their essential properties and also examples of their use in the removal of singularities. We shall demonstrate their use on (15.13) and (15.14) and on another more complicated integral below. But first we briefly list the properties that we shall need.

Jacobian elliptic function properties

There are three basic functions: $\text{sn}(v)$ is rather like $\sin\theta$, $\text{cn}\, v$ like $\cos\theta$, and $\text{dn}\, v$ something like $\cos 2\theta$ except that it is shifted upward and never goes negative (Figure 15.2). Most important for our purposes are the Pythagorean identities

$$\begin{aligned} \text{sn}^2 v + \text{cn}^2 v &= 1 \\ \text{dn}^2 v &= 1 - k^2 \cdot \text{sn}^2 v \end{aligned} \tag{15.15}$$

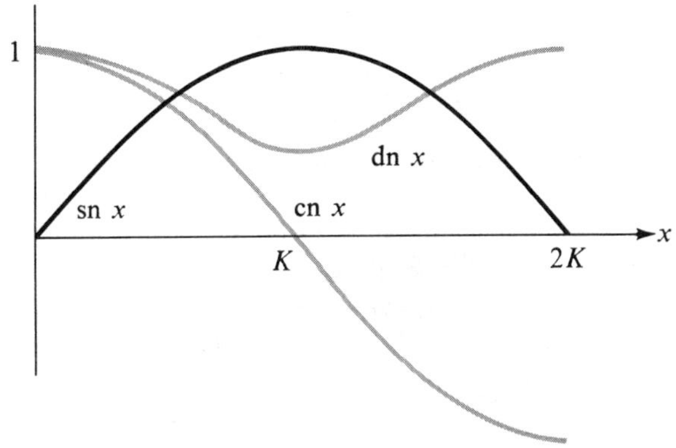

Figure 15.2. Jacobian elliptic function; $k = 1/\sqrt{2}$

and the differentiation formulas

$$\begin{aligned} d(\text{sn } v) &= (\text{cn } v)(\text{dn } v)\, dv \\ d(\text{dn } v) &= -k^2(\text{sn } v)(\text{cn } v)\, dv \end{aligned} \tag{15.16}$$

The reader by now has noticed the appearance of a parameter k. The Jacobian elliptic functions really have two arguments, v and k, the latter traditionally being called a *parameter* to help distinguish it. (We shall use it to abolish the second factor in our integrals.) When v is zero $\text{sn}(v)$ is zero, while both $\text{cn } v$ and $\text{dn } v$ are unity, regardless of k.

In (15.13) we make the formal substitution

$$t = \text{sn}(v|k) = \text{sn}(v)$$

the shorter form usually being used when only one k exists in the problem or whenever no confusion is apt to arise. We obtain

$$\begin{aligned} I(b) &= \int_0^1 \frac{dt}{\sqrt{(1 - t^2)(b^2 - t^2)}} = \int_0^{K(k)} \frac{\text{cn } v \cdot \text{dn } v \cdot dv}{\sqrt{(1 - \text{sn}^2 v)(b^2 - \text{sn}^2 v)}} \\ &= \frac{1}{b} \int_0^{K(k)} \frac{\text{dn } v \cdot dv}{\sqrt{1 - \dfrac{1}{b^2}\,\text{sn}^2 v}} \end{aligned} \tag{15.17}$$

where the first factor in the denominator now disappears through use of the first Pythagorean identity. The upper limit $K(k)$ is the complete elliptic integral of the first kind and need not particularly bother us here. It is merely that value of v that makes $\text{sn } v$ equal to unity—analogous to $\frac{1}{2}\pi$ for $\sin \theta$. Once k is fixed, K is available from tables or may easily be computed. We now observe that if k is identified with $1/b$, then the second Pythagorean identity removes the second singularity and we have the trivial integral

$$I(b) = \frac{1}{b} \int_0^K dv = \frac{K(k)}{b} \tag{15.18}$$

with no quadrature left to perform. In any real problem there would be some function $f(t)$ in the numerator of (15.13) sufficiently complicated to prevent analytic integration. This function would persist into (15.18). Thus we would remove the two singularities but be left with a well-behaved quadrature (always assuming that f is polynomic in its local behaviors).

Turning to our second example (15.14) we have $1 \leq t \leq b$. Here we may achieve our purpose either by

$$t = \frac{1}{\text{dn } v} \qquad \text{or} \qquad t = b\, \text{dn } v \tag{15.19}$$

We choose the first transformation of our dummy variable, requesting the student to verify that the other is equally effective. We have

$$dt = +\frac{k^2 \cdot \text{sn } v \cdot \text{cn } v}{\text{dn}^2 v} dv$$

so (15.14) becomes

$$\int_0^? \frac{k^2 \cdot \text{sn } v \cdot \text{cn } v \cdot dv}{\text{dn}^2 v \sqrt{\left(\frac{1}{\text{dn}^2 v} - 1\right)\left(b^2 - \frac{1}{\text{dn}^2 v}\right)}} = \int_0^? \frac{k^2 \cdot \text{sn } v \cdot \text{cn } v \cdot dv}{\sqrt{(1 - \text{dn}^2 v)(b^2 \cdot \text{dn}^2 v - 1)}} \tag{15.20}$$

The first factor under the radical immediately collapses to be $k^2 \cdot \text{sn}^2 v$ while the second, on applying the Pythagorean identity, is only a little more stubborn. We now have

$$\int_0^? \frac{k \cdot \text{cn } v \cdot dv}{\sqrt{b^2 - 1 - b^2 k^2 \cdot \text{sn}^2 v}} = \frac{1}{b}\int_0^? \frac{k \cdot \text{cn } v \cdot dv}{\sqrt{\left(1 - \frac{1}{b^2}\right) - k^2 \cdot \text{sn}^2 v}}$$

If we are to cancel the cn v of the numerator we must this time identify k^2 with $1 - 1/b^2$, at which point the integral collapses completely into

$$\frac{1}{b}\int_0^? dv = \frac{?}{b}$$

where the upper limit is to be found from

$$b = \frac{1}{\text{dn}(?)} \qquad \text{or} \qquad \text{dn}(?) = \frac{1}{b}$$

Careful examination of Table 16.5 of AMS 55 shows that if

$$k^2 = 1 - \frac{1}{b^2} = m$$

then, when v is $K(k)$, dn v takes on the value $1/b$ ($\sqrt{m_1}$ in AMS notation)—a result that is comfortably analogous to that of the previous example. The final value of the integral is simply $K(k)/b$. Formally, it is the same as (15.18) except that the dependence of K, through k, on b is different.

The computation of elliptic functions and integrals

Lest the reader wonder just how he may evaluate an integral containing a $\text{cn}^2 v$ or an upper limit of K, we here reassure him that several algorithms have been published [see Bibliography]. Further, AMS 55 (16.4, 17.6) gives examples and formulas based on the arithmetic-geometric-mean (AGM) schema that can be encoded in about 10 lines of FORTRAN. They work remarkably accurately, though some of the algorithms based on series may be more efficient. We merely set up

$$a_0 = 1 \qquad b_0 = \sqrt{1 - k^2} \qquad c_0 = k$$

then compute

$$a_{i+1} = \tfrac{1}{2}(a_i + b_i) \qquad b_{i+1} = \sqrt{a_i b_i} \qquad c_{i+1} = \tfrac{1}{2}(a_i - b_i)$$

stopping when c_N is negligible – usually four to six iterations. The value of $K(k)$ is $\pi/(2a_N)$. Once the AGM table for a particular value of k is set up, then sn v may be computed from the recurrence

$$\sin(2\varphi_{n-1} - \varphi_n) = \frac{c_n}{a_n} \sin \varphi_n$$

starting with

$$\varphi_N = 2^N \cdot a_N \cdot v$$

and concluding with

$$\text{sn}(v|k) = \sin \varphi_0$$

The functions $\text{cn}(v|k)$ and $\text{dn}(v|k)$ follow from the Pythagorean identities (15.15). One must watch with some care the branch of the sine during this recurrence, since φ_n must decrease monotonically with n. Also note that the parameter m of AMS 55 is our k^2. Both notations are widely used. For more extensive examples of the use of elliptic functions and convenient compendium of identities thereof, see Bowman [1961].

Integrals similar to (15.13) and (15.14) occur frequently in conformal mapping of polyogonal regions. There the parameter k is associated with the essential geometry of the region, while v, usually complex, is the location of an arbitrary interior point. Thus, typically, we evaluate our elliptic functions for many v's while holding k fixed. This means that the AGM table need not be recomputed until a different geometry is contemplated.

Three "removable" singularities

Riddle: When is a removable singularity not removable?

Answer: When there are three of them.

More interesting, and frustrating, is the integral

$$\int_0^1 \frac{dt}{\sqrt{(1 - t^2)(b^2 - t^2)(a^2 - t^2)}} \qquad 1 < b < a \tag{15.21}$$

where a and b are both larger than, but close to, unity. The same treatment accorded to (15.13) will remove two of the three singularities—leaving one to trouble us. Common sense dictates that we remove the singularity at 1, being on the range of integration, and the factor at b, since it is nearer than the one at a. and hence more damaging. We let

$$t = \operatorname{sn}(v|k) \qquad k = \frac{1}{b}$$

as before, to get

$$I(a, b) = \frac{1}{b}\int_0^{K(k)} \frac{dv}{\sqrt{a^2 - \operatorname{sn}^2 v}} \tag{15.22}$$

an integral that leaves us somewhat apprehensive at the upper limit, where $\operatorname{sn} v$ is equal to unity and hence very near to a. In any quadrature we may avoid the subtractions by rewriting (15.22) through the Pythagorean identity to obtain

$$I(a, b) = \frac{1}{b}\int_0^{K} \frac{dv}{\sqrt{(a^2 - 1) + \operatorname{cn}^2 v}} \tag{15.23}$$

where the quantity $(a^2 - 1)$ may be computed once and for all. (In integrals of this type, which arise constantly in conformal mapping problems employing the Schwartz–Christoffel transformation, the effective computational parameters turn out to be the small quantities $(b - 1)$, $(a - 1)$, and $(a - b)$—which should be stored directly for frequent use. The formal manipulations of the Jacobian functions, however, appear simpler with a and b.

The major contribution to (15.23) arises at the upper limit since $\operatorname{cn} K$ is zero. (We said it was like the cosine, with K playing the role of $\frac{1}{2}\pi$.) If we are worried about the adequacy of (15.23) as the computational form when $\operatorname{cn} v$ is small, we may divide the range of integration at some intermediate point d, transforming the questionable integral via

$$u = K - v$$

so that we now have

$$b \cdot I(a, b) = \int_0^{d} \frac{dv}{\sqrt{(a^2 - 1) + \operatorname{cn}^2 v}} + \int_0^{K-d} \frac{du}{\sqrt{(a^2 - 1) + \left(1 - \dfrac{1}{b^2}\right) \cdot \dfrac{\operatorname{sn}^2 u}{\operatorname{dn}^2 u}}}$$

Since both b and a are slightly larger than unity, both parenthesized expressions in the integrals are small. We factor out the second one, leaving our integral represented by

$$b \cdot I(a, b) = \int_0^d \frac{dv}{\sqrt{(a^2 - 1) + \text{cn}^2 v}} + \frac{1}{\sqrt{1 - \dfrac{1}{b^2}}} \int_0^{K-d} \frac{du}{\sqrt{\dfrac{a^2 - 1}{1 - 1/b^2} + \dfrac{\text{sn}^2 u}{\text{dn}^2 u}}} \tag{15.24}$$

as its expedient computational form. The second integral is the major contributor because of the smallness of $(1 - 1/b^2)^{1/2}$ in the denominator out in front. The quantity $(a^2 - 1)/(1 - 1/b^2)$ is usually the order of unity or larger in typical physical problems. [Some may prefer to restrict $K - d$ to be small, whence they may replace dn u by unity and sn u by u. Then the second integral of (15.24) can be integrated analytically.]

A singular integral equation

In a direct solution for the potential around a parallel-plate condenser we must solve a singular integral equation (Chapter 18) for the unknown charge $c(x)$ on the plate. This equation,

$$100 = \int_0^b c(\xi) \cdot \ln \frac{[(x - \xi)^2][(x + \xi)^2]}{[(x - \xi)^2 + 4][(x + \xi)^2 + 4]} \cdot d\xi \qquad 0 \leq x \leq b \tag{15.25}$$

must hold for all values of x on the range $(0, b)$. When ξ equals x the argument of the logarithm becomes zero and hence the integrand is infinite. The troublesome factor may be separated from its better-behaved companions by employing the properties of the logarithm to rewrite (15.25) as

$$\begin{aligned} 100 = 2\int_0^b c(\xi) \cdot \ln|x - \xi| \cdot d\xi + 2\int_0^b c(\xi) \cdot \ln(x + \xi) \cdot d\xi \\ - \int_0^b c(\xi) \cdot \ln[(x - \xi)^2 + 4] \cdot d\xi - \int_0^b c(\xi) \cdot \ln[(x + \xi)^2 + 4] \cdot d\xi \end{aligned} \tag{15.26}$$

where now our obvious troubles arise from the first integral. For any value of x we may temporarily contemplate, the dummy ξ will equal it at one instant during the performance of the quadrature. A moment's reflection, however, discloses that the second integrand is also singular, though only once – for x and ξ can both simultaneously be zero. The last two integrals seem quite safe with their irreducible deposit of 4 in the logarithmic bank.

Integral equations with this degree of complexity must usually be solved numerically, although occasionally analytic devices such as dual Fourier series [Sneddon, 1966, p. 246] suffice to produce infinite series solutions. The numerical strategy is simple: We replace the unknown continuous function $c(\xi)$ by a set of unknown constants (c_i) on a set of points (ξ_i) covering the range $(0, b)$. If we now set x equal to *one* of the ξ_i, each integral in (15.26) may be expressed by some quadrature formula over the n points. Since the unknown c_i's enter linearly, we now have *one* equation for the n unknowns. But the value of x may be set, in turn, to each of the ξ_i values, thus generating a new equation each time. We therefore find that we have n linear algebraic equations for the n unknown c_i's – which may presumably be solved. So much for the strategy – but the tactics are more complex!

We have already pointed out that some of the integrals we must evaluate have logarithmic infinities in their integrands. Since most of the contribution of area inevitably occurs because of the singularity, we must take considerable care to represent these integrands accurately there. In particular, we must *not* try to ignore these singularities by some such device as choosing our x's on a grid that interlocks with the ξ_i's but never equals them – so that "the infinities never occur" – or some other such form of wishful avoidance. The singularities occur in the *integrals* whether we look carefully at them or not, and the chief contributions to the values of those integrals come from the singularities. Our replacement of the integrals over a finite grid had better preserve those contributions and preserve them accurately. Avoidance is a disastrous tactic; confrontation is required.

Before returning to the details of integral replacement, we must raise another crucial problem. Although it is not obvious from the form of (15.25), it is, nevertheless, a fact that the unknown function $c(x)$ must itself be singular at b, the upper end of its range of definition. Further, it can be determined a priori that $c(x)$ near b behaves like

$$\frac{g(x)}{\sqrt{b^2 - x^2}}$$

where $g(x)$ is finite and slowly varying. This hidden singularity of $c(x)$ at b therefore introduces a second difficulty that we must accommodate while replacing the integrals of (15.26) – and accommodate in all four of these integrals. Naturally the worst troubles occur in the first integral, which now is seen to possess two singularities whenever x is *at or near b*. (For remember that we need not be exactly at a singularity for its influence to be felt decisively.) The remainder of our discussion will deal with some of the integrals we must replace, together with several appropriate replacement tactics.

The distribution of points

We shall take our n points evenly spaced on $(0, b)$ and shall include both end points. Other grids are possible; indeed, some may even be preferable if we wish to maximize our computational efficiency. But for expository purposes this simplest of grids is adequate. It produces all the troubles characteristic of these problems and it also suffices for the ultimate solution of the integral equation – a process we shall not pursue to completion in this book. If we take 11 points we may depict our line as in Figure 15.3. We see that we need to represent each integral over small regions that each include one point in their

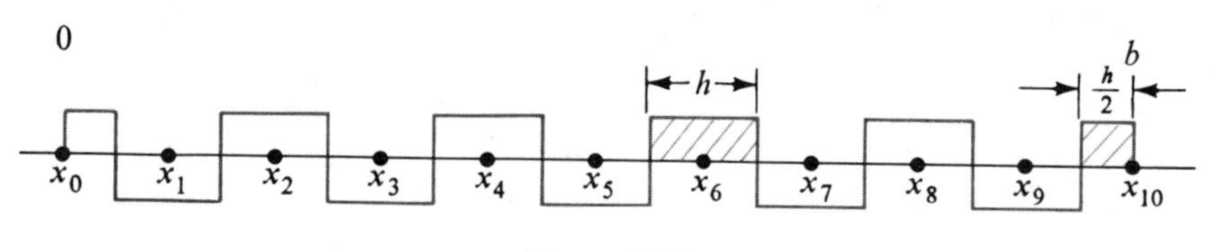

Figure 15.3

interiors. The extreme regions are exceptional, being half as long and having their point at one end. Thus we need to evaluate integrals of which the following are typical:

$$\int_{x_6-h/2}^{x_6+h/2} c_6 \cdot \ln|x_6 - \xi| \cdot d\xi \qquad \text{(A)}$$

$$\int_{x_6-h/2}^{x_6+h/2} c_6 \cdot \ln(x_8 - \xi) \cdot d\xi \qquad \text{(B)}$$

$$\int_{b-h/2}^{b} g_{10} \frac{\ln(b - \xi)}{\sqrt{b^2 - \xi^2}} d\xi \qquad \text{(C)}$$

$$\int_{b-h/2}^{b} g_{10} \frac{\ln[(b - \xi)^2 + 4]}{\sqrt{b^2 - \xi^2}} d\xi \qquad \text{(D)}$$

$$\int_{x_6-h/2}^{x_6+h/2} c_6 \cdot \ln[(x_4 - \xi)^2 + 4] \cdot d\xi \qquad \text{(E)}$$

Here the expressions c_6 and g_{10} are taken to be unknown but constant over the limited range of integration in which they occur. Naturally the more points in our grid, the smaller these ranges and the better this assumption. All combinations of the 11 x_j and the 11 ranges of integration occur for each of the four integrals in (15.26). Most of these 484 integrals are well behaved [for example, (E)] and may be approximated in several straightforward obvious ways. The simplest is to take the value of the integrand at the midpoint of the range and multiply it by h. A precise replacement, however, is often possible through analytic integration. For the integral (E) we obtain, respectively,

$$c_6 \cdot h \cdot \ln(4h^2 + 4)$$

and

$$c_6\left[\frac{5h}{2}\ln\left(4 + \frac{25h^2}{2}\right) - \frac{3h}{2}\ln\left(4 + \frac{9h^2}{4}\right) - 2h + 4\left(\tan^{-1}\frac{5h}{4} - \tan^{-1}\frac{3h}{4}\right)\right] \tag{15.27}$$

by these two philosophies. Whether or not the increased computational complexities of the second form are worth their trouble is a moot point. Since the singular integrals that give the major contributions to our equation system cannot be analytically integrated and since the c_i are not actually constant, we necessarily are stuck with some degree of imprecision elsewhere. Hence we wonder about the wisdom of building too carefully a marble palace upon foundations of mere wood – if not sand. We should, however, do no worse than elsewhere, and since the exact integral can often relieve us of such worries, we do not dismiss it out of hand. But note also that evaluation of the exact form in (15.27) will suffer from precision problems through subtraction of nearly equal quantities when h is very small. The choice in (15.27) remains both judicious and personal.

Turning to the more difficult integrals, we next consider (A). Fortunately it can be integrated analytically – a piece of good luck that we hurriedly accept since here the alternatives are distinctly unpleasant. We have

$$c_6\int_{x_6-h/2}^{x_6+h/2} \ln|x_6 - \xi| \cdot d\xi = c_6\int_{x_6-h/2}^{x_6} \ln(x_6 - \xi)\cdot d\xi + c_6\int_{x_6}^{x_6+h/2} \ln(\xi - x_6)\cdot d\xi$$

$$= 2c_6\int_0^{h/2} \ln u \cdot du \tag{15.28}$$

where the obvious changes in the dummy variables of integration happen to lead to identical forms which still have integrands singular at the lower limit. A final integration by parts removes all our troubles, giving the expression

$$A = h\cdot\ln\frac{h}{2} - h = h\cdot\ln h - h(\ln 2 + 1) \tag{15.29}$$

We note that the first term goes to zero with h, though in our integral replacement we shall never increase the number of points beyond some computationally reasonable number (50 or perhaps 100) and hence h remains distinctly finite.

The integral (B) is well behaved in the sense that analytical integration is again possible. Since the integrand is not singular on the range of the quadrature we may incline toward the use of the simpler midpoint replacement – as with (E). But here such replacement could be dangerous, for the integrand *is* singular just a little off-stage – $2h$ units off stage to be precise. If h is small, this can be quite close indeed and our simplest assumption, that thc integrand is effectively constant over $(x_6 - h/2, x_6 + h/2)$, will no longer be good. The analytic solution is much safer.

When we turn to (C) we come to our most serious difficulties. The blasted thing has *two* singularities in its integrand, both at b. Separately, each can be managed analytically. The logarithmic singularity yields to integration by parts as in (A); the $(b^2 - \xi^2)^{-1/2}$ factor may be removed by the standard trigonometric cosine substitution. But together they are intractable; each maneuver combines with the other to introduce insurmountable analytic complexities in the desingularized integrand. Thus we are forced to complete our quadrature by approximate numerical methods. Several are available and the relative precisions of the many approximate representations for (C) have not been fully investigated by your author. He merely gives two among some eight he has tried. They are not bad.

The first approximation

We begin by making the cosine transformation to rid ourselves of the denominator factor. We have

$$I(b, k) = \int_{b-k}^{b} \frac{\ln(b - \xi)}{\sqrt{b^2 - \xi^2}} d\xi = \int_{0}^{\theta_k} \ln(b - b\cos\theta) \cdot d\theta \qquad \begin{cases} \xi = b\cos\theta \\ b - k = b\cos\theta_k \end{cases} \tag{15.30}$$

The logarithm is, of course, still singular at the lower limit. Integration by parts improves the condition of our integral. It becomes

$$I(b, k) = \Big[\theta \cdot \ln(b - b\cos\theta)\Big]_{0}^{\theta_k} - \int_{0}^{\theta_k} \frac{\theta \cdot \sin\theta}{1 - \cos\theta} d\theta \tag{15.31}$$

and the term in brackets disappears at the lower limit. The integrand of the second term (15.31) is now at least finite at zero. It is computationally indeterminate there, but L'Hospital's rule or series replacement of the trigonometric functions disclose it to have the value 2 there. The integrand also has all derivatives finite at the origin so we may now take considerable liberties with it.

Our manipulations thus far have been exact; henceforth we shall – must – do violence to the integrand. Thus the value of the upper limit becomes pertinent. For typical geometries of physical interest b is the order of unity – say 0.7 to 1.5. If we are using 11 points, then k is $b/20$, so that θ_k is approximately 0.32 (rather larger than your author's intuition expected). Increasing the points to 21 reduces θ_k to 0.224. By now it is apparent that we must be prepared to accommodate values of θ up to 0.2 or 0.3 in any approximation we choose. One such possibility has been to replace $\sin\theta$ and $\cos\theta$ by their series—but the size of θ is discouraging for this tactic, too many terms being required for analytic convenience. (The availability of algebraic manipulation of series by computer will alter our ideas of what is convenient, but we are assuming that such facilities are still awkward. So we seek elsewhere.)

Since the integrand is now distinctly polynomic in its behavior, we should consider applying one of our standard quadrature formulas. In particular, we might apply Simpson's rule to generate an explicit *formula* for the integral in (15.31). Thus we may write

$$\int_0^{\theta_k} \frac{\theta \cdot \sin\theta}{1 - \cos\theta}\, d\theta \approx \frac{\theta_k}{6}\left[\frac{\theta_k \cdot \sin\theta_k}{1 - \cos\theta_k} + 4\frac{\bar{\theta}\sin\bar{\theta}}{1 - \cos\bar{\theta}} + 2\right] \qquad \text{where } \bar{\theta} = \frac{\theta_k}{2}$$

Obviously, better approximations may be written immediately by subdividing the range $(0, \theta_k)$ and applying Simpson's rule multiply. In the same spirit we may use any of the standard quadrature formulas, Gaussian or Newton–Cotes, with any required degree of subdivision and replication. This class of techniques seems the most promising to your author, subdivision being stopped when the required precision has been obtained.

The second approximation

A different approach to the integral of (15.31) lies in first rewriting the denominator as $2\sin^2(\theta/2)$ and then introducing the functions $\sin\theta/\theta$—which vary slowly over the range $(0, \theta_k)$. Thus we have exactly

$$\frac{\theta \cdot \sin\theta}{1 - \cos\theta} = \frac{2(\sin\theta/\theta)}{\left(\dfrac{\sin\theta/2}{\theta/2}\right)^2} \tag{15.32}$$

and now any adequate approximation for $\sin\theta/\theta$ may be applied. Since the largest value of θ will be 0.32, we may try replacing $\sin\theta/\theta$ by a truncation of its Maclaurin series. We have

$$\frac{\sin\theta}{\theta} = 1 - \frac{\theta^2}{3!} + \frac{\theta^4}{5!} - \frac{\theta^6}{7!} \tag{15.33}$$

where the last term will be no larger than 2.1×10^{-7}. We may dispense with this term by Chebyshev economization (Chapter 12), but in so doing it seems desirable to retain the property that the function be exactly unity for θ equal to zero. Thus we first rewrite the series as

$$\frac{\sin\theta}{\theta} = 1 - \frac{\theta^2}{3!}\left(1 - \frac{\theta^2}{20} + \frac{\theta^4}{840}\right)$$

and economize the $\theta^4/840$. Letting $\theta = 0.32y$ and noting that

$$T_4(y) = 1 - 8y^2 + 8y^4 \qquad -1 \le y \le 1$$

we may write

$$\begin{aligned}\frac{\sin\theta}{\theta} &= 1 - \frac{\theta^2}{3!}\left(1 - \frac{\theta^2}{20} + \frac{\theta^4}{840}\right)\\ &= 1 - \frac{\theta^2}{6}\left(1 - \frac{(0.32)^2}{20}y^2 + \frac{(0.32)^4}{840}y^4\right)\\ &= 1 - \frac{\theta^2}{6}\left[1 - \frac{(0.32)^2}{20y}y^2 + \frac{(0.32)^4}{840}\left(-\frac{1}{8} + y^2 + \frac{T_4}{8}\right)\right]\end{aligned} \quad (15.34)$$

Now we may drop the last term, thereby committing a maximum error *inside* the brackets of (15.34) of

$$\frac{(0.32)^4}{840}\cdot\frac{1}{8} = 1.6 \times 10^{-6}$$

or a total maximum error of 2.7×10^{-8} – considerably better than simple dropping of the term. The approximation is

$$\frac{\sin\theta}{\theta} \approx 1 - \frac{\theta^2}{6}(0.99999\,8440 - 0.04987\,810\theta^2) \qquad \text{for } |\theta| \le 0.32 \quad (15.35)$$

In the denominator of (15.32) we have an argument that is only $\frac{1}{2}\theta$, hence less than 0.16, so a different approximation would allow greater precision – or perhaps the same precision with fewer terms. This last remark is important, for our ultimate purpose is to integrate analytically some approximation of (15.32). Thus the presence of a denominator in the form of $(1 - b\theta^2)^2$ would be much more tractable than $(1 - c_1\theta^2 + c_2\theta^4)^2$. Repeating the manipulations of the previous paragraph but with the limit 0.16 we obtain

$$\frac{\sin\theta}{\theta} = 1 - \frac{\theta^2}{1}(0.99999\,99025 - 0.4996\,952\theta^2)$$

by economizing the $\theta^4/840$ term. A second economization will replace θ^2 inside the parentheses by its median value, $(0.16)^2/2$, leading to the simpler approximation

$$\frac{\sin\varphi}{\varphi} \approx 1 - 0.16656\,00488\varphi^2 \qquad |\varphi| \le 0.16$$

with maximum error less than 2.8×10^{-6}.

On substituting into (15.32) we now must integrate

$$I = 2\int_0^{\theta_k} \frac{(1 - a_1\theta^2 + a_2\theta^4)}{(1 - b\theta^2)^2}\,d\theta \qquad \text{with} \begin{cases} a_1 = 1.66666\,407 \\ a_2 = 0.00831\,3017 \\ b = 0.04160\,0122 \end{cases}$$

Long division and partial fraction separations lead directly to the form

$$\frac{1 - a_1\theta^2 + a_2\theta^4}{(1 - b\theta^2)^2} = \frac{a_2}{b^2} + \frac{\dfrac{a_1}{b} - 2\dfrac{a_2}{b^2}}{1 - b\theta^2} + \frac{1 + \dfrac{a_2}{b^2} - \dfrac{a_1}{b}}{(1 - b\theta^2)^2}$$

where each integration may be performed analytically. Our final formula is

$$\frac{I}{2} = \frac{a_2}{b^2}\theta_k + \left(\frac{a_1}{b} - \frac{2a_2}{b^2}\right) \cdot \frac{1}{2\sqrt{b}} \ln\left|\frac{1 + \theta_k\sqrt{b}}{1 - \theta_k\sqrt{b}}\right|$$

$$- \left(\frac{a_1}{b} - \frac{a_2}{b^2} - 1\right)\left(\frac{1}{4\sqrt{b}} \ln\left|\frac{1 + \theta_k\sqrt{b}}{1 - \theta_k\sqrt{b}}\right| - \frac{\theta_k}{2(\theta_k{}^2 b - 1)}\right)$$

The well-behaved integral

Finally we take a look at (D), an integral that has the trigonometrically removable singularity at b but whose logarithm is well behaved. Dropping the constant factor g_{10} and making the usual cosine transformation we have

$$D = \int_{b-h/2}^{b} \frac{\ln[4 + (b - \xi)^2]}{\sqrt{b^2 - \xi^2}}\,d\xi = \int_0^{\theta} \ln[4 + b^2(1 - \cos\theta)^2] \cdot d\theta$$

Using the half-angle identity for $\sin(\theta/2)$, D may be written

$$D = \int_0^{\theta} \ln\left[4 + 4b^2\sin^4\frac{\theta}{2}\right] \cdot d\theta = \theta_1 \cdot \ln 4 + \int_0^{\theta_1} \ln\left(1 + b^2\sin^4\frac{\theta}{2}\right) \cdot d\theta \tag{15.36}$$

and we are now concerned with approximating the last integral. Since we are considering a problem in which b is near unity and θ_1 is apt to be less than 0.32, we may quickly estimate the first term of (15.36) to be

$$(0.32)(1.39) = 0.45$$

while the integral is, crudely,

$$\int_0^{0.32} \ln\left[1 + \left(\frac{\theta}{2}\right)^4 b^2\right] \cdot d\theta \approx \int_0^{0.32} b^2\left(\frac{\theta}{2}\right)^4 \cdot d\theta = 0.00004194b^2 \quad (15.37)$$

Clearly, what we do to estimate this integral is of almost no importance unless we need more than five significant figures. Our crudest technique, exhibited here, is adequate. [We replaced $\sin y$ by y, then replaced $\ln(1 + x)$ by x.] Returning to (15.36) to evaluate the first term adequately we add the contribution from (15.37) to give

$$D_1 = 0.4436\,5613$$

An alternative, though cruder, treatment is to evalue the integrand in (15.36) at the midpoint of the range of integration, at $\theta_1/2$, to give

$$D_2 = \theta_1 \cdot \ln \cdot 4 + \theta_1 \ln\left(1 + b^2 \sin^4\frac{\theta_1}{4}\right) = 0.4436\,272$$

If we want more accuracy we may integrate by parts in (15.36) to get

$$D = \theta_1 \cdot \ln \cdot 4 + \theta_1 \cdot \ln\left(1 + b^2 \sin^4\frac{\theta_1}{2}\right) - 8b^2 \int_0^{\theta_1} \frac{\frac{\theta}{2} \cdot \sin^3\frac{\theta}{2} \cdot \cos\frac{\theta}{2} \cdot \frac{d\theta}{2}}{1 + b^2 \sin^4\frac{\theta}{2}} \quad (15.38)$$

and then replace the $\theta/2$ factor in the numerator by $\sin(\theta/2)$, a quite minor piece of larceny in view of the smallness of $\theta/2$. We get

$$D \approx \theta_1 \ln \cdot 4 + \theta_1 \ln\left(1 + b^2 \sin^4\frac{\theta_1}{2}\right) - 8b^2 \int \frac{y^4}{1 + b^2 y^4} dy \quad (15.39)$$

where

$$y = \sin\frac{\theta}{2}$$

The last form of the integral may, with considerable difficulty, be integrated analytically to yield a quite complicated expression[1] that succeeds in computing the quantity 0.000164 as the difference between 1.27456 and 1.27440. Not good computational tactics at all! Better would be to freeze the complicating

[1] It is

$$8\sin\frac{\theta}{2} - \frac{2}{\sqrt{b}}\left\{\frac{1}{\sqrt{2}} \ln\frac{by^2 + y\sqrt{2b} + 1}{by^2 - y\sqrt{2b} + 1} + \sqrt{2}[\tan^{-1}(y\sqrt{2b} + 1) + \tan^{-1}(y\sqrt{2b} - 1)]\right\}$$

denominator of (15.39) at the midrange of θ, thereby permitting analytic integration of the numerator to give the small contribution of this integral directly. We have, finally,

$$D \approx \theta_1 \cdot \ln \cdot 4 + \theta_1 \cdot \ln\left(1 + b^2 \sin^4 \frac{\theta_1}{2}\right) - \frac{1.6b^2 \sin^5 \dfrac{\theta_1}{2}}{1 + b^2 \sin^4 \dfrac{\theta_1}{4}}$$

$$D_3 = 0.4436\,5607$$

The accuracy of our several approximations may be judged in Table 15.1, where the correct value comes from evaluating the integral of (15.36) via Simpson's rule. When θ_1 is smaller than 0.32, the accuracy of all the approximations increases. For most work the simplest, D_1, is adequate.

Table 15.1

Correct value	0.4436 556	$10^5 \times$ error
$\theta_1 \ln \cdot 4$	0.4436 142	4.0
D_1	0.4436 561 +	0.05
D_2	0.4436 272	3.0
D_3	0.4436 561 –	0.05

Problems

1. Evaluate

$$\int_{0.8}^{1.0} \frac{\ln(1 - x)}{\sqrt{4 - x^2}}\,dx$$

by at least three approximate methods. Also find the correct value to at least six significant figures.

2. Evaluate

$$\int_1^b \frac{dt}{\sqrt{(t^2 - 1)(b^2 - t^2)}} \qquad b = 1.005, 1.1, 1.5, 3.0$$

both exactly, through use of elliptic functions, and also by suitable approximate techniques.

3. Integrate

$$\int_{0.8}^{1.0} \frac{\ln(1 - x)}{\sqrt{1 - x^2}}\, dx$$

after suitably removing the two singularities, showing its value to be approximately

$$-(1.965 + \ln \cdot 5) \cos^{-1}(0.8)$$

How much error still remains in this approximation?

4. Evaluate

$$F(a) = \int_{0}^{10.0} \frac{e^{-at}}{\sqrt{t}}\, dt \qquad a = 0.2(0.2)2$$

without resorting to brute-force use of Simpson's rule over the range (0, 10). Note that if the upper limit were ∞, a change of the dummy variable permits analytic integration in closed form as $\sqrt{\pi/a}$. Check your results against tabled values for accuracy, first having tried to develop some internal estimates of the accuracy you are achieving.

5. Examine integral (B), evaluating it by several techniques of your own devising. Try to estimate their probable error before checking your answers against the exact values from analytic integration.

CHAPTER

16

INSTABILITY IN EXTRAPOLATION

Unstable numerical processes must be ranked high on our list of computational booby traps. They probably cause as much trouble as unanticipated singularities. Instabilities can arise to plague us in Newton's method for solving most transcendental equations; they abound within the plethora of recurrence relations that we may use for evaluating classical functions; and our fear of them often causes our solution process for nonlinear partial differential equations to be woefully inefficient. Their sources are too varied to permit a unified treatment; indeed, many potentially unstable processes are too complex for precise analysis. In this chapter we merely sketch the general properties of one large class of instability: the instability that frequently arises in *extrapolative* processes. We discuss recurrence formulas, the integration of ordinary differential equations from initial conditions, and the solution of the diffusion (parabolic partial differential) equation. The first two subjects have been touched on earlier (Chapters 1 and 5); the last is new. Our hope in combining and partially repeating some of these topics is to convey a qualitative feeling for a phenomenon that is both important and elusive.

"Extrapolative process" is itself not a precise term. We use it here as a generic name for processes that build out from one end—creating their next datum principally from data that they have recently created by earlier cycles of the same process. The purest example of an extrapolative process is the integration from initial conditions of a differential equation. Here each cycle of the process produces a new datum—not a succession of values that is trying to converge to a single value, but rather a progression of values that represents a function of the independent variable. Specifically, we rule out the iterative root-finding algorithms for transcendental equations. The common quadrature rules are not extrapolative within our meaning, nor are the eigenvalues algorithms. A borderline case is the forward elimination process in the solution of simultaneous linear algebraic equations. And, of course, calculation of cosine and Bessel functions by their recurrence relations are definitely extrapolative. We begin with them.

Recurrence calculation

In Chapter 1 we saw that we could evaluate $\cos 3\theta$ from $\cos\theta$ and $\cos 2\theta$, then get $\cos 4\theta$ from $\cos 3\theta$ and $\cos 2\theta$, and so on. We also found, empirically, that the process had some accuracy problems, though they seemed not too severe. The basic formula was the recurrence relation

$$\cos(n+1)\theta = 2\cos\theta\cdot\cos n\theta - \cos(n-1)\theta \tag{16.1}$$

which may be abbreviated as

$$T_{n+1} = 2\cos\theta\cdot T_n - T_{n-1} \tag{16.2}$$

For a specified θ this is a linear *difference* equation of second order with constant coefficients. We are concerned with what it does to errors that are fed into it attached to the T_n's. If we let C_n be the actual numbers we get from the computational process (16.2), then the error in C_n is given by

$$\varepsilon_n = C_n - T_n$$

Since the C's are generated by successive application of

$$C_{n+1} = 2\cos\theta\cdot C_n - C_{n-1}$$

we may subtract the precise algorithm (16.2) to generate the difference equation for our errors. We have, of course the same equation as for the function itself,

$$\varepsilon_{n+1} = 2\cos\theta\cdot\varepsilon_n - \varepsilon_{n-1} \tag{16.3}$$

Equations of this form have elementary solutions

$$\varepsilon_n = \rho^n \tag{16.4}$$

and, being of second order, have two of them. The general solution is a linear combination of the two elementary solutions. Substituting (16.4) into (16.3) we get

$$\rho^{n+1} = 2\cos\theta\cdot\rho^n - \rho^{n-1}$$

or

$$\rho^2 - 2\cos\theta \cdot \rho + 1 = 0 \tag{16.5}$$

a quadratic with two complex roots

$$\rho = \cos\theta \pm i\sin\theta = e^{\pm i\theta}$$

The error ε_n is therefore

$$\varepsilon_n = Ae^{i\theta n} + Be^{-i\theta n} \tag{16.6}$$

As n increases the magnitude of ε_n does not change, because $e^{\pm i\theta n}$ remains on the unit circle. Thus we do not expect an error committed early in our use of the recurrence (16.2) to grow spectacularly, but neither do we expect it quietly to disappear. And this analysis is for the propagation of *one* error—presumably created by rounding off T_0. Later independent errors caused by later roundings intrude themselves to produce a statistical error that tends to grow, though not quickly. For any one specific example we might, of course, be lucky and find that accidental cancellations have given us a smaller than expected error, but it is not likely. *Slow but bearable error growth* is our expectation, the statistical growth being proportional to the square root of the number of iterations.

Turning to the recurrence relation for the Bessel function $J_n(x)$, we find the error equation to be

$$\varepsilon_{n+1} = \frac{2n}{x}\varepsilon_n - \varepsilon_{n-1} \tag{16.7}$$

This also is a linear second-order difference equation, but one coefficient contains n and hence is not constant. Its precise solution is, of course, a Bessel function—or rather a linear combination of the two Bessel functions

$$\varepsilon_n = A \cdot J_n(x) + B \cdot Y_n(x)$$

and, for fixed x, $J_n(x)$ goes to zero with increasing n, but $Y_n(x)$ goes through a minimum for n approximately equal to $x/2$, then ultimately grows large. Hence the error will grow.

If we are not that familiar with the growth properties of Bessel functions we may still get a qualitative idea of the error if we imagine n/x to be frozen at some fixed value and use the solution of this constant-coefficient difference equation. We have

$$\rho^2 - \frac{2n}{x}\rho + 1 = 0 \qquad \frac{n}{x} \text{ constant} \tag{16.8}$$

so

$$\rho = \frac{n}{x} \pm \sqrt{\left(\frac{n}{x}\right)^2 - 1} \tag{16.9}$$

and

$$\varepsilon_n = A\left[\frac{n}{x} + \sqrt{\left(\frac{n}{x}\right)^2 - 1}\right]^n + B\left[\frac{n}{x} - \sqrt{\left(\frac{n}{x}\right)^2 - 1}\right]^n \tag{16.10}$$

If $|n/x| < 1$, then both terms are complex and $|\varepsilon_n|$ is unity, so that ε_n does not grow. But if $n/x > 1$, then the first term grows in unbounded exponential fashion. Thus our recurrence is clearly unstable if $n > x$. This error growth is doubly disastrous for $J_n(x)$ since the function itself becomes small as n increases, so the error quickly overtakes the function.

The Legendre polynomials have the three-term recurrence

$$(n + 1)P_{n+1}(x) = (2n + 1) \cdot x \cdot P_n(x) - n \cdot P_{n-1}(x) \tag{16.11}$$

Again the difference equation for the local error, being the same as that for the function, is linear but with coefficients that, because of n, are variable. Our technique of pretending that n is constant gives the quadratic

$$n(n + 1)\rho^2 - (2n + 1)x\rho + n = 0 \tag{16.12}$$

The solutions are

$$\rho = x\left(\frac{n + \frac{1}{2}}{n + 1}\right) \pm \sqrt{\left(\frac{n + \frac{1}{2}}{n + 1}\right)^2 x^2 - \frac{n}{n + 1}} \approx x \pm \sqrt{x^2 - 1} \tag{16.13}$$

When x is greater than unity the error will behave like

$$\varepsilon_n = A(2x)^n + B(0)$$

which is distinctly unpleasant. However, the normal range of x for the Legendre polynomials is $|x| < 1$, and here the error is complex with magnitude

$$|\varepsilon_n| = \frac{n}{n + 1} < 1$$

slightly less than unity—thus the individual errors in the recurrence will damp out. Because these rounding errors made at each iteration combine statistically, slow erosion of the result can be expected. The student desiring a more thorough treatment of this topic would do well to begin with an article by Oliver [1967].

Ordinary differential equations—initial-value problems

In Chapter 5 we have seen that instabilities can arise to plague some predictor-corrector schemes for integrating ordinary differential equations from initial conditions. All initial-value problems in ordinary differential equations are necessarily extrapolative. We must build out into the unknown from our

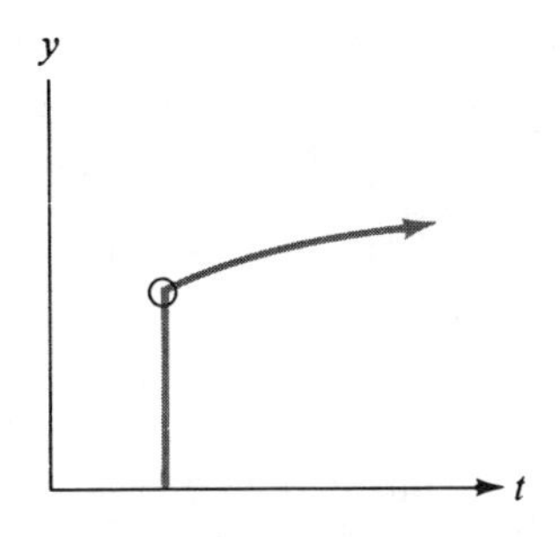

Figure 16.1

starting value(s) (Figure 16.1) without the safe rock on which to rest the far end of our finished solution curve that we have in the boundary-value problem (Chapter 6). But the extrapolative schemes available to us divide into those that gather all their information at the one current point versus those that utilize the past history of the solution that sits available in several previous points. Computational efficiency generally favors the several-point schemes, but their uncritical use carries with it the danger of occasional disasters through instability.

The several-point schemes, generally known as predictor-corrector integration methods, compute their next value, y_{n+1}, of the solution from some small number of previous values and their derivatives. Often y_{n+1} also depends on the derivative at the current point. Formally we may express these facts by

$$y_{n+1} = F(y_n, y_{n-1}, \ldots ; y'_{n+1}, y'_n, y'_{n-1}, \ldots) \tag{16.14}$$

but such formalisms are not particularly revealing. Indeed, the functional dependence of y_{n+1} on the several y'_i and y_i's is usually nonlinear and quite complicated, although as computational procedures they are rather simple to carry out. A person wishing to study the stability of a particular predictor-corrector method thus has two courses open to him: he may decide on an experimental approach, actually using the scheme on a number of typical standard problems with known solutions to see what happens; or he may make simplifying assumptions sufficient to linearize (16.14) into

$$y_{n+1} = a_n y_n + a_{n-1} + \cdots + b_{n+1} y'_{n+1} + b_n y'_n + \cdots \tag{16.15}$$

whence some rigorous analyses become possible—for very simple differential equations such as

$$y' + ky = 0 \tag{16.16}$$

Lest the reader conclude that such analyses are too removed from real problems to be useful, we should point out that this analytic approach at least permits us to detect and reject *unstable* schemes. For if a predictor-corrector system is unstable for the simple exponential differential equation (16.16), we certainly should think twice before adopting it as our general tool for integrating all sorts of complicated differential equation systems. The choice of an integration scheme depends on several conflicting *desiderata*, but stability is clearly one of the more important. Most experienced problem solvers would rank it first, ahead of low truncation and round-off errors, numerical convenience, number of derivative evaluations, and so on.

We note that (16.15) is a linear difference equation. For predictor-corrector systems this equation is of higher order than the order of the differential equation it represents. Thus it has *extraneous* solutions, solutions that are not solutions to the differential equation and hence are unwanted by us. For a predictor-corrector scheme to be stable these extraneous solutions must disappear—if not absolutely, at least they must not become large relative to the solution of the differential equation. For the simple differential equation (16.16) with the solution e^{-kx}, this requirement translates into a requirement that the indicial (that is, characteristic) equation of (16.15) must have roots that are all inside the unit circle except for one which lies on it to give rise to the wanted solution. The analysis is instructive but tangential to our interests. We refer the curious student to the excellent treatment accorded the subject by Ralston [1965] and, earlier, by Hamming [1962]. Our readers with initial-value differential equation problems to solve would do well to use one of the standard package schemes that are certified stable and reasonably efficient. Construction of a new integration routine is by now a highly specialized art demanding both judgment and experience. It should not be entered upon casually by the novice. Fortunately it seems no longer necessary, as the standard techniques are by now quite good. The difficulties that remain for the user are the troubles that may be inherent in his differential equation itself rather than those introduced by the integration scheme. For example, while moving in the direction of increasing x, no standard extrapolative scheme, however stable, will follow the e^{-kx} solution of

$$y'' - k^2 y = 0$$

for very long. The growing solution e^{kx} will enter and soon dominate the numerical computation. The problem, rather than the method, is poorly posed. It is with such considerations that the engineer and other users of computers must be principally concerned. But this is the business of Chapter 6, to which we refer the reader.

Parabolic partial differential equations

Parabolic partial differential equations differ essentially from elliptic equations in that they are necessarily extrapolative. As with Laplace's equation, this book attempts only to present the basic numerical properties of parabolic differential equations. For this purpose the diffusion equation

$$\frac{\partial^2 u}{\partial x^2} = \frac{1}{c^2}\frac{\partial u}{\partial t} \tag{16.17}$$

will usually suffice. This equation describes the flow of heat in a long, thin insulated rod (Figure 16.2) that has constant temperature at each cross section. We are given the initial temperature distribution along its length, $f(x)$, and are also told what the temperatures are at both ends of the rod; that is, we are given boundary conditions, possibly as functions of time, as well as initial conditions. The problem is to discover the temperature $u(x, t)$ at each point in the rod at all subsequent times.

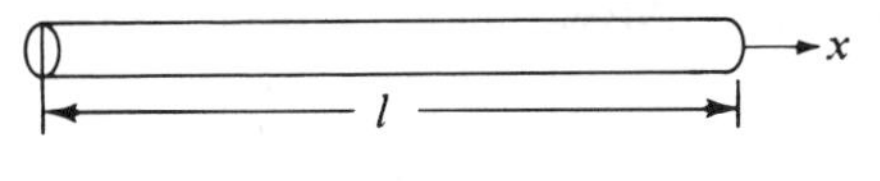

Figure 16.2

For both physical and mathematical reasons this is an initial-value problem. We know the value of u at some initial time and our job is to extrapolate forward in time to find later temperature distributions. Entropy forbids that the diffusion process run backward, and fundamental mathematical difficulties prevent us from trying to solve the problem backward, too. In terms of the (x, t) plane we know u on three boundaries (Figure 16.3), and our problem is to build out in the direction of increasing t from the given conditions and the differential equation.

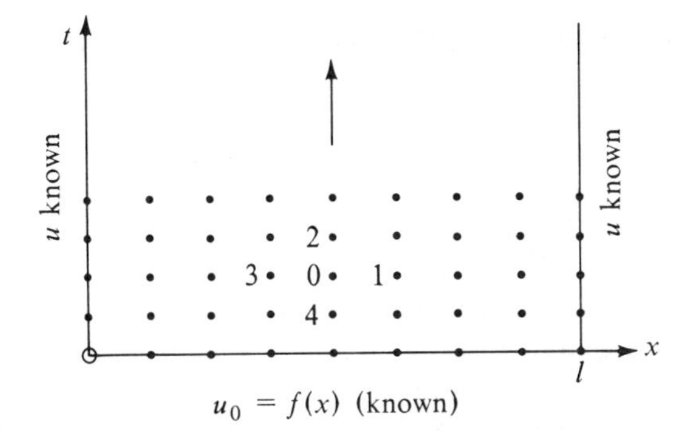

Figure 16.3

No essential points are lost if we nondimensionalize our equation, absorbing the diffusion parameter c into a new time variable. We have

$$\frac{\partial^2 u}{\partial x^2} = \frac{\partial u}{\partial t} \tag{16.18}$$

If we now make the standard replacements for the partial derivatives in (16.18) on the grid of Figure 16.3, we obtain

$$\frac{u_1 - 2u_0 + u_3}{h^2} = \frac{u_2 - u_0}{k} \tag{16.19}$$

where the horizontal spacing h and the vertical (time) spacing k may be different—there being no reason of symmetry or similarity of dimension that might imply the desirability of a square grid. While our replacement of $\partial^2 u/\partial x^2$ is standard enough, we have made a somewhat arbitrary choice in (16.19) for $\partial u/\partial t$. With equal plausibility we might have used the symmetric approximation

$$\frac{u_2 - u_4}{2k} \approx \frac{\partial u}{\partial t} \tag{16.20}$$

or, with perhaps somewhat less intuitive appeal, the backwards approximation

$$\frac{u_0 - u_4}{k} \approx \frac{\partial u}{\partial t} \tag{16.21}$$

All three approximations converge to $\partial u/\partial t$ in the limit as k goes to zero—but of course we never make k that small, so the question remains about the relative advantages of these three replacements. The answer to this question comprises the remainder of this chapter.

The forward difference

Taking the *forward* difference replacement (16.19) for the time derivative our difference equation reads

$$u_2 = u_0 + \frac{k}{h^2}(u_1 - 2u_0 + u_3) \tag{16.22}$$

This single computational rule links three adjacent temperatures at one time with one temperature at the next time step forward. The four points of the computational star lie in a triangle (Figure 16.4) that points forward. Once we know the temperatures along one time line in our (x, t) picture, then (16.22) permits us to calculate the temperatures, point by point, on the next line. We have one application of (16.22) for each interior point to be calculated; the boundary values enter only in the calculation of the extreme interior points.

The algorithm is extremely simple; no simultaneous equations need be solved, only one line of temperatures needs to be stored at any instant, the computation moves in the proper direction – what could go wrong?

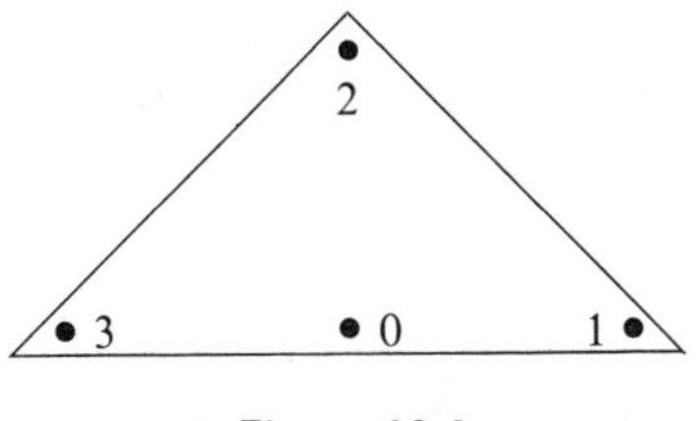

Figure 16.4

Such euphoria is common, but it is dangerous. Our pleasure in the simplicities of the algorithm may blind us to its essential extrapolatory character. The future values depend directly on the present values; errors currently present are certainly transmitted forward. The question is whether they are *amplified* or *diminished* during the transmission. If these errors are amplified, the simplicity of our algorithm is useless for we shall not be able to employ it for long. We must examine what happens to errors as they pass through (16.22).

We shall conduct the error investigation in an heuristic fashion, being content to refer the reader to the rigorous treatments that, we feel, are too cumbersome to serve the average problem well. It is sufficient to demonstrate one unstable configuration to damn any iterative extrapolatory algorithm, for Murphy's law guarantees that: If, in an experimental situation, anything *can* go wrong, it will!

We have one parameter, k/h^2, in (16.22) still at our disposal. As a place to begin we shall consider setting k/h^2 to unity, the value being pleasant, but arbitrary. Our algorithm now reads

$$u_2 = u_1 - u_0 + u_3 \qquad \frac{k}{h^2} = 1 \tag{16.23}$$

Let us now consider what would happen if all the values of u along one time row in (x, t) plane had identical positive errors of size ε (Figure 16.5). Applying (16.23) it is clear that the next row of u's will also exhibit the same ε errors, having

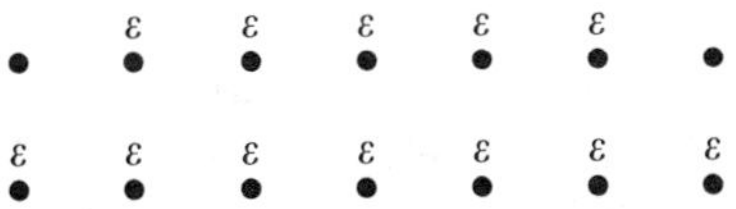

$$\begin{array}{ccccccc} \bullet & \overset{\varepsilon}{\bullet} & \overset{\varepsilon}{\bullet} & \overset{\varepsilon}{\bullet} & \overset{\varepsilon}{\bullet} & \overset{\varepsilon}{\bullet} & \bullet \\ \overset{\varepsilon}{\bullet} & \overset{\varepsilon}{\bullet} & \overset{\varepsilon}{\bullet} & \overset{\varepsilon}{\bullet} & \overset{\varepsilon}{\bullet} & \overset{\varepsilon}{\bullet} & \overset{\varepsilon}{\bullet} \end{array}$$

Figure 16.5

inherited them unchanged by being pumped through (16.23) since

$$\varepsilon_2 = \varepsilon_1 - \varepsilon_0 + \varepsilon_3 = \varepsilon - \varepsilon + \varepsilon = \varepsilon$$

Not good—but not catastropic either. The uniform error pattern is simply reproduced.

If we now propose an error pattern with alternating signs on one row the story is less pleasant. Now (Figure 16.6) one application of (16.23) produces an

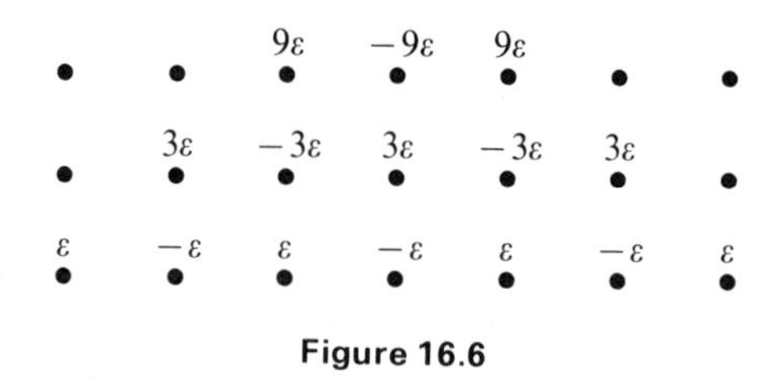

Figure 16.6

error of $+3\varepsilon$ or -3ε, and these errors again alternate in sign. The next row will again be amplified by 3 to give alternating errors of $\pm 9\varepsilon$, and so on. This error pattern is unstable under algorithm (16.23) and hence the algorithm is unstable. Any row of errors can be broken up into segments that are either uniform in sign or are alternating in sign. Since the uniform pattern is propagated undiminished while the alternating pattern is exponentially amplified, the average behavior (in some approximate sense) is amplification by a *factor* that is between 1 and 3. The rate is really unimportant—the mere fact of exponential amplification suffices to make us look for a different algorithm.

We now consider a value for k/h^2 of $\frac{1}{4}$. The general forward difference algorithm (16.22) becomes

$$u_2 = \tfrac{1}{4}u_1 + \tfrac{1}{2}u_0 + \tfrac{1}{4}u_3 \qquad \frac{k}{h^2} = \tfrac{1}{4} \tag{16.24}$$

We try our two patterns of errors on this algorithm and again the uniform pattern is transmitted undiminished. The alternating-sign pattern, however, this time *vanishes completely*. We have

$$\varepsilon_2 = \tfrac{1}{4}(+\varepsilon) + \tfrac{1}{2}(-\varepsilon) + \tfrac{1}{4}(+\varepsilon) = 0$$

Thus the average behavior is somewhere between an amplification of one and zero—and amplification factors *less* than unity are better termed suppression factors. For our algorithm (16.24) is a *noise suppressor*. It is stable in the sense that any one rounding error will be suppressed. We still must worry about the fact that we are continually making new rounding errors and that they combine statistically to erode our results—but such growth is never exponential and usually is proportional to the square root of the number of arithmetic operations. Since all extrapolatory algorithms suffer from this statistical erosion, it is not the crucial factor in our choice between them.

The discovery that (16.24) is exponentially stable, while cheering to the analyst, is less satisfying to the bill payer. Normally we would like to take h fairly small in order to represent adequately the various bumps in the initial temperature profile. Likewise, we would usually prefer then to step along in time with fairly good size steps – being anxious to find out how rapidly the diffusion process is approaching its stable asymptotic configuration. But if h is small, then h^2 is very small, and our choice of $\frac{1}{4}$ for k/h^2 makes k one quarter of h^2, thereby causing us to take many very small steps in time. This is both annoying and costly. Accordingly, we shall examine the other replacements for the time derivative, (16.20) and (16.21), in our parabolic partial differential equation to see if they offer any better prospects.

Before turning to them, however, we would finally look at (16.22) with k/h^2 set to $\frac{1}{2}$. This value has been recommended by otherwise respectable texts on the grounds that it gives the simplest possible algorithm

$$u_2 = \tfrac{1}{2}(u_1 + u_3) \qquad \textbf{(16.25)}$$

each point being the average of two points in the previous time line. Unfortunately, this rule creates a checkerboard of temperature values containing red (⊙) and black (×) temperatures (Figure 16.7) that know nothing about each other.

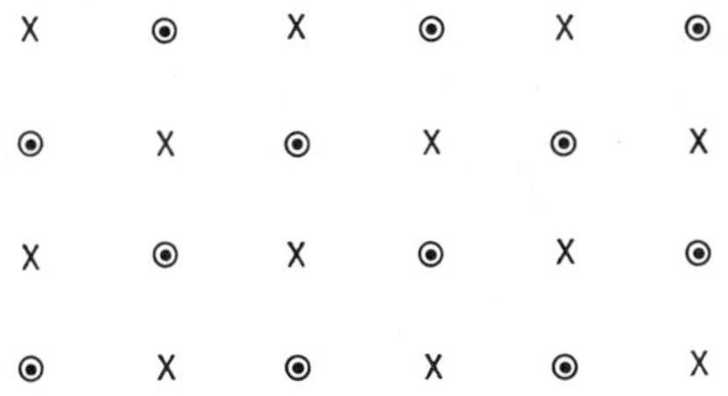

X ⊙ X ⊙ X ⊙

⊙ X ⊙ X ⊙ X

X ⊙ X ⊙ X ⊙

⊙ X ⊙ X ⊙ X

Figure 16.7

Each red temperature is the average of two previous red temperatures, while each black temperature is the average of two black. If we think of the physical problem that this process is supposedly imitating, we see that the temperature at x_0 now depends [according to (16.25)] only on the previous temperatures at *adjacent portions* of the rod and not at all on the temperature at x_0 itself a moment ago. This is patent physical nonsense! We need no stability analyses to reject the proposed algorithm as absurd. (For the record, the value $\frac{1}{2}$ turns out to be the critical value for k/h^2 that divides the unstable algorithms from the stable ones. For practical computation various considerations of efficiency have caused authors [Milne, 1953] to recommend the stable values $\frac{1}{4}$ and $\frac{1}{6}$ for actual work.)

We may formalize the preceding error analysis by letting E_n represent the maximum absolute error on time row n. Then, using the triangular inequality and letting λ represent the grid parameter k/h^2, the error propagation equation for (16.22)

$$\begin{aligned} e_{j,n+1} &= e_{j,n} + \lambda(e_{j+1,n} - 2e_{j,n} + e_{j-1,n}) \\ &= e_{j,n} \cdot (1 - 2\lambda) + \lambda(e_{j+1,n} + e_{j-1,n}) \end{aligned} \tag{16.26}$$

gives

$$e_{j,n+1} \leq (1 - 2\lambda)E_n + 2\lambda E_n = E_n \quad \text{if} \quad 1 - 2\lambda > 0 \tag{16.27}$$

and

$$e_{j,n+1} \leq (2\lambda - 1)E_n + 2\lambda E_n = (4\lambda - 1)E_n \qquad \text{if } 1 - 2\lambda < 0 \tag{16.28}$$

(Worse bounds may easily be obtained, but they do not interest us.) If λ is less than one half, we see from (16.27) that *any* error on time row $n + 1$ is no worse than the worst error on time row n—hence the algorithm is stable. When λ is greater than one half, however, (16.28) only tells us that the error on time row $n + 1$ is no worse than the worst error of time row n *multiplied* by the factor $(4\lambda - 1)$—which is always greater than unity. Thus inequality (16.28) cannot guarantee stability. By the same token it does not rule stability out. To show that instability can actually occur we have to use another tool. But we have already exhibited an error pattern that is unstable, so we shall not pursue the same matter analytically. The student desiring to see the analytic treatment may peruse Isaacson and Keller [1966, pp. 503–504].

Backward difference

We now consider replacing the time derivative of (16.18) by the *backward difference* (16.21). The algorithm reads

$$\frac{u_0 - u_4}{k} = \frac{u_1 - 2u_0 + u_3}{h^2}$$

or

$$\frac{k}{h^2}(u_1 - 2u_0 + u_3) - u_0 = -u_4 \tag{16.29}$$

where we have placed all the temperatures from the new time line on the left, the old one on the right. Our computational star is Figure 16.8, and it points backward. We must, of course, use it in the forward direction, however, so we see that (16.29) thus connects three unknown temperatures with one that is known and hence is not immediately useful. This contrasts sharply with the forward algorithm (16.22), which permits immediate computation of a new temperature in terms of temperatures already known.

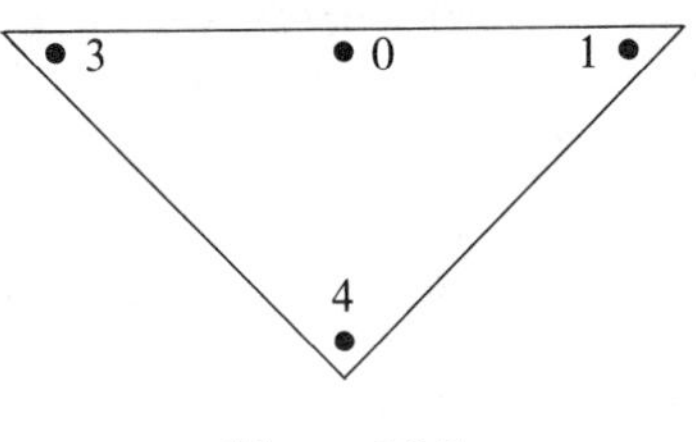

Figure 16.8

It is only when we consider an entire row of temperatures that we see how to use our backward equation. We have one backward equation at each interior grid point (Figure 16.9) on the forward row and hence, having also exactly one unknown temperature at each interior point, we have a system of n linear algebraic equations in n unknowns that may be solved. The boundary points

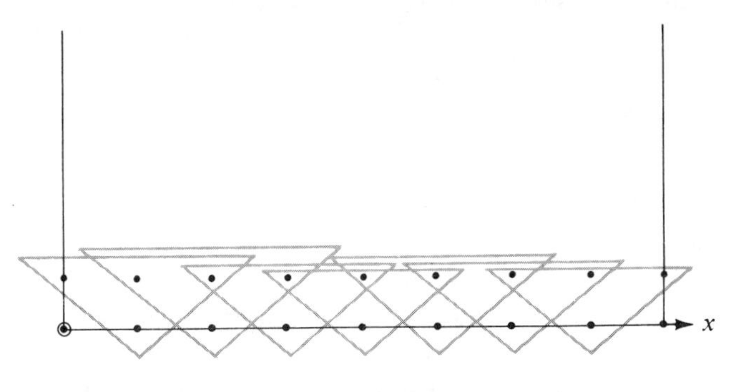

Figure 16.9

also enter into our system of equations, but the temperatures there are known from the boundary conditions. Thus we are able to compute temperatures via (16.29) *a line at a time* instead of a point at a time as before. We must pay the price of solving a system of linear equations, albeit a system of a very special form that lends itself readily to efficient evaluation – but of that, more later. For the moment we merely admit that the backward algorithm seems to require more computation per line than does our forward algorithm. We therefore shall not be particularly enthusiastic about using it unless it has some compensatory advantages—like stability. Fortunately, it does.

The errors committed by using a system of equations (16.29) from one time row of Figure 16.9 satisfy the same algebraic equations. But if we now postulate the error patterns that disclosed the behavior of our point-at-a-time algorithm, we cannot immediately discover their behavior for the backward algorithm. The simultaneous equation system obstructs our vision and calculations. We must reason via a different route. The basic error equation is

$$e_{i,n} = e_{i,n-1} + \frac{k}{h^2}(e_{i+1,n} - 2e_{i,n} + e_{i-1,n})$$

which may be written

$$e_{i,n} + 2\lambda e_{i,n} = e_{i,n-1} + \lambda(e_{i+1,n} + e_{i-1,n}) \qquad \lambda = \frac{k}{h^2}$$

where we have arranged our equation so as to have positive coefficients throughout. Considering all the errors on row n, let us designate the *largest* of them by E_n. We then have

$$(1 + 2\lambda)e_{i,n} \leq E_{n-1} + 2\lambda E_n \tag{16.30}$$

This inequality shows that no error on the nth row, when multiplied by $(1 + 2\lambda)$, is worse than the expression on the right of (16.30). In particular, this is true for the worst error of the $e_{i,n}$ in the nth row—which we have called E_n. Thus

$$(1 + 2\lambda)E_n \leq E_{n-1} + 2\lambda E_n$$

or

$$E_n \leq E_{n-1}$$

and λ has disappeared. We thus have shown that the worst error on the nth row is no worse than the worst error on the previous row—that is, the algorithm is stable. Thus the individual errors are not amplified *no matter what value of the grid spacing parameter, λ, we choose.* This result opens up the possibility of employing a rather large λ that would permit us to march along with a much larger time step via the backward difference algorithm than would be permitted by (16.22). Thus, although one time step may involve more arithmetic (we have to solve a system of equations) we may be able to take sufficiently fewer time steps to achieve a net saving of computational labor.

Whether or not this potential saving is actually realized depends on the efficiency with which we solve our system and also on the size of λ we may use—for λ is far from free. We have only shown that large λ's do not imply exponential instability; nothing has been said about how well the stable solution of the difference equation (16.29) conforms to the solution of our differential equation

(16.18). We are still bound by the requirement for small cumulative errors. As with all reasonable replacements of continuous problems by finite-difference approximations on a grid, the agreement between the continuous problem and its finite-difference replacement grows better as the grid is refined, approaching exact replication in the limit as (h, k) go to zero. Conversely, as the grid is coarsened, the agreement becomes worse. Thus, although we may still be able to solve (16.29), we rapidly lose interest in the solution merely because it no longer happens to represent our problem.

These two concepts, *truncation error* and *stability*, are essentially different. Both are important. The student should make sure he understands them and does not confuse their effects. The forward-difference algorithm may have a quite small truncation error, especially if both h and k are small, but if k/h^2 is greater than $\frac{1}{2}$ the algorithm is unstable and hence nearly useless. This remains true even as h and k approach zero provided only that λ remains greater than $\frac{1}{2}$. On the other hand, the forward difference algorithm remains stable for λ less than $\frac{1}{2}$ and the backward difference algorithm remains stable for all λ as (h, k) are made larger, but the truncation error grows with such change until neither of these finite-difference algorithms is an adequate representation of the original problem.

When we look at the form of the equations we must solve in the line at a time backward-difference algorithm we see that the matrix is *tridiagonal.* Thus the amount of labor is proportional to r, the number of points in the row, rather than to r^3 as a general $r \times r$ system would require. Furthermore, for any constant coefficient parabolic partial differential equation the matrix is always the same, so it may be factored once and for all and then each row requires only the computation of the right-hand side followed by multiplication by the two factored matrices (see p. 481 in Chapter 18). The labor is more per row than for the point-at-a-time algorithm, but it is far from prohibitive, being about twice as many arithmetic operations. Thus if we can take time steps that are more than twice as long by using the backward algorithm we have a net profit.

The symmetric difference

If we replace the time derivative in the diffusion equation (16.18) by the symmetric difference (16.20) we have

$$\frac{u_2 - u_4}{2k} = \frac{u_1 - 2u_0 + u_3}{h^2}$$

or

$$u_2 = u_4 + 2\frac{k}{h^2}(u_1 - 2u_0 + u_3) \tag{16.31}$$

and our computational star now involves all five points (Figure 16.10). Clearly this algorithm, requiring solution values from two preceding time rows, could never be used to begin our solution; nevertheless we can still wonder whether it might not be useful in continuing after starting by another algorithm. What are its computational characteristics? It certainly does not require a solution of a system of equations, a favorable point. What about stability?

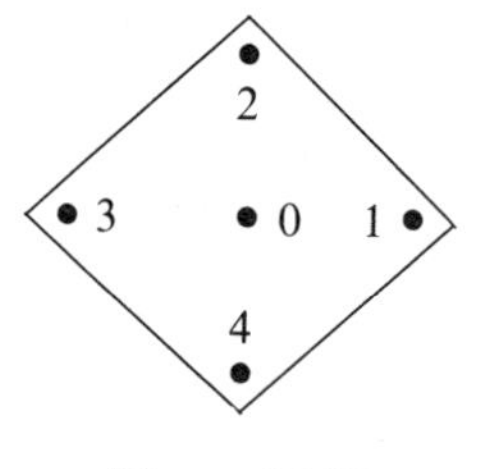

Figure 16.10

We may postulate our two standard patterns. If the errors are uniform in size, we obtain our standard result; the errors propagate unchanged. Errors of uniform size but checkerboarded in sign, however, yield errors that are

$$e_{n+1} = e_n \cdot (1 + 8\lambda)$$

an unstable growth for all λ. Thus (16.31) is a uniformly *unstable* algorithm and should never be used. Ironically, the *truncation* error of (16.31) is an order of magnitude *smaller* than those of the two useful algorithms. Thus a person who considered only the accuracy of his differential equation's replacement would conclude that (16.31) gives a better solution method than (16.24) or (16.29). Unfortunately, in extrapolatory problems exponential instability will defeat all minor virtues!

Another symmetric replacement

Instead of replacing our parabolic differential equation (16.18) on a symmetric star centered on a grid point, as in Figure 16.10, we can use the star Figure 16.11, again symmetric but centered between two points with the same space-direction coordinate. Noting the new numbering of the points, we have

$$\frac{u_2 - u_5}{k} = \frac{1}{2}\left(\frac{u_1 - 2u_2 + u_3}{h^2} + \frac{u_4 - 2u_5 + u_6}{h^2}\right)$$

or

$$u_2 - \frac{\lambda}{2}(u_1 - 2u_2 + u_3) = u_5 + \frac{\lambda}{2}(u_4 - 2u_5 + u_6) \tag{16.32}$$

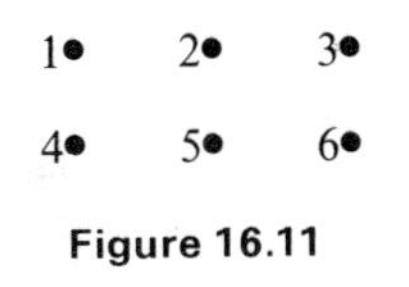

Figure 16.11

which is

$$(1 + \lambda)u_2 - \frac{\lambda}{2}(u_1 + u_3) = (1 - \lambda)u_5 + \frac{\lambda}{2}(u_4 + u_6) \qquad \textbf{(16.33)}$$

Again we have several unknowns on the left, requiring solution of the algorithm a line at a time. A stability analysis which may be carried out by the method used for the backward time-difference algorithm shows (16.33) to be stable for all λ less than or equal to one. Actually it is stable for *all* λ, though nonelementary techniques are required to establish this fact. See Isaacson and Keller [1966]. This algorithm, called the Crank–Nicholson method, has smaller truncation errors (they are proportional to $\Delta t^2 + \Delta x^2$) than any of the other stable algorithms discussed in this chapter. This fact constitutes a possible computational advantage over the backward difference algorithm. But from the stability point of view there is no difference between them.

Problems

1. Consider the functions

$$I_n = \int_0^1 x^n \cdot e^{x-1} \cdot dx \qquad n = 0, 1, \ldots$$

Derive the recurrence

$$I_n = 1 - nI_{n-1} \qquad n = 1, \ldots$$

and examine its stability characteristics.

2. Start with I_0 to six significant figures and evaluate $I_1, I_2, \ldots, I_q$ by a desk calculator – which involves no rounding. Compare your value of I_9 with its correct value of 0.0916123.

3. Start with the correct value of I_9 and recurse your way back to I_0. Is the experimental evidence in agreement with your analysis of Problem 1?

4. Consider how you might use the cylinder function recurrence

$$C_{n+1} + C_{n-1} = \frac{2n}{x} \cdot C_n$$

to evaluate the "other" Bessel function $Y_n(x)$.

CHAPTER 17

MINIMUM METHODS

It is with a sense of reluctance that your author introduces this topic, for minimum-seeking methods are often used when a modicum of thought would disclose more appropriate techniques. They are the first refuge of the computational scoundrel, and one feels at times that the world would be a better place if they were quietly abandoned. But even if these techniques are frequently misused, it is equally true that there are problems for which no alternative solution method is known—and so we shall discuss them. A better title for this chapter might be "How to Find Minima—If You Must!"

Minimum methods are popular because one may easily take any set of complicated equations and produce from them a positive function whose minimum coincides with the solution to the equations. We may then seek the minimum—comfortable in the feeling that if we are going downhill we must be getting closer to the answer. This philosophy provides a simple idea capable of being simply implemented by an iterative program that may even be self-correcting. The unpleasant fact that the approach can well require 10 to 100 times as much computation as methods more specific to the problem is ignored—for who can tell what is being done by the computer? Minimum methods do not even possess the reliability their users impute to them. They may oscillate indefinitely, producing no answer at all, or they may converge on minima that do not correspond to a solution of the original equations. But every person who has learned formal differentiation can visualize the problem of finding a minimum in two dimensions and can devise some scheme that has a fair chance of getting there, eventually. We therefore face a steady stream of visitors to the computer who would seek minima and our principal purpose is to help find them efficiently—unless, of course, we can suggest better ways to solve their problems.

Before trying to catalogue the various techniques, we shall consider a problem for which we know the answer. We take the simultaneous equations

$$y = \frac{3.5}{x}$$

$$x = \tan^{-1} y \text{ (principal branch)}$$

which we have already solved by iterations that depended rather strongly on a knowledge of the geometry of these two curves in the (x, y) plane. We construct

$$f = \left(y - \frac{3.5}{x}\right)^2 + (x - \tan^{-1} y)^2$$

which is clearly positive except at the solution to the set, where it is zero. Figure 17.1 shows a rough plot of contours along which f is constant. From an arbitrary point, $A(x_0, y_0)$ we may easily compute the direction **grad** f and take a step *backward* along it (the gradient points *uphill*). If our step is moderate

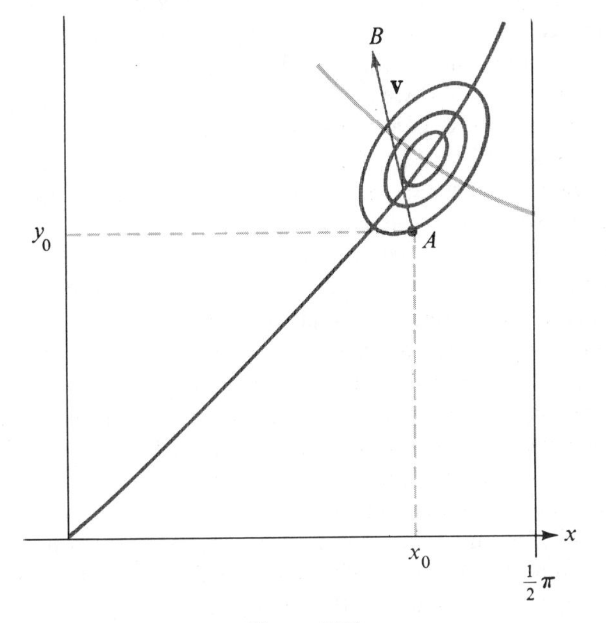

Figure 17.1

enough, we will achieve some progress downhill and may then repeat the process. Too large a step will obviously carry us across the valley—possibly to a point, B, higher than we currently occupy, so some care is required. Too small a step wastes time; too large a step produces instability. These are two horns of the dilemma.

A further difficulty arises whenever the contours are long and narrow (Figure 17.2), for most gradients now point the quickest way to the valley floor,

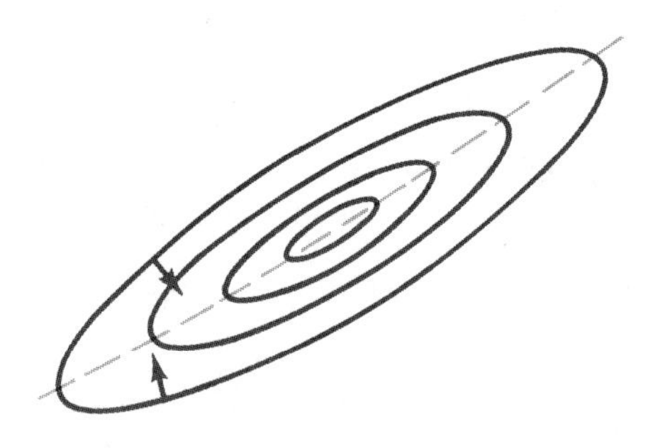

Figure 17.2

but *not* to the center. Indeed, the center is poorly defined along the valley since the f surface declines so slowly as we approach it. We are merely stating the obvious: that quadratic minima are flat for quite a distance—but if we seek a solution point, should we not seek it with a metric that defines it clearly rather than with one that depends critically on variations in the fifth significant figure? Thus f functions that are formed from a sum of absolute values might work more efficiently. (There are, however, some virtues to the squared geometry. We shall have more to say on this subject.)

The gradient vector is clearly zero at the minimum, but it does no good to equate our gradient components separately to zero, for we thereby generate a set of simultaneous equations that is harder to solve than the original set and has no compensatory virtues. More important is the fact that the level lines are ellipses, approximately, and we may often use the properties of conic sections to help find the center. Our doubts here concern how close we must be to a minimum before we may rely on the ellipticity of the level lines, and the answer is never known except in the vaguest of terms: We must be close in comparison to the distance to the nearest saddle point. Since we usually have less information about saddle points than about roots, this observation really doesn't help! All too often we choose a technique that depends more on the personality of the programmer than on the characteristics of the problem. In an attempt to introduce some order in a field that borders on the chaotic we now discuss some of the more useful techniques, stressing their dependence on the geometric assumptions and their implications for the amount of computing we must do.

Downhill methods

With the uncertainties that surround the values of higher derivatives of complicated functions, we are driven to consider methods that do not depend directly upon those derivatives. We recall how Taylor's series in one dimension represented a function quite well *locally* but how badly it could do once we had left the neighborhood of the point where the expansion was made. If we want to represent some function like an arch of a cosine curve by a parabola, we always do better to insist on matching the two curves at three discrete points rather than fit the function and two derivatives at a single point. The fit-at-several-points philosophy might be called an *interpolative* philosophy, while the fit-everything-at-one-point (Taylor's series) can be called an *extrapolative* philosophy. The extrapolative philosophy gives quite good fits very close to *the* point, but extrapolation for any real distance is distinctly dangerous. Accordingly, when we are not in the immediate vicinity of a minimum, we shall usually prefer methods that are *interpolative*, that is, do not depend too strongly on the properties of the surface f at one point.

One such technique, admittedly crude, is to evaluate f at a cluster of points and then movc to that point of the cluster at which f is least, there to repeat the process. The cluster, often called a *star*, usually consists of a central point and two additional points for each coordinate axis, one on the positive and one on the negative side of the central point. In two dimensions we have Figure 17.3.

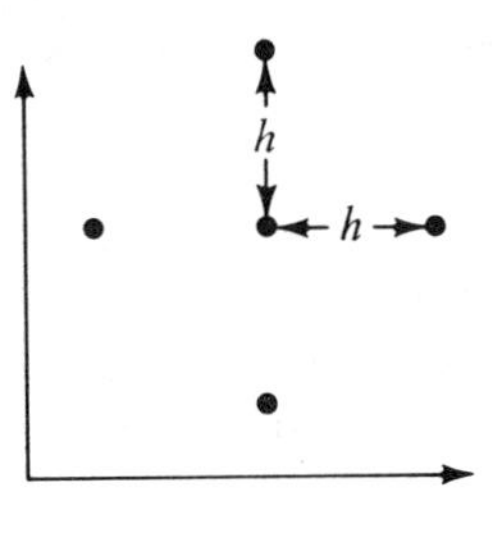

Figure 17.3

After an often considerable number of iterations, the minimum value of f will be at the central point, and the size of star may then be reduced and the routine re-entered – this entire cycle being repeated until the central point is bounded by the other points to as many significant figures as are needed. This process is the n-dimensional analogue of the binary chop, and it is atrociously inefficient, especially after we are close to a minimum which we have good reason to believe is isolated. On the other hand, it is an extremely stable searching technique that is seldom distracted by local bobbles in f. In higher dimensions, where such distractions abound, this persistent single-mindedness is a virtue that cannot be ignored and we must present the downhill star search as the basic tool to be used when other techniques fail.

We can dress up our basic technique quite a bit by noting that after we know f at each point we have enough information available to compute an approximate gradient from our star, so it is possible to move it in several (or all) the coordinates at once instead of one coordinate at a time. We can even estimate some of the second partial derivatives of f fairly well. We must now decide whether we shall:

1. Move the star exactly one increment in each of these coordinates where f_{min} is not the central point. (This will keep the star anchored to one of the regular grid of points spaced h apart in every dimension.) Or
2. Move the star along the gradient direction, even though this will take the star off the regular grid of points. Or
3. Move the star along the gradient direction a distance which depends on the curvature of the f surface as estimated from the approximate second derivatives of f at the star points.

Options 1 and 2 do not change the spirit of the downhill search. They make it somewhat more efficient without a commensurate increase in programming complexity and are thus usually worthwhile. Option 3 introduces a new element—the stepsize is now a function of the values of f, and a rather delicate function: the curvature. The step size can become infinite and will do so whenever the curvature becomes *zero*, as at inflection points. If we elect option 3 we reintroduce the nervousness that always accompanies the use of second derivatives in minimum seekers, and we can make this use acceptable only by limiting the amount of nervousness that we will permit—by limiting the maximum step size, for example. If we omit such safeguards, we will find our star happily stepping out to East Limbo when nobody is looking because it wandered quietly downhill into an inflection point. Option 3 is not bad, but it is different and must be implemented carefully.

Ray minimum methods

Some grand strategies require, as a substrategy, that we somehow select a direction in which to move and then move to the minimum value of f in that direction. In this section we confine our attention to the problem of finding the minimum along this single direction without worrying how the direction was selected or, indeed, whether the minimum along it is really the point we ought to be seeking. In Figure 17.4 we have plotted f against distance along the vector $\mathbf{v}$ of Figure 17.1 and, if our usual assumptions about ellipses are satisfied, we know that f is a parabola with a minimum that is generally greater than zero (and never less than zero). Obviously, we can march to the minimum in a downhill pedestrian fashion, but such a procedure ignores far too much information to be justified. It will be chosen only by the lazy and irresponsible programmer.

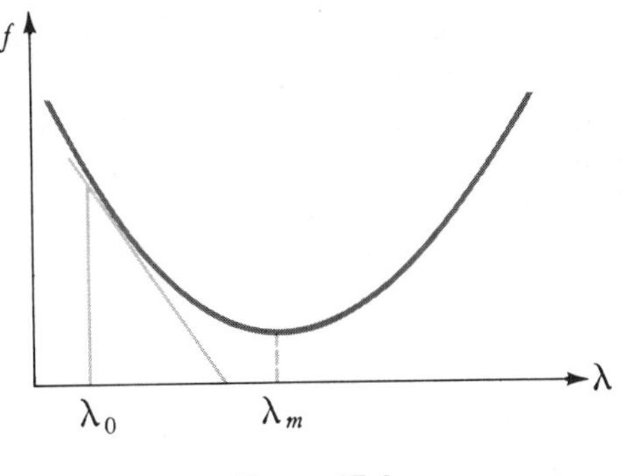

Figure 17.4

Two geometrics must be distinguished: Either the minimum is known to be zero – which is less usual – or it is not. If it *is* zero, then we may repeatedly slide down a tangent, though the process is somewhat slow toward the end. Such a technique is Newton's method, but the root being sought is double and hence dangerous. A proper Newtonian technique is to use *double* the normal correction (p. 55). Another Newtonian technique requires the calculation of the second derivative but this may be difficult or even impossible. To

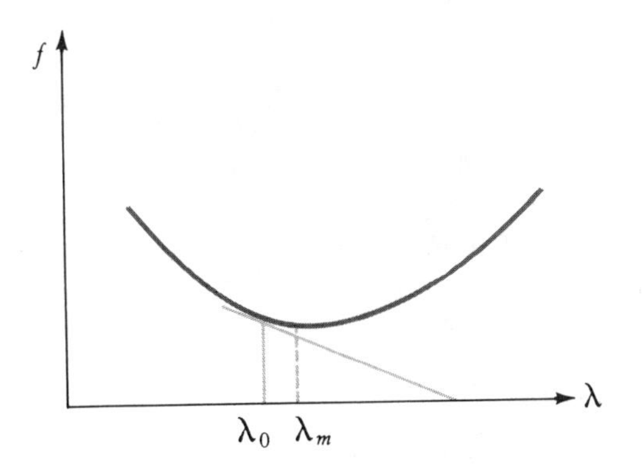

Figure 17.5

estimate the second derivative by some second difference is possible, but then we are moving into our next technique. If the minimum f must be presumed to be positive, as is usual, then sliding down a tangent is obviously silly, since it can lead to the right location only by the rarest of accidents and it can lead to a lot of less agreeable places in an awful hurry! (Figures 17.4 and 17.5.)

The reasonable technique for finding the positive minimum of an almost-parabola depends on whether or not it is practical to evaluate first derivatives. If it is, then we seek the zero of the first derivative curve, which is nearly a straight line. False Position is a reasonable method, although a modified Newton approach is preferable. Under the modified Newton philosophy we find the approximate slope of our derivative curve by evaluating it at two points. We then slide along this line of constant slope to the axis to get our next estimate of the zero. Figure 17.6 shows a somewhat unfavorable geometry with the slope determined at points 1 and 2, all diagonal lines being parallel to this first one.

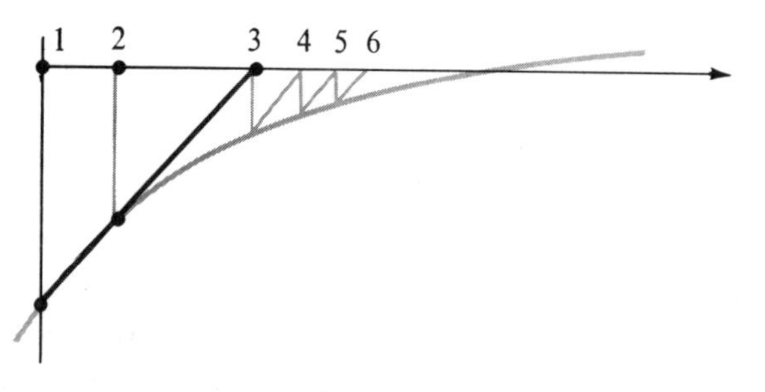

Figure 17.6

A true Newton method would require the evaluation of the slope at each cycle, and since this picture is already the first derivative, we would be faced with computing the second directional derivative of our function – which we emphatically wish to avoid. In more than two dimensions the evaluation of *second* derivatives is usually prohibitively laborious, the number of the second partial derivatives in n dimensions being $(n)(n + 1)/2$. The directional second derivative is then the quadratic form $\mathbf{v}^T Q \mathbf{v}$, where Q is the symmetric matrix of second partial derivatives and $\mathbf{v}$ the unit vector in the required direction. Fortunately an almost-parabola has an almost-constant second derivative, so a single evaluation, even if quite approximate, is often sufficient, and this can be obtained by evaluating the *first* derivative at two points on the line – a much more economical computation.

Cubic and parabolic fits

An alternative philosophy that uses both functional *and* derivative information at two points is to fit a cubic and step immediately to its minimum. Geometrically, we are fitting a parabola with a cubic perturbation, four parameters being available. Typically we find ourselves at some point x_0 in n-space with a direction specified by a unit vector $\mathbf{v}$. We may immediately compute

$$f_0 \quad \text{and} \quad g_0 = \mathbf{g}^T \cdot \mathbf{v}$$

(the dot product of the gradient of f and $\mathbf{v}$) (see Figure 17.7). Then we move

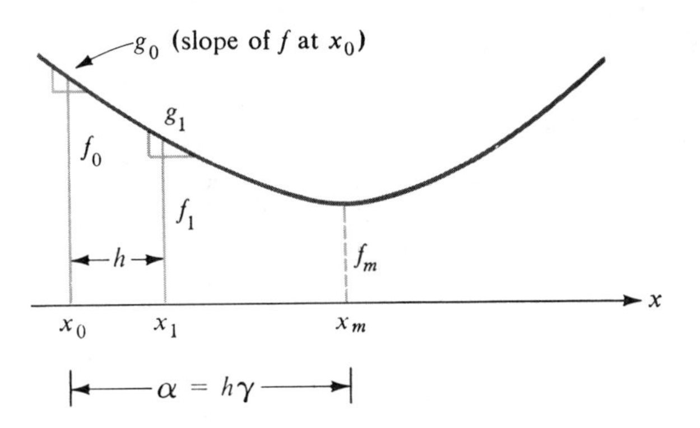

Figure 17.7. A plot of f in the direction of $\mathbf{v}$

some convenient distance h along $\mathbf{v}$ to another point x_1 and there compute

$$f_1 \quad \text{and} \quad g_1 = \mathbf{g}_1{}^T \cdot \mathbf{v}$$

[If we know enough about the probable local geometry to place x_1 close to the minimum, so much the better. In the quadratic algorithm (p. 457) an h of $|s|$ is often about right, as is the double Newton correction

$$-\frac{2(f_0 - f_m)}{g_0}$$

When these differ, the conservative choice is the smaller.]

Our cubic in the one dimensional variable x may be expressed in terms of the location of the minimum x_m as

$$f = f_m + \frac{b}{h^2}(x - x_m)^2 + \frac{c}{h^3}(x - x_m)^3 \tag{17.1}$$

where, to make b and c dimensionless, we include h defined by

$$\boxed{h = x_1 - x_0}$$

Defining α and γ by

$$\alpha = h\gamma = x_m - x_0$$

and noting that

$$g = \frac{2b}{h^2}(x - x_m) + \frac{3c}{h^3}(x - x_m)^2 \tag{17.2}$$

we may easily show that

$$\boxed{F = f_1 - f_0} = b(1 - 2\gamma) + c(1 - 3\gamma + 3\gamma^2) \tag{17.3}$$

$$\boxed{G = (g_1 - g_0)h} = 2b + 3c(1 - 2\gamma) \tag{17.4}$$

$$g_0 h = -2b\gamma + 3c\gamma^2 \tag{17.5}$$

and hence may compute c directly via

$$\boxed{c = G - 2(F - g_0 h)} \tag{17.6}$$

We expect c to be small, since it measures the departure of our curve from parabolic geometry.

Eliminating b between (17.4) and (17.5) we get a quadratic equation for γ, the required location of the minimum:

$$3c\gamma^2 + (G - 3c)\gamma + g_0 h = 0 \tag{17.7}$$

Since c is assumed to be small, and since we want the smaller root, the appropriate solution to (17.7) is

$$\boxed{\gamma = \frac{-2g_0 h}{(G - 3c) + \sqrt{(G - 3c)^2 - 12c \cdot g_0 h}}} \tag{17.8}$$

There are shorter computational forms that are algebraically equivalent to ours (which consists of the boxed formulas here) but they involve possible losses of accuracy by requiring small numbers to be calculated as differences of large, nearly equal quantities.

Occasionally, through automatic application of the algorithm near a saddle point, the cubic fitted to f_0, f_1, g_0, and g_1 may not have a minimum. This condition causes the argument of the square root in (17.8) to be negative, so it should be tested. Under such circumstances, one usually must abandon the algorithm and restart in another, more favorable, part of the domain, but one less drastic possibility may be tried: to fit a parabola to f_0, f_1, and g_1. This is easily accomplished by

$$\boxed{F = f_1 - f_0} = b(1 - 2\gamma)$$

$$g_1 h = 2b(1 - \gamma)$$

so that

$$b = g_1 h - F$$

and

$$\gamma = \frac{1 - F/b}{2}$$

Although this parabola may always be fitted, its minimum may not have very much to do with the point we seek, especially if the cubic fitting has already failed.

If computation even of first derivatives is impossible, then we must pick three points on our curve, fit a parabola through them, and move to the minimum of the fitted parabola. This point may then be refined by a repetition of the procedure, though a simple binary chop may do as well – now that the minimum is at hand. The absence of derivative information hurts here, and one must be careful to distinguish between problems for which the derivative computation is awkward or troublesome and those for which it is impossible. It must be *very* laborious before it is more efficient to ignore it! One practical difficulty with the three-point method is deciding on the spacing of the three points. Clearly we would prefer to span the minimum or at least choose our points so as to include some portion of the parabola commensurate with the distance to the minimum, and we usually do not have that information readily available. At worst, however, we may pick three points, compute a second derivative estimate, and, if this shows our points to be chosen on a clearly unsuitable scale, alter it before proceeding to fit the parabola.

Formulas for the minimum of a fitted parabola, expressed in terms of the three ordinates and abscissas, are easily found to be

$$x_m = x_3 - \frac{h}{2} - \frac{f_2 - f_3}{f_1 - 2f_2 + f_3} \cdot h$$

and

$$x_m = x_2 + \frac{f_1 - f_3}{f_1 - 2f_2 + f_3} \cdot \frac{h}{2}$$

where the definitions of the various terms are obvious from Figure 17.8. The first formula places heavier stress on x_3, which is probably a good idea when we know that it is the nearest of the three points to the minimum. Otherwise the symmetric formula based around x_2 seems more natural. Except for the way errors in the several entries affect their values, the formulas are identical.

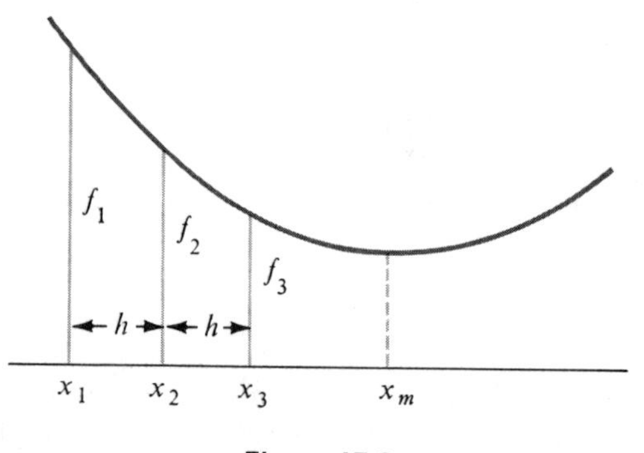

Figure 17.8

For a function $f(\mathbf{x})$ of n variables with a gradient $\mathbf{g}$ evaluated along a line $\mathbf{v}$ we obtain the quantities of Figure 17.7 from

$$g_1 = \mathbf{g}_1 \cdot \mathbf{v}$$

$$g_0 = \mathbf{g}_0 \cdot \mathbf{v}$$

where $\mathbf{v}$ is a unit vector and

$$h = |\mathbf{x}_1 - \mathbf{x}_0|; \qquad \text{i.e., } (\mathbf{x}_1 - \mathbf{x}_0) = h\mathbf{v}$$

Ellipsoid center seeking

One of the grander strategies for attacking the minimization problem was conceived by Powell [1962]. He observed that if we have nested similar ellipses, then there is an easy way to find a line that passes through their common center. In Figure 17.9 we are at P and we wish to find a point M such that PM passes through the center C. We carry out two steps:

1. Proceed an arbitrary distance down the gradient from P to the point R. (In practice it will be neater to proceed to a *minimum* along this gradient, but this is *not* essential to our purpose.)
2. Proceed perpendicular to PR to a minimum, which is the required point, M. (This minimum in the *space* perpendicular to PR is necessary.)

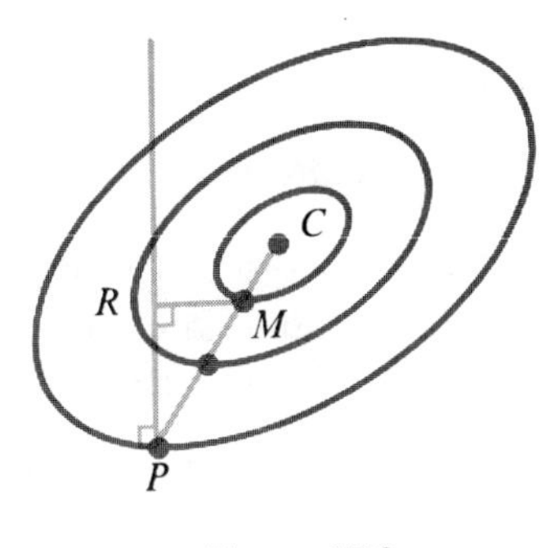

Figure 17.9

To implement this method in two dimensions, we need only find M and then proceed along PM to a minimum, which would be the center C, if all our assumptions were satisfied and our work precise. In the real world, whether neither condition is usually true, we shall probably have to repeat the entire cycle. The entire cycle consists of determining the gradient direction, then proceeding to a minimum, along three lines by any expedient technique.

Now consider the same algorithm in three independent variables. Our level surfaces are concentric ellipsoids and our *gradient* in step 1 is a line perpendicular to a tangent *plane*. Proceeding along the gradient from P to R (which may, though need not be, at a minimum) we consider the *plane* Q, at R, perpendicular to PR and observe that the level surfaces cut it in a nested set of ellipses (Figure 17.10) and the desired point M is their center. (Remember, M must be at the minimum value of f in the *space* perpendicular to PR.) So we are now sitting at R, faced with the same problem we had at P, but in one fewer dimensions.

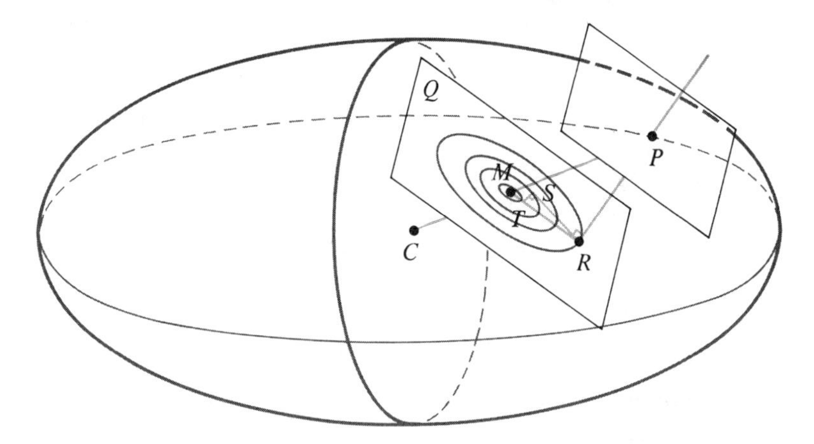

Figure 17.10

Operationally:

1. We find the gradient at P and proceed to R.
2. We find the gradient at R and take the component of it that is perpendicular to PR – that is, the part that lies in Q. We then proceed along this gradient to S (which may, though need not be, a minimum).
   a. Proceed perpendicular to RS in the plane Q (a unique direction) to a minimum that we call T.
   b. Proceed along RT to the minimum which will be M, the center of the ellipses in Q.
3. Proceed along PM to a minimum which will be the center, C.

This procedure generalizes to higher dimensions easily, although we have difficulty visualizing it. Step 2 gets repeated $n - 2$ times, where n is the number of independent variables. Each time, we obtain the gradient and then extract that part of it which is perpendicular to all the preceding step 2 lines along which we have moved, and then move along this line, which is unique, to a convenient point, probably a minimum. Finally we are down to the two-dimensional problem and the last execution of step 2 yields a point T. Steps b and 3 above are really the same type of operation, and there will be $n - 1$ of them, the last one leading to the center.

The completely systematic nature of this procedure becomes clearer if we switch to a subscript notation, replacing R by P_{n-1}, f by P_1, and M by C_{n-1}. Then we

1. Go from P_n to P_{n-1} along the gradient.
2. Go along the perpendicular projection of the local gradient at P_i. The perpendicular projection is what is left of the gradient after it is orthogonalized with respect to the lines

$$(P_n, P_{n-1}), (P_{n-1}, P_{n-2}), (P_{n-2}, P_{n-3}), \ldots, (P_{j+1}, P_j).$$

   (Repeat step 2 until reaching P_1.)
3. Go along P_1, P_0, which is perpendicular to all (P_{i+1}, P_i) lines, to the *minimum* to establish P_0. (This is the only essential minimum thus far.)
4. Go along P_2P_0 to a minimum to be called C_2. (Repeat step 4, going along P_{i+1}, C_i to produce C_{i+1} at the minimum, finally producing C_n, which is the center.)

Those who enjoy programming will see that this process invites an elegant encoding as a recursive procedure – but we will not pursue this further lest the FORTRAN enthusiast become unhappy.

Contour analytic geometry

The analytic basis for the geometry we have been discussing is simply Taylor's series in several dimensions. We illustrate by our two-dimensional example. Expanding $f(x, y)$ about (x_0, y_0) we obtain

$$f(x, y) = f(x_0, y_0) + \Delta x \frac{\partial f}{\partial x}\bigg|_0 + \Delta y \frac{\partial f}{\partial y}\bigg|_0 + \frac{(\Delta x)^2}{2} \cdot \frac{\partial^2 f}{\partial x^2}\bigg|_0$$

$$+ \frac{2(\Delta x)(\Delta y)}{2} \cdot \frac{\partial^2 f}{\partial x\, \partial y}\bigg|_0 + \frac{(\Delta y)^2}{2} \cdot \frac{\partial^2 f}{\partial y^2}\bigg|_0 + \text{higher terms} \qquad (17.9)$$

If we replace our point (x_0, y_0) by the position vector $\mathbf{x}$ and our increment $(\Delta x, \Delta y)$ by a distance λ along a unit vector $\mathbf{v}$ (see Figure 17.11), and if we denote

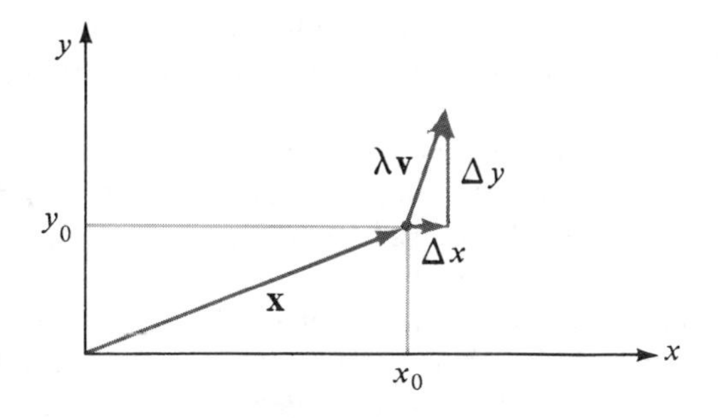

Figure 17.11

the symmetric matrix of all possible second derivatives of f by Q [these are all evaluated at (x_0, y_0) and hence Q is a matrix of numbers], we may rewrite our expansion in the somewhat more convenient form

$$f(\mathbf{x} + \lambda\mathbf{v}) = f(\mathbf{x}) + \lambda(\mathbf{grad}\, f(\mathbf{x}) \cdot \mathbf{v}) + \frac{\lambda^2}{2}(\mathbf{v}^T Q \mathbf{v}) + \text{higher terms in } \lambda \qquad (17.10)$$

If, for the moment, we imagine the point $\mathbf{x}$ to be the minimum point in the f surface, then $\mathbf{grad} f(\mathbf{x})$ is identically zero (this is also true at a maximum point and another kind of point. Why? Draw the contours near this third kind of point) and the term in λ vanishes. If in addition we demand that the minimum $f(\mathbf{x})$ actually be *zero*, then our series expansion has been cut down to the λ^2 terms – we neglected terms in λ^3 and higher long ago – and we have

$$f(\mathbf{x} + \lambda\mathbf{v}) = \frac{\lambda^2}{2} \cdot (\mathbf{v}^T Q \mathbf{v}) \qquad (17.11)$$

or

$$f(x, y) = (\Delta x)^2 \cdot \frac{\partial^2 f}{\partial x^2} + 2(\Delta x)(\Delta y)\frac{\partial^2 f}{\partial x\, \partial y} + (\Delta y)^2 \cdot \frac{\partial^2 f}{\partial y^2}$$

which is clearly the equation of a general conic section when we demand that $f(x, y)$ be a constant, f_0. We have

$$f_0 = a(\Delta x)^2 + 2b(\Delta x)(\Delta y) + c(\Delta y)^2$$

The symmetry of Q arises from the fact that $\partial^2 f/\partial x\, \partial y$ is equal to $\partial^2 f/\partial y\, \partial x$. The further condition on the determinant $|Q|$

$$\frac{\partial^2 f}{\partial x^2} \cdot \frac{\partial^2 f}{\partial y^2} > \left(\frac{\partial^2 f}{\partial x\, \partial y}\right)^2 \qquad \text{or} \qquad |Q| > 0 \tag{17.12}$$

suffices to make the contours ellipses rather than hyperbolas or parabolas, and this inequality is always satisfied at a minimum or maximum in two dimensions. In higher dimensions the criterion is the positivity or negativity of all the eigenvalues of Q.

Even when the minimum value of f is not zero, the shape of the contour is not affected—a constant value is simply added to each f. We therefore see that the *ellipses are always with us precisely to the extent that the terms of the order of λ^3 and higher are unimportant relative to the terms in λ^2*. As λ becomes very small, the dominance of λ^2 is clear, but we may have to wait until λ is quite small indeed if various third and fourth partial derivatives are large. Sometimes our intuition will tell us that these terms are neglectable, but usually we do not have that good a feeling for the complicated f function that we are trying to minimize and we must take appropriate precautions lest the higher terms louse up our plans.

When we feel that we are working in a region where the contours are really ellipses, we have several facts available that are useful. Any straight line, for example, through the contours gives values of f along it which are a *parabolic* function. That is, if we plot f as a function of distance along any vector cutting through the ellipses in any direction, this plot will be an upward-opening parabola. It will be a different parabola for different lines, but the functional form will always be the same. This fact may be easily demonstrated by replacing Δy by $c\,\Delta x$ in (17.9), a straight line in the (x, y) plane requiring that $\Delta y/\Delta x$ be constant. f is then a quadratic function of Δx and nothing else.

Another fact, possibly annoying, is that passing to a minimum value of f along a succession of gradients will cause one to traverse a series of straight lines which are mutually perpendicular (see Figure 17.12) and hence the recomputation of the gradient at each minimum point is, in theory, unnecessary. Actually, of course, this fact is *very* sensitive to the curves being geometrically similar, as well as to their having a common center, and good computational practice would seldom dispense with the safeguard of computing the gradient at each turning point. Still, the fact that the information is just possibly redundant is annoying, for one feels that one's computational effort should be used to get the maximum information each time, and not merely to check up on a process that should be working smoothly.

Figure 17.12

A third fact about nested ellipses is that their center is easily expressible in terms of the various first and second derivatives that are the coefficients of (17.9). We are thus tempted to compute various derivatives and try to leap to the center – but this technique is more dependent on the absence of higher terms than are the previous suggestions and must usually be rejected as sheer wishful thinking so far as practical problems with multiple minima are concerned.

Once we pick a direction **v** we may wish to move to the local minimum along this line. At this minimum the derivative along **v** is zero, so that

$$0 = \frac{df}{d\lambda} = \mathbf{grad}\, f \cdot \mathbf{v} + \lambda \quad \mathbf{v}^T Q \mathbf{v} + \cdots$$

and hence

$$\lambda_{\min} = -\frac{\mathbf{grad}\, f \cdot \mathbf{v}}{\mathbf{v}^T Q \mathbf{v}} \tag{17.13}$$

Formula (17.13) gives us a way to move directly to a local minimum along **v** provided $\mathbf{v}^T Q \mathbf{v}$ and the first derivatives of f are known. To the extent that our function is quadratic, Q is a matrix of constants and hence $\mathbf{v}^T Q \mathbf{v}$ is a scalar constant that may be estimated quite accurately by the usual second difference of f along **v**. Thus, when first derivatives are calculable and when we also believe cubic and higher terms to be unimportant, movement along a given **v** to an approximate local minimum is a one-step computation.

More generally, all our remarks about two-dimensional strategies hold in higher dimensions and our vector formulas are valid. But the booby traps increase rapidly with the number of dimensions, for there are more ways a function can be "flat" without also being a minimum. Also the number of higher derivatives increases sharply, so that there is – in some vague statistical sense – a greater chance of some of these higher terms being important and we are less able to depend on the ellipsoids really existing as we would like to imagine them.

Minima without derivatives

Probably the most effective strategy when derivatives cannot be calculated is a sophistication of the most elementary – proceeding to a minimum along each of the n coordinate directions in turn. In fact, this is exactly the first iteration of our process, but at its conclusion we now possess information that can improve the efficiency of the next iteration. We now know the *average direction that we have moved during this iteration.* It seems desirable to make this average direction one of the n directions in which we shall next move to a minimum. In the absence of a better idea about which direction to discard (n linearly independent directions per iteration *are* sufficient!) we might throw away the first one, adding the newest one to the far end of our list. In practice, this discard philosophy turns out to be bad, especially if more than five dimensions are involved, since near linear dependence tends to build up within the list of directions. It seems better to drop the direction in which we made the *largest decrease* in our function, since this is usually a major component of the new direction we are adding. Occasionally it is better not to add the new direction at all – just sticking with the old ones for another iteration.

Powell's algorithm [1964] stated more precisely is:

1. Along each of the n directions ξ_r, find the minimum of f along that direction and its position – that is, from our current position $\mathbf{p}_{r-1}$ find

$$\min_{\lambda_r} f(\mathbf{p}_{r-1} + \lambda_r \xi_r)$$

and call the new location $\mathbf{p}_r$ so that

$$\mathbf{p} = \mathbf{p}_{r-1} + \lambda_r \xi_r$$

Since the n directions are traversed in a definite order, we always start from our current minimum and proceed downhill, noting the amount by which we decrease f at each stage and identifying the stage in which this decrease is the largest. Having minimized once in each of our current n directions, we then conduct two tests to see if we should retain our old set of directions intact and simply repeat step 1 again. These tests use three quantities:

$$f_1 = f(\mathbf{p}_0)$$

$$f_2 = f(\mathbf{p}_n)$$

$$f_3 = f(2\mathbf{p}_n - \mathbf{p}_0)$$

The points $\mathbf{p}_0$ and $\mathbf{p}_n$ are simply the places we were at the beginning and end of step 1. The third point may be written

$$\mathbf{p}_n + (\mathbf{p}_n - \mathbf{p}_0)$$

and is clearly colinear with the first two at an equal distance farther down the line that connects them. Thus we have computed f at three equally spaced points. Denoting the largest decrease during one direction within step 1 by Δ we now

2. Test whether

$$f_3 \geq f_1$$

and/or

$$2[f_1 - 2f_2 + f_3][(f_1 - f_2) - \Delta]^2 \geq \Delta[f_1 - f_3]^2$$

If either of these conditions holds, we retain our old set of directions and use $\mathbf{p}_n$ from step 1 as the new $\mathbf{p}_0$ – that is, we simply iterate the entire step 1 again.

The first condition says that the average direction $(\mathbf{p}_n - \mathbf{p}_0)$ we moved during step 1 is rather thoroughly exhausted, since a further step of the same size in that direction would carry us farther up the opposite wall of the valley than we were at $\mathbf{p}_0$. Thus we see no need to include $\mathbf{p}_n - \mathbf{p}_0$ as a new direction.

The second condition implies that the progress toward a minimum during step 1 did not depend primarily on the one direction in which the largest decrease Δ occurred, or else that the cross section in the average direction has a good curvature and that f_2 is near the minimum of it. Again, this average direction seems exhausted.

3. If neither of the tests in step 2 hold, then the average direction we moved in step 1

$$\xi = \mathbf{p}_n - \mathbf{p}_0$$

is not yet exhausted. Thus we move immediately to the minimum in this direction by finding

$$\min f(\mathbf{p}_n + \lambda\xi)$$

and take this point as the starting location $\mathbf{p}_0$ for our next iteration of step 1. We also add ξ to the list of directions to be used by step 1, dropping the direction ξ_r in which the largest decrease was made last time.

As with most minimization strategies, an essential subroutine is one which moves to a local minimum along a given line, ξ_i. And, as usual, this step better be efficient, since it is the innermost loop of a lengthy algorithm. When explicit calculation of the functional derivatives is denied us, we must compensate by deducing as much as we usefully can about their values from the functional evaluations themselves. The essential geometry of f along any line that passes near a minimum is parabolic; that is, if the level lines of f are nested ellipses, then along any cross section f is parabolic. Since the *second* derivative of a parabola is constant, we might reasonably expect it to be no worse than slowly varying on any one direction in our problem. It seems a good tactic, therefore, to estimate the second derivative of f along ξ_i as early in our algorithm as we can and to store it for use in subsequent moves along that direction. Since parabolas require three pieces of information and since we always possess the value of f at the beginning of each move along a direction, we need only our stored value of f'' plus one additional evaluation of f in order to determine the parabola and hence its minimum. Therefore, if the major calculational cost is the functional evaluation, we reduce the basic iteration to one evaluation of f plus minor bookkeeping. Since the direction ξ_i is stored as a vector that must be normalized under some convention, we might as well normalize in such a manner as to store the second derivative datum.

We already have available, from a previous step, the value of

$$f(\mathbf{p}_0) = u_0$$

The basic new calculation for the minimum along the direction ξ_i is to evaluate f at $\mathbf{p}_0 + \xi_i$. If the direction ξ_i has been normalized so that its length is $(f'')^{-1/2}$ taken along ξ_i, then our next estimate for the location of $f_{\min}$ is at

$$\mathbf{p}_1 - [\tfrac{1}{2} + (u_1 - u_0)]\xi_i$$

The quantity in the brackets is dimensionless and is applied directly to ξ_i. If we have not normalized ξ_i, then the bracket is

$$\left[\frac{1}{2} + \frac{u_1 - u_0}{h^2 u''}\right]$$

where the h^2 is the square of the distance from $\mathbf{p}_0$ to $\mathbf{p}_1$—that is, the squared length of $\mathbf{p}_1 - \mathbf{p}_0$—and u'' is presumably a stored quantity that has been previously estimated.

The derivation is instructive. We have a parabola whose minimum is u_m at s_m. Thus

$$u = b(s - s_m)^2 + u_m \qquad b > 0$$

Then

$$u_1 - u_0 = b(s_1 - s_0)(s_1 + s_0 - 2s_m)$$

or

$$s_m = \frac{s_0 + s_1}{2} - \frac{1}{2b}\frac{u_1 - u_0}{s_1 - s_0}$$

or, letting $s_1 - s_0 = h$ and $s_m - s_0 = \delta_0$,

$$\delta_0 = \frac{s_1 - s_0}{2} - \frac{1}{2b}\cdot\frac{u_1 - u_0}{h}$$

but

$$u'' = 2b$$

so

$$\delta_0 = \left[\frac{1}{2} - \frac{u_1 - u_0}{u''h^2}\right]$$

Since h is the step from s_0 to s_1, δ_0 is expressed as a fraction of that h—thus the bracket is to be multiplied into each component of ξ_i.

An estimate of h^2u'' is obtained from the standard second difference of u. This requires evaluating u (that is, f) at $\mathbf{p}_0$, $\mathbf{p}_0 + \mathbf{h}$, and then either at $\mathbf{p}_0 + 2\mathbf{h}$ or $\mathbf{p}_0 - \mathbf{h}$, the choice being inclined toward the minimum. In the beginning h is more or less arbitrary—probably unity—but if it seems desirable to reevaluate u'' later, we will have an h around from the steps we have been taking. An incorrect value for u'' will slow convergence of the algorithm but will not usually stop it. [*Problem*: If an error of δ is made in u'' and if the true curve is actually a parabola, show that the error in x_2 (the distance from $x_{\min}$) is the average of the errors in x_0 and x_1 multiplied by $\delta/(u'' + \delta)$.]

Minima with first derivatives available

The geometric rationale for this method is not clear, but its authors [Fletcher and Powell, 1963] claim for it greater efficiency than for any of the other methods we have quoted. We give it as a cookbook procedure, followed by a few comments concerning its properties. We use $f(\mathbf{x})$ for the function to be minimized. The independent variables constitute the vector $\mathbf{x}$. The *gradient* of $f(\mathbf{x})$ is $\mathbf{g}(\mathbf{x})$, usually abbreviated $\mathbf{g}$. Subscripts refer to iterations, not to components of vectors. H, A, and B are symmetric matrices. We assume that whenever an $\mathbf{x}$ is given, f and $\mathbf{g}$ can be computed directly.

$i = 0$

$\mathbf{x}_0$ is given a priori, hence $f_0, \mathbf{g}_0$

$H_0 = I$ (the unit matrix)

→ $\mathbf{s}_i = -H_i\, \mathbf{g}_i$ (a direction, always downhill, and approximately the distance to the minimum along that direction)

C: Execute a subroutine that minimizes f in the direction $\mathbf{s}_i$ – that is, find the (positive) value of λ that minimizes $f(\mathbf{x}_i + \lambda \mathbf{s}_i)$ and call that critical value of λ by the name α_i. The nature of this subroutine is irrelevant to the principal iteration, although for efficient operation a somewhat sophisticated procedure is desirable. We have given one earlier.

C: $\mathbf{d}_i = \alpha_i \mathbf{s}_i$ (the incremental distance we will move along $\mathbf{s}_i$)

C: $\mathbf{x}_{i+1} = \mathbf{x}_i + \mathbf{d}_i$ (our new position near the minimum along $\mathbf{s}_i$)

C: $f(\mathbf{x}_{i+1}), \mathbf{g}_{i+1}$ (note that $\mathbf{d}_i^T \cdot \mathbf{g}_{i+1} = 0$, except for rounding errors)

C: $\mathbf{y}_i = \mathbf{g}_{i+1} - \mathbf{g}_i$

C: $A_i = \dfrac{\mathbf{d}_i \mathbf{d}_i^T}{\mathbf{d}_i^T \mathbf{y}_i}$ (the numerator is the *backward*, or matrix, product of two vectors; the denominator is an ordinary dot product)

C: $B_i = \dfrac{(H_i \mathbf{y}_i)(H_i \mathbf{y}_i)^T}{\mathbf{y}_i^T H_i \mathbf{y}_i}$ (again, the numerator is the matrix product of two vectors; the denominator is a quadratic form, a scalar)

C: $H_{i+1} = H_i + A_i - B_i$ (it can be proved that H_i, having started positive definite, will always remain positive definite)

↓

Is $i < n$? —no→ Exit, too many iterations!

↓ yes

Test for smallness of $\mathbf{d}_i$ and $\mathbf{s}_i$ —yes, small!→ Stop (at minimum)

no → $i = i + 1$ (return to $\mathbf{s}_i = -H_i\,\mathbf{g}_i$)

This method uses first-derivative information explicitly in the gradient vector, $\mathbf{g}_i$. Second-derivative information is estimated by $\mathbf{y}_i$ and incorporated into H_i, which gradually evolves to become $Q_{\min}^{-1}$ – the inverse of the matrix of the second partial derivatives, evaluated at the minimum. We can write the Taylor's series of f as

$$f(\mathbf{x} + \mathbf{v}) = f(\mathbf{x}) + \mathbf{g}(\mathbf{x})^T\mathbf{v} + \frac{1}{2!}\mathbf{v}^T Q(\mathbf{x})\mathbf{v} + \cdots$$

where $\mathbf{v}$ is the vector displacement from the arbitrary point of expansion. At a minimum $\mathbf{g}$ becomes zero. Also we see that the gradient at $\mathbf{v}$ is

$$\mathbf{g}(\mathbf{v}) = \mathbf{g}(\mathbf{x}) + Q(\mathbf{x}) \cdot \mathbf{v}$$

If we consider the problem of getting to the point where $\mathbf{g}$ is zero from the point $\mathbf{x}$ where we currently are, we have

$$0 = \mathbf{g}(\mathbf{x}) + Q(\mathbf{x}) \cdot \mathbf{v} \qquad \text{or} \qquad \mathbf{v} = -Q(\mathbf{x})^{-1} \cdot \mathbf{g}(\mathbf{x})$$

Thus if H_i is approaching $Q(\mathbf{x})^{-1}$, then $\mathbf{s}_i$ is approximately the vector that takes us to the minimum and hence α_i should be close to unity.

Second derivative geometries again

If we are prepared to believe that a Taylor's series through quadratic terms represents our function more or less adequately, at least over the region we would be willing to traverse in one step toward a minimum we seek, then most of the previous algorithms are inefficient. They do not make full use of the geometric information available in the second derivatives at the point of expansion. In this section we review in more detail the pertinent geometric facts regarding quadratic surfaces in order to profess a philosophy of minimum seeking that has recently arrived on the computational scene. As with most philosophies, there are several variants between whose relative merits we do not prefer to adjudicate. More experience is needed. But since the class of methods seems sensible, is headed in the proper direction, we prefer to include it here even if our description is necessarily incomplete.

If we plot the most general quadratic in two independent variables,

$$ax^2 + 2hxy + by^2 + 2fx + 2gy + c = 0 \tag{17.14}$$

we find that most of the information about the shape of the curve is contained in the three coefficients of the quadratic terms. Thus a, b, and h determine whether our curve is an ellipse, hyperbola, circle, or a parabola. The coefficients f and g are principally concerned with where the center of the figure might be. Thus

$$4x^2 + 9y^2 - 36 = 0$$

is an ellipse, centered at the origin, with semiaxes of three and two (Figure 17.13) while

$$4(x - 1)^2 + 9(y - 2)^2 - 36 = 0$$

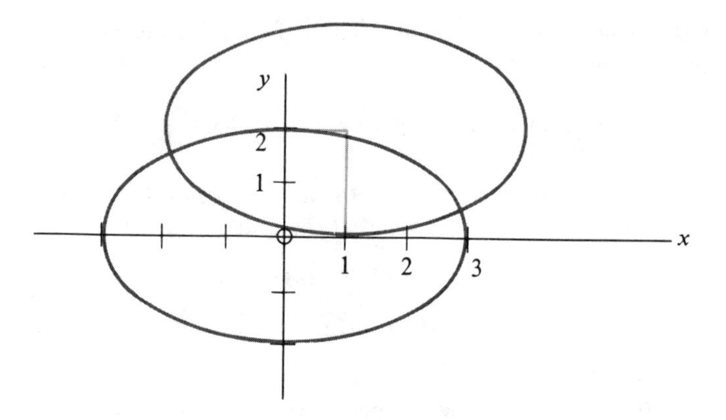

Figure 17.13

which clearly has the same quadratic terms, is the same ellipse simply shifted so that its center sits at (1, 2). If we rotate the ellipse, an xy quadratic term appears and the coefficients of x^2 and y^2 change in (17.14)—but the combination $(ab - h^2)$ retains its same value. If we write the coefficients as the symmetric matrix

$$D = \begin{bmatrix} a & h & f \\ h & b & g \\ f & g & c \end{bmatrix}$$

then (17.14) is seen to be the quadratic form

$$\mathbf{v}^T D \mathbf{v} = 0$$

where

$$\mathbf{v}^T = [x \quad y \quad 1]$$

Although not computationally useful, this form has the virtue that it expands directly to quadratic systems in three and more independent variables. More important, this form displays the algebraic quantities that do not change when the coordinate axes are rotated, the *invariants* of the system. Thus the crucial quantity that tells whether (17.14) is an ellipse, parabola, or hyperbola is simply the principal minor of the determinant of the matrix—itself being the determinant of the quadratic-term coefficients. We have

$$S = \begin{vmatrix} a & h \\ h & b \end{vmatrix} = ab - h^2 \begin{cases} > 0 & \text{ellipse} \\ = 0 & \text{parabola} \\ < 0 & \text{hyperbola} \end{cases}$$

From this quantity we may tell immediately which kind of figure we shall have if we plot (17.14). (There are also some degenerate geometries that can be identified.) Furthermore, the coordinates of the center of the figure are nearly visible in D, being solutions to the linear equations

$$ax_c + hy_c + f = 0$$
$$hx_c + by_c + g = 0$$

equations that are directly available by deleting the last row of D. Again, this observation generalizes to quadratic forms in higher dimensions.

In three independent variables the general quadratic equation is

$$ax^2 + by^2 + cz^2 + 2hxy + 2fxz + 2gyz + 2lx + 2my + 2nz + d = 0 \tag{17.15}$$

or, once again,

$$\mathbf{v}^T D \mathbf{v} = 0$$

where this time

$$D = \begin{bmatrix} a & h & f & l \\ h & b & g & m \\ f & g & c & n \\ l & m & n & d \end{bmatrix} \quad \text{and} \quad \mathbf{v} = \begin{bmatrix} x \\ y \\ z \\ 1 \end{bmatrix}$$

The center of the figure occurs at the solution to

$$\begin{aligned} ax + hy + fz + l &= 0 \\ hx + by + gz + m &= 0 \\ fx + gy + cz + n &= 0 \end{aligned} \tag{17.16}$$

Whether the figure is an ellipsoid or hyperboloid is found by solving the eigenvalue problem

$$S\mathbf{u} = \lambda\mathbf{u} \tag{17.17}$$

where

$$S = \begin{bmatrix} a & h & f \\ h & b & g \\ f & g & c \end{bmatrix}$$

We examine whether all three values of λ that permit solution of (17.17) have the same sign (ellipsoid) or different signs (hyperboloid). The quantities a, h, and so on, are the several second partial derivatives of (17.15) multiplied by a factor of one half.

Looking further at the geometry of the ellipse, we note that the *lengths* of the axes are related to the coefficients of the quadratic terms. Writing the equation of the standard two-dimensional ellipse

$$\left(\frac{x}{r_1}\right)^2 + \left(\frac{y}{r_2}\right)^2 = 1$$

in which r_1 and r_2 are the lengths of the semiaxes, we see that

$$a = \frac{1}{r_1^{\,2}} \qquad b = \frac{1}{r_2^{\,2}}$$

But these facts hold only when the ellipse is sitting resolutely parallel to the coordinate axes, that is, having no xy quadratic cross term in the equation. To find the more generally useful relation we must turn to the matrix representation of the quadratic terms, which have their own quadratic form

$$\mathbf{u}^T S \mathbf{u}$$

and further to the eigenvalue equation

$$S\mathbf{u} = \lambda \mathbf{u} \qquad \mathbf{u}^T = [x \quad y]$$

The vectors $\mathbf{u}$ that satisfy this equation are the *directions* of the principal axes of the quadratic form—a fact that will soon be useful to us. Further, the corresponding λ's bear quantitative relations to the lengths of these principal axes. For our standard ellipse we have

$$S = \begin{bmatrix} \dfrac{1}{r_1^{\,2}} & 0 \\ 0 & \dfrac{1}{r_2^{\,2}} \end{bmatrix}$$

whose eigensolutions are

$$\lambda_1 = \frac{1}{r_1^{\,2}} \qquad \mathbf{u}_1^{\,T} = [1 \quad 0]$$

and

$$\lambda_2 = \frac{1}{r_2^{\,2}} \qquad \mathbf{u}_2^{\,T} = [0 \quad 1]$$

so we see that each eigenvalue λ_i of the matrix S is the inverse square of the length of semiaxis whose direction is given by the corresponding eigenvector,

$\mathbf{u}_i$. We thus gain complete information about our ellipsoid by solving the eigenvalue problem for the second derivative matrix. For a small number of dimensions this is not a formidable task.

By contrast, the standard hyperbola

$$\left(\frac{x}{r_1}\right)^2 - \left(\frac{y}{r_2}\right)^2 = 1$$

gives the matrix

$$S = \begin{bmatrix} \frac{1}{r_1^2} & 0 \\ 0 & -\frac{1}{r_2^2} \end{bmatrix}$$

whose eigensolutions are

$$\lambda_1 = \frac{1}{r_1^2} \qquad \mathbf{u}_1^T = [1 \quad 0]$$

$$\lambda_2 = -\frac{1}{r_2^2} \qquad \mathbf{u}_2^T = [0 \quad 1]$$

and we see that a matrix with both positive and negative eigenvalues characterizes hyperbolic geometries. In the context of minimization problems, level lines with hyperbolic geometries are found in the vicinity of saddlepoints. Ellipses are found around both minima and maxima, their only difference being found in the signs of their eigenvalues, those of minima being all positive while those of maxima are all negative.

Returning to the function we wish to minimize, $H(x)$, we expanded it into a Taylor series at the point $\mathbf{b}$. We have

$$H(\mathbf{x}) = H(\mathbf{b}) + (\mathbf{x} - \mathbf{b})^T\mathbf{G}_\mathbf{b} + \tfrac{1}{2}(\mathbf{x} - \mathbf{b})^T S_\mathbf{b}(\mathbf{x} - \mathbf{b}) + \text{cubic and higher terms} \tag{17.18}$$

with $\mathbf{G}_\mathbf{b}$ the gradient vector (the vector of first derivatives of H) and $S_\mathbf{b}$ the matrix of second derivatives, both evaluated at $\mathbf{b}$. To determine a minimum of H, we may adopt the slightly indirect strategy of seeking a *zero* of its gradient. Since the gradient may be expressed, by differentiating (17.18) we have

$$\mathbf{G}(\mathbf{x}) = \mathbf{G}_\mathbf{b} + S_\mathbf{b}(\mathbf{x} - \mathbf{b}) + \cdots \tag{17.19}$$

and we may find its zero by equating (17.19) to zero and solving, first having abandoned the higher-order terms as hopefully irrelevant. This algorithm clearly is Newton's method in higher dimensions. We have

$$0 = \mathbf{G}_\mathbf{b} + S_\mathbf{b}(\mathbf{x} - \mathbf{b}) \tag{17.20}$$

that gives the correction $(\mathbf{x} - \mathbf{b})$ as

$$\mathbf{x} - \mathbf{b} = -S_{\mathbf{b}}^{-1} \cdot \mathbf{G}_{\mathbf{b}}$$

which is to be evaluated at the current location **b**. We note in (17.15) that the coefficients $[l, m, n]$ constitute, except for the factor of one half, the gradient vector. Thus (17.20) is the same equation as (17.16) which is thus seen to be an implementation of Newton's method for finding a place where the gradient becomes zero. In using this strategy for minimization of functions we must of course remember that zero gradients also occur at maxima and at saddle points of quadratic surfaces. Thus we should test the geometry of our surface before plunging ahead with the solution of Newton's method. Fortunately, this is easily done through the signs of the eigenvalues. (For two-dimensional problems the decision is simpler since positivity of the discriminant guarantees two eigenvalues of like sign – hence ellipses.) If all eigenvalues are positive, the figure is an ellipsoid, and we know that its center is the point where the quadratic function takes on its minimum value.

The ellipsoidal strategy for minimization

When we evaluate the first and second derivatives in a typical minimization iteration we may then determine whether all the eigenvalues are positive. If so, S is positive definite, the level surfaces of our form are closed ellipsoids, and their center is the minimum of the quadratic form. To the extent that the third and higher derivatives of our function are unimportant, these ellipsoids are also the level surfaces of our function and the center is its minimum. Since this is what we seek, there is no reason not to go there – immediately. Once there we may evaluate our actual function at the new location and we may compare it with an evaluation of our quadratic form that was expanded about the old location. Their difference is a measure of the inadequacy of the quadratic expansion. In most instances we shall wish to repeat this iteration until motion is negligible. This case, being the well-behaved one, poses no particular problems. It gets us to the nearby minimum very efficiently.

When we examine the largest and the smallest eigenvalues (for which it is *not* necessary to find all the eigenvalues – see Chapter 8) it is quite likely that one will be positive and the other negative, implying that we are nearer to a saddle point than to a minimum. Further, if our current location is near a maximum rather than a minimum, all the eigenvalues will be negative. In either event, a minimum is not immediately within our grasp, so the best we can hope to do is to move in a promising direction, as far as we dare. A general procedure is to add a constant to all the eigenvalues of the matrix S sufficient to make them all at least positive. Operationally this is simple: We add the constant to each of the principal diagonal elements of S. This causes the level lines to cease being hyperbolic and to become ellipsoids again — in effect, forgetting about the saddle point. Then one Newton step with the positivized matrix will move us to that point M in a sphere of radius r_α around our current location where the quadratic form is minimum. Since no global minimum exists in this region, this minimum is necessarily on the boundary of the sphere (Figure 17.14), so we have done as well as we could in one step, given that we limited ourselves to a step of length r_α.

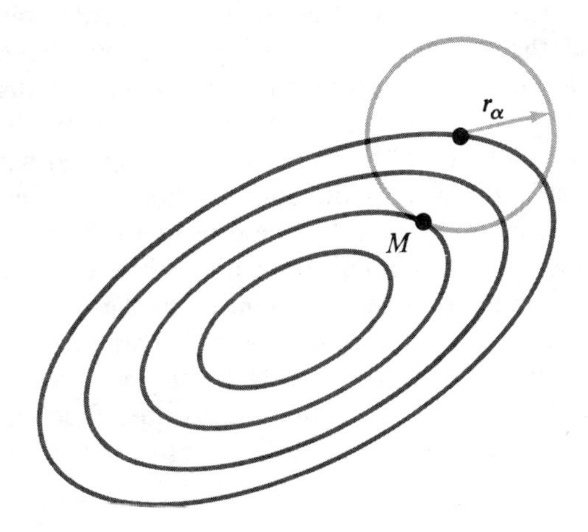

Figure 17.14

To a considerable extent the radius r_α is under our control. It depends on just how positive we make the smallest eigenvalue λ_n of our modified matrix S. Although there are pathologic cases, usually r_α is infinite if λ_n is 0 and it decreases as λ_n is made increasingly positive. So we may choose our own radius of risk by the α we add to the principal diagonal of the S matrix — that is, by deciding just how positive we wish λ_n to be. The several tactics available under the general strategy group into two areas:

1. We may choose α through a crude or sophisticated analysis of previous steps.
2. We may evaluate the elements of S, being the local second partial derivatives of our function, either numerically or analytically.

Numerical evaluation of the derivatives is often less laborious, but is apt to be quite imprecise — especially near the minimum. Such a choice can delay convergence and hence be "penny-wise but pound-foolish."

A reasonable choice for α is usually of the form

$$\alpha = \lambda_n + R \cdot \|\mathbf{G}\|$$

with λ_n the algebraically smallest eigenvalue of S. $\mathbf{G}$ is the gradient, [see (17.18)] and R is a nervousness factor that can vary according to the evidence of over- or under-shooting on the previous step. We determine R by the recent history of agreement between the change in the quadratic approximation and the change in the actual function being minimized. A large R produces a large α and hence a smaller r_α, producing a more conservative step. Conversely, a small R permits a larger r_α and step. Any complete algorithm must also handle exceptional cases such as accidental landing on a saddle point, whence $\|\mathbf{G}\|$ becomes zero and the resultant r_α could be disastrous. Thus $\|\mathbf{G}\|$ must be tested and suitable alternatives provided. For a complete discussion of two versions of this strategy we refer to the original papers by Goldfeld, Quandt, and Trotter [1966, 1968] where the methods are compared with others in the recent literature on a number of difficult functions. As they are careful to point out, comparative evaluations of these techniques are tricky, for many *desiderata* exist and no one minimization algorithm achieves superiority in them all. Prudent comment currently is limited to the observation that these ellipsoidal center seekers are more stable and rather slower than the algorithms given earlier in this chapter. In their defense it must be noted that the reported tests used a standard package that found eigenvalues much more precisely—and laboriously—than was needed. Additional tests are in order.

CHAPTER 18

LAPLACE'S EQUATION—AN OVERVIEW

Many books discuss Laplace's equation, often within the context of an application. Thus we find potential theory, hydrodynamic flow, elasticity, steady-state heat transmission, and subsonic aerodynamics all concerned with solutions to this same equation. The language frequently differs, the boundary conditions vary considerably, and the emphases are variously placed—but the subject remains the solution of this, the commonest partial differential equation. In our first treatment of the topic we put aside every distraction, every complication, restricting our discussion to the basic character of Laplace's equation as it affects its numerical solution. Accordingly, we shall here only talk about solving

$$\frac{\partial^2 u}{\partial x^2} + \frac{\partial^2 u}{\partial y^2} = 0 \tag{18.1}$$

over a rectangular region (Figure 18.1) where u is completely specified on the

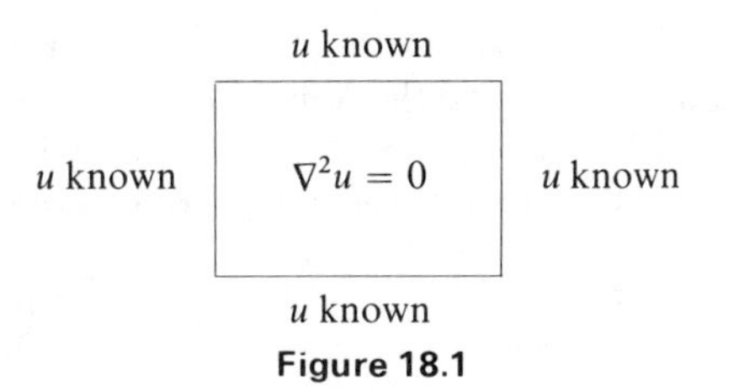

Figure 18.1

boundary. Curved boundaries, polar coordinates, first derivative boundary conditions—all these topics will come later. They are important because they are necessary if we would represent physical problems realistically, but they do not affect the essential philosophy of numerical solution.

The fundamental device is the replacement of our partial differential equation by a set of linear algebraic equations on a grid of points. To this end we introduce such a grid (Figure 18.2) and, in line with our doctrine of expository

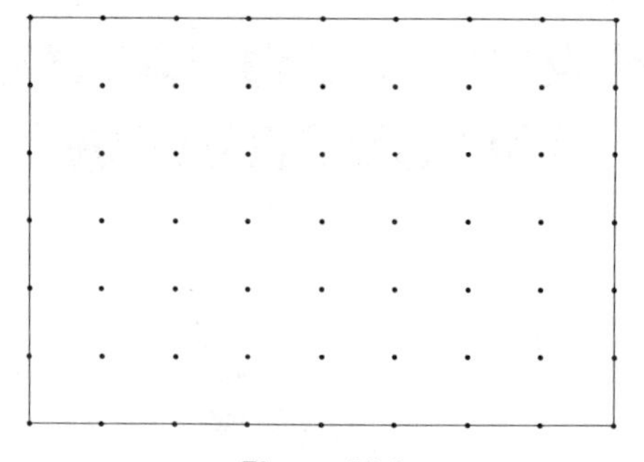

Figure 18.2

simplicity, we shall assume that the grid is square and that our boundaries fall on it. We shall need to talk about groups of these points, most commonly about a "central" point and the four "adjacent" points, by which we refer to a *computational star* or cluster arranged and numbered as in Figure 18.3. This

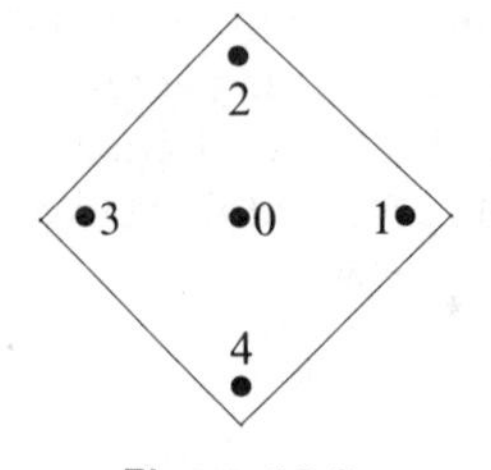

Figure 18.3

notation is local and relative: the central (0) point could be any interior point in Figure 18.2 and then (2) would be the point *above* it. The dependent variable, u, at (3) will be denoted by u_3, and so on. In this notation we have the standard simplest approximation to the second partial derivatives given by

$$\left.\frac{\partial^2 u}{\partial x^2}\right|_0 \approx \frac{u_1 - 2u_0 + u_3}{h^2} \qquad \left.\frac{\partial^2 u}{\partial y^2}\right|_0 \approx \frac{u_2 - 2u_0 + u_4}{h^2} \tag{18.2}$$

where h has been used to represent the grid spacing. Thus at point (0) Laplace's equation becomes the algebraic equation

$$u_1 + u_2 + u_3 + u_4 - 4u_0 = 0 \qquad \text{or} \qquad u_0 = \tfrac{1}{4}(u_1 + u_2 + u_3 + u_4) \tag{18.3}$$

Equation (18.3) says that the variable u at the central point is the arithmetic average of the u's at the four surrounding points. This statement agrees with our knowledge about the averaging properties of Laplace's equation. [The largest and smallest values of u must occur on the boundary of the region, for example, in the exact solution of (18.1).]

Looking at Figure 18.2 we see that the u's are known on the boundary points but are unknown at the interior points. Each interior point thus contributes exactly one unknown. But it also contributes exactly one equation. Thus for n interior points we possess a set of n linear algebraic simultaneous equations in n unknowns—the standard high school problem. What is *not* standard is the size of the system. It is large! Even for our modest problem of Figure 18.2 we have 35 equations. If the variation of u along the boundaries is at all rapid, we might need to halve our spacing in order to represent this variation adequately. This would up the ante to 165 equations. And if the problem should really be three dimensional so that our equation becomes

$$\frac{\partial^2 u}{\partial x^2} + \frac{\partial^2 u}{\partial y^2} + \frac{\partial^2 u}{\partial z^2} = 0$$

we get a three-dimensional grid that could easily include 500 to 1000 points! High school techniques will not suffice. Indeed, unless the equations have some special properties that can be used to facilitate their solution, the problem is nearly hopeless. Accordingly, we shall set up our equations, examining them for regularities that suggest solution techniques.

We now number our points in Figure 18.2 in some regular way. The specific system is not crucial so long as a reasonable order is used. We show ours in Figure 18.4. The equation that comes from replacing Laplace's equation at point 3, for example, involves the unknowns u_2 and u_4, as well as u_{10}. It also involves the known value of u at the boundary point directly above 3. It is

$$u_2 - 4u_3 + u_4 + u_{10} = -u_b$$

·1	·2	·3	·4	·5	·6	·7
·8	·9	·10	·11	·12	·13	·14
·15	etc.					

Figure 18.4

In the same way, the equations centered at points 1 through 7 all involve one or two boundary points while those centered on completely interior points such as 9 through 13 consist only of unknowns. Expressing our equations in matrix form we find that part of it looks like

	1	2	3	4	5	6	7	8	9	10	11	12	13	14	15	16	17				
1	-4	1						1										$\cdots$	u_1		$-b_0\ -b_1$
2	1	-4	1						1										u_2		$-b_2$
3		1	-4	1						1									u_3		$-b_3$
4			1	-4	1						1								u_4		$-b_4$
5				1	-4	1						1							u_5		$-b_5$
6					1	-4	1						1						u_6		$-b_6$
7						1	-4	·						1					u_7	$=$	$-b_7\ -b_9$
8	1						·	-4	1						1				u_8		$-b_8$
9		1						1	-4	1						1			u_9		0
10			1						1	-4	1						1	$\cdots$	u_{10}		0

Figure 18.5

This form is far from a random set of coefficients. In particular:

1. The diagonal coefficients dominate the rows and are all equal to -4.
2. Most of the nonzero coefficients lie in a tridiagonal pattern while the remainder lie in diagonal bands farther afield. (The matrix is "tri-diagonal with fringes.")
3. No row has more than five nonzero coefficients.

These facts immediately suggest the Jacobi or Gauss–Seidel iteration techniques —a suggestion that will be somewhat tempered by further examination. In those techniques, each equation is solved for its diagonal variable as a linear combination of the other variables, plus the constant term, if any. In Jacobi a trial solution is then substituted into the right-hand sides of all the equations to produce a new set of trial values—the iteration being continued until sufficient agreement between successive iterations occurs. (Gauss–Seidel uses the new value of each variable as soon as it is available. It is preferred over Jacobi both because it uses less storage on a computer and, when both methods converge, converges slightly faster.) The dominance of the diagonal coefficients guarantees convergence, the large percentage of zero coefficients ensures a small amount of arithmetic per iteration, and the iterative nature of the process permits efficient utilization of any good approximation we might possess from a priori knowledge. These impressive advantages effectively rule out any attempt to solve these equations by general elimination of variables techniques. Competition for Gauss–Seidel iterations would occur only via methods that use *more* of the structure of our matrix to produce a still more efficient algorithm. A glance at our matrix with its nearly tridiagonal form suggests that we ought to try to invoke that form in some essential way.

Solution of tridiagonal systems

Systems that are *strictly* tridiagonal may be solved directly via an efficient algorithm whose arithmetic labor is proportional to n rather than to n^3. Further, they require storage of only $4n$ elements instead of the more usual $n^2 + n$. For both of these reasons we can entertain the prospect of solving really large linear systems much more calmly when they are tridiagonal. The cost in time and space may still be reasonable even if there are 5000 equations.

The central idea is the factoring of the tridiagonal matrix A into two bidiagonal matrices, L and R. We have

$$Ax = b = LRx$$

If we now write Rx as g we get the two bidiagonal systems

$$Lg = b \qquad Rx = g$$

each of which may be expediently solved, the first from the top downward, the second from the bottom up. Further, the diagonal elements of R can all be taken as unity, thus avoiding their storage (Figure 18.6). This factorization sometimes breaks down when one of the α_i becomes zero, but for the tridiagonal

$$\begin{bmatrix} l_1 & & & \\ \gamma_2 & l_2 & & \\ & \gamma_3 & l_3 & \\ & & \gamma_4 & l_4 \end{bmatrix} \begin{bmatrix} 1 & r_1 & & \\ & 1 & r_2 & \\ & & 1 & r_3 \\ & & & 1 \end{bmatrix} = \begin{bmatrix} \alpha_1 & \beta_1 & & \\ \gamma_2 & \alpha_2 & \beta_2 & \\ & \gamma_3 & \alpha_3 & \beta_3 \\ & & \gamma_4 & \alpha_4 \end{bmatrix}$$

$$L \qquad\qquad R \qquad = \qquad A$$

$$\begin{array}{llll} l_1 = \alpha_1 & l_2 + \gamma_2 r_1 = \alpha_2 & l_3 + \gamma_3 r_2 = \alpha_3 & l_4 + \gamma_4 r_3 = \alpha_4 \\ l_1 r_1 = \beta_1 & r_2 l_2 = \beta_2 & r_3 l_3 = \beta_3 & \end{array}$$

Figure 18.6

systems arising from Laplace's equation on a systematically numbered grid, this trouble will not occur. The total number of arithmetic operations is about $5n$.

Block tridiagonal systems

Returning to our actual tridiagonal fringed system (Figure 18.5) we see that it has the general form of Figure 18.7, where the T_{ii} are tridiagonal matrices and

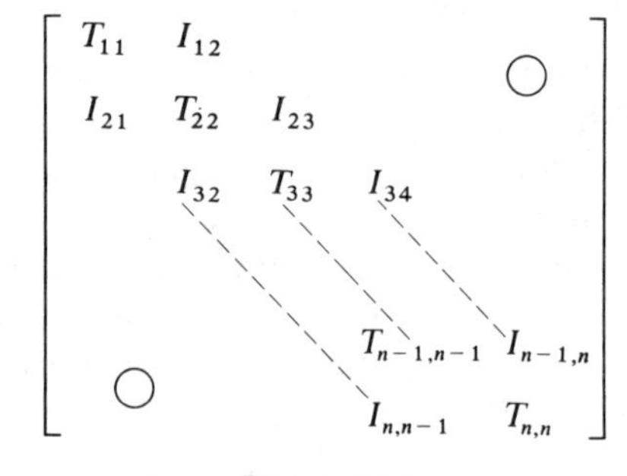

Figure 18.7

the I_{ij} are unit matrices. Further, in terms of these matrix blocks the system is *block tridiagonal* in that the nonzero elements are grouped in a triple diagonal band of matrices. Because of the form of the T's and the I's the elements within this band are still largely zeros, but at least we have managed to cut out large chunks of zeros from our consideration by this grouping. Block tridiagonal matrices can be factored in a way that formally resembles the LR factorization for the strictly tridiagonal matrices discussed above, but unfortunately not all of the analogous efficiencies come along with the factorization. Still, the method is far from unattractive. If, for example, our main fringed matrix had partitioned into a 3×3 block system, we could have written it as

$$\begin{bmatrix} A_1 & & \\ I & A_2 & \\ & I & A_3 \end{bmatrix} \cdot \begin{bmatrix} I & \Gamma_1 & \\ & I & \Gamma_2 \\ & & I \end{bmatrix} = \begin{bmatrix} T_{11} & I & \\ I & T_{22} & I \\ & I & T_{33} \end{bmatrix}$$

and then solved

$$\begin{aligned} A_1 &= T_{11} & \Gamma_1 &= T_{11}^{-1} \\ A_2 &= T_{22} - \Gamma_1 & \Gamma_2 &= A_2^{-1} \\ A_3 &= T_{33} - \Gamma_2 \end{aligned}$$

Although the T_{ii} are ordinary tridiagonal matrices, computational complexities arise at the inversion of T_{11} for the *inverse* of a tridiagonal matrix unfortunately is *not* tridiagonal. In this instance it is a full matrix. Thus we lose the efficiencies of tridiagonal form in the small while preserving them in the large—that is, the factorization of our big $n \times n$ matrix requires only $2(m - 1)$ *matrix* operations even though those operations are now on square matrices each of size n/m. (Here m is 3.)

Alternating directions

A quite different strategy designed to regain tridiagonal efficiencies for fringed tridiagonal systems is the *method of alternating directions.* This method, unlike the preceding one, is iterative. It thus rewards one for beginning with a good approximate answer, but it also usually requires some kind of acceleration device if efficient performance is to be achieved. The rather sophisticated details are set forth in the specialized literature, to which we refer the reader—Forsythe and Wasow [1960] being a good place to begin. Here we confine our remarks to the general ideas of the method, emphasizing its place in a family of methods applicable to Laplace's equation.

The usual replacement for Laplace's equation on a square grid gives highly regular systems of algebraic equations such as we depict on p. 480 (Figure 18.5). Furthermore, one can easily show that the matrix A of that equation system $A\mathbf{u} = \mathbf{b}$ is *negative* definite—which is a positive definite system that has been multiplied through by -1. For all practical geometries the common finite difference Laplacian operator gives rise to these, the best of all possible matrices. Just about any standard solution method will succeed, and many theorems are available for your pleasure. For example, it is a theorem that the Gauss–Seidel iteration will converge for a positive definite equation system from *any* starting vector $\mathbf{u}_0$. In practice, however, it is the computing bill that you are trying to minimize, and this theorem does not concern itself with that. A poor choice for $\mathbf{u}_0$ may not upset Gauss–Seidel, but it will still upset you at the end of the month. For efficient iterative computation you must start with a $\mathbf{u}_0$ that is at least a reasonable approximation to the correct answer. Otherwise the number of operations can become depressingly large.

A fairly general iterative algorithm that succeeds in solving positive definite equation systems computes the $k + 1$st iteration of the solution vector $\mathbf{u}_{k+1}$ from the kth by the formula

$$\mathbf{u}_{k+1} = \mathbf{u}_k - \alpha(A\mathbf{u}_k - \mathbf{b}) \tag{18.4}$$

This process modifies the old vector $\mathbf{u}_k$ by some factor α times the *residual* by which $\mathbf{u}_k$ did not satisfy the system. The choice of 1/4 for α gives us the Gauss–Seidel algorithm (remember that A is positive definite; our example of Figure 18.5 would have $+4$'s on the principal diagonal and -1's scattered elsewhere.)

A different choice for α will give faster convergence – indeed, the best of the known strategies employs different α's on different iterations. The subject is interesting but complicated and we refer you to Forsythe [1960].

The two methods thus far proposed for solving our positive definite equation system lie at two extremes of a spectrum of methods. The first is direct elimination of variables on the system $A\mathbf{u} = \mathbf{b}$, breaking up the system into submatrices so as to avoid treating blocks of coefficients known to be zeros. The second, given in the preceding paragraph, is a pure iteration that invokes no simultaneous equation-solving since the correction for each component of the new $\mathbf{u}_{k+1}$ can be computed directly from the old information. We cannot directly compare the computational efficiency of these two methods for we still have the awkward question of how to assess the contribution of the goodness of the starting vector for the iterative algorithm. It is your author's guess, however, that a sensible direct elimination method is preferable whenever one can fit it into one's computer memory unless one has a quite good initial approximation such as would be available if one were solving a whole series of nearly identical problems and could thus use the solution of the previous problem to begin the next.

We can easily produce iterative algorithms that lie between these two extremes in the sense that we transfer some, but not all, of the components of the $\mathbf{u}$ employed in calculating the residual in (18.4) to the other side by making them $\mathbf{u}_{k+1}$'s. (If we make every $\mathbf{u}_k$ on the right of (18.4) into a $\mathbf{u}_{k+1}$, we get the original equation system for solution – presumably by elimination of variables.) In matrix language, we propose splitting the matrix A into the sum of two matrices to give

$$\mathbf{u}_{k+1} = \mathbf{u}_k - \alpha(A_L\mathbf{u}_{k+1} + A_R\mathbf{u}_k - b)$$

and then transfer the A_L (left) terms over to the left side of the equation system giving

$$(\alpha A_L + I)\mathbf{u}_{k+1} = \mathbf{u}_k - \alpha(A_R\mathbf{u}_k - b) \tag{18.5}$$

If A_L is not diagonal, this is a simultaneous equation system that must be solved by an elimination scheme. Since its solution is not a final answer to our problem but merely the next iterate in a sequence of vectors $\mathbf{u}_k$ that are hopefully to converge to that answer, we shall not be happy about this complication of our iterative process unless it buys us a commensurate increase in the speed of convergence. In particular, there is no point in generating an iterative set of simultaneous equations as complicated as the non-iterative original set. Since the original set was tridiagonal with fringes, the only reasonable simpler set seems to be plain tridiagonal (or, just possibly, bi-diagonal). Several splittings of the matrix A will give us a pure tridiagonal system to solve. One could, for example, choose A_L to be the tridiagonal part of A, thus transferring everything except those annoying fringes of Figure 18.5 to the left. This choice, however, makes the remaining part of the "residual" in (18.5) no longer small since in each equation we have moved only 3 of the 5 elements that represent the 2-dimensional Laplacian operator. Intuitively, a better choice might be those elements that represent the entire 1-dimensional Laplacian, either u_{xx} or u_{yy}. In terms of the local computational star (Figure 18.3), we would move *either* $(u_1 - 2u_0 + u_3)$ *or* $(u_2 - 2u_0 + u_4)$ to the left in each equation. The first choice gives a tridiagonal system immediately from our matrix (p. 480) while the second choice requires that we *reorder* the equations to regain tridiagonal form. (Our matrix arose from equations ordered by numbering the grid points by rows in Figure 18.4. The other system corresponds to numbering the points by columns.)

The *method of alternating directions* is a two-stage iteration. In the first stage we solve the tridiagonal system that results from moving the terms representing u_{xx} to the left; in the second we solve the tridiagonal system that results from having moved the u_{yy} terms to the left and then rearranging the equations. To avoid the rearrangement, practical computational logistics demand that we keep both versions of our equation set in memory, but both are pure tridiagonal and hence each requires only about $4n$ cells. More important, their solutions require a number of arithmetic operations that is proportional to n. Since direct eliminations with even fringed tridiagonal systems require labor that grows as n^3, one can afford a fair number of iterations with alternating directions whenever n is large—and we do not usually go to iterative solutions unless n is large. For many practical problems this alternating directions scheme holds the current record as the most efficient method for solving the large fringed tridiagonal systems that arise from replacing Laplace's equation on a regular grid.

Less regular boundaries

Whenever possible, we should impose a grid for Laplace's equation that includes the boundaries, for our equation system is thereby simplified. If a square grid will not fit a rectangular region, the last row of points will have to have a different vertical spacing, k. This makes the typical equation for a bottom point (Figure 18.8)

$$\frac{u_1 - 2u_0 + u_3}{h^2} + \frac{2}{k(h+k)}\left[b_4 + \frac{k}{h}u_2 - \left(1 + \frac{k}{h}\right)u_0\right] = 0$$

or

$$\frac{k}{2h}\left(1 + \frac{k}{h}\right)(u_1 + u_3) + \frac{k}{h}u_2 - \left(1 + \frac{k}{h}\right)^2 u_0 = -b_4$$

More complicated treatments are demanded by curved boundaries. Of course, if our region has some regular shape for which there is a recognized

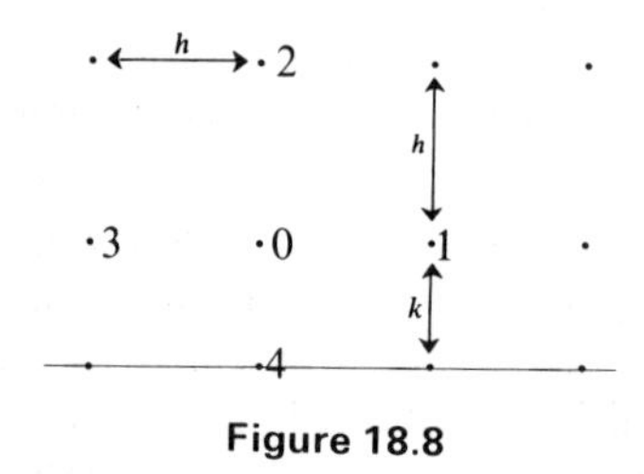

Figure 18.8

coordinate system, we should switch to that system with its corresponding form of Laplace's equation. Thus a finite cylinder with circular symmetry requires cylindrical coordinates and Laplace's equation becomes

$$\frac{\partial^2 u}{\partial r^2} + \frac{1}{r}\frac{\partial u}{\partial r} + \frac{\partial^2 u}{\partial z^2} = 0 \tag{18.6}$$

while an oblate spheroid takes oblate spheroidal coordinates with the equation

$$\frac{\partial}{\partial \xi}\left[(\xi^2 + 1)\frac{\partial u}{\partial \xi}\right] + \frac{\partial}{\partial \eta}\left[(1 - \eta^2)\frac{\partial u}{\partial \eta}\right] + \frac{\xi^2 + \eta^2}{(\xi^2 + 1)(1 - \eta^2)}\frac{\partial^2 u}{\partial \varphi^2} = 0$$

Such systems raise small annoyances concerning grid points where terms in the equation become indeterminate, but they are resolved by use of L'Hospital's rule. Thus in the cylindrical form, the second term of (18.6) seems to blow up at the center. But the condition of circular symmetry requires that, at the center

$$\frac{\partial u}{\partial r} = 0$$

so that on differentiating both numerator and denominator of this term, we find that *along the axis of the cylinder only* Laplace's equation is

$$\frac{\partial^2 u}{\partial r^2} + \frac{\partial^2 u}{\partial r^2} + \frac{\partial^2 u}{\partial z^2} = 2\frac{\partial^2 u}{\partial r^2} + \frac{\partial^2 u}{\partial z^2} = 0$$

Symmetry or boundary conditions will similarly resolve all other apparent infinities in the various forms of Laplace's equation—or the problem has no finite solution and probably does not represent the original physics very well!

When the boundaries are curves that do not lie easily on any of the usual coordinate systems we must devise special replacements for Laplace's equation at each interior point that lies adjacent to the boundary. We give one example in Figure 18.9. Here points 0, 3, and 4 are interior, but points 1 and 2 on the regular grid would fall outside. We want a replacement for Laplace's equation at point 0 that includes the known boundary values u_A and u_B at the points A and B which lie at distances fh and gh, respectively, from point 0. Using a Taylor

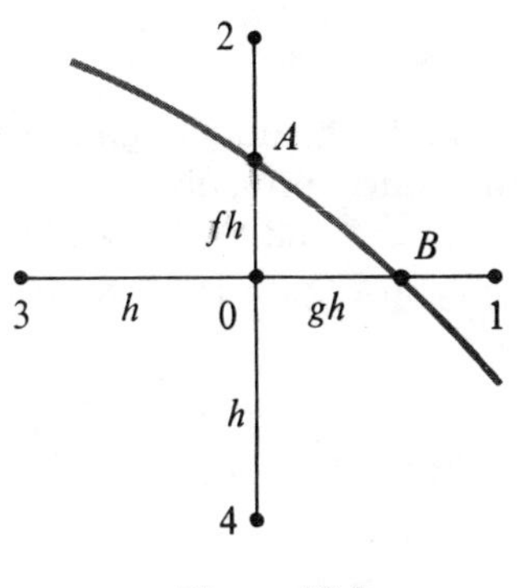

Figure 18.9

expansion about point 0, we may express the potential at A up through second-order terms as

$$u_A = u_0 + fh \cdot u_y + \frac{(fh)^2}{2} \cdot u_{yy}$$

and that at the point 4 by

$$u_4 = u_0 - hu_y + \frac{h^2}{2} \cdot u_{yy}$$

so that we may eliminate the first partial derivative u_y—giving the expression

$$u_A + fu_4 = (1 + f)\, u_0 + \frac{fh^2}{2}(1 + f)\, u_{yy} \tag{18.7}$$

Similarly, the potentials at B and 3 give

$$u_B + gu_3 = (1 + g)\, u_0 + \frac{gh^2}{2}(1 + g)\, u_{xx} \tag{18.8}$$

If we now weight (18.8) by $2/[gh^2(1 + g)]$ and (18.7) by $2/[fh^2(1 + f)]$ and add them, we get

$$\frac{2}{h^2}\left[\frac{u_B + gu_3}{g(1 + g)} + \frac{u_A + fu_4}{f(1 + f)} - \left(\frac{1}{g} + \frac{1}{f}\right)u_0\right] = u_{xx} + u_{yy} = 0 \tag{18.9}$$

which is a reasonable replacement for Laplace's equation at point 0. It contains the unknowns u_0, u_3, and u_4, all linearly, as well as the known constants u_A and u_B and the known geometric parameters f, g. The general grid spacing h^2 will disappear. If the mesh is rather fine, there will be many equations of this type, so it is worthwhile writing a program to set up these equations, as well as to solve the resulting system.

Normal derivative conditions

Many potential flow problems have some boundaries on which the normal derivative is prescribed—and often prescribed as *zero*. If the boundary lies on our grid system such boundary conditions are easily incorporated into our equation system. For example, in Figure 18.10 we may express the derivative

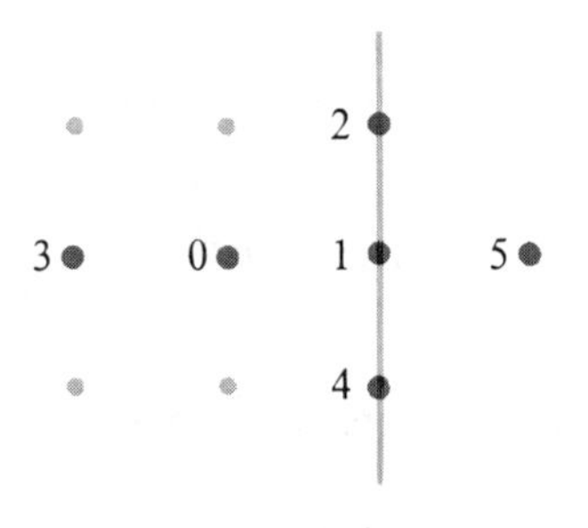

Figure 18.10

at point 1 in terms of u_1, u_0, and u_3 via

$$\left.\frac{\partial u}{\partial x}\right|_1 \approx \frac{1}{2h}(3u_1 - 4u_0 + u_3) \tag{18.10}$$

or, more crudely, by

$$\left.\frac{\partial u}{\partial x}\right|_1 \approx \frac{u_1 - u_0}{h}$$

The price for using the cruder expression is the need for a smaller h—although it may not be enough of an effect to justify the more precise replacement. One generally uses more precise finite difference approximations in the expectation of being able to get away with a larger mesh size h and hence a smaller number of equations while still achieving some desired accuracy in the solution. But this critical size of h depends, in some sense, on the *average* precision with which the differential equation and boundary conditions have been replaced. Since there are usually many more interior points than boundary points, good computational logistics requires us to spend most of our efforts in replacing Laplace's equation well at the interior points and permits us to relax somewhat (but not too much!) when replacing the boundary conditions. Recent research suggests that the boundary replacements can be one order of magnitude in h less accurate for the Dirichlet boundary conditions without substantially affecting the accuracy of the solution.

An alternative philosophy for incorporating normal derivative conditions along straight boundaries is to add an extra row of grid points outside the boundary (point 5 in Figure 18.10). We then set up Laplace's equation at the point on the boundary and finally eliminate the extra point by the derivative condition. Thus we write

$$-4u_1 + u_2 + u_0 + u_4 + u_5 = 0$$

and use the symmetric derivative estimate

$$2h \cdot \left.\frac{\partial u}{\partial x}\right|_1 = u_5 - u_0 \tag{18.11}$$

to produce

$$-4u_1 + u_2 + 2u_0 + u_4 = -2h\left.\frac{\partial u}{\partial x}\right|_1 \tag{18.12}$$

This device is especially satisfying when the normal derivative is zero, but your author has no feeling for the relative efficiency of these two methods. He merely observes that the representations (18.10) and (18.11) of the normal derivative are both of the order h^2 in their error terms.

When the boundary is curved there seems to be no nice method for incorporating the normal boundary conditions. Several can easily be constructed, but none of them exhibits that neatness and aura of inevitability that commend so many difference expressions to their users. Consider in Figure 18.11 the boundary point A and the normal to the boundary there, which intersects the grid at B and C. We may write Laplace's equation at 0 in terms of the potentials at 1, A, 3, 4 just as in the previous section. We have

$$\frac{2}{f(1+f)} \cdot (u_A + fu_4) + (u_1 + u_3) - 2\left(1 + \frac{1}{f}\right) \cdot u_0 = 0 \tag{18.13}$$

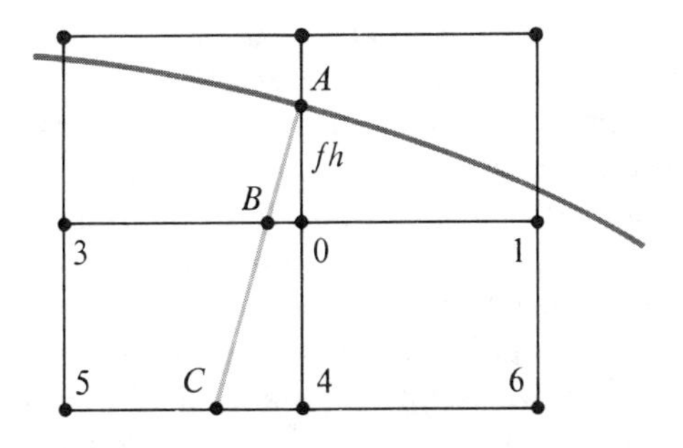

Figure 18.11

while we may also express the normal derivative at A as a linear expression in u_A, u_B, and u_C. Thus we can eliminate u_A in (18.13) at the cost of introducing an expression linear in u_B and u_C. These variables in turn can be expressed by interpolation formulas of varying degrees of complexity but always linear in the potentials at the grid points. Thus u_B can be written in terms of u_3, u_0, and u_1; u_C in terms of u_5, u_4, and u_6. The final equation involves the value of the derivative at A (the given boundary condition) and the unknowns at points 0, 1, 3, 4, 5, and 6 – as well as coefficients formed from the fixed geometric factors such as f. Clearly, such an equation does not fit neatly into our fringed tridiagonal matrix but, provided the percentage of such equations is small, the alternating directions strategem will still produce a solution.

The principal difficulty is the setting up of the equations, for the large geometric variety of curved boundaries prevents easy mechanization. Several schemes are to be found in various computer centers, but no consensus has emerged that can confidently be recommended. Part of the problem is the mechanism by which the boundary is to be described to the program. Should it be a small number of analytic expressions in (x, y) devised by the customer, or should it be a collection of points which the program then connects with "smooth" arcs of some type to create its own boundary? With graphic input perhaps the customer might even sketch his boundaries!

Further complications

Most realistic potential problems, especially those with irregular boundaries, have regions where the potential changes rapidly and others, often in the majority, where variations are obviously going to be slight. In an effort to reduce the large number of grid points, and hence the large number of equations, we shall often wish to use a coarse grid in some regions, a fine one elsewhere. At the transition boundaries our replacement of Laplace's equation obviously takes on non-standard forms which we shall leave to the reader to devise. The difficulties are minor but they cannot be ignored.

More serious is the problem of *accelerating the convergence* of the solution. Most algorithms for Laplace's equation are iterative and are characteristically slow in converging – especially as they near the correct answer. For practical computation, we must accelerate their convergence. Two general devices are used: *Aitken extrapolation* (p. 216) and *overrelaxation*. In the first, two or more potentials from successive iterations at the same grid point are used to yield an extrapolated value there. This device is applied at every grid point to yield an approximate solution that, hopefully, skips a large number of iterations. It also introduces local aberrations which must be smoothed by a number of ordinary iterations before attempting another such Aitken extrapolation – otherwise we shall end up extrapolating the aberrations rather than the converging potentials.

Unfortunately it requires a large amount of storage. Overrelaxation is similar, though somewhat more *ad hoc*. In this device the change between two successive ordinary iterations is obtained, multiplied by a somewhat arbitrary[1] amplification factor to produce an overcorrection that is then actually applied. Again, some choices for the amplification can be shown to be safe, if inefficient, while others might produce instability or rapid convergence – depending on just how they are used. But their proper use is a dimly understood art.

As a final comment we observe that having solved a potential problem to convergence on a grid spacing of h, we may quadruple the number of points and solve again at a spacing of $h/2$. Since we already possess rather good approximate values, the convergence should not take too long. Finally, if the region is convex we can usefully apply Richardsonian extrapolation (page 135) at the points common to both grids to get values that could only otherwise be obtained by solving over a prohibitively fine grid. This Richardsonian extrapolation, please note, is an extrapolation as h is made small. The other extrapolative devices mentioned here are applied on a constant-sized grid and are only extrapolations across iterations that speed the iterative convergence of the solution on that one grid.

Laplace's equation with partial open boundaries

Our previous discussion of Laplace's equation supposed that we wanted the potential inside a closed region with potentials given on the boundary. We now turn to a problem where the region is partially open. Consider the potential in the vicinity of the finite parallel plate condenser. We shall two-dimensionalize our problem by supposing the two plates (Figure 18.12) are infinitely long in the

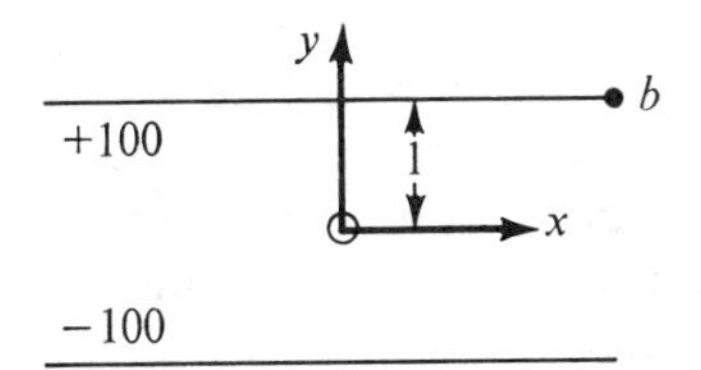

Figure 18.12. A parallel-plate condenser

direction perpendicular to the surface of the page. The difficulties introduced by the finiteness in the x direction are bad enough for our immediate purposes. We are given the potential on the strips, say $+100$ on the top and -100 on the

[1] After you have solved the problem several times you find out what factor you ought to have used.

bottom, and we desire to calculate the potential at other reasonable places, especially near the ends of the plates. Mathematically the problem is defined by these potentials and by the condition that the potential at infinity is zero — but this condition at infinity is not numerically very useful. One can, of course, put the condenser into a large box on which the potential is speci-

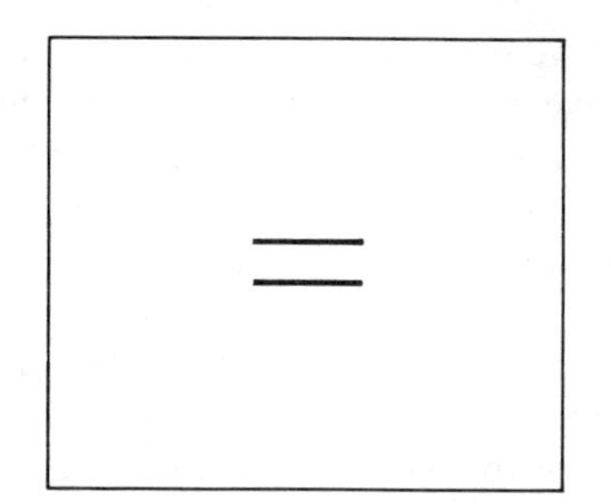

Figure 18.13. A small condenser in a large box

fied to be zero (Figure 18.13), thereby producing the type of problem we have previously considered, but such a strategy is unattractive. It makes us impose most of our grid points in regions that hold no interest for us. Since the grid approach leads to a system of simultaneous equations and since the amount of labor for solving these equations can easily be of the order of N^3, where N is the number of grid points, we are clearly going to expend most of our computational energies calculating potentials about which we could not care less. In fact, this approach rapidly leads to systems that tax the capacity of our largest computers. We can, of course, ameliorate the difficulties by using coarser grids out far from the condenser; we can even determine the asymptotic form of the solution, thereby permitting use of calculated nonzero potentials on a somewhat reduced box — but these are palliatives that partially compensate for a fundamentally unsound strategy. The box must go.

There is another difficulty with the straightforward box approach that remains hidden until we think of the pattern of constant potential and flow lines that describe the solution. We then realize that the flow lines will be strongly bunched at the ends of the plate, implying the need for special treatment there. Actually the charge on the plates becomes infinite at the end points and this creates a computational singularity that must be removed before any simple replacement of Laplace's equation on a grid will succeed. Thus the grid in the box turns out to be not so simple after all. It may be an easy way to program the problem, but unfortunately the answers will be wrong. The student should try it, together with other methods, to get a feeling for the magnitudes of the errors.

Fortunately Laplace's equation, especially for simple geometries in two dimensions, permits a number of alternative approaches. In the remainder of this chapter we explore several other ways to compute the parallel-plate condenser potentials. We do not carry these other methods through in full detail, for to do so would extend this chapter far beyond its fair proportions. Our purpose here is merely to indicate thc general approaches that are possible and to display the nature of their difficulties, for difficulties they all have.

The integral equation

Probably the most general alternative technique for potential problems with partially open boundaries is to set up the integral equation for the potential in terms of the charge on the plates. In two-dimensional potential theory we learn that potential at any point (x, y) is given by

$$\varphi(x, y) = -Cf \ln r^2$$

where r is the distance from the point to the elementary charge of strength f. (In three dimensions it is, of course, merely r^{-1}.) C is a proportionality constant that depends on units. It will disappear. Taking into account the x symmetry and the y antisymmetry of our condenser problem (Figure 18.12), we see that the total potential at any point is given by

$$\varphi(x, y) = -C \int_0^b f(\xi) \cdot \ln \frac{[(x-\xi)^2 + (y-1)^2][(x+\xi)^2 + (y-1)^2]}{[(x-\xi)^2 + (y+1)^2][(x+\xi)^2 + (y+1)^2]} \, d\xi \tag{18.14}$$

where ξ is a dummy variable that runs along the plate and we have normalized our geometry by taking the spacing between the plates to be 2. The charge $f(\xi)$ is unknown. Since (18.14) must hold *on* the plate, as well as everywhere else, we obtain

$$100 = -C \int_0^b f(\xi) \cdot \ln \frac{(x-\xi)^2 \cdot (x+\xi)^2}{[(x-\xi)^2 + 4][(x+\xi)^2 + 4]} \, d\xi \tag{18.15}$$

and it is this integral equation that we would solve for the unknown $f(\xi)$. Once $f(\xi)$ is available, it may be substituted back into (18.14) and the potential $\varphi(x, y)$ evaluated at any point we wish.

A standard approach to integral equations of this type is to replace the plate with a set of points $\{x_i\}$, the function $f(\xi)$ with a set of unknown $\{f_i\}$ at these points, and the equation (18.15) by a set of algebraic equations linear in the unknown f_i. We obtain only one algebraic equation at each x_i, but the quadrature over the plate introduces all the f_i into each equation. Thus for n points we have a set of $n \times n$ linear simultaneous algebraic equations to solve for the f_i. Obviously

n, being the number of points required to represent a function along a line, is smaller by an order of magnitude than N, the number of grid points in that box of Figure 18.13. We are thus prepared to put up with some additional complexities in the equations generated by this integral equation approach.

Since we already discussed in Chapter 15 the principal difficulties encountered in solving equation (18.15) we shall here merely point out that the function $f(\xi)$ is singular at b, having the form

$$\frac{g(\xi)}{\sqrt{b^2 - \xi^2}} \tag{18.16}$$

there, $g(\xi)$ being well behaved. This singularity must be recognized and explicitly removed in the replacement of (18.15) by an algebraic system. Likewise, for some combinations of (x_i, ξ_j)—and all possible combinations occur—a logarithm becomes infinite. Again we must take appropriate measures to remove the singularity quantitatively and carefully or we shall find that the algebraic system has very little to do with the physical problem. This is true because the singularities are the principal contributors to the integral in (18.15). We refer the reader to page 424 for typical details.

The method is quite flexible. Unlike methods discussed below, it does not depend on fortuitous geometries. Rather it embodies directly a basic physical principle that survives even when the condenser plates are badly warped. The disadvantages of this integral equation lie in the analytic care required to accommodate the singularities and the organizational care to ensure that all the minus signs are in the right places in one's computer program. The grid in the box has the virtue of a programmable simplicity that here is totally lacking. If you have a high error rate in writing computer programs or in finding mistakes in complicated formal integrations, then this method may turn out to be unusable *for you.* Another person with demonstrated ability to carry out such operations precisely will find the technique highly useful. There seems to be no practical way to check for minor errors in the output of the integral equation method short of having two different persons do it.

Fourier series—conformal map

Our parallel-plate condenser problem can also be solved by a conformal-map-plus-a-Fourier-series method. The central idea is to map one of the plates onto the unit circle with the rest of the plane, including the other plate, going into its interior. In effect we have put the problem into a very finite box where we may now deal with it in one of several ways. The method we discuss here continues to use mapping techniques, although a grid approach presumably could be applied instead.

We note that classical Fourier series and separation of variable techniques allow us to represent the solution to Laplace's equation analytically in the interior of the unit circle in terms of a potential known on the boundary. Thus we can write the interior potential at (r, θ) as

$$\varphi(r, \theta) = \sum_{n=0}^{\infty} A_n \rho^n \cos 2n\theta \qquad 0 \leq \rho \leq 1 \tag{18.17}$$

while on the unit circle it becomes the *known* potential $f(\theta)$, so that

$$f(\theta) = \sum_{n=0}^{\infty} A_n \cos 2n\theta \tag{18.18}$$

The A_n can therefore be evaluated either by analytic quadratures of a continuous known $f(\theta)$ or, numerically, by summations over values of $f(\theta_i)$ on a grid of points uniformly spaced around the circumference of the unit circle. For either technique there is a convenient orthogonality relation between the cosine functions that permits individual evaluation of the A_n's – a highly desirable though not essential simplification that reduces the amount of computation labor. These boundary-value-problem solution devices are treated in many books, one of the clearer being Churchill [1941]. Once we have evaluated the A_n's for the boundary potential $f(\theta)$, (18.17) permits calculation of the resultant potential at any interior point.

The simple mapping function

$$w = \tfrac{1}{2} \cdot \left(z + \frac{1}{z}\right) \tag{18.19}$$

takes the top plate of Figure 18.14 onto the unit circle, the bottom one to its interior, as shown. We may use this tool to transfer potentials from the one picture to the other whenever it suits our convenience. We may also decide which of the two plates is to be the unit circle and which shall have the interior image. The z plane has the advantage that a specified potential on the boundary permits calculation of its resultant potential on the other plate.

The grand strategy, then, is to set up an iterative mapping procedure. We assume an approximate solution to the problem: the potential $\ln \rho$ in the z plane giving *zero* on the boundary and a nonzero but *nearly constant* potential (DC is nearly a circular arc) on the interior plate CD – which we immediately map onto $C'D'$. At the same time we apply our mapping function (18.19) to the lower plate but use the negative potential $-\ln \rho$. Adding these two potentials gives us nearly constant antisymmetric potentials on the two plates – and zero potential at infinity. If these induced potentials were indeed constant our problem would be solved, for a linear transformation can turn any two different constant potentials on the two plates into the ± 100 that we want. But they are not

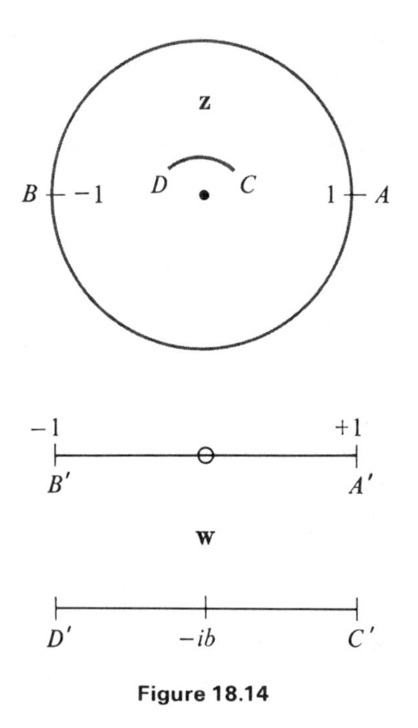

Figure 18.14

constant and so we must make some modifications. Since we continue to treat the plates antisymmetrically, we describe the process for only one.

We accept an average potential on $C'D'$, leaving a residual potential there that we would like to remove. We now cancel this residual by applying its negative to $C'D'$, but this will produce a small potential on $A'B'$ – thus disturbing the just created constant potential there. To find out the precise effect we map $C'D'$ onto the unit circle and let $A'B'$ go into the interior, then use our Fourier series solution to get the answer. We now undertake to balance this residual on $A'B'$ by applying *its* negative there, calculating the disturbance produced on $C'D'$ by the same mapping and Fourier series method. Because the effect of any potential falls off rapidly with distance, each effect is considerably smaller than its cause, and we need bother only about the deviations of each effect from constancy along the plate, thus further reducing each magnitude. In this way the successive corrections fall off rapidly to insignificance and we generate the solution to our condenser problem as the sum of several successively smaller potentials, each with its Fourier series representation. The combined Fourier series and the one logarithmic term permit us to compute the potential at any interior point within the unit circle and the mapping (18.19) transfers this potential to the corresponding point in the w plane.

The calculations are highly repetitive and efficiently programmed for a computer. The difficulties are the practical ones of getting Fourier series that converge rapidly enough for accurate representations of the highly varying potentials near the plates. Stubborn geometries (this one is not) may require more efficient representation, such as rational functions (continued fractions) if we are to avoid series with several hundred terms, but these are questions of additional *finesse* rather than of fundamental changes in the strategy. The method is quite general and may be applied to rather arbitrary configurations of several finite charged plates. It is briefly discussed in Kantorovich and Krylov [1958].

Another conformal map

The parallel-plate condenser geometry permits another conformal transformation that solves the problem "exactly." We note that the complex transformation

$$w = \int_0^t \frac{(c^2 - t^2) \cdot dt}{\sqrt{(1 - t^2)(a^2 - t^2)}} \tag{18.20}$$

maps the first quadrant of Figure 18.15 onto the first quadrant of Figure 18.16.

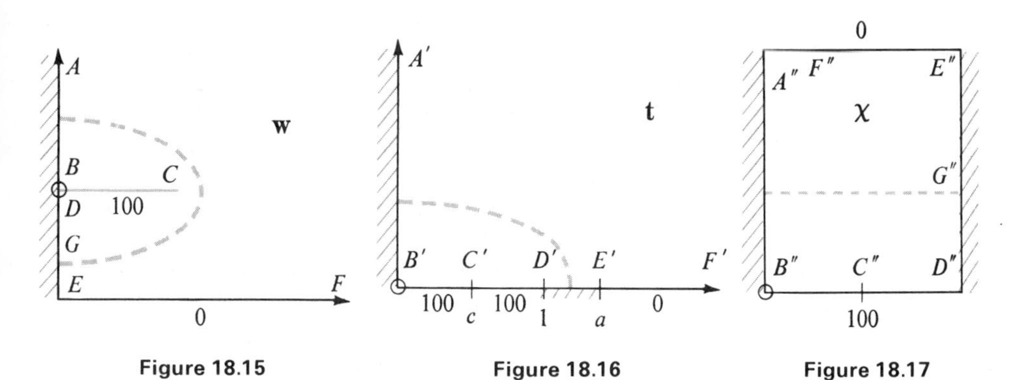

Figure 18.15

Figure 18.16

Figure 18.17

A second rather similar transformation

$$\chi = \int_0^t \frac{dt}{\sqrt{(1 - t^2)(a^2 - t^2)}} \tag{18.21}$$

maps Figure 18.17 onto Figure 18.16. Each transformation maps a solution of Laplace's equation onto the corresponding solution of Laplace's equation in the other geometry. Figure 18.15 is one quarter of our parallel-plate condenser, with BC being half the upper plate. ABE is an axis of symmetry, requiring the reflective boundary condition (a zero normal derivative), while EA is the central horizontal line of antisymmetry along which the potential must be zero. Figure 18.17 is a trivial potential problem, solvable by inspection. Thus if we can map a point from Figure 18.15 through Figure 18.16 to Figure 18.17 we will know its potential to be simply 100 times its fractional distance from $A''E''$ toward $B''D''$.

Both integrals have two singularities. These can, and must, be removed by a change in the dummy variable of integration. If there had been only one singularity a trigonometric substitution would have been appropriate. Here, to remove two singularities, we must use Jacobian elliptic functions. We have discussed the technique in Chapter 15.

There is an analytic neatness to this solution that is attractive. We may go farther: A line of constant potential, horizontal in Figure 18.17, may be traced in Figure 18.15 quite simply by noting that

$$\frac{dw}{d\chi} = \frac{dw/dt}{d\chi/dt} = c^2 - t^2$$

hence we may integrate this two-equation system (t, w, and χ are complex) in its real and imaginary components—starting from any boundary point G. The boundary points are easily expressed in terms of real integrals or real elliptic functions for which adequate algorithms exist.

Why would anybody bother with other methods? Is there some hidden difficulty? Of course there is, though it happens not to be too bad. For reasonable condenser geometries such as the one pictured in Figure 18.15 with BC/DE equal to unity, the corresponding values of c and a are 0.8405 and 1.0104, respectively—both much closer to *one* than we would prefer. This fact is more of an annoyance than an obstacle, for we know what to do. We store $(1 - c)$ and $(a - 1)$ as our fundamental parameters and use them in the relevant algorithms for evaluating the Jacobian elliptic functions and quadratures. If we blindly use c and a, giving no thought to their magnitudes, we shall experience crippling losses of significant figures that will make a mockery of our calculations. (Hence our quotes on "exactly" in the first sentence of this section.) The difficulty here is simply a booby trap for the unwary; the experienced analyst will not be deterred. Finally, however, we must point out that very few geometric potential problems lend themselves to this type of mapping. The earlier techniques took expositional precedence because they are more generally useful.

CHAPTER 19

NETWORK PROBLEMS

In this final chapter we turn away from numerical methods to glance briefly at a class of nonnumeric problems involving networks. In so doing your author hopes to whet an interest in what might have been called logical problem solving were it not for the inference that numerical methods are thereby illogical! Obviously the solution of nonnumeric problems is a large subject, so he here restricts himself to sampling from the network group as being at least indicative of the variety that awaits the explorer.

The network, frequently called a *graph*, has nodes and edges – or junctions and lines. It may arise as a picture of a highway system or of a radio circuit. It may represent a somewhat arbitrary idea such as a construction schedule or it may simply describe a Family Tree – hopefully without too many closed loops. We shall discuss four problems:

1. Finding the quickest way between two points in Baltimore at 5 P.M.
2. Paving the least amount of road surface in Alabama that will still provide paved access to all its towns.
3. Finding the probable bottlenecks in the construction schedule of a new airport.
4. Finding the maximum flow through a network of pipes, given their separate capacities.

All these problems are trivial if the network is small. Since many real problems with large networks exist, however, it follows that we shall be principally concerned with the speed of our algorithms and the efficiency of our storage.

Storage of network information

The network consists of *nodes*, which we shall name with integers, and *edges* which we shall describe by the pairs of nodes they connect. Thus in Figure 19.1 nodes 3 and 1 are connected by the edge (3, 1), while nodes 5 and 6 are not directly

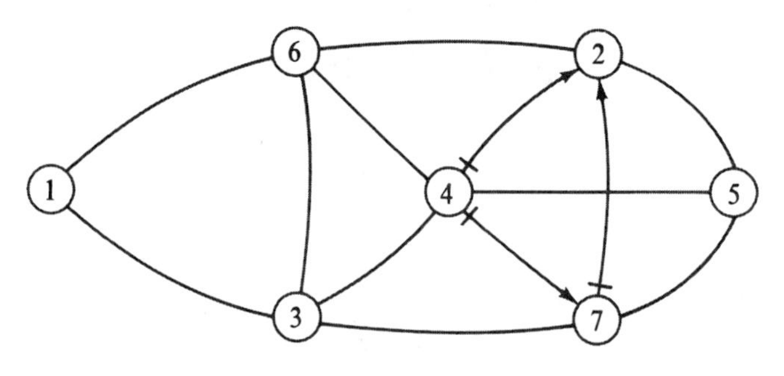

Figure 19.1

connected by an edge. Both nodes and edges may carry *values.* Thus if an edge represents a pipe it might have a capacity of 7 ft^3/min of sulfuric acid, while a node representing a person in a family tree might have values equal to the number of children he had and the size of his estate. The first value would be redundant, since the number of children would be the number of edges eminating upward from the node, while the second value would be additional information not implicit in the network geometry. Edges may be *directed.* Thus the edge (3, 1) may be distinct from (1, 3) in some problems, while in others no directional information will be needed. Cities with one-way streets leap to mind. The values associated with edges are frequently called *distances*, for obvious reasons. There is some danger in such mnemonics for we might wish to refer to the distance between two cities on a map that were not directly connected by a road (edge)—but the nomenclature still seems useful.

Embedded in any graph or network is a myriad of *trees*—defined as graphs with no closed circuits. Again, the nomenclature often becomes specialized for particular applications. We shall try to avoid such specializations, relying on the intuitive geometric concepts that gave birth to the terms. Thus once we identify a node as the *root* of a tree, we may systemmatically proceed to explore its *branches.*

A *nodal matrix* is the most voluminous and least efficient way to store a network. It may, however, give the best accessibility and hence the fastest algorithms. So if we have sufficient memory space in our computer, the nodal matrix should not be lightly dismissed. A nodal, or incidence, matrix is a square matrix of size n, the number of nodes, having a nonzero entry at (i, j) if the edge (i, j) exists. If our only information is the existence of the (i, j) edge, our matrix may be binary, a 1 indicating the presence of the edge. If we are storing values or distances, then this is the item that appears. In either event, *zero* implies no connection between nodes i and j. The obvious problem with the nodal matrix is the large number of nodal pairs that probably are *not* connected in a large network. This information could be implied rather than explicitly stored as zeros, thereby presenting the same information in a smaller compass.

The *neighborhood array* stores network connective information as a collection of vectors, one per node. The components of the vectors are the neighbor node numbers, the nodes that can be reached directly from this one. While reasonably compact this notation is still not completely dense except for the peculiar graphs where every node has exactly k neighbors. It is, however, apt to be more compact than the incidence matrix. For manipulative convenience each vector is filled out with zeros to the length of the longest vector, thus permitting them to be referenced as an $N \times K$ array, where N is the total number of nodes and K is the largest number of neighbors enjoyed by any node. Although not logically necessary, a vector that shows the actual number of neighbors enjoyed by each node is convenient. The information could of course be retrieved by counting the number of nonzero entries in the pertinent row of the neighborhood matrix. (But note that an incidence matrix is *binary*, so that size comparisons are not direct.)

A *list* gives the greatest practical compression of network connective information. As the name implies, it is simply a consecutive list of node pairs, one for each edge. If the ordering of these edges is unimportant, the list can be compressed by grouping together all the edges that eminate from one node. The common node name may then be omitted, saving nearly a factor of 2 in memory space. Of course, the deleted information has to somehow be supplied – usually by a nodal vector that points to the first entry of the list associated with that node. As with the neighborhood array, values must be stored in a similar but separate list.

By way of example, we give all three representations for the network of Figure 19.1, a graph in which three edges are directed (one-way streets), while the other 10 are not. We double the undirected edges, replacing each of them by *two* directed edges – giving a total of 23 to be listed. The nodes are numbered randomly. The incidence matrix has zeros everywhere it does not have 1's, although we have here omitted all but the diagonal zeros for visual clarity. We have also emphasized the 1's that represent the one-way edges. Clearly if all edges were undirected, the matrix would be symmetric and its upper half would suffice. The total storage required is thus either N^2 or $N(N-1)/2$ bits.

Node n \ Node m	1	2	3	4	5	6	7
1	0		1			1	
2		0			1	1	
3	1		0	1		1	1
4		1	1	0	1	1	1
5				1	0		1
6	1	1	1	1		0	
7		1	1		1		0

Incidence Matrix

The neighborhood array for Figure 19.1 is

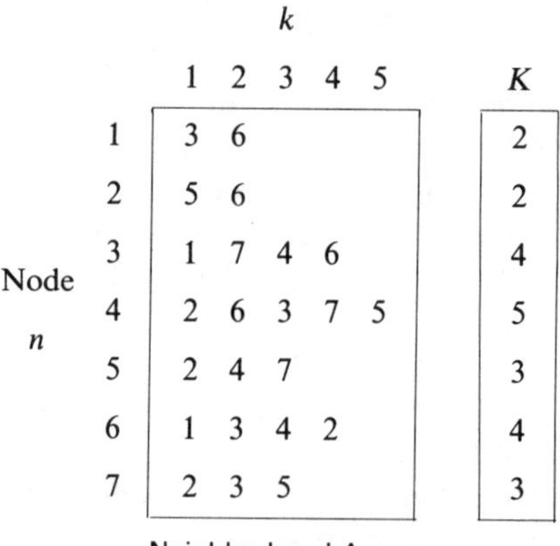

Node n \ k	1	2	3	4	5	K
1	3	6				2
2	5	6				2
3	1	7	4	6		4
4	2	6	3	7	5	5
5	2	4	7			3
6	1	3	4	2		4
7	2	3	5			3

Neighborhood Array

Again, zeros have been omitted in our diagram. Some information about the relative ordering among the neighbor nodes can be stored in the neighborhood array. Thus, in each vector, we have listed the smallest numbered neighbor node first and continued counterclockwise in our diagram. This choice obviously has nothing to do with the connections of our graph but only with the particular way it happens to be drawn in Figure 19.1. Storage for the neighborhood matrix requires at least NK symbols or $NK\{1 + [\log_2 N]\}$ bits.

Finally, we have the compressed *list*. The edges are grouped according to the node from which they originate, are numbered sequentially in the order they happen to be written down.

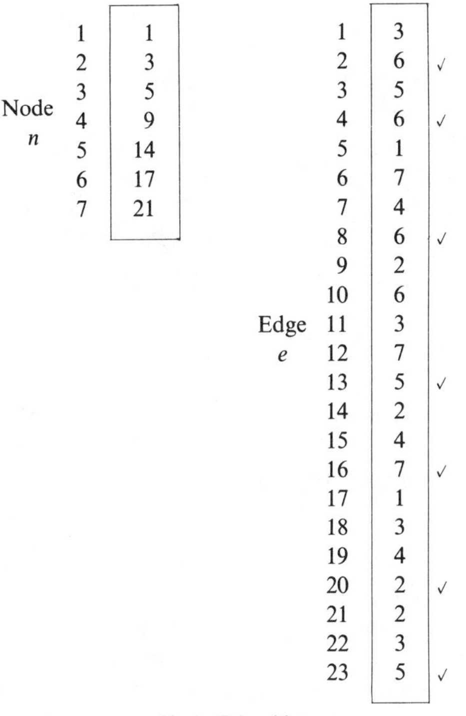

Node n	
1	1
2	3
3	5
4	9
5	14
6	17
7	21

Edge e		
1	3	
2	6	✓
3	5	
4	6	✓
5	1	
6	7	
7	4	
8	6	✓
9	2	
10	6	
11	3	
12	7	
13	5	✓
14	2	
15	4	
16	7	✓
17	1	
18	3	
19	4	
20	2	✓
21	2	
22	3	
23	5	✓

Node Edge List

The small nodal vector points to the *first* edge that originates from each node. Thus edges emerging from node 4 begin with the ninth edge in the edge vector, which says that it is the edge that goes to node 2 – hence (4, 2). Since the node 5 edges do not start until 14, all the edges from 9 through 13 must originate from node 4. This notation is too compressed for many uses because it is difficult to find where a group of edges ends. Frequently the addition of a binary flag (✓) on the last edge of a group will more than justify the small additional storage by greatly decreasing the running time of algorithms that must traverse the network. (Any values or distances would be contained in a second edge vector, in effect turning it into a 23×2 matrix.)

Note that this list depends on its order to achieve its considerable compression. If we wish to reorder the edges – sorting them by length, for example – we must use a list that includes both nodes in each edge designation, thereby becoming twice as large, though more flexible.

Traffic in Baltimore—the best path through a network

Suppose we have a map of a large city like Baltimore (Figure 19.2) on which we are given the times required to traverse each block at rush hours. We are now

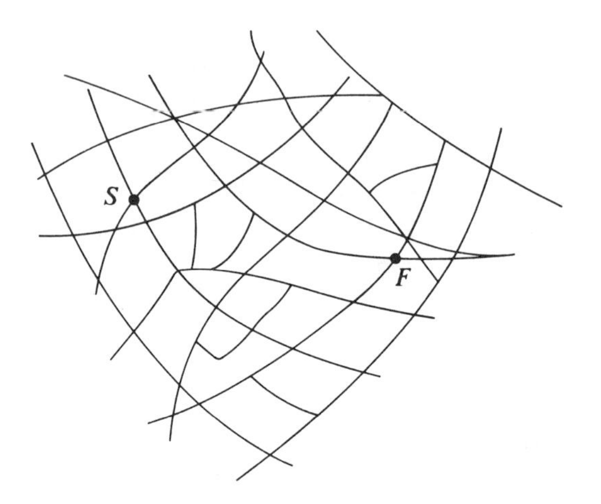

Figure 19.2. Streets of "Baltimore"

given two points, S and F, and are asked to find the quickest route from S to F through the city. One simple-minded approach would be to start at S and pass to all nodes (that is, intersections) that can be reached directly from S, tagging each of the neighboring nodes with the time required to get there from S. We then take each of these nodes in turn, passing to their next neighbors, labeling each with the *total* time required to reach it from S. In this and later stages we shall often reach a node that has already been reached previously by another route. On such occasions we substitute the new travel time only if it is *smaller* than the one already there. After processing all the neighbors of one node we must flag that node lest we reprocess it upon reaching it by some circular route. When all the nodes have been processed, each has been labeled with the *minimum time* required to reach it from S.

We may now find the minimum time *route* from S by starting at F and working backward, always choosing those nodes that can be reached from the neighbors in the times indicated on the route—that is, the number along the street must equal the difference between the times that label the nodes bounding that street—otherwise this street is not in a minimum path from S. In this way we shall ultimately reach S, possibly by more than one route since ties have not been excluded in our problem. If some of the streets are one-way, then we can make them all one-way, replacing each two-way street by two one-way streets. This may drastically enlarge the problem but introduces no qualitative differences.

The algorithm just described will solve our problem but is not very efficient. The first part gives us more information than we want by giving the minimum times from S to *all* nodes in our graph rather than confining itself to the node F.

The second part, the reverse tracing of the minimum route, is a small fraction of the total work and is relatively efficient since it usually eliminates irrelevant streets rather quickly – pursuing very few unprofitable paths.

Two improvements in the strategy suggest themselves. If we can obtain the time required by any one path from S to F, we may then stop processing nodes that show travel times in excess of the feasible route that we already know. This technique will greatly improve the efficiency of our algorithm, but depends on a priori knowledge that we may not possess or easily obtain.

A different approach, suggested by Nicholson [1966], would carry the minimum time algorithm out simultaneously from both S and F, testing after each node is processed to see if the two trees have yet met. A meeting establishes a feasible time (and route) but does not guarantee a minimum – which could arise from a yet-unprocessed node that has more efficient streets connected to it. Thus we continue to process the nodes with *smallest* access times, testing after each to see if the shortest travel time to a branch end in the S tree plus the shortest travel time to a branch end from F is still less than the shortest travel time over the feasible S, F paths already established. When this test fails we are then certain that we cannot find any paths that are faster than those already found, which must therefore include the minimal paths. Again, our algorithm can be split, confining the first part to times while relegating the route finding to a second operation, or the two may be combined and carried out in parallel. Under the sequential philosophy, we trace the paths backward to S and F from their discovery points where the two trees met. For such tracing, one-way streets must be reversed.

As a coding technique we may construct two matrices: an intersection matrix and a street matrix. We show a sample problem thus encoded below, the basic data being presented in Figure 19.3. Note that three of the streets are one-way, as indicated by arrow heads. The numbers on the streets are the times for traversal in arbitrary units. The numbering of the intersections is random.

In the street matrix, we form our listings by grouping together all the streets

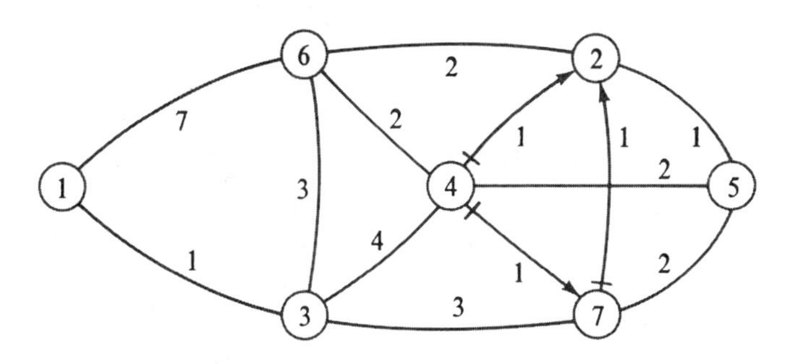

Figure 19.3. Streets with traversal times

which *originate at the same intersection.* This is essential to our coding. The order within these groupings is unimportant as is the order of the groupings themselves. A two-way street will appear twice; a one-way street only once.

Note that the intersection matrix gives the row of the street matrix in which will be found the beginning of the group of streets that originate from this intersection. The end of that group is signaled by the flag f and the far ends of the several streets in the group are given by the column D. The column V gives traversal time for the street. If we wish to drive through the city we clearly can

1. Pick an intersection, I, finding the group of streets originating there.
2. From the street matrix choose a street, finding its terminal intersection.
3. From the I matrix we can locate the group of streets that originate at the second intersection, and so on. If we total the V's as we drive, we know how long it has taken. There is nothing to prevent our looping.

The I matrix is shown in its initial condition for finding fastest times from intersection 1 to anywhere. In the less efficient algorithm we pick

I	S	T	f
1	1	0	1
2	3	99	0
3	5	99	0
4	9	99	0
5	14	99	0
6	17	99	0
7	21	99	0

Intersection Matrix

S	D	V	f
1	3	1	0
2	6	7	1
3	5	1	0
4	6	2	1
5	1	1	0
6	7	3	0
7	4	4	0
8	6	3	1
9	2	1	0
10	6	2	0
11	3	4	0
12	7	1	0
13	5	2	1
14	2	1	0
15	4	2	0
16	7	2	1
17	1	7	0
18	3	3	0
19	4	2	0
20	2	2	1
21	2	1	0
22	3	3	0
23	5	2	1

Street Matrix

any intersection having a flag of 1 and compute the time to the several intersections that can be reached from it. If these times are less than those already computed, they are substituted for their predecessors and the flags for those intersections are set to 1. (If the computer times are equal to or greater than the values already sitting in the T slots for the appropriate intersections, no action is taken.) The algorithm begins by having the flag of the S node being set to 1, its T to 0 – all other flags being 0 and all the T's to 99 or some other impossibly large number. It finishes when all flags become zero. We display in the table the history of the algorithm for our example, showing the various values of T and f

I	T	f	T	f	T	f	T	f	T	f	T	f	T	f	T	f	T	f	T	f
1	0_x	1	0	0	0	0	0	0	0	0	0	0	0	0	0	0	0	0	0	0
2	99	0	99	0	99	0	6_x	1	6	0	6	0	6	0	5_x	1	5	0	5	0
3	99	0	1_x	1	1	0	1	0	1	0	1	0	1	0	1	0	1	0	1	0
4	99	0	99	0	5_x	1	5	0	5	0	5	0	5	0	5	0	5	0	5	0
5	99	0	99	0	99	0	7	1	7_x	1	7	0	7	0	6	1	6_x	1	6	0
6	99	0	7	1	4	1	4	1	4	1	4_x	1	4	0	4	0	4	0	4	0
7	99	0	99	0	4	1	4	1	4	1	4	1	4_x	1	4	0	4	0	4	0

that were recorded and often replaced. The subscript x indicates the node whose streets are being traversed to give the times in the next stage of the algorithm. In Figure 19.4 we show the final values, the minimum times to reach the

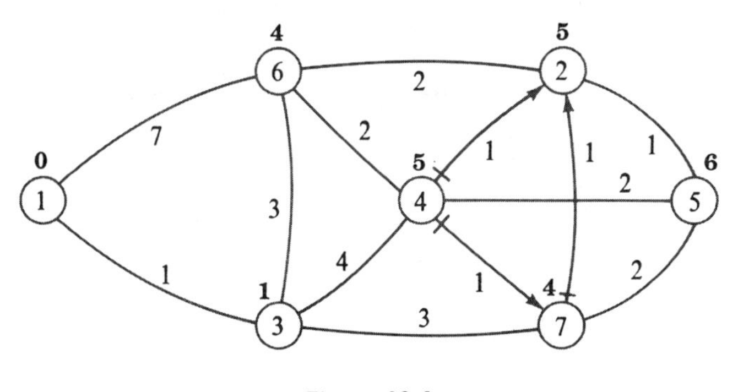

Figure 19.4

respective intersections from 1, adjacent to the intersection. If we wish to find the minimal route to an intersection, say 2, after the minimal T data are available, we can do so either:

1. By working backward from each intersection in turn, always going to the

predecessor intersection whose T differs from the T where we currently are by exactly the V of the street; or

2. By tracing the entire minimal tree forward from the origin – a difficult program to write without a list-processing language.

In order to pursue the first option we must reverse our one-way streets, but otherwise leave our lists unchanged. The algorithm is simply a subtraction and comparison to find the predecessor intersection which is then listed and moved to. But since the minimal path is not always unique (see the path to 5) provision must be made for branching.

The more efficient Nicholson algorithm uses this simple algorithm as an essential part, modified to produce two sets of paths radiating from S to F, respectively – checking after each extension to see if a path has yet been found and, if so, whether other more efficient paths are still logically possible.

PERT, critical paths and networks

Suppose we consider a construction job where a number of interrelated processes must be carried out, some necessarily following others in strict precedence. By way of concrete example we suppose 12 *activities* and 8 recognizable *events* that are related as in Figure 19.5. An event may be the completion of

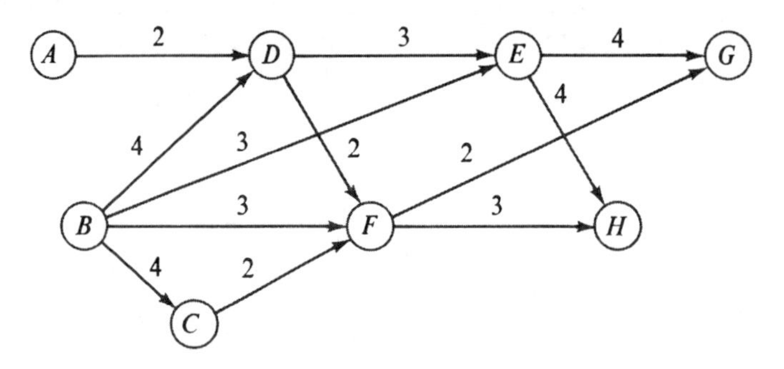

Figure 19.5

the pouring of concrete or the letting of the electrical wiring contract, and so on, and is one of the *nodes* of our graph – lettered $A, B, \ldots, H$. An *activity* such as the pouring of concrete extends from one node to the next and is represented in our graph by an edge or branch. Associated with each activity is a *value*, that is, the length of time it goes on, which appears in Figure 19.5 explicitly on each branch. The questions of interest are:

1. How rapidly can the entire project be completed? In graphical terms, what is the *longest*-valued path through the directed network?

2. What are the earliest dates each event can possibly be completed? (This is a subquestion that has to be answered in answering question 1.)
3. What are the latest dates each event may be completed without affecting the answer to question 1?

If our graph is small, the answers are easily found by inspection and a little pencil work. We display the results for our sample problem in Figure 19.6 with

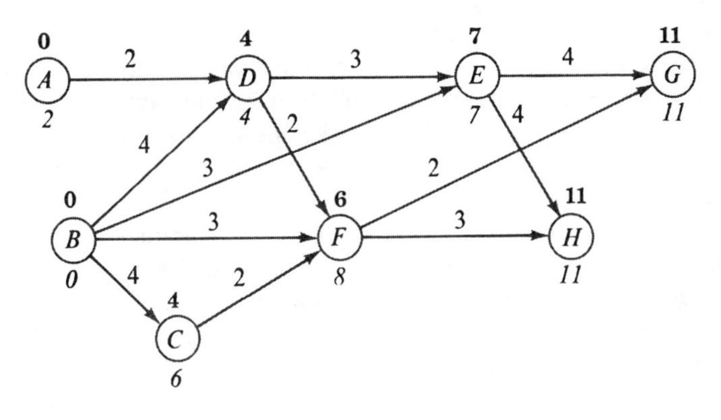

Figure 19.6

the earliest dates above each event and the latest permissible date below. In real construction projects or national economic planning projects, however, we may find several thousand activities instead of 12, and the answers are not so easily obtained. So we seek computer algorithms.

In this example we have already performed one important operation: The nodes are labeled in topological order—that is, the activities all proceed

A, D	2
B, E	3
B, D	4
B, C	4
B, F	3
C, F	2
D, E	3
D, F	2
E, G	4
E, H	4
F, G	2
F, H	3

List 1

from an earlier to a later alphabetical symbol. Thus if we try writing down our facts as a list, the algorithm to compute the upper numbers in Figure 19.6 is easily seen to be quite simple. We merely assign the value *zero* to A and B, then proceed down list 1 (see page 509) systematically, computing a score for the second node of each pair by adding the current score to the score already held by the first node. If the resulting total is *greater* than the score already assigned to the second node, we replace it with the bigger one. Otherwise we pass on. A single pass through the list gives the correct scores.

The latest permissible dates are found by sorting our list of activities into order on the second node, starting with H and working backward. The same algorithm is now used, starting with G and H at 11 and subtracting the values, recording the new value if it is *smaller* than the one already there. On a computer we would assign impossibly large values to all the nodes at the start of this part, excepting only G and H.

Thus we see that the topological ordering is very useful—either in putting our activity information into convenient physical order for the other algorithms or in constructing linkages for the list in topological order, which may then be perused at leisure without physically relocating any data. We give below two different algorithms for ordering the nodes of a directed network. Finally, we give (p. 515) a scheduling algorithm that does not require prior topological ordering, although it must then do some work of its own that is equivalent.

Topological ordering of nodes in a network

We shall rename the nodes of our previous network more or less randomly in order to display two algorithms for finding a topological ordering. Since the algorithms are to be performed ultimately on a computer, it is convenient to let our node names be *numbers*, though letters will do as well if relative ordering and sorting is feasible with them on your computer.

The first algorithm given by Lasser [1961] starts by sorting the activity list into order on the source node and, within each group, on the sink node. Thereafter it proceeds according to the following rules (and see the table):

1. Move the first labeled grouping into the middle field, deleting from the first field.
2. Move any groupings beginning with labels that are successor labels to the first group into the middle field.
3. When no more can be moved under rule 2, move the entire contents of the middle field to the *bottom* of the final field. Then draw a line *above* these items to define a new bottom for the final field. Then go to rule 1.

First field	Middle field	Final field
1, 5	1, 5	6, 2
1, 7	1, 7	
		3, 1
2, 1	2, 1	3, 2
2, 8	2, 8	3, 4
	8, 5	3, 8
3, 1	8, 7	4, 1
3, 2		
3, 4	3, 1	2, 1
3, 8	3, 2	2, 8
	3, 4	8, 5
4, 1	3, 8	8, 7
	4, 1	
6, 2		1, 5
	6, 2	1, 7
8, 5		
8, 7		

Rules 1 through 3 are repeated until all items have been moved to the final field. The first nodes in the final field are now in topological order except for the nodes that have no successors. The student should follow through our example and

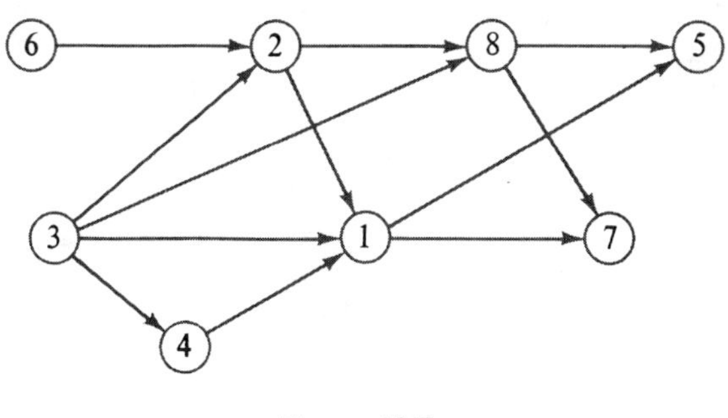

Figure 19.7

verify by comparing Figure 19.7 and 19.6 that the algorithm has indeed succeeded. The strokes indicate groups that have already been moved at the moment that the grouping headed by (2.1) gets moved into the final field.

A second algorithm, given by Kahn [1962], wanders through the network placing, altering, and removing quantitative flags. We shall describe the algorithm twice, first on the network, then on the computer.

1. Place zeros on the nodes with no precedents: (6) and (3).
2. Label each node with the number of its immediate predecessors.
3. Form a list of ordered nodes, beginning with those having zero predecessors. As each node is added to this list:
   a. Reduce its predecessor count to a canonical -1.
   b. Reduce the predecessor count of all its successor nodes by unity.
4. Repeat stage 3 until -1 sits on all the nodes. The ordered list is then in topological order.

On the computer we can encode the Kahn algorithm by using the two matrices already familiar to Baltimore drivers. The activity matrix is ordered so that all activities with the same source are grouped. Only the last two columns of our two matrices are actually stored, the rest of the information being implicitly presented. For example, event 3 is the third item in the event matrix and the activity column says that the activities originating at event 3 begin at the fifth location in that matrix. The successor column says that these activities terminate at 1, 2, 4, and 8 – the flag of 1 saying that this is the last activity of that

Activity	Edge	Flag	Successor event
1	1, 5	0	5
2	1, 7	1	7
3	2, 1	0	1
4	2, 8	1	8
5	3, 1	0	1
6	3, 2	0	2
7	3, 4	0	4
8	3, 8	1	8
9	4, 1	1	1
10	6, 2	1	2
11	8, 5	0	5
12	8, 7	1	7

Activity Matrix

Event No.	Activity No.	No. of predecessors
1	1	3
2	3	2
3	5	0
4	9	1
5	·	2
6	10	0
7	·	2
8	11	2

Event Matrix

group. We are thus in a position to carry out Kahn's algorithm, reducing the predecessor count as we add the various events to an ordered list – beginning with those having zero predecessors.

The occurrence of a loop in the directed PERT graph is, of course, an error. In a small system it is usually obvious, but in large machine-analyzed nets we must rely on mechanical detection. Discovery that a loop exists comes from failure of the ordering algorithm, but pinpointing the nodes that comprise it is not simple. Kahn suggests three techniques:

1. List all events which have not been ordered but which have had their predecessor count reduced by at least 1. These events either are in a loop or have a predecessor in a loop.
2. Run the ordering routine a second time from the beginning reversing the roles of predecessor and successor. Events remaining unordered in both forward and reverse applications of the algorithm are either in loops or lie between two loops.
3. Use a precedence matrix to specify the loops exactly.

The mechanism seems cumbersome, a recent algorithm by Knuth [1968] being neater. But see below for an implementation.

Critical path networks—continued

An alternative method of storing our edge and node data facilitates the detection and identification of loops. It is given by Klein [1967], but note that his Fig. 2 is numbered inconsistently with his worked example. We rework our example here by his system—and then do it again after introducing a loop.

We write down our activities, in any order, in an activity table and our nodes in a node table.

Activity		Common Father	Common Son	V
1	A, D	·	*5*	2
2	B, E	**5**	*3*	3
3	D, E	**4**	·	3
4	D, F	·	*7*	2
5	B, D	**6**	·	2
6	B, C	**8**	·	4
7	C, F	·	*8*	3
8	B, F	·	·	3
9	F, G	**10**	*11*	2
10	F, H	·	*12*	3
11	E, G	**12**	·	4
12	E, H	·	·	4

Node		Start of chain Father	Start of chain Son	Counts Father	Counts Son	Early times	Work count	Z list
1	A	**1**	·	*0*	**1**	0		~~A~~
2	B	**2**	·	*0*	**4**	0		~~B~~
3	D	**3**	*1*	*2*	**2**	~~2~~ 4	‖	
4	C	**7**	*6*	*1*	**1**	4	|	
5	F	**9**	*4*	*3*	**2**	3	|	
6	G	·	*9*	*2*	**0**			
7	E	**11**	*2*	*2*	**2**	3	|	
8	H	·	*10*	*2*	**0**			

We then construct linkage information (**bold face** and *italic*). In the node table we list the first lines of the activity table in which each node is mentioned as a father (**bold**) and as a son (*italic*). In the activity table we provide chain links downward to the next occurrence of each node as a **father** and as a *son*. Thus F as a son is seen to first occur on line 4 of the activity table and its next occurrence is on line 7, the next is line 8, which is the last such occurrence. Thus F has a father count of 3.

The algorithm for the early time for completion requires a list, the Z list, of nodes all ready to process (and later, already processed). We also have a work count which counts the number of times we have computed an early time for that node. When the work count equals the father count we have visited the node once from each source that leads to it so it is finished and its current early time has reached its final value. We can now add this node to our Z list to be used as a source node for the computation of early times of its successors. We begin by entering into the Z list all nodes with zero fathers (which equals their current work count). In our example, these are A and B, to which we assign the beginning date—here taken as zero. After processing A and B our node table appears as shown above, with A and B checked off the Z list—but not removed from it, since we may need to know which nodes have been processed.

At this point we look for new nodes having work counts equal to father counts, finding D and C, which are thus added to the Z list. The algorithm proceeds as shown.

Node	Father Count	Early times	Work count	Z list
A	*0*	0		~~*A*~~
B	*0*	0		~~*B*~~
D	*2*	~~2~~ 4	‖	~~*D*~~
C	*1*	4	|	~~*C*~~
F	*3*	~~3~~ 6	‖|	
G	*2*			
E	*2*	~~3~~ 7	‖	
H	*2*			

Processing D we find (D, E) requiring three time units, which must be added to the now final early time of $D(4)$ to give 7 for E. F may also be reached as a son of D, causing $4 + 2$ or 6 to be applied to F. Since this is larger than the 3 that is already there, it is substituted for it.

In processing C we can only reach F with an early time of 6. Since this is the same as the current early time, no change is made but the work count is increased to 3.

Now both *E* and *F* may be added to our *Z* list, and we finish the algorithm.

Node	Father count	Early times	Work count	*Z* list
A	*0*	0		~~*A*~~
B	*0*	0		~~*B*~~
D	*2*	~~2~~ 4	‖	~~*D*~~
C	*1*	4	|	~~*C*~~
F	*3*	~~3~~ 6	‖|	~~*F*~~
G	*2*	~~8~~ 12	‖	~~*E*~~
E	*2*	~~3~~ 7	‖	*G*
H	*2*	~~9~~ 11	‖	*H*

The *Z* list holds the terminal nodes *G* and *H* because they have equal work and father counts, but they have no sons.

Loop detection

If we carry out our early time algorithm on a network that has a *loop* in it (logically impossible, but transcription errors will occur and must be detected and rectified in any large system) it will break down before completion. The usual breakdown, caused by an interior loop, is the exhaustion of the *Z* list with no new nodes ready for admission, because none have yet been reached from all possible fathers. (One of their fathers is also a son and has not been reached.)

	Act.	*F*	*S*	*V*
1	*A, D*	·	*5*	2
2	*B, E*	**5**	*3*	·
3	*D, E*	**4**	·	·
4	*D, F*	·	*7*	·
5	*B, D*	**6**	·	·
6	*B, C*	·	·	
7	*C, F*	·	·	
8	*F, B*	**9**	·	
9	*F, G*	**10**	*11*	
10	*F, H*	·	*12*	
11	*E, G*	**12**	·	
12	*E, H*		·	

	Node	*F*	*S*	Counts *F*	Counts *S*	Early times	Work count	*Z*	*L*
1	*A*	**1**	·	*0*	**1**	0		~~*A*~~	*D*
2	*B*	**2**	*8*	*1*	**3**				*B*
3	*D*	**3**	*1*	*2*	**2**	2	1		*F*
4	*C*	**7**	*6*	*1*	**1**				*D* √
5	*F*	**8**	*4*	*2*	**3**				
6	*G*	·	*9*	*2*	**0**				
7	*E*	**11**	*2*	*2*	**2**				
8	*H*	·	*10*	*2*	**0**				

Having found a node that shows this sign of being a loop (for others, see Klein)—we may trace the loop explicitly by going *backward*, from son to father, remaining among the nodes that have not yet been stricken from the Z list—that is, the unfulfilled nodes. A loop has been traced when the first node appears a second time. We illustrate with our same table, in which the activity BF has been reversed.

After processing A we run out of our Z list. D has a work count of only 1 but has two fathers. Thus we enter D in our L list and proceed backward, avoiding A. We quickly obtain DBF as our loop.

Minimum state highway net for Alabama

Suppose that we decided to widen and apply a high-quality surface to enough roads in Alabama so that every village can be reached by a modern road. Suppose further that none of the existing road surfaces is considered adequate (Alabama was chosen alphabetically!), so we can avoid the requirement that specific pieces of the existing system be included in the new minimum state highway net. [The term "net" is poor, for the minimum system will be a *tree*; there will be no closed loops (why?).]

The minimum tree network is the tree that includes all the network nodes and minimizes the *sum* of the values of the included edges. Clearly it is our minimal highway system—the tree of minimum total length. (Note that this is not the same tree that will usually be generated in our first approach to the Baltimore traffic problem, which was the tree of minimum distance to each node from one root, S. In that tree, the distance from the root to the tip of *each* branch was minimized.)

The most efficient algorithm is also the most direct: to construct the tree by always adding the shortest branch that will join on to the existing tree without completing a circuit. This algorithm is easily effected by first sorting the edges in order of increasing size, taking the smallest as the first branch of our tree, then repeatedly adding the smallest remaining branch that has one end (but not both) already in the tree. When all nodes have been included, the algorithm stops.

As an example consider the network of roads shown in Figure 19.8 and represented by a list of edges that is ordered according to increasing values. We begin by placing the shortest edge (4, 5) in our tree, checking it off our list and checking the two nodes as being included in our tree. The next edge (4, 7) satisfies our requirement of one node in the tree and one node in our new node list. The next edge (8, 6) has neither node currently in our growing tree, so it must be skipped. Thus we next add (8, 7). Now we may add (8, 6), showing the need for provision for backing up in our edge list. Finally (3, 8), (1, 4), and (6, 2) complete our minimal tree, shown in heavy lines in Figure 19.8.

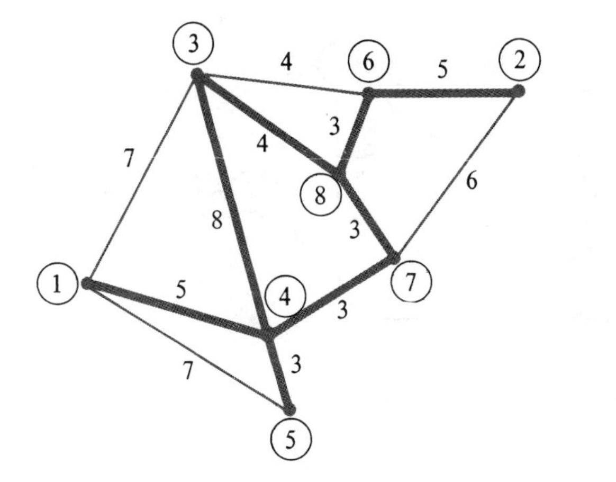

Figure 19.8

Edge	Distance		
4–5	3	√	
4–7	3	√	
8–6	3		√
8–7	3	√	
3–8	4		√
3–6	4		
1–4	5		√
6–2	5		√
7–2	6		
1–3	7		
4–3	8		

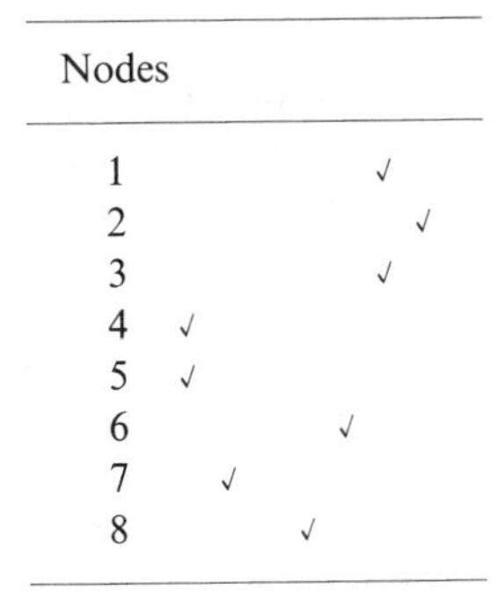

Nodes						
1					√	
2						√
3					√	
4	√					
5	√					
6				√		
7		√				
8			√			

Flow through a network of pipes

Suppose we are given the diagram, Figure 19.9, of a network of pipes that shows the flow admittances, as well as the physical interconnections. A pump is asserted to be pushing 17 gallons of water per minute through this system and we are concerned to know the flows through the several pipes, as well as the pressures at the various nodes. (Alternatively, the pressure supplied by the pump could be given without significantly altering the problem.)

We shall adopt as our physical law the proposition that the pressure drop through a pipe is proportional to the square of the flow rate. Thus we have

$$p_i = [\operatorname{sgn} f]c_i f_i^2 \tag{19.1}$$

where p is the pressure drop and the constant of proportionality c_i is the flow resistance. This c_i for the ith pipe depends on its length, cross section, roughness,

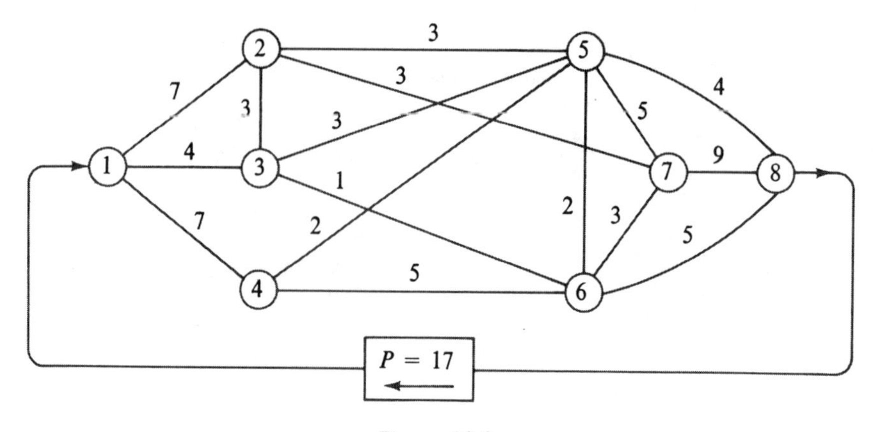

Figure 19.9

and some fluid characteristics such as viscosity and density—but we are not concerned with these details and lump them together in order to concentrate on our central purpose, how to set up the equations systematically from the network. The sign-of-f term, (sng f) is necessary because both the pressure drop and the flow bear signs indicating direction, which unfortunately will not survive squaring. Using (19.1) we may calculate p_i if we know f_i, and vice versa.

We now choose a *spanning tree* for our network—a tree that includes all the nodes. It is indicated by the heavy branches in Figure 19.10. Many other

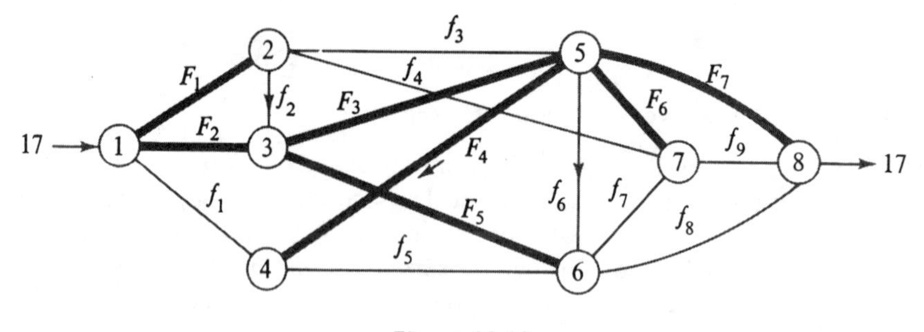

Figure 19.10

spanning trees are possible, some of which are even preferable to this one. We chose the present example to emphasize that special cautions are not necessary, any standard method for obtaining a spanning tree from a network being adequate. If we happen to choose an unrealistic direction as positive (F_4), our final numerical value for this variable will simply turn out to be negative.

We have assigned flow variables to our in-tree branches (F_k) and to our out-of-tree branches (f_k). The subscripting is arbitrary, the direction being generally positive toward the right unless an arrow indicates differently. The nodal equations express the fact that all liquid entering a node must also leave it.

$$
\begin{aligned}
&(1) \quad F_1 + F_2 + f_1 = 17 \\
&(2) \quad f_2 + f_3 + f_4 = F_1 \\
&(3) \quad F_3 + F_5 = F_2 + f_2 \\
&(4) \quad f_5 = f_1 + F_4 \\
&(5) \quad f_6 + F_6 + F_7 + F_4 = F_3 + f_3 \qquad \textit{nodal equations} \\
&(6) \quad f_7 + f_8 = f_5 + f_6 + F_5 \\
&(7) \quad f_9 = f_4 + f_7 + F_6 \\
&(8) \quad 17 = f_8 + f_9 + F_7
\end{aligned}
\tag{19.2}
$$

Either the first or the last equation of this nodal set is redundant, since all liquid entering at node (1) must exit at (8).

In order to emphasize the fact that a pressure drop is only a function of the corresponding flow rate, we shall let the expression $[f_i]$ stand for the pressure drop p_i. We note that each out-of-tree pipe closes a unique circuit through the tree. We shall use these circuits for our pressure drop equations, which must sum to zero. We obtain

$$
\begin{aligned}
[f_1] &= [F_2] + [F_3] + [F_4] \\
[f_2] &= [F_2] - [F_1] \\
[f_3] &= [F_2] + [F_3] - [F_1] \\
[f_4] &= [F_2] + [F_3] + [F_6] - [F_1] \\
[f_5] &= [F_5] - [F_3] - [F_4] \qquad \textit{circuit equations} \\
[f_6] &= [F_5] - [F_3] \\
[f_7] &= [F_3] + [F_6] - [F_5] \\
[f_8] &= [F_3] + [F_7] - [F_5] \\
[f_9] &= [F_7] - [F_6]
\end{aligned}
\tag{19.3}
$$

The equation for $[f_5]$. for example, shows that the pressure drop between nodes 3 and 6 is equal to $[F_3] + [F_4] + f_5$ or to $[F_5]$, depending on which route one takes (see Figure 19.10).

Note immediately that if all the $[F_i]$'s are known, the f_k's are trivially evaluated from the circuit equations. Likewise, if all the f_k's are known, the nodal equations (the first or last being omitted) constitute a linear system that determines the F_i. Somewhat less obvious is the fact that this system, too, may be *evaluated*, one F at a time, rather than having to be solved as a simultaneous set. Thus the amount of arithmetic labor is really very small. Note that nodal equations 2, 4, 6, 7, and 8 contain only one F_i, hence may be evaluated immediately – giving F_1, F_4, F_5, F_6, and F_7. These equations contain only one F because they arise from the nodes that terminate the tree branches in Figure 19.10. Having evaluated these F's we may now trim back our tree by removing those branches. We see that the node 5 equation now contains only one unknown, F_3, which may thus be evaluated and its branch lopped off. This leaves node 3 exposed, so the node 3 equation may be used to evaluate F_2, thereby completing our evaluation of the F_i's.

We now have a strategy for setting up and solving, iteratively, for the pressure drops and flows throughout our equation system. We begin with some set of values for the F's, solve for the f's through the circuit equations, then apply the nodal equations to get a new set of F's. To guard against unstable iterations we probably do *not* use these new F's directly but rather use some weighted average of the old F's and the new. The details are of little interest here, the principles having been discussed in Chapters 2 and 14.

The important point about this network formulation of the flow problem is that it permits a computer algorithm to be written that is *not* specific to this geometry. In effect, we can cause the program to set up and then evaluate its own equations from the network geometry. Both the incidence matrix and the list forms of our network can be useful here.

Let us now examine the matrix, Figure 19.11, in which the flow variables appear to see if our equations are contained therein. We see immediately that the first nodal equation is merely the sum of the elements of the first row of the matrix equated to zero. The second equation is the sum of the second row equated to the sum of the second column, and so on. If we add the items as we go down a column, then consider the diagonal element to be an equality sign, turn and add the elements as we proceed across the row, we get our nodal equation. The only additional requirement is that any subdiagonal element (F_4) have first been reflected with changed sign into the upper half of the matrix. Alternatively, the entire matrix could have been skew-symmetrized by reflection and sign change, whence each nodal equation is simply a row sum.

The order of evaluation for the nodal equations can be ascertained by finding first those equations containing only one F_k. Since any *column* contains exactly one F_k (why?), this implies finding rows that contain no F_k's. Having evaluated those equations, we blank out the newly found F's and examine other rows for absence of F's.

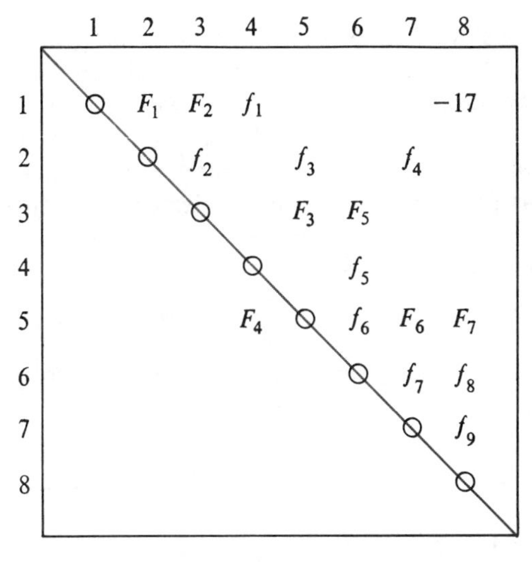

Figure 19.11

Finally, we wonder if the circuit equations are lurking within the structure of our incidence matrix. Clearly (Figure 19.10) the potential drop in each out-of-tree branch is simply the difference between the potentials at the ends of the two tree branches it connects. Thus we need only start at each branch end and work our way back to the root, accumulating potential drops as we go. Considering a matrix of tree elements and only one out-of-tree element, Figure 19.12,

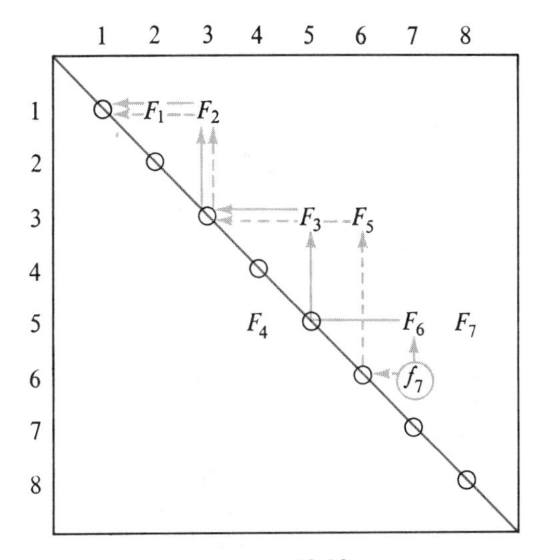

Figure 19.12

we see that pursuing a branch toward the root consists of *moving vertically to an F, then horizontally to the principal diagonal*, vertically to the next F, and so on – stopping when we reach the top left corner. Two paths are possible from any starting point, depending on whether the first move is vertical or horizontal, but vertical moves always go to F's, horizontal moves to the diagonal.

The two paths for f_7 both include the element F_2, causing that item to cancel when the accumulated pressure drops for the two paths are finally subtracted. Thus we have

$$[f_7] = \{[F_6] + [F_3] + [F_2]\} - \{[F_5] + [F_2]\}$$

The f_1 path includes a subdiagonal element, but the rule for pursuing the tree

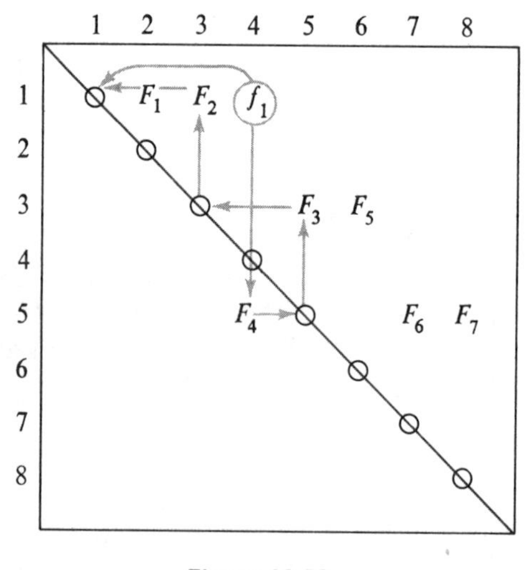

Figure 19.13

root is unaffected (Figure 19.13). We get

$$[f_1] = [F_4] + [F_3] + [F_2]$$

Our aim is to provide an encodable algorithm that evaluates F's and f's for *any* network, as specified at run time on the computer by, say, the resistance matrix. Sufficient information is contained therein both about the topology of the network and about the impedance characteristics of its individual pipes to determine the flows. We should not have to prepare a separate computer routine for each network. Using the incidence matrix permits such encoding – but so does a *list* representation. If the computer language is FORTRAN or ALGOL, some persons may feel the list form preferable, since it matches well with a subscripted variable type of thinking. Direct machine-language encoding may cause different preferences.

Double subscript notation

Once we actually turn the programming in a subscripted language, we shall find a double-subscript notation for our variables much more revealing than the single subscript we have been using. We reexpress our nodal and circuit equations using F_{jk} to represent the flow rate from node j to node k (Figure 19.14).

$$
\begin{array}{llr}
(1)\quad F_{12} + F_{13} + f_{14} = 17 & [f_{14}] = [F_{13}] + [F_{35}] + [F_{54}] & \\
(2)\quad f_{23} + f_{25} + f_{27} = F_{12} & [f_{23}] = [F_{13}] - [F_{12}] & \\
(3)\quad F_{35} + F_{36} = f_{23} + F_{13} & [f_{25}] = [F_{13}] + [F_{35}] - [F_{12}] & (19.4) \\
(4)\quad f_{46} = f_{14} + F_{54} & [f_{27}] = [F_{13}] + [F_{35}] + [F_{57}] - [F_{12}] & \\
\quad\vdots & [f_{46}] = [F_{36}] - [F_{35}] - [F_{54}] & \\
 & \vdots & \\
\textit{nodal equations} & \textit{circuit equations} &
\end{array}
$$

The structure of the nodal equations is again transparent: the kth equation has all the elements with k for the first subscript on its left, all those with k for the second subscript on its right. Each circuit equation is composed of a positive

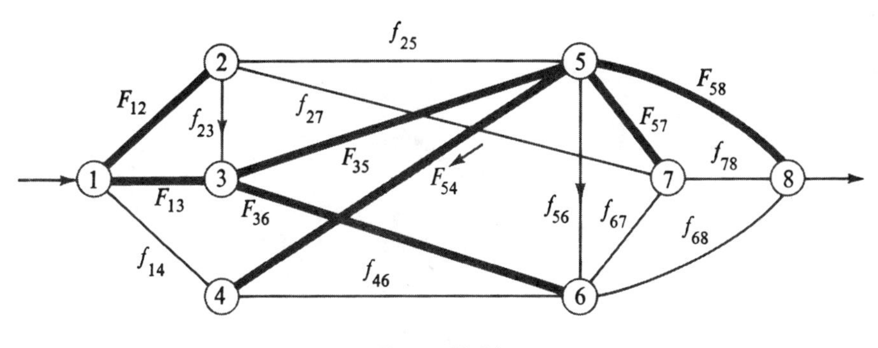

Figure 19.14

and a negative branch, with a chaining of the second and first subscripts through successive elements. The right-hand element of each chain has its right subscript in agreement with a subscript of f_{jk} term. If we include the canceling elements that get us back to the root via both branches, then both start with 1 for their first subscript. Under this convention the equation for $[f_{46}]$ would read

$$[f_{46}] = \{[F_{13}] + [F_{36}]\} - \{[F_{13}] + [F_{35}] + [F_{54}]\}$$

and the general rule is easily encoded in FORTRAN—aided by a *predecessor vector* that gives the predecessor node for every node. For our tree it is

Node	1	2	3	4	5	6	7	8
Predecessor	0	1	1	5	3	3	5	5

While a clever programmer could implement the general set of equations that are represented by (19.4), a much simpler structure appears if we can find a spanning tree that is one continuous line from source to sink—all trunk and no branches. If we then number the nodes in order along this linear tree as in Figure 19.15, the structure of the equations becomes particularly simple—including a simplicity in the order of evaluating the nodal equations. The student should write the equations out and consider their implementation. Finding a spanning linear tree by automatic computer techniques is, however, far from trivial. The human eye is much more efficient at this sort of pattern recognition. But if the programmer is willing to put that part of the job onto the customer's shoulders, his own task will be much easier. Since the simplification is unnecessary, your author feels it tantamount to an abnegation of programming responsibility—but then *he* doesn't have to write the program!

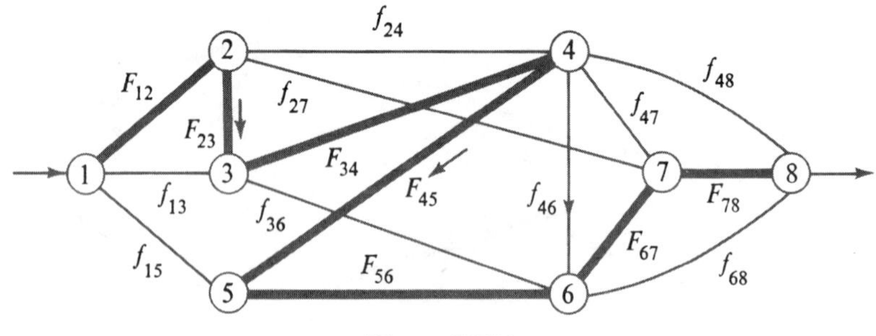

Figure 19.15

Maximal network flows—the bottleneck problem

Another, and simpler, problem that can be posed in terms of finding the maximum flow through a network of pipes has been solved by Ford and Fulkerson [1955]. In this problem we are given "capacities" of every pipe in the network

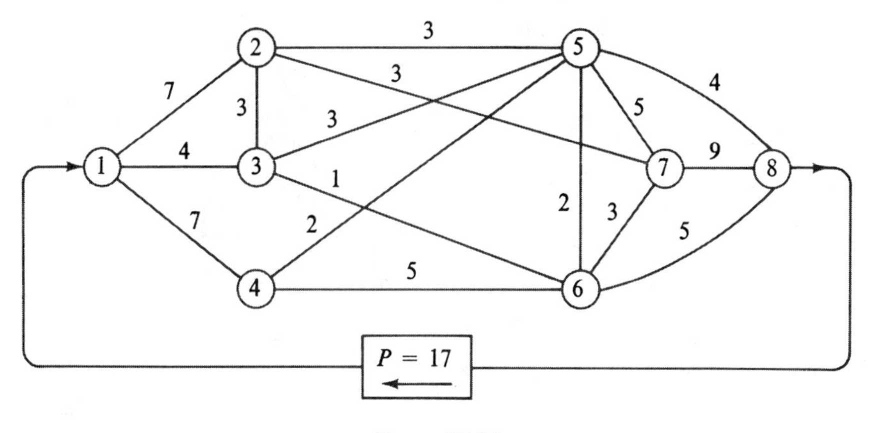

Figure 19.16

and are asked what is the total amount that passes through from source to sink under the assumption that every pipe will carry its exact capacity provided the incompressible fluid can keep on going through the subsequent pipes. This is, then, a *bottleneck* problem – somewhat unrealistic when expressed in terms of fluid flow but probably useful in examining conveyor systems and railway nets where "pumping a little harder" is not apt to have much influence. We shall illustrate the algorithm using the same network topology as in Figure 19.9. We reproduce it here as Figure 19.16 but with the interpretation of the numbers on each pipe as maximum flow rates that will be achieved provided only that the fluid can get out the far end of the pipe. The pipes are presumed to be directional and so we shall duplicate the pipes (2, 3) and (5, 6) in our example with two one-way pipes since we are not quite sure in which direction the fluid here will flow.

To arithmetize our problems we set up a square matrix whose rows and columns are the nodes of the network and the entries are the capacities. We have

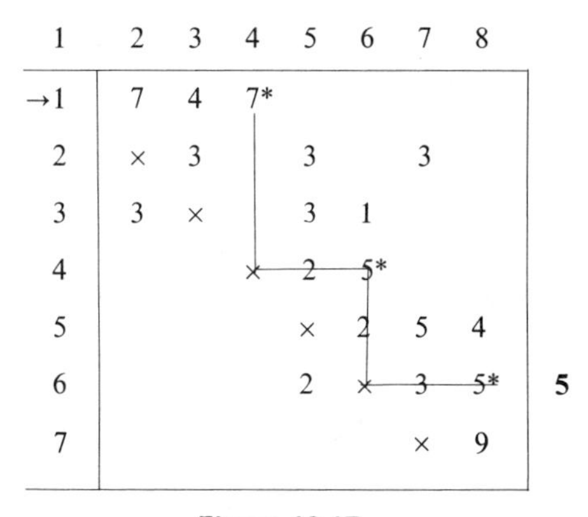

1	2	3	4	5	6	7	8	
→1	7	4	7*					
2	×	3		3		3		
3	3	×		3	1			
4			×	2	5*			
5				×	2	5	4	
6				2	×	3	5*	**5**
7						×	9	

Figure 19.17

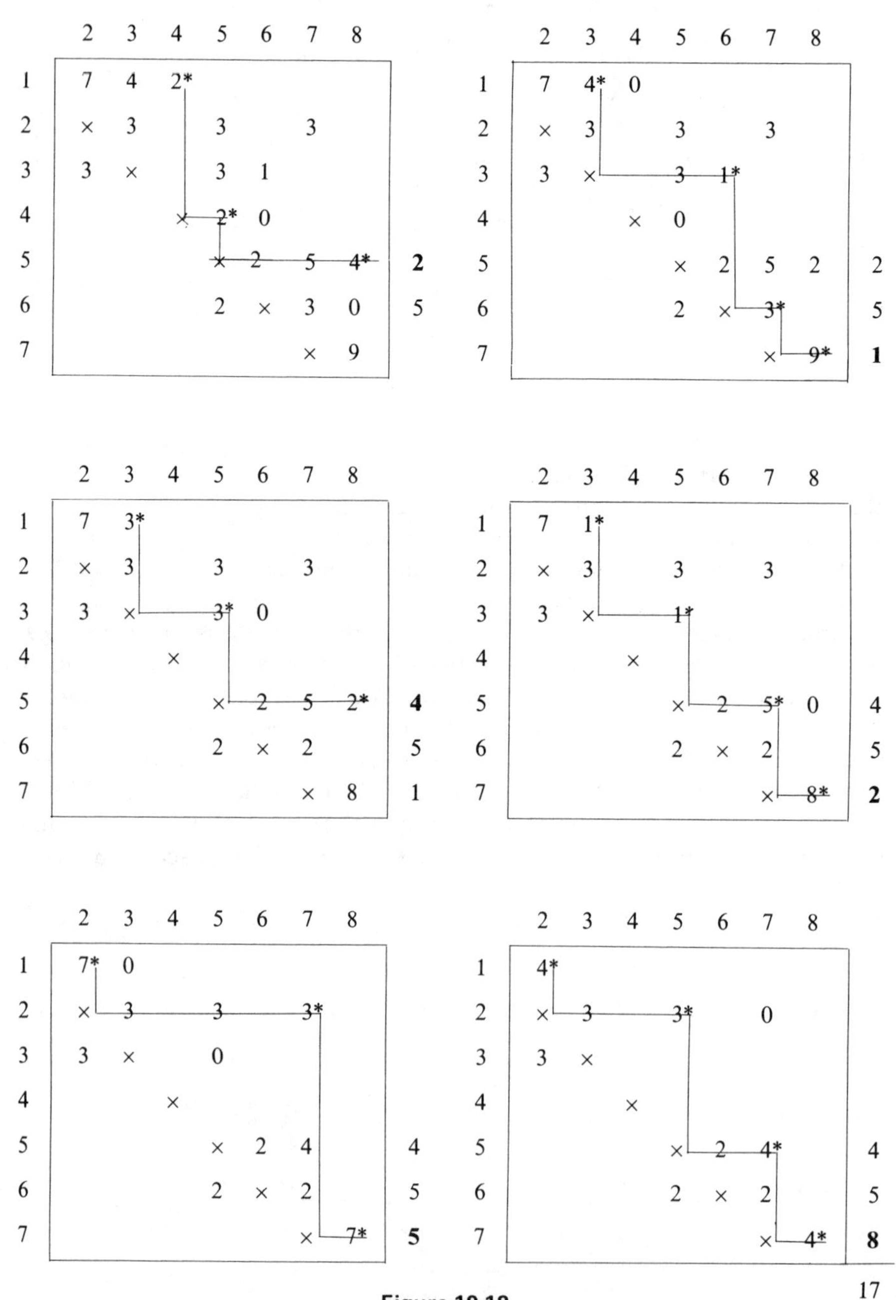
2 3 4 5 6 7 8
1 7 4 2*
2 × 3 3 3
3 3 × 3 1
4 × 2* 0
5 × 2 5 4* 2
6 2 × 3 0 5
7 × 9
2 3 4 5 6 7 8
1 7 4* 0
2 × 3 3 3
3 3 × 3 1*
4 × 0
5 × 2 5 2 2
6 2 × 3* 5
7 × 9* 1
2 3 4 5 6 7 8
1 7 3*
2 × 3 3 3
3 3 × 3* 0
4 ×
5 × 2 5 2* 4
6 2 × 2 5
7 × 8 1
2 3 4 5 6 7 8
1 7 1*
2 × 3 3 3
3 3 × 1*
4 ×
5 × 2 5* 0 4
6 2 × 2 5
7 × 8* 2
2 3 4 5 6 7 8
1 7* 0
2 × 3 3 3*
3 3 × 0
4 ×
5 × 2 4 4
6 2 × 2 5
7 × 7* 5
2 3 4 5 6 7 8
1 4*
2 × 3 3* 0
3 3 ×
4 ×
5 × 2 4* 4
6 2 × 2 5
7 × 4* 8
17

Figure 19.18

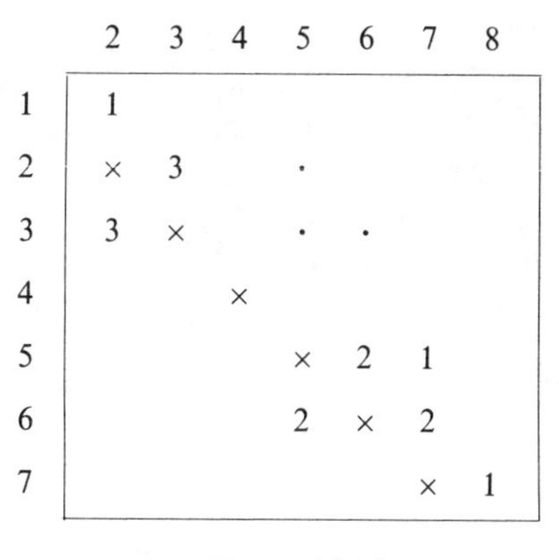

	2	3	4	5	6	7	8
1	1						
2	×	3		·			
3	3	×		·	·		
4			×				
5				×	2	1	
6				2	×	2	
7						×	1

Figure 19.19

The precise arrangement is irrelevant to the ultimate success of the algorithm, but its efficiency may be enhanced if the nodes are listed in some sort of natural progression from the source to the sink as here. Thus nodes farther to the right are nearer the sump and the pipes tend to cluster close to the diagonal of the matrix.

The algorithm of Ford and Fulkeron is iterative, each iteration consisting of two parts:

1. Starting with 1 we seek a path to 8 using pipes with as-yet-unused capacity. (From 3 we may go over to 2, 5, or 6; from 4 we may go to 5 or 6; from 7 we find only 8.)
2. Having found a path and noted the maximum amount which can be pumped through it, we reduce by that amount the capacities of every pipe in our matrix that we used on this path.

When we can no longer carry out our algorithm in a sense that no free path exists from 1 to 8, we are through. Some remarks seem useful:

1. We reach the next pipe by passing downward from the pipe just traversed until we encounter the diagonal ×, then move either right (preferably) or left to any nonzero entry.
2. For automatic searching, we shall take the rightmost option when several are available.
3. When pipes to the *left* of the diagonal are used in step 1, we then rule out the subsequent use of the symmetrically placed element for part of the current path. (It is the parallel pipe running in the opposite direction.)

We have marked the first route in Figure 19.17. Its capacity is 5. In Figure 19.18 we show subsequent developments of the algorithm, disclosing a total capacity of 17 in the final diagram. In Figure 19.19 there is no path leading from 1 through 8. The numbers remaining in the matrix are unused capacities. We also see that the remaining capacity of one unit could be used if we augmented the original matrix in any of several places—(2, 5), for example, or (3, 5), or (3, 6)—as this would permit one additional path of capacity unity. This algorithm is simple to automate and carry out on a digital computer. We leave it as a final exercise.

AFTERTHOUGHTS

The students in my Numerical Methods course usually have been Juniors—with a sprinkling of graduates and the occasional rather sophisticated Sophomore. Mostly they have been Engineers and Scientists. (Mathematicians at Princeton are proudly Pure while most Computer Scientists find an obligatory decimal point to be slightly demeaning.) Thus the student interest has centered on solving 'typical' physical problems that in practice devolve to simplified versions such as those treated in the text. Since some instructors have requested more exercises I have included some more below.

My recent practice has been to go rapidly through the first 6 chapters in as many weeks with weekly computation laboratories in which simple (pencil and pocket-calculator) problems are worked with an instructor present to get people out of mudholes onto productive paths. (These sessions are partly remedial as at least half of my students have had some exposure to these early topics in a first computer programming course that contained considerable numerical methods.)

In the second half of the semester I cover topics from the rest of the book according to the perceived class interests but the computational homework has been devoted entirely to one half-term project—each student's problem being unique. I feel strongly that somewhere in the undergraduate scientific curriculum the student should be asked to work a problem to which he does not know the answer and for which the answer, when found, should not proclaim its correctness. In most underclass courses the answers to the exercises are known *a priori*: they are in the back of the book or they turn out to be π or $\sqrt{2}$—or one's roommate did the problem last term. For such problems the effort goes entirely into *getting* the answer. But in realistic engineering problems somewhere between 50 and 90 percent of the effort goes into *verifying that the 'answer' obtained is, indeed, correct*. The student needs this kind of experience at least once, somewhere, and I choose to supply it here.

A problem type that has been very effective for these purposes is the preparation of an efficient computer *function* (i.e., a working, thoroughly debugged program) for an integral like

$$H(b) = \int_0^\infty \frac{e^{-bx}}{\sqrt{x(1+x)}}dx.$$

For large b an asymptotic series is useful; for small b a singularity has to be propitiated, analytically, before a series in b is possible—and these two approaches leave a gap that the student has to struggle to fill. One of the ground rules is that numerical quadrature is *not* to be part of the final function (too slow) although it is a necessary tool during the development. Likewise, covering the gap by interpolation in a large stored table lies outside the aesthetic framework of the exercise, although I have reluctantly accepted it for part credit when submitted as a last-ditch effort by a weak student. A mature mathematician can perform this exercise in a couple of afternoons, but a student often needs a half-semester of a course, even with considerable help. Weekly conferences with the instructor are essential; without them, there is the danger that work either is postponed into impossibility or proceeds forever down ultimately unproductive paths.

There is a whole family of similar integrals that can be assigned; generally

$$G(b) = \int_0^\infty \frac{e^{-bx^j}}{x^k(1+x^l)^m}dx$$

with

$$j = 1, 2 \qquad k = 0, 1/2 \qquad l = 1, 2 \qquad m = 1/2, 1$$

that offer varying degrees of difficulty so that the problem can be tailored to the student's abilities—or ambitions. In addition to the techniques covered in the book, most of these integrals can be evaluated efficiently by rather sophisticated recurrences. When time has permitted I have derived the recurrence for the integral of page 261*—thereby tempting the better student to explore this new tool for his gap-covering algorithm.

Tests and examinations

I have usually given two 50-minute tests and a final 3-hour examination. These, together with observations available from the labs and weekly project conferences have provided a more-than-adequate basis for grading. The tests have been 'closed book' except for Abramowitz & Stegun and a pocket calculator —neither of which are usually necessary but knowing that they will be available usually keeps the students from spending time memorizing formulae. Hopefully they will have concentrated on strategic algorithmic principles. Here is a 50-minute test that is difficult to do well but one that provides good opportunities for exhibiting mastery of the course concepts.

Test

1. It is necessary to compute the sequence of 50 functions

$$F_k(b) = \int_0^1 e^{bt} t^k \, dt \qquad k = 0, 1, \ldots, 49$$

for various b that are not known at the time you must write the computer program. Integration by parts gives the recurrence

$$bF_k(b) + kF_{k-1}(b) = e^b \qquad k = 1, 2, \ldots$$

from

$$F_0(b) = (e^b - 1)/b$$

---

*Run

$$I_{n-1} = I_n + cK_n/(2n) \qquad J_{n-1} = (2nJ_n + cI_{n-1})/(2n-1) \qquad K_{n-1} = K_n + J_{n-1}$$

from largish N down to 1, whence $F(b) = J_0/c/K_0$. The starting values of I_n, J_n and K_n are arbitrary but should not all be zero.

The recurrence is economical, if usable. For what values of (b, k) can the recurrence be used to deliver 10 significant figures (or 10 decimals, if you prefer) on a 13 digit machine?

2\. a) How would you integrate

$$b\frac{d^2u}{dt^2} + \frac{a}{t}\frac{du}{dt} + u^3 = 0$$

from the initial conditions

$$u(0) = 1 \qquad \frac{du}{dt}(0) = 0$$

for various (a, b)?

b) If the coefficient b is large (say, 1000), what effect will it have on your favored process?

3\. A large number of peculiar quintic polynomials are to be solved for all their real roots that lie on the range $(-10, 10)$. These polynomials are peculiar because the coefficient of their x^5 term is typically smaller than the other coefficients by at least a factor of 100. Five algorithms are listed below.

a) Write 2 or 3 sentences about each, explaining how it can be applied to this problem (if at all) and comment on its usefulness.

b) Give *your* recommendation for handling this problem—not necessarily from the list—and defend it.

1. Forward-Backward Division
2. Newton's Method
3. Binary Chop
4. Lin's Method
5. Lehmer's (concentric circles) Method

Exercises

These exercises are intended to be performed with pencil, paper and a pocket calculator. Most are 'remedial' in the sense that, ideally, students taking this course should be able to do them in relatively short order—say, 5 to 20 minutes each—on the basis of prior experience. Since many lack that experience, I have given them in supervised laboratories where individual guidance is available before too much frustration sets in.

Precision Problems

1\. How would you evaluate

$$\ln\sqrt{x+1} - \ln\sqrt{x}$$

for large x without losing significant figures? Compare your method with direct evaluation for $x = 123456.7$.

2. Find the behavior of

$$\ln\left[\frac{\sqrt{1 + x^2} - x}{\sqrt{1 + x^2} + x}\right]$$

   a) for large x
   b) for x near zero

   and then suggest how the function should be evaluated.

3. Evaluate

$$w(x) = 1 + \frac{1}{2}\tanh x^2 - \cosh x$$

   for $x = 0.1$
   a) as expressed here
   b) from a series expansion

   Keep all arithmetic to 10 significant figures if possible. Where and how does error arise?

4. Consider evaluating

$$\frac{\cosh x}{2 - \cos x} \qquad x \geq 0$$

   and then decide how you would solve for

$$\frac{\cosh x}{2 - \cos x} = 1 + \epsilon$$

   for small ϵ.

5. How should one solve $\tan\theta = B$ for θ if B is greater than 10? Assume you do not have specialized tables.

6. Devise iterative computational algorithms for solving

$$\frac{\tanh x}{\arctan x} = b$$

for x when

a) $b = s$ near $2/\pi$

b) b is near 1

without serious instability or loss of significance. Estimate the range of applicability of your algorithms.

7. Consider the quadratic equation obtained by eliminating either the $\sin^2 x$ or $\sinh^2 y$ term from the equation pair

$$\beta = \sin^2 x + \sinh^2 y \qquad \beta = u^2 + v^2$$

$$\alpha \tanh^2 y = \tan^2 x \qquad \alpha = u^2/v^2$$

as a method for determining x and y when (u, v) are given. Where are the trouble spots? What has to be done to avoid significant figure loss? How many separate algorithms do you end up with via this approach?

8. Examine the limiting behaviors of

$$\frac{\sin x}{1 + \cos x} \quad \text{and} \quad \frac{\sinh x}{\cosh x - 1}$$

as x approaches zero (from each side).

Conformal mapping

[The student should have seen complex variables in their rectangular and polar forms, but no course in the subject is presumed. The instructor should do a mapping first—perhaps $z = \sin t$ using the first quadrant of t.]

9. Find the complex z-map of the figure under

$z = \tanh t$

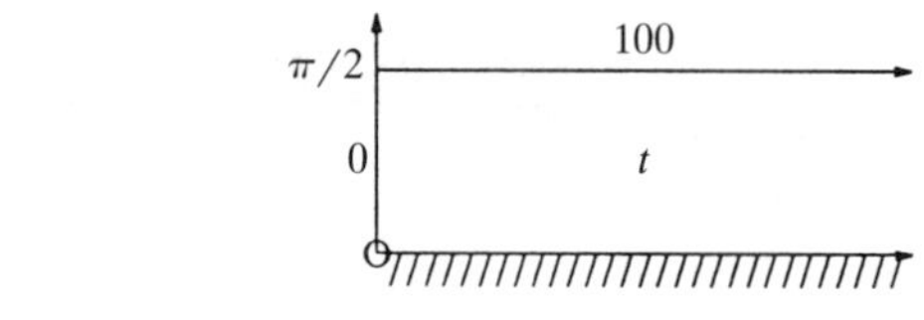

labelling the critical points.

10. Consider the mapping from t to w of the figure in Problem 9 via

$$\sin w = \tanh t.$$

Compare the final potential problem with the original one.

11. Devise an algorithm to evaluate a point w that corresponds to a *complex* value of t near $(0, i\pi/2)$.

Singularity removal

12. Find a series in b for

$$F(b) = \int_0^1 \frac{dx}{\sqrt{1 - bx^3}} \tag{a}$$

when b is small. How many terms are needed to obtain approximately 8 significant figures when b is 0.5? For what ranges of b values is the series practical? Then transform

$$\int_0^1 \frac{dx}{\sqrt{1 - x^3}}$$

to remove the singularity, sketch, estimate its value. Devise a method for evaluating $F(1 - \beta)$ when β is small. [*Hint*: Transform the dummy variable in (a) so that the parameter is in the *limit* rather than in the integrand. Then consider an intergral that is related to the difference between $F(1)$ and $F(b)$.]

BIBLIOGRAPHY

Abramowitz, M., and I. Stegun, *Handbook of Mathematical Functions* (AMS 55), National Bureau of Standards, Washington, D.C., U.S. Government Printing Office, 1964.

Acton, F. S., *Analysis of Straight-Line Data*, New York, Dover, 1966.

"Algorithms for elliptic functions and elliptic integrals," *Numerische Mathematik*, *5*; also see *CACM*, *4* (1961), 180, 543; *CACM*, *6* (1963), 166; *CACM*, *9* (1966), 12.

Bowman, F., *Introduction to Elliptic Functions*, New York, Dover, 1961.

Bulirsch, R., and S. Stoer, "Numerical treatment of ordinary differential equations by extrapolation methods," *Numerische Mathematik*, *8* (1966), 1–13, 93–104.

Churchill, R. V., *Fourier Series and Boundary Value Problems*, New York, McGraw-Hill, 1941.

Clenshaw, C. W., "A note on the summation of Chebyshev series," *MTAC*, *9* (1955), 118–120.

Comrie, L. J., *Chambers's Shorter Six-Figure Mathematical Tables*, London and Edinburgh, W. and R. Chambers, 1966.

Comrie, L. J., *Chambers's Six-Figure Mathematical Tables*, Vols. I and II, London and Edinburgh, W. and R. Chambers, 1949.

Fletcher, R., and M. J. D. Powell, "A rapidly convergent descent method for minimization," *Computer J.*, *6* (1963), 163–168.

Ford, L. R., Jr., "Maximal flows through a network," *Can. J. Math.*, *8* (1956), 399–404.

Ford, L. R., Jr., and D. R. Fulkerson, "A simple algorithm for finding maximal network flows and an application to the Hitchcock problem," Santa Monica, RAND Corp., P-743, Sept. 1955.

Forsythe, G. E., and W. R. Wasow, *Finite-Difference Methods for Partial Differential Equations*, New York, Wiley, 1960.

Gentleman, W. M., and G. Sande, "Fast Fourier transforms—for fun and profit," *Proc. Fall Joint Computer Conf.* (1966), 563–578.

Goldfeld, S. M., R. E. Quandt, and H. F. Trotter, "Maximization by improved quadratic hill-climbing and other methods," *Econometrics Res. Program Res. Mem. 95*, Princeton University, April 1968.

Goldfeld, S. M., R. E. Quandt, and H. F. Trotter, "Maximization by quadratic hill-climbing," *Econometrica*, *34* (1966), 541–551.

Graffe, C. H., *Die Auflösung der höheren numerischen Gleichungen*, Zurich, F. Schilthess, 1837.

Hamming, R. W., *Numerical Methods for Engineers and Scientists*, New York, McGraw-Hill, 1962.

Hartree, D. R., *Numerical Analysis*, New York, Oxford University Press, 1958.

Hastings, C., *Approximations for Digital Computers*, Princeton, N.J., Princeton University Press, 1955.

Isaacson, E., and H. B. Keller, *Analysis of Numerical Methods*, New York, Wiley, 1966.

Jahnke-Emde, *Tables of Functions with Formulae and Curves*, 4th ed., New York, Dover, 1951.

Kahn, A. B., "Topological sorting of large networks," *CACM*, *5* (1962), 558–562.

Kantorovich, L. V., and V. I. Krylov, *Approximate Methods of Higher Analysis* (translated by C. D. Benster), Groningen, Noordhoff, 1958.

Klein, M., "Scheduling project networks," *CACM*, *10* (1967), 225–230.

Knuth, D. E., *Fundamental Algorithms*, Reading, Mass., Addison-Wesley, 1968.

Kogbetliantz, E. G., "Generation of elementary functions," in *Mathematical Methods for Digital Computers*, Vol. 1, A. Ralston and H. F. Wilf, eds., New York, Wiley, 1960.

Kuiken, H. K., "Determination of the intersection points of two plane curves by means of differential equations," *CACM*, *11* (1968), 502–506.

Lance, G. N., *Numerical Methods for High Speed Computers*, London, Iliffe and Sons, 1960.

Lanczos, C., *Applied Analysis*, Princeton, N.J., Van Nostrand, 1956.

Lasser, D. J., "Topologically ordering of a list of randomly numbered elements of a network," *CACM*, *4* (1961), 167–168.

Lehmer, D. H., "A machine method for solving polynomial equations," *JACM*, *8* (1961), 151–162.

Lin, S. N., "Numerical solution of complex roots of quartic equations," *J. Math. and Phys.*, *26* (1947), 279–283.

Maehly, H., "Methods for fitting rational approximations. Part I: Telescoping procedures for continued fractions," *JACM*, *7* (1960), 150–162.

Maehly, H., "Methods for fitting rational approximations. Parts II and III," *JACM*, *10* (1963), 257–277 . . . (equation 11.5 on p. 269 is in error).

Maehly, H., "Rational approximations for transcendental functions," internal report, Mathematics Department, Princeton University, 1956.

Milne, W. E., *Numerical Calculus*, Princeton, N.J., Princeton University Press, 1949.

Milne, W. E., *Numerical Solution of Differential Equations*, New York, Wiley, 1953.

Nicholson, T. A. J., "Finding the shortest route between two points in a network," *Computer J.*, *9* (1966), 275–280.

Oliver, J., "Relative error propagation in the recursive solution of linear recurrence relations," *Numerische Mathematik*, *9* (1967), 321–340.

Olver, F. W. J., "The evaluation of zeros of high-degree polynomials," *Phil. Trans. Roy. Soc. London, A*, *244* (1952), 385–415.

Powell, M. J. D., "An efficient method for finding the minimum of a function of several variables without calculating derivatives," *Computer J.*, *7* (1964), 155–162.

Powell, M. J. D., "An iterative method for finding stationary values of a function of several variables," *Computer J.*, *5* (1962), 147–151.

Ralston, A., *A First Course in Numerical Analysis*, New York, McGraw-Hill, 1965.

Ralston, A., and H. F. Wilf, eds., *Mathematical Methods for Digital Computers*, Vol. 1, New York, Wiley, 1960.

Rosser, J. B., "A Runge–Kutta for all seasons," *SIAM Rev.*, *9* (1967), 417–452.

Rutihauser, H., "Deflation bei Bandmatrizen," *Z. Angew. Math. Phys.*, *10* (1959), 314–319.

Rutishauser, H., "Der quotienten-differenzen Algorithmus," *Z. Angew. Math. Phys.*, *5* (1954), 496–507.

Sneddon, I. N., *Mixed Boundary Value Problems in Potential Theory*, Amsterdam, North-Holland, 1966 (and New York, Wiley).

Spielberg, K., "Representation of power series in terms of polynomials, rational approximations, and continued fractions," *JACM*, *8* (1961), 613–627.

Tukey, J. W., and J. W. Cooley, "An algorithm for the machine calculation of complex Fourier series," *Math. of Comp. 19* (1965), 297–301.

Wilkinson, J. H., "*The Algebraic Eigenvalue Problem*," London, Oxford, 1965.

Wilkinson, J. H., "The evaluation of the zeros of ill-conditioned polynomials, part 1," *Numerische Mathematik*, *1* (1959), 150–166.

Wilkinson, J. H., "Householder's method for the solution of the algebraic eigenproblem," *Computer J.*, *3* (1960), 23–27.

Wynn, P., "Transformations to accelerate the convergence of Fourier series," *MRC Tech. Sum. Rep. 673*, U.S. Army Mathematics Research Center, University of Wisconsin, July 1966.

INDEX